T0202789

# Lecture Notes in Artificial Intelligence 13726

Subseries of Lecture Notes in Computer Science

More information about this subseries at https://link.springer.com/bookseries/1244

Weitong Chen · Lina Yao · Taotao Cai ·
Shirui Pan · Tao Shen · Xue Li (Eds.)

# Advanced Data Mining
# and Applications

18th International Conference, ADMA 2022
Brisbane, QLD, Australia, November 28–30, 2022
Proceedings, Part II

Springer

*Editors*
Weitong Chen
The University of Adelaide
Adelaide, SA, Australia

Lina Yao
The University of New South Wales
Sydney, NSW, Australia

Taotao Cai 🄳
Macquarie University
Sydney, NSW, Australia

Shirui Pan
Griffith University
Brisbane, QLD, Australia

Tao Shen
Microsoft
Beijing, China

Xue Li
The University of Queensland
Brisbane, QLD, Australia

ISSN 0302-9743        ISSN 1611-3349 (electronic)
Lecture Notes in Artificial Intelligence
ISBN 978-3-031-22136-1        ISBN 978-3-031-22137-8 (eBook)
https://doi.org/10.1007/978-3-031-22137-8

LNCS Sublibrary: SL7 – Artificial Intelligence

This Springer imprint is published by the registered company Springer Nature Switzerland AG
The registered company address is: Gewerbestrasse 11, 6330 Cham, Switzerland

# Preface

The 18th International Conference on Advanced Data Mining and Applications (ADMA 2022) was held in Brisbane, Australia, during November 28–30, 2022. Researchers and practitioners from around the world came together at this leading international forum to share innovative ideas, original research findings, case study results, and experienced insights into advanced data mining and its applications. With the ever-growing importance of appropriate methods in these data-rich times, ADMA has become a flagship conference in this field.

ADMA 2022 received a total of 198 submissions. After a rigorous single-blind review process 192 reviewers, 76 regular papers were accepted to be published in the proceedings, 39 were selected to be delivered as oral presentations at the conference and 37 were selected as poster presentations. This corresponds to a full oral paper acceptance rate of 19.6%. The Program Committee (PC), composed of international experts in relevant fields, did a thorough and professional job of reviewing the papers submitted to ADMA 2022, and each paper was reviewed by at least three PC members. With the growing importance of data in this digital age, papers accepted in ADMA 2022 covered a wide range of research topics in the field of data mining, including machine learning, text mining, graph mining, predictive data analytics, recommender systems, query processing, analytics-based applications, and privacy and security analytics. It is worth mentioning that, firstly, ADMA 2022 organized a physical event, allowing for in-person gatherings and networking, secondly, a special inclusive workshop has been organized to enhance the experience of women non-binary and gender non-conforming in the data mining community.

We thank the PC members for completing the review process and providing valuable comments within tight schedules. The high-quality program would not have been possible without the expertise and dedication of our PC members. Moreover, we would like to take this valuable opportunity to thank all authors who submitted technical papers and contributed to the tradition of excellence at ADMA. We firmly believe that many colleagues will find the papers in this proceedings exciting and beneficial for advancing their research. We would like to thank Microsoft for providing the CMT system that is free to use for conference organization, Springer for the long-term, support and the University of Queensland and ARC Training Centre for Information Resilience (CIRES) sponsorship of the conference.

We are grateful for the guidance of the steering committee members, Osmar R. Zaiane, Jianxin Li, and Guodong Long. With their leadership and support, the conference run smoothly. We also would like to acknowledge the support of the other members of the organizing committee. All of them helped to make ADMA 2022 a success. We appreciate local arrangements from the local co-chairs, Guangdong Bai and Henry Nguyen, the time and effort of the publication co-chairs, Taotao Cai, Shirui Pan and Tao Shen, the effort in advertising the conference by the publicity co-chairs, Ji Zhang, Philippe Fournier-Viger and Grigorios Loukides, the effort on managing the Tutorial sessions by tutorial co-chairs, Tianyi Zhou and Can Wang, We would like to give very special thanks to the web chair, Shaofei Shen, Hao Yang and Ruiqing Li, for creating a beautiful website and maintaining the information. We also thank Kathleen Williamson for her contribution to managing the registration system and financial matters. Finally, we would like to thank all the other co-chairs who have contributed to the conference.

November 2022

Xue Li
Lina Yao
Weitong Chen

# Organization

## Steering Committee

Xue Li      University of Queensland, Australia
Osmar R. Zaiane      University of Alberta, Canada
Jianxin Li      Deakin University, Australia
Guodong Long      University of Technology Sydney, Australia

## Program Committee Co-chairs

Lina Yao      University of New South Wales, Australia
Weitong Chen      University of Adelaide, Australia

## Local Chairs

Guangdong Bai      University of Queensland, Australia
Henry Nguyen      Griffith University, Australia

## Publicity Co-chairs

Ji Zhang      University of Southern Queensland, Australia
Philippe Fournier-Viger      Shenzhen University, China
Grigorios Loukides      King's College London, UK

## Publication Chairs

Shirui Pan      Griffith University, Australia
Tao Shen      Microsoft, China
Taotao Cai      Macquarie University, Australia

## Tutorial Co-chairs

Tianyi Zhou      University of Maryland, USA
Can Wang      Griffith University, Australia

## Web Co-chairs

Shaofei Shen      University of Queensland, Australia
Hao Yang      University of Queensland, Australia
Ruiqing Li      University of Queensland, Australia

## Industry Track Co-chairs

| | |
|---|---|
| Lu Liu | Google, USA |
| Jiajun Liu | CSIRO, Australia |
| Sen Wang | University of Queensland, Australia |

## Program Committee

| | |
|---|---|
| Abdulwahab Aljubairy | Macquarie University, Australia |
| Adita Kulkarni | SUNY Brockport, USA |
| Ahoud Alhazmi | Macquarie University, Australia |
| Akshay Peshave | GE Research, USA |
| Alan Liew | Griffith University, Australia |
| Alex Delis | National and Kapodistrian University of Athens, Greece |
| Ali Abbasi Tadi | University of Windsor, Canada |
| Anbumunee Ponniah | BITS Pilani, India |
| Atreju Tauschinsky | SAP SE, Germany |
| Bin Guo | Northwestern Polytechnical University, China |
| Bin Xia | Nanjing University of Posts and Telecommunications, China |
| Bin Zhao | Nanjing Normal University, China |
| Bo Ning | Dalian Maritime University, China |
| Bo Tang | Southern University of Science and Technology, China |
| Carson Leung | University of Manitoba, Canada |
| Chang-Dong Wang | Sun Yat-sen University, China |
| Chaoran Huang | University of New South Wales, Australia |
| Chen Wang | Chongqing University, China |
| Claudia Antunes | Universidade de Lisboa, Portugal |
| Clemence Magnien | Centre national de la recherche scientifique, France |
| David Broneske | German Centre for Higher Education Research and Science Studies, Germany |
| Dechang Pi | Nanjing University of Aeronautics and Astronautics, China |
| Dima Alhadidi | University of New Brunswick, Canada |
| Dong Li | Liaoning University, China |
| Dong Huang | South China Agricultural University, China |
| Donghai Guan | Nanjing University of Aeronautics and Astronautics, China |
| Eiji Uchino | Yamaguchi University, Japan |
| Ellouze Mourad | Université de Sfax, Tunisia |

| | |
|---|---|
| Elsa Negre | LAMSADE, Paris-Dauphine University, France |
| Farid Nouioua | University of Souk Ahras, Algeria |
| Fatma Najar | Concordia University, Canada |
| Genoveva Vargas-Solar | CNRS, France |
| Guanfeng Liu | Macquarie University, Australia |
| Guangdong Bai | University of Queensland, Australia |
| Guangquan Lu | Guangxi Normal University, China |
| Guangyan Huang | Deakin University, Australia |
| Guillaume Guerard | ESILV, France |
| Guodong Long | University of Technology Sydney, Australia |
| Haïfa Nakouri | ISG Tunis, Tunisia |
| Hailong Liu | Northwestern Polytechnical University, China |
| Hantao Zhao | Southeast University, China |
| Haoran Yang | University of Technology Sydney, Australia |
| Harry Kai-Ho Chan | Roskilde University, Denmark |
| Hongzhi Wang | Harbin Institute of Technology, China |
| Hongzhi Yin | University of Queensland, Australia |
| Hui Yin | Deakin University, Australia |
| Indika Priyantha Kumara Dewage | Tilburg University, The Netherlands |
| Jerry Chun-Wei Lin | Western Norway University of Applied Sciences, Norway |
| Jiali Mao | East China Normal University, China |
| Jian Yin | Sun Yat-sen University, China |
| Jiang Zhong | Chongqing University, China |
| Jianqiu Xu | Nanjing University of Aeronautics and Astronautics, China |
| Jing Du | University of New South Wales, Australia |
| Jizhou Luo | Harbin Institute of Technology, China |
| Jules-Raymond Tapamo | University of KwaZulu-Natal, South Africa |
| Junchang Xin | Northeastern University, China |
| Junhu Wang | Griffith University, Australia |
| Junjie Yao | East China Normal University, China |
| Ke Deng | RMIT University, Australia |
| Khanh Van Nguyen | University of Technology Sydney, Australia |
| Lei Duan | Sichuan University, China |
| Lei Li | The Hong Kong University of Science and Technology (Guangzhou), China |
| Li Li | Southwest University, China |
| Liang Hong | Wuhan University, China |
| Lin Yue | University of Queensland, Australia |
| Lizhen Cui | Shandong University, China |
| Lu Chen | Swinburne University of Technology, Australia |

| | |
|---|---|
| Lu Chen | Zhejiang University, China |
| Lu Jiang | Northeast Normal University, China |
| Lukui Shi | Hebei University of Technology, China |
| Lutz Schubert | Universität Ulm, Germany |
| Madalina Raschip | Alexandru Ioan Cuza University of Iasi, Romania |
| Maneet Singh | Indian Institute of Technology Ropar, India |
| Manqing Dong | University of New South Wales, Australia |
| Mariusz Bajger | Flinders University, Australia |
| Markus Endres | University of Augsburg, Germany |
| Mehmet Ali Kaygusuz | Middle East Technical University, Turkey |
| Meng Wang | Southeast University, China |
| Miao Xu | University of Queensland, Australia |
| Mirco Nanni | ISTI-CNR Pisa, Italy |
| Moomal Farhad | United Arab Emirates University, United Arab Emirates |
| Mourad Nouioua | Harbin Institute of Technology (Shenzhen), China |
| Mukesh Mohania | Indian Institute of Technology - Bombay, India |
| Nenggan Zheng | Zhejiang University, China |
| Nicolas Travers | Léonard de Vinci Pôle Universitaire, Research Center, France |
| Nizar Bouguila | Concordia University, Canada |
| Noha Alduaiji | Majmaah University, Saudi Arabia |
| Omar Al-Janabi | Universiti Sains Malaysia, Malaysia |
| Paul Grant | Charles Sturt University, Australia |
| Peiquan Jin | University of Science and Technology of China, China |
| Peisen Yuan | Nanjing Agricultural University, China |
| Peng Peng | Hunan University, China |
| Philippe Fournier-Viger | Shenzhen University, China |
| Pragya Prakash | Indraprastha Institute of Information Technology, India |
| Priyamvada Bhardwaj | Otto von Guericke University Magdeburg, Germany |
| Prof. Feng Yaokai | Kyushu University, Japan |
| Qing Xie | Wuhan University of Technology, China |
| Quan Z. Sheng | Macquarie University, Australia |
| Qun Chen | Northwestern Polytechnical University, China |
| Quoc Viet Hung Nguyen | Griffith University, Australia |
| Rania Boukhriss | MIRACL-FSEG, Sfax University, Tunisia |
| Rogério Luís Costa | Polytechnic of Leiria, Portugal |
| Rong-Hua Li | Beijing Institute of Technology, China |
| Sadeq Darrab | Otto von Guericke University Magdeburg, Germany |

| | |
|---|---|
| Sai Abhishek Sara | Indian Institute of Technology, Bombay, India |
| Saiful Islam | Griffith University, Australia |
| Salim Sazzed | Old Dominion University, USA |
| Sanjit Kumar Saha | Brandenburg University of Technology Cottbus – Senftenberg, Germany |
| Sayan Unankard | Maejo University, Thailand |
| Sen Wang | Griffith University, Australia |
| Senzhang Wang | Central South University, China |
| Shan Xue | University of Wollongong, Australia |
| Sheng Wang | Wuhan University, China |
| Shi Feng | Northeastern University, China |
| Shiyu Yang | Guangzhou University, China |
| Shutong Chen | Xidian University, China |
| Sonia Djebali | ESILV, France |
| Suman Banerjee | IIT Jammu, India |
| Sutharshan Rajasegarar | Deakin University, Australia |
| Tao Shen | University of Technology Sydney, Australia |
| Tarique Anwar | University of York, UK |
| Thanh Tam Nguyen | Griffith University, Australia |
| Tianchi Sha | Beijing Institute of Technology, China |
| Tianrui Li | Southwest Jiaotong University, China |
| Tiexin Wang | Nanjing University of Aeronautics and Astronautics, China |
| Tim Oates | University of Maryland Baltimore County, USA |
| Tung Kieu | Aalborg University, Danish |
| Uno Fang | Deakin University, Australia |
| Wei Chen | University of Auckland, New Zealand |
| Wei Hu | Nanjing University, China |
| Wei Emma Zhang | University of Adelaide, Australia |
| Weijun Wang | University of Goettingen, Germany |
| Weiwei Yuan | Nanjing University of Aeronautics and Astronautics, China |
| Wen Zhang | Wuhan University, China, China |
| Xiang Lian | Kent State University, USA |
| Xiangfu Meng | Liaoning Technical University, China |
| Xiangguo Sun | The Chinese University of Hong Kong, China |
| Xiangmin Zhou | RMIT University, Australia |
| Xiangyu Song | Deakin University, Australia |
| Xianzhi Wang | University of Technology Sydney, Australia |
| Xiao Pan | Shijiazhuang Tiedao University, China |
| Xiaocong Chen | University of New South Wales, Australia |
| Xiaohui (Daniel) Tao | University of Southern Queensland, Australia |

Xiaowang Zhang                 Tianjin University, China
Xie Xiaojun                    Nanjing Agricultural University, China
Xin Cao                        University of New South Wales, Australia
Xingquan Zhu                   Florida Atlantic University, USA
Xiujuan Xu                     Dalian University of Technology, China
Xueping Peng                   University of Technology Sydney, Australia
Xuyun Zhang                    Macquarie University, Australia
Yajun Yang                     Tianjin University, China
Yanda Wang                     Nanjing University of Aeronautics and
                               Astronautics, China
Yanfeng Zhang                  Northeastern University, China
Yang Li                        University of Technology Sydney, Australia
Yang-Sae Moon                  Kangwon National University, South Korea
Yanhui Gu                      Nanjing Normal University, China
Yanjun Zhang                   Deakin University, Australia
Yao Liu                        University of New South Wales, Australia
Yasuhiko Morimoto              Hiroshima University, Japan
Ye Yuan                        Beijing Institute of Technology, China
Ye Zhu                         Deakin University, Australia
Yicong Li                      University of Technology Sydney, Australia
Yixuan Qiu                     University of Queensland, Australia
Yong Zhang                     Tsinghua University, China
Yong Tang                      South China Normal University, China
Yongpan Sheng                  Chongqing University, China
Yongqing Zhang                 Chengdu University of Information Technology,
                               Australia
Youwen Zhu                     Nanjing University of Aeronautics and
                               Astronautics, Australia
Youxi Wu                       Hebei University of Technology, China
Yu Liu                         Huazhong University of Science and Technology,
                               China
Yuanbo Xu                      Jilin University, China
Yucheng Zhou                   University of Technology Sydney, Australia
Yue Tan                        University of Technology Sydney, Australia
Yuhai Zhao                     Northeastern University, China
Yunjun Gao                     Zhejiang University, China
Yurong Cheng                   Beijing Institute of Technology, China
Yuwei Peng                     Wuhan University, China
Yuxiang Zhang                  Civil Aviation University of China, China
Zesheng Ye                     University of New South Wales, Sydney, Australia
Zheng Zhang                    Harbin Institute of Technology, Shenzhen
Zhi Cai                        Beijing University of Technology, China

| Zhihui Wang | Fudan University, China |
| Zhiqiang Zhang | Zhejiang University of Finance and Economics, China |
| Zhixin Li | Guangxi Normal University, China |
| Zhixu Li | Soochow University, China |
| Zhuowei Wang | University of Technology Sydney, Australia |
| Zijiang Yang | York University, Canada |
| Zongmin Ma | Nanjing University of Aeronautics and Astronautics, China |

# Contents – Part II

## Classification, Clustering and Recommendation

## Multi-objective, Optimization, Augmentation, and Database

## Others

# Contents – Part I

## On-Device Application

## Other Application

## Pattern Mining

**Graph Mining**

# Text Mining

# Towards Idea Mining: Problem-Solution Phrase Extraction from Text

Haixia Liu[1(✉)], Tim Brailsford[1], James Goulding[2], Tomas Maul[3], Tao Tan[4], and Debanjan Chaudhuri[5]

[1] Computer Science and Creative Technology, University of the West of England, Bristol, UK
haixia.liu@uwe.ac.uk
[2] N/LAB, Nottingham University Business School, Nottingham, UK
[3] School of Computer Science, University of Nottingham Malaysia, Semenyih, Malaysia
[4] The Faculty of Applied Science, Macao Polytechnic University, Macao, China
[5] Computer Science, The University of Bonn, Bonn, Germany

**Abstract.** This paper investigates the feasibility of problem-solution phrases extraction from scientific publications using neural network approaches. Bidirectional Long Short-Term Memory with Conditional Random Fields (Bi-LSTM-CRFs) and Bidirectional Encoder Representations from Transformers (BERT) were evaluated on two datasets, one of which was created by University of Cambridge Computer Laboratory containing 1000 positive examples of problems and solutions (UCCL1000) with the corresponding phrases annotated. The F1-scores computed on the UCCL1000 dataset indicate that BERT is an effective approach to extract solution phrases (with an F1-score of 97%) and problem phrases (with an F1-score of 83%). To test the model's robustness on a different corpus with a different annotation scheme, a dataset consisting of 488 problem-solution samples from the Conference on Neural Information Processing Systems (NIPS488) was collected and annotated by human readers. Both Bi-LSTM-CRFs and BERT performances were dramatically lower for NIPS488 in comparison with UCCL1000.

**Keywords:** Text mining · Problem-solution extraction · NLP

## 1 Introduction

The discovery of original and new scientific ideas is a key phase of research innovation. This process usually starts with a literature review. Apart from researchers who are working in academia, scientists from industry and government also need to keep track of new trends. Given increasing publication rates, and the diversification of the literature into ever more specialized fields it is becoming increasingly challenging for both academic and industry researchers to decide how to most productively spend their time on selecting the important parts of a text. It is also difficult for government officers to pick up the most useful pieces of information that are available. The main goal of an abstract includes

W. Chen et al. (Eds.): ADMA 2022, LNAI 13726, pp. 3–14, 2022.
https://doi.org/10.1007/978-3-031-22137-8_1

distilling the main purpose of the corresponding paper. A paper's novel ideas are embedded in its abstract along with the problems it is solving and these can be extracted using pattern recognition. Mining scientific ideas by manually extracting them from a large body of literature tends to be massively time consuming. People can easily get lost in thousands of abstracts. Scientists in academia are trying to discover state-of-the-art methods for specific problems within their research area and are hoping to invent novel methods that are better than the existing ones; while, researchers in industry are looking for practical solutions that can be implemented and are working effectively in real scenarios. Instructors who are assessing essays online need an assistant that can automatically analyze essays [1].

In order to perform idea mining from text, a functional idea definition is crucial. Liu *et al.* [2] explored idea definition from a technical perspective, where ideas were represented by <problem, solution> pairs. How to extract the important information automatically from the text and make it structured is becoming increasingly important. In this paper, for the first time (to the best of our knowledge), two machine learning methods were compared to extract problem-solution phrases.

## 2   Related Work

A variety of methods for idea mining from text have been experimented by researchers. Thorleuchter *et al.* [3] introduced an approach for extracting ideas from unstructured text based on the length and the term weights of stop and non-stop words. The extracted ideas are represented by the retrieved words using text patterns, which are built around each targeted term in the new text. The represented words should occur on the left and right side of the non-stop words. The outputs using this method are a list of words and therefore the relations between the extracted words are lost, which makes the pattern less understandable. Some researchers investigated idea mining from the perspective of text classification rather than idea extraction. Christensen *et al.* [4] focused on classifying online community texts into Idea Text and Non-Idea Text using a supervised learning approach. They concluded that it is possible to automatically identify ideas written as text in online communities, however, their study did not provide methods for extracting ideas from text. Liu *et al.* [2] explored idea definition from a technical perspective, where ideas were represented by <problem, solution> pairs. It's stated in the paper [5] that the most important parts of the abstract are the *document problem* and *problem solution*. Liu *et al.* [2] used a part-of-speech tagging technique to extract noun-phrases from scientific publication abstracts. A rule based method was adopted to classify the noun-phrases into problems and solutions. Although <problem, solution> pairs embody an effective definition of ideas, the representation of problems and solutions is not easy to define. While the primary concepts are predominantly carried by the noun-phrases, simply using noun-phrases to represent problems and solutions is not enough. For example, from the sentence researchers have developed a computational method to predict the function of unknown yeast genes, simply using the noun phrase yeast genes to represent the research problem is not as clear as using a span of consecutive text predict the function of unknown yeast genes. In order to make the expression of the <problem, solution> more comprehensible and understandable, a span of consecutive text to represent problems and solutions are worth study. Heffernan and Teufel [6] created a new corpus containing ground truth for problem-solution strings. They also present an automatic classifier to make a binary decision about problemhood and solutionhood of a given phrase. The classifier was based on supervised machine learning methods that intake a set of 8 features. However, their experiments were focusing on distinguishing problems from non-problems and solutions from non-solutions. Moreover, the 8 features being used were hand-crafted, which is time consuming. This paper will utilize the annotated corpus by Heffernan and Teufel [6] for the task of problem-solution phrases extraction from a given sentence using neural networks.

### 2.1 Problem Formation

Considering a single-labeled sentence $T$ represented as an ordered set of $N$ words, where $T = \langle w_1, w_2, \ldots, w_N \rangle$, then the functional definitions used to extract our representation of problem-solution phrases are as follows:

**Problem-phrase:** is an ordered subset of the text determined to be a *problem* extracted from $T$:

$$\phi = \langle wp_1, wp_2, \ldots, wp_n \rangle. \tag{1}$$

**Solution-phrase:** is an ordered subset of the text determined to be a *Solution* extracted from $T$:

$$\psi = \langle ws_1, ws_2, \ldots, ws_n \rangle. \tag{2}$$

Our goal is to extract $\phi$ or $\psi$ given $T$.

As stated in the paper [7], the ground truth for problem-solution strings were defined to be at most one sentence long. The parsed dependencies were examined and some target words such as *problem* and *solution* were used as the seeds to identify subject position. Then, the syntactic arguments were chosen as the candidate Problem-Solution phrase. Semantically similar words of the target words were used to increase the variations. Examples of problem-solution phrases are shown in Fig. 1.

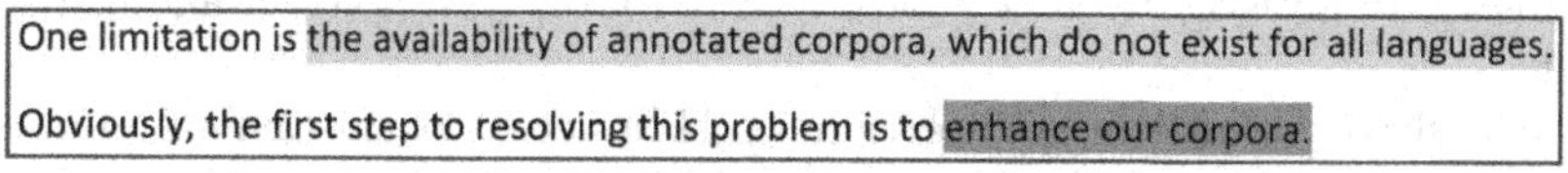

One limitation is the availability of annotated corpora, which do not exist for all languages.

Obviously, the first step to resolving this problem is to enhance our corpora.

**Fig. 1.** An example of annotated problem phrase was highlighted in yellow shown on the top. An example of annotated solution phrase was highlighted in green shown on the bottom. (Color figure online)

## 3  Methodology

Detecting Problem-Solution phrases is a form of Named Entity Recognition (NER) [8] since only parts of the sentence are considered as the target to be tagged. In order to detect Problem-Solution phrases, a classification scheme is determined based on the IO [9] format. Here, I is a token inside a chunk and O is a token outside a chunk. Although IO format cannot distinguish between adjacent chunks of the same named entity, it's suitable for our study due to our problem formation: the prediction is based on a single sentence with a single label. The goal of detecting Problem-Solution phrases is to correctly label every word in a sentence as one of the three categories: outside of the chunk (O), inside of the problem (I-P) or inside of the solution (I-S). Therefore, the three classes to be predicted are **I-P, I-S** and **O**.

### 3.1  Models for Extracting Problem-Solution Phrases

Existing models for sequence labeling are linear statistical models, such as Maximum Entropy Markov models (MEMMs) [10] and Conditional Random Fields (CRFs) [11]. Research findings have shown that the model combining bidirectional LSTM (Bi-LSTM) networks and CRF is robust and it can produce accurate tagging performance without resorting to word embedding [12,13]. Considering that the task belongs to the tagging problem category and the words surrounding the problem-solution tags have certain patterns, it is hypothesized that using Bi-LSTM-CRF to detect problem-solution phrases can give better results.

*Bidirectional LSTM.* Long Short-term Memory Networks (LSTM) [14] belong to recurrent neural network (RNN) [15]. LSTM networks are good at learning long-term dependencies. The LSTM had the ability to erase and add information to the cell states and it has regulated gates.

*Bidirectional LSTM and CRF tagging (Bi-LSTM-CRFs).* Bi-LSTM-CRFs [12] were explored in this study since the advantages of Bi-LSTM-CRFs are: (1) Bi-LSTM takes into account the information from both of the left and right side of the current word; (2) instead of predicting the label of the individual word independently, CRF has the transition matrix connecting the context with the current word. Research findings proved that Bi-LSTM-CRFs have achieved state-of-the-art performance in the task of NER [16]. While most literature focuses on extracting a relatively short span of text such as Location, Person, Organization etc., this study investigates how good Bi-LSTM-CRFs is to extract a longer span of text.

   The workflow of utilizing bidirectional LSTM networks (bi-LSTMs) and conditional random fields (CRFs) to extract problem-solution phrases closely follows the steps described in the paper [12].

   The widely used transformer based model BERT [17] was also explored on the two datasets.

## 4 Experiment

### 4.1 Dataset UCCL1000

UCCL1000 dataset was created by Heffernan and Teufel [7] on a subset of the ACL anthology[1] released in March 2016 containing 22,878 publications. A random subset of 2,500 papers was selected across the entire ACL timeline. Only documents having abstracts were considered. A ground truth for problem-solution strings was defined on the corpus. The annotated samples were independently validated for correctness by two annotators (the two authors of this paper). Correctness was defined by two criteria, which were detailed in the paper Heffernan and Teufel [7].

From the annotated sentences that passed the quality test for both independent assessors, 500 samples of positive problems, 500 samples of negative problems, 500 samples of positive solutions and 500 samples of negative solutions were randomly selected. The resulting 1000 positive samples (500 positive problems and 500 positive solutions) were used in this study.

### 4.2 Dataset NIPS488

In order to evaluate the neural networks performance on a different corpus with a different annotation scheme, a human annotated dataset is needed. Compared with the contents of a paper, abstracts have fewer licensing issues, resulting in more easily accessible data. Therefore, it's a good decision to obtain problem-solution phrases from the abstracts. A guiding principle underlying the annotation scheme was proposed: keep the sequence as short as possible, while retaining enough information to distinguish the novel contribution of the paper. Four hundred and fifty abstracts were obtained and analysed[2] from the Proceedings of the Neural Information Processing Systems conference (NIPS).

*Guidance for Annotation.* The annotation task was conducted using the abstract of the corpus. The annotation rules were as follows:

- A Problem (Solution) sequence might be a word, a list of words or an entire sentence. However, the sequence should not be separated by other words.
- For each abstract, only one problem and one solution were expected to be identified. If there was more than one problem (or solution) in an abstract, the most important one was chosen. When there are multiple Problem-Solution phrases in a paper, then the main problem (or solution) should be the one that is most related to the title.
- The chosen sequence should reflect the novelty of the paper.
- The chosen sequence should be as short as possible.
- The distance between the chosen sequence and the root of the sentence should be as close as possible.[3]

---

[1] https://aclanthology.org/.
[2] Publication years: 2008–2016.
[3] The distance could be measured by the depth level on the parsed dependency tree.

– If no Problem-Solution phrase were identified, the abstract was excluded from the analysis.

The annotations were collected by seven computer science researchers, who manually highlighted Problem-Solution phrases on printed abstracts with coloured highlighter pens. Four-hundred and fifty abstracts were examined, out of which there were 244 abstracts having the problem-solution phrases clearly stated. An example of a human annotated problem/solution phrases is shown in Examples of problem-solution phrases were shown in Fig. 2.

> The paper presents and evaluates the power of parallel search for exact MAP inference in graphical models. We introduce a new parallel shared-memory recursive best-first AND/OR search algorithm, called SPRBFAOO, that explores the search space in a best-first manner while operating with restricted memory.

**Fig. 2.** An example of human annotated abstract. The *problem* (*solution*) phrases were highlighted in yellow (blue). (Color figure online)

### 4.3 Dataset Summary

UCCL1000 dataset contains 7920 O (Outside), 7389 I-S and 6792 I-P entities. NIPS488 contains 6797 O (Outside), 2120 I-S and 1946 I-P entities. In comparison, NIPS488 is imbalanced and smaller.

### 4.4 Text Preprocessing

After basic text preprocessing such as noise removal, the next step is to make the raw text structured, which includes sentence segmentation, tokenization and token-label assignment.

Take the first sentence shown in Fig. 2 for example, let $x_1$ represent the span of text The paper presents and evaluates the power of parallel search for exact MAP inference in graphical models and $y_1$ represent their labels. Part of the output of the first step is shown in Table 1.

The second step is to build dictionaries for tokens and tags respectively by converting tokens and tags to numerical values. An uncased tokenizing mechanism was adopted, meaning that all the letters were converted into lower-cased letters. The reason to use an uncased tokenized model is that the problem-solution statements are usually case-insensitive. Each token was assigned with a unique integer, also known as index, such that a sentence was represented by a list of integers. The tokens in the pre-trained embeddings were merged to the token dictionary.

After the second step, the sentences in Fig. 2 were converted to a list of lists: [[6965, ..., 139], [15, ..., 1]], where each list represented a sentence. Similarly, each tag was represented by a unique index:

$$y = [0, 1, 2] \tag{3}$$

**Table 1.** An example of outputs after the first step of text preprocessing. The words in the second column were excerpted from the first sentence shown in Fig. 2.

| Sentence ID | Word | Tag |
| --- | --- | --- |
| Sentence:1 | For | O |
| Sentence:1 | Exact | I-P |
| Sentence:1 | Map | I-P |
| Sentence:1 | Inference | I-P |
| Sentence:1 | In | I-P |
| Sentence:1 | Graphical | I-P |
| Sentence:1 | Models | I-P |

### 4.5  Input Representations

Before training the models, all unique token indexes should be converted to meaningful input features. There are several options to represent the features, such as using one-hot-vector [18] and word embeddings [19]. Muneeb et al. [20] pointed out two major drawbacks with one-hot-vector representations: first, the length of the vector is huge and second, there is no notion of similarity between words. Word embeddings [19] have proved to be an effective representation in some NLP tasks, such as sentence classification [21] and sentiment detection [22]. In comparison with randomly initialized word embeddings, pre-trained ones carry semantic information. A lot of researchers found that a good initialization of the input layer can improve the performance of models significantly [13]. Chung *et al.* [23] explained that the learned vectors contain semantic information pertaining to the underlying spoken words, and are close to other vectors in the embedding space if their corresponding underlying spoken words are semantically similar. Song *et al.* [24] found out that pre-trained embeddings are more effective than randomized ones. Cases et al. [25] also demonstrated that pre-trained word2vec embeddings significantly outperformed random one as long as the network is properly configured.

Word embeddings are learned from raw text. A projection matrix is derived using unsupervised learning, which means, the values in the matrix are learned by maximizing the likelihood that words are predicted from their context. Each word can be represented by the corresponding row in the matrix, which is called word vector or word embedding. The dimensionality of the word vectors determines the size of the input layer. Although some researchers claimed [26] that the dimension of the word vectors should be chosen based on corpus statistics as well as NLP tasks, the empirical dimension is usually set between 50 to 300. Chung et al. [23] found out that increasing the embedding size does not always result in improved performance for their experiment of learning word embeddings from speech and they further emphasised that word embeddings of 50 dimensions are able to capture enough semantic information of the words, as the best result was obtained by them. Bairong et al. [27] investigated the different embedding vector sizes for the End-to-End Conversation Modeling task. In this experiment, pre-trained word embeddings were used as the input features. The word embedding dimension was set to 300. A vectorized representation of the input data is needed for training the models. Sequences with variable length need transformation to make sure each sequence has the same length. A post-sequence truncation method was adopted in this study, where the values were removed from the end of the sequence if it was larger then maxlen, which was set to 75/128 for Bi-LSTM-CRF/BERT respectively.

## 4.6   Training and Evaluation

*Training Bi-LSTM-CRF and BERT.* BI-LSTM-CRF models were trained for each dataset separately. Word embedding vectors trained on GoogleNews were used to initialize the embedding layer since it outperforms randomly initialized embedding vectors in the embedding layer. The hidden unit size in the BiLSTM network was set to 50 because researchers found that model performance is not sensitive to hidden layer sizes [12] and 50 units were shown to be a good option [28]. The recurrent dropout rate was set to 0.1. Default parameters for the CRF layer were adopted[4]. Each model was trained for 20 epochs with batch size 32. The Embedding layer and BiLSTM network implementations were based on keras library[5].

BERT models were trained for 20 epochs using Huggingface Bert-base-uncased pretrained model[6]. Comparisons were done between 32 (train), 32 (validation) and 4 (train), 2 (validation) batch sizes, which were named as BS32-32 and BS4-2 respectively.

*Evaluation.* k-fold cross-validation is a popular form of model validation [29]. Typically, researchers perform k-fold cross-validation using k = 5 or k = 10, as these values have been shown empirically to yield test error rate estimates that suffer neither from excessively high bias nor from very high variance [30]. Therefore, 5-fold cross-validation was adopted in this study.

F1-scores were reported in this study since the F1-score is a widely used measurement for most NER systems [31]. Because this study focused on extracting problem-solution phrases, the evaluation emphasised F1-scores for problem-solution entity recognition. In addition to F1-scores, precision and recall were reported.

## 4.7   Result Analysis

The results are shown in Table 2. The F1-scores for problem-solution phrase extraction on UCCL1000 dataset were 0.68/0.91 using Bi-LSTM-CRFs (BLC)

---

[4] https://github.com/keras-team/keras-contrib/blob/master/keras_contrib/layers/crf.py.

[5] https://keras.io/api/.

[6] https://huggingface.co/bert-base-uncased.

and 0.83/0.97 using BERT. However, the results on NIPS488 dataset are very low. Batch size strategy comparison indicated that B4-2 outperforms B32-32 on NIPS488 dataset.

**Table 2.** Results (Precision/Recall/F1-score) generated by the model Bi-LSTM-CRFs (BLC) and BERT. UCCL1000 and NIPS488 indicated the corresponding dataset that the experiments were carried out on. 33–33 and 4–2 indicated the batch-sizes for train and validation (train-validation) datasets.

| Tag-Model | Precision/Recall/F1(UCCL1000) | Precision/Recall/F1(NIPS488) |
|---|---|---|
| P-BLC32-32 | 0.64/0.72/0.68 | 0.09/0.12/0.10 |
| P-BERT32-32 | 0.79/0.85/0.82 | 0.15/0.26/0.18 |
| P-BERT4-2 | 0.81/0.85/0.83 | 0.17/0.29/0.22 |
| S-BLC32-32 | 0.89/0.94/0.91 | 0.06/0.08/0.07 |
| S-BERT32-32 | 0.97/0.98/0.97 | 0.10/0.18/0.13 |
| S-BERT4-2 | 0.95/0.98/0.97 | 0.14/0.23/0.17 |

Examples of error analysis on UCCL1000 and NIPS488 datasets are shown in Figs. 3 and 4 respectively.

```
Word                ||True ||Pred
================================
Another        : O      O
limitation     : O      O
that           : O      O
was            : O      O
discovered     : O      O
after          : O      O
the            : O      I-P
data           : O      I-P
analysis       : O      I-P
was            : O      I-P
that           : O      I-P
a              : I-P    I-P
data           : I-P    I-P
input          : I-P    I-P
error          : I-P    I-P
caused         : I-P    I-P
All            : I-P    I-P
Negative       : I-P    I-P
Chinese        : I-P    I-P
n-gram         : I-P    I-P
category       : I-P    I-P
```

**Fig. 3.** An example of problem phrase extraction error analysis on UCCL1000 dataset. The predicted problem phrases indicated that extra words were recognized as part of the problem entities but the ground truth showed that only the words in the clause are considered to be correct in this particular case.

```
Word              ||True  ||Pred
================================
We              : O       O
propose         : O       O
multiplicative  : I-S     I-S
approximation   : I-S     I-S
scheme          : I-S     I-S
MAS             : I-S     I-S
for             : O       I-S
inference       : O       O
problems        : O       O
```

**Fig. 4.** An example of solution phrase extraction error analysis on NIPS488 dataset. The word *for* was wrongly detected as part of the solution phrases.

## 5   Discussion

A problem for training and evaluation in experiments of this nature is that it is difficult to enforce consistent annotation rules due to the differing subjective perceptions of the annotators. For the NIPS488 dataset, some of the annotators highlighted the problem (solution) explanations stated in a clause, rather than selecting the actual problem (solution) names. One of the biggest challenges when extracting only one main problem-solution phrase from an abstract is dealing with multiple problem-solution phrases that exist in the same abstract. This challenge might be the reason that the model could not achieve good result on the NIPS488 dataset.

## 6   Future Work

In the future, several aspects could be improved:

*Adding a Sentence Classification Stage.* To overcome the challenge caused by the second rule in the *Guidance for annotation,* labeling each sentence with one of the labels: main-problem, main-solution, main-ps (the examined sentence contains both main problem and main solution), non-main-ps (the examined sentence is neither main problem nor main solution sentence) before extracting problem-solution phases could be useful. Seventy nine abstracts were obtained from the Journal of Machine Learning Research (JMLR79). Each sentence was annotated with one of the labels described above. The annotation was done by one computer science researcher.

*Annotation Tool for Collecting More Data.* In the future, it is possible to use crowd sourcing techniques to get the same abstracts annotated by many different people. Many more annotations from authors should be collected.

*Novel Idea Computation.* Potential useful ideas can be discovered by analyzing the problem-solution phrases that are not seen together in one abstract. Using a similar method in the paper [2], it is possible to accelerate the ideation process using a collaborative filtering algorithm, where problem phrases are considered as users and solution phrases as the items to be recommended.

# 7  Conclusion

The idea to extract problem-solution phrases from a given sentence using neural network techniques is new, to the best of our knowledge. With high quality dataset, the model Bi-LSTM-CRFs can spot meaningful patterns in text, which is intriguing and potentially valuable. It is hoped in the future, the work may contribute to novel idea computation and information retrieval (IR) in such a way that based on users' problems, the IR system can retrieve the papers that contain the solutions that can potentially solve these problems. However, although this work is promising, it needs to be repeated with larger high quality datasets.

# References

1. Azman, A., Alksher, M., Doraisamy, S., Yaakob, R., Alshari, E.: A framework for automatic analysis of essays based on idea mining. In: Alfred, R., Lim, Y., Haviluddin, H., On, C.K. (eds.) Computational Science and Technology. LNEE, vol. 603, pp. 639–648. Springer, Singapore (2020). https://doi.org/10.1007/978-981-15-0058-9_61
2. Liu, H., Goulding, J., Brailsford, T.: Towards computation of novel ideas from corpora of scientific text. In: Appice, A., Rodrigues, P.P., Santos Costa, V., Gama, J., Jorge, A., Soares, C. (eds.) ECML PKDD 2015. LNCS (LNAI), vol. 9285, pp. 541–556. Springer, Cham (2015). https://doi.org/10.1007/978-3-319-23525-7_33
3. Thorleuchter, D., Van den Poel, D., Prinzie, A.: Mining ideas from textual information. Expert Syst. Appl. **37**(10), 7182–7188 (2010)
4. Christensen, K., Nørskov, S., Frederiksen, L., Scholderer, J.: In search of new product ideas: identifying ideas in online communities by machine learning and text mining, Creativity & Innovation Management (2016)
5. Trawiński, B.: A methodology for writing problem structured abstracts. Inf. Process. Manage. **25**(6), 693–702 (1989)
6. Heffernan, K., Teufel, S.: Identifying problem statements in scientific text. Patrick Saint-Dizier (2018)
7. Heffernan, K., Teufel, S.: Identifying problems and solutions in scientific text. Scientometrics **116**(2), 1367–1382 (2018). https://doi.org/10.1007/s11192-018-2718-6
8. Sang, E.F., De Meulder, F.: Introduction to the CoNLL-2003 shared task: language-independent named entity recognition. Arxiv **21**(08), 142–147 (2003)
9. Krishnan, V., Ganapathy, V.: Named Entity Recognition. Stanford Lecture CS229 (2005)
10. Mccallum, A., Freitag, D., Pereira, F.C.N.: Maximum entropy Markov models for information extraction and segmentation. In: Proceedings of ICML, pp. 591–598 (2000)
11. Lafferty, J., et al.: Conditional random fields: probabilistic models for segmenting and labeling sequence data. In: Proceedings of the Eighteenth International Conference on Machine Learning, ICML, vol. 1, pp. 282–289 (2001)
12. Huang, Z., Xu, W., Yu, K.: Bidirectional LSTM-CRF models for sequence tagging. Comput. Sci. (2015)
13. Lample, G., Ballesteros, M., Subramanian, S., Kawakami, K., Dyer, C.: Neural architectures for named entity recognition, arXiv preprint arXiv:1603.01360 (2016)

14. Hochreiter, S., Schmidhuber, J.: Long short-term memory. Neural Comput. **9**(8), 1735–1780 (1997)
15. Mikolov, T., Karafiát, M., Burget, L., Černocký, J., Khudanpur, S.: Recurrent neural network based language model. In: Eleventh Annual Conference of the International Speech Communication Association (2010)
16. Mai, K., et al.: An empirical study on fine-grained named entity recognition. In: Proceedings of the 27th International Conference on Computational Linguistics, pp. 711–722 (2018)
17. Devlin, J., Chang, M.-W., Lee, K., Toutanova, K.: BERT: pre-training of deep bidirectional transformers for language understanding, arXiv preprint arXiv:1810.04805 (2018)
18. Uriartearcia, A.V.: One-hot vector hybrid associative classifier for medical data classification. PLoS ONE **9**(4), e95715 (2014)
19. Mikolov, T., Chen, K., Corrado, G., Dean, J.: Efficient estimation of word representations in vector space, arXiv preprint arXiv:1301.3781 (2013)
20. Muneeb, T., Sahu, S., Anand, A.: Evaluating distributed word representations for capturing semantics of biomedical concepts. In: Proceedings of BioNLP 2015, pp. 158–163 (2015)
21. Komninos, A., Manandhar, S.: Dependency based embeddings for sentence classification tasks. In: Conference of the North American Chapter of the Association for Computational Linguistics: Human Language Technologies, pp. 1490–1500 (2016)
22. Xue, B., Fu, C., Zhan, S.: A study on sentiment computing and classification of Sina Weibo with word2vec. In: IEEE International Congress on Big Data, pp. 358–363 (2014)
23. Chung, Y.-A., Glass, J.: Speech2vec: a sequence-to-sequence framework for learning word embeddings from speech. In: Proc. Interspeech **2018**, pp. 811–815 (2018)
24. Song, Y., Shi, S.: Complementary learning of word embeddings. In: IJCAI, pp. 4368–4374 (2018)
25. Cases, I., Luong, M.-T., Potts, C.: On the effective use of pretraining for natural language inference, arXiv preprint arXiv:1710.02076 (2017)
26. Patel, K., Bhattacharyya, P.: Towards lower bounds on number of dimensions for word embeddings. In: Proceedings of the Eighth International Joint Conference on Natural Language Processing (Volume 2: Short Papers), pp. 31–36 (2017)
27. Bairong, Z., Wenbo, W., Zhiyu, L., Chonghui, Z., Shinozaki, T.: Comparative analysis of word embedding methods for DSTC6 end-to-end conversation modeling track. In: Proceedings of the 6th Dialog System Technology Challenges (DSTC6) Workshop (2017)
28. Trivedi, S., Rangwani, H., Singh, A.K.: IIT (BHU) submission for the ACL shared task on named entity recognition on code-switched data. In: Proceedings of the Third Workshop on Computational Approaches to Linguistic Code-Switching, pp. 148–153 (2018)
29. Geisser, S.: The predictive sample reuse method with applications. J. Am. Stat. Assoc. **70**(350), 320–328 (1975)
30. James, G., Witten, D., Hastie, T., Tibshirani, R.: An Introduction to Statistical Learning. STS, vol. 103. Springer, New York (2013). https://doi.org/10.1007/978-1-4614-7138-7
31. Pal, B., Tarafder, A.K., Rahman, M.S.: Synthetic samples generation for imbalance class distribution with LSTM recurrent neural networks. In: Proceedings of the International Conference on Computing Advancements, pp. 1–5 (2020)

# Spam Email Categorization with NLP and Using Federated Deep Learning

Ikram Ul Haq[1(✉)], Paul Black[1], Iqbal Gondal[1], Joarder Kamruzzaman[1], Paul Watters[2], and A. S. M. Kayes[2]

[1] School of Science, Engineering and Information Technology, ICSL, Federation University, Melbourne, Australia
{i.ulhaq,p.black,iqbal.gondal, joarder.kamruzzaman}@federation.edu.au
[2] Department of Computer Science and Information Technology, La Trobe University, Ballarat, Australia
{p.watters,a.kayes}@latrobe.edu.au

**Abstract.** Emails are the most popular and efficient communication method that makes them vulnerable to misuse. Federated learning (FL) provides a decentralized machine learning (ML) model, where a central server coordinates clients that collaboratively train a shared ML model. This paper proposes Federated Phishing Filtering (FPF) technique based on federated learning, natural language processing, and deep learning. FL for intelligent algorithms fuses trained models of ML algorithms from multiple sites for collective learning. This approach improves ML performance by utilizing large collective training data sets across the corporate client base, resulting in higher phishing email detection accuracy. FPF techniques preserve email privacy using local feature extraction on client email servers. Thus, the contents of emails do not need to be transmitted across the network or stored on third-party servers. We have applied FL and Natural Language Processing (NLP) for email phishing detection. This technique provides four training modes that perform FL without sharing email content. Our research categorizes emails as benign, spam, and phishing. Empirical evaluations with publicly available datasets show that accuracy is improved by the use of our Federated Deep Learning model.

**Keywords:** Spam detection · Phishing detection · Federated learning · Model averaging · Deep learning · Privacy-preserving · TF/IDF · Incremental learning

## 1 Introduction

Emails are a common and effective communication tool, which makes them susceptible to misuse. Among the cyberattacks propagating through email, phishing emails exploit users for financial gain. Phishing is "a scalable act of deception whereby impersonation is used to obtain information from a target" [1]. Phishing emails can be used for fraud involving the spoofing of reputable companies, creating a plausible premise, collecting information, linking to websites that gather information, hiding links or hostnames,

W. Chen et al. (Eds.): ADMA 2022, LNAI 13726, pp. 15–27, 2022.
https://doi.org/10.1007/978-3-031-22137-8_2

switching ports, and having incorrect sender address [2]. While spam emails may contain malicious software, illegal advertising, and fraud schemes. The business sector is increasingly a major victim of ransomware, and phishing attacks. While safety precautions are taken, humans often fail to recognize such attacks, resulting in substantial financial losses. This paper presents a new approach to combat phishing campaigns, by federating the knowledge of these phishing campaigns for all of the organization's clients. For example, banks have millions of net banking customers, which could share their phishing email data with banks to implement effective phishing filtering schemes. As FL extracts features using the customer's computer, the privacy of banking customers is protected.

FL provides a decentralized ML model, where a central server coordinates clients that collaboratively train a shared ML model [3]. FL allows corporations to improve client email systems' security by creating a collective phishing detection model utilizing email intelligence from a large customer base. Although various spam detection techniques are already proposed, these techniques mostly use conventional ML approaches for spam categorization and are not using FL models in real-time environments. The key tasks undertaken by the proposed methodology include: Email classification in a real-time FL environment. The accurate classification of client emails requires preparation of the training model using the selected features from client emails. As this is a real-time system, there is a need for real-time model retraining using the new email data. Customer input is used to highlight undetected phishing emails. One of the strategies proposed in this paper is to fuze the trained models of all the clients. So there is a need for data fusion when the training data size is known and unknown.

This paper's objective is to classify emails and feature extraction has been performed using NLP techniques to reveal the distinct inherent structural characteristics that are commonly present in the emails of various categories. ML techniques will explore and enhance the capability of deep neural networks with FL for spam email detection and classification. As a result, users will be better protected against financial fraud resulting from email-based malware.

## 2    Related Work

Phishing and spam detection remain a huge challenge for the email systems despite ML-based approaches' development. Key prior works are presented below to give context to this paper's contribution.

[4] propose spam detection and there are two variants of the proposed technique: one is the "Basic MailRank", which computes a global reputation score for each email address, whereas in the 'Personalized MailRank' the score of each email address is different for each MailRank user. MailRank is highly resistant to spam attacks and performs well for sparse networks. Chi-Yao et al. [5] offered Incremental SVM model for spam detection on dynamic social networks. The authors' technique is an incremental work to efficiently retrain the SVM model. The technique was evaluated on a live data set from a university email server and proved to be efficient and effective. A text clustering-based spam detection system was developed by [6]. This spam detection technique uses text clustering based on a vector space model. Results have shown that the technique is effective in spam detection for unsupervised clustering.

Another spam detection is proposed by Damiani et al. [7], which is an open digest-based technique for spam detection. Its main purpose was to overcome digest-based spam detection issues. Research by Mohammad et al. [8] has streamlined the important features that influence phishing prediction. Spam identification based on the header features of emails is proposed by [9]. B. et al. [10] have proposed a spam detection technique, however it's a binary classifier and they have not considered email header features.

Verma et al. [11] highlight that NLP is challenging and there is very limited research on NLP for phishing detection. One of the attempts was by [12], however, authors have focused on features like stopwords, words, punctuation, and counts of stopwords. Seymour & Tully [13] highlight that NLP is used to extract patterns from raw-text and recurring patterns can be identified from phishing emails so NPL can be helpful in phishing detection.

Mehta et al. [14] highlight concept drift (CD) challenges that arise when ML is used. Gepperth & Hammer [15] indicate that when data samples' temporal structure is considered, changes occur over time. [16] Zliobaite mentions that CD refers to a problem of non-stationary learning over time or changes in the data distribution over time.

## 3   Federated Phishing Filter (FPF)

The detection and categorization of malicious emails requires an effective feature extraction mechanism. In this paper, the relevance and importance of the extracted features have been investigated for the development of an accurate phishing filtering system. Once the features are extracted and selected, then a multi-class classifier is trained using a deep learning algorithm for the real-time FL environment.

Since emails are textual, the text in the email contains inherent ambiguity, but the rich structure of natural languages can provide meaningful features for better classification. Hence, NLP is a good candidate for email feature extraction. In this research, we consider email headers, links, and attachments in addition to email body text.

### 3.1   Natural Language Processing

To map the selected features to the input data format for the training of a ML model, we use the Term-Frequency Inverse Document Frequency (TF/IDF) vectorization technique [17]. The email body is first summarized in this technique before converting to a TF/IDF vector. TF/IDF can be explained with Eq. 1 [18].

$$w_{ij} = tf_{ij} \times \log(\frac{N}{df_i}) \tag{1}$$

where $tf_{ij}$ = the number of occurrences of $i$ in $j$, $df_i$ is the number of documents containing $i$, and $N$ is the total number of the documents in the corpus. TF/IDF score in Eq. (1), can be explained with an example. Consider an email that contains 1,000 words, where the word "money" appears 3 times. The term frequency (TF) for "money" is then (3/1000) = 0.003. Now, in 10,000 emails, the word "money" appears in 1,000 of them. Therefore, the inverse document frequency (IDF) is calculated as log (10,000/1000) = 1. Thus, the TF-IDF weight is the product of these quantities is 0.003 * 1 = 0.12, and is used as a feature.

### 3.2 Deep Learning Model for Spam Categorization

Heaton et al. [19] explains that Deep learning is a branch of ML in which an Artificial Neural Network (ANN) learns from a huge volume of data. Deep learning does not require careful engineering and domain knowledge to perform feature extraction, because it automatically extracts important features from the raw data [20].

### 3.3 Spam Detection and Categorization Model

Our proposed model applies NLP to email body data and uses TF/IDF to perform the feature extraction and encoding. The deep learning spam detection and categorization model has been trained in four different ways to achieve a high spam detection accuracy. Our proposed model is explained in Fig. 1.

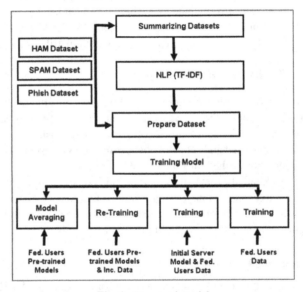

**Fig. 1.** Proposed model

### 3.4 Federated Learning

Unlike traditional ML, an FL approach generates a shared model based on decentralized training of user data. FL allows users' local models to learn from the shared and federated prediction model while retaining the training data locally, thus avoiding the need for centralized training [21]. Figure 2 shows a generic FL environment diagram.

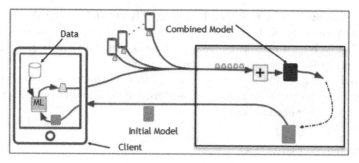

**Fig. 2.** Federated learning

FL provides a decentralized collective ML model [33]. Update to model weights in FL can be expressed with Eq. 2.

$$\theta \leftarrow \theta - \sum_{i=1}^{K} \frac{n_i}{N} H_i \tag{2}$$

where $\theta$ are the weights of the model and k are the selected federated users and $N = \sum_{i=1}^{K} n_i$ is the total number of data points used in this iteration.

In this paper, Differential Privacy (DP) has been used to preserve the data of participating clients. DP in ML fits a model without knowing details of an individual's data. A learning algorithm computes an estimate in each of T iterations. The data used to compute the estimate is sampled using a probability q. The sensitivity of the estimate is bounded by a constant d, and noise sampled from N (0, $\sigma2$, $\delta2$) is added to the estimate in each iteration. To compute the weights of the next iteration, the estimate is subtracted from the current weights. Then, constants c1, c2 exist so that the algorithm is ($\varepsilon$, $\delta$)—differentially private for any $\varepsilon < c1q2T$ and $\delta > 0$ if noise is added. FL is differentially private and can be expressed with Eq. 3.

$$\sigma < c2 \frac{\sqrt[q]{T log(\frac{1}{\delta})}}{\varepsilon} \tag{3}$$

### 3.5 Federated Training Models

We have proposed four different models for the FPF. Any of these training scenarios may be used, depending on the availability of the data samples or pre-trained models.

**Training from Server Model (TSM):** This model assumes a pre-trained model. All federated users provide their extracted features data to the server. At initialization, FPF uses pre-trained model, then retrains incrementally during steady-state operation, using the federated users' data.

**Training from New Data (TND):** IN TND, all federated users provide their extracted features to the server, and the FPF consolidates the user's features and then trains the users' models.

In **Re-Training with Incremental Learning (TIL)** model, federated users provide their pre-trained models and new data. The system re-trains user models with incremental learning. [21] highlights that incremental learning is required in interactive scenarios where training examples are given based on human input over time.

In **Model Averaging (MA)** method, federated users are not required to supply their data. Instead, they provide their pre-trained models, and the federation system calculates the mean of the pre-trained models.

### 3.6  Federated Averaging (FA)

In addition to being suitable for environments with sensitive data, FA helps to minimize client-to-server and server-to-client communication cost. Users train the generic neural network model with an FA algorithm, and the trained weights are sent back to the server. To return the final weights, the server then takes the average of the updates. A typical FA diagram is shown in Fig. 3.

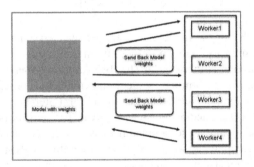

**Fig. 3.**  Federated averaging

### 3.7  Federated Averaging Strategies

FA uses the averaging equation provided by [35] in Eq. 4.

$$f(w) = \sum_{k=1}^{K} \frac{nk}{n} (Fk(w)) \tag{4}$$

where

$$Fk(w) = \frac{1}{nk} \sum_{i \in P_k} f_i(w) \tag{5}$$

If the partition $Pk$ was formed by distributing the training examples over the clients uniformly at random, then we would have $EP_k[F_k(w)] = f(w)$, where the expectation is over the set of examples assigned to a fixed client $k$. Two federated model averaging strategies were developed in this research. The Weighted Average Strategy (WAS) is based on the training model dataset size, while the Equal Weighting Strategy (EWS) is independent of the dataset size on which model was training.

### 3.8   Equal Weighting (EWS)

Using EWS, models are averaged regardless of the dataset size on which the model was trained. This strategy can be used, if the training dataset size is unknown or, a model is trained with the datasets of the same size. In this strategy, there are no extra over-heads to calculate a weighted score based on the training model size. It is less time-consuming and more ideal for models of the same size datasets. Model averaging in EWS can be explained in Eq. 6.

$$avg_{ew} = \int_{k}^{N} avg(l \in y^m)$$   (6)

where $f$ is an average weighting function for a given federated network $N$ for $k$ federated users for all layers $l$ on federated users models $m$.

### 3.9   Weighted Average (WAS)

With the use of the WAS, models are averaged with a weight based on the size of the training dataset. The model averaged by this strategy has higher accuracy as compared to the model averaged by EWS strategy. Model averaging in was can be explained in Eq. 7.

$$avg_{wa} = \int_{k}^{N} avg\Delta(l \in y^m)$$   (7)

where $f$ is a weighting average function, for a given federated network $N$ for $k$ federated users for all layers $l$ on federated users models, $m$. $\Delta$ is a weight score, which can be further explained with Eq. 8.

$$\Delta = \frac{s^m}{\sum D_t}$$   (8)

where $D_t$ is the total training data size of all the models, and $s^m$ is the size of a particular model $m$.

### 3.10   Datasets

The publicly available Ham/Spam & Phishing datasets from CEAS2008, Spam Assassin, Jose Phishing, and Untroubled.org, data-sources were used in this research. These datasets were combined to avoid class imbalance problems. Two datasets were prepared from the combined dataset for the empirical evaluation. One dataset was prepared using email header fields only (Dataset-H), while the other dataset was developed using a combination of email headers field and NLP features of email contents (Dataset-NLP).

## 4   Empirical Evaluation

An empirical evaluation of the deep learning model was performed using the datasets described in Sect. 3.8. As the characteristics of phishing emails change over time, phishing detection systems need to be trained with new features on an ongoing basis. This is the concept which underpins our proposed approach in developing an FPF. We have validated our models using average accuracy.

A loss function in supervised learning quantities how close a prediction of a model is to the correct answer $y_i$. The full loss for a prediction function $f$ is given by Eq. 9.

$$L = \frac{1}{n} \sum_{i=1}^{n} loss\left( \int (x_i), y_i \right)$$ (9)

where $f(x_i)$ is prediction function for a correct answer $y_i$.

The templates of Spam and Phishing emails are thought to change over time. In real-time systems, it's natural that due to changing dynamics of the environment of a system, the relationship between inputs and outputs changes, and this CD phenomena can be captured in form of datasets. In this paper, it is anticipated that there will be CD in the federated phishing detection system, so experiments have been conducted to study the phenomena.

In the CD experiment, the FPF was trained on Dataset1, and then testing was performed on a dataset created by mixing an increasing amount of a second independent dataset (Dataset2). This experiment aims to investigate the ability of the FPF to perform well in the presence of CD in the test dataset. Dataset1 was created from CEAS2008 Ham and Spam data; and the Jose phishing dataset. Dataset2 was created from the Spam Assassin Ham and Spam dataset, and the Jose phishing dataset.

The CD experiment investigates the response of federated ML to a test dataset that is changing over time. In conventional ML, this CD may result in reduced classification accuracy. In this experiment, Data1 provides a baseline dataset. An independent dataset (Dataset 2) is progressively mixed with Dataset 1 to simulate CD in the test dataset. Table 1 shows the split of the datasets used for training purposes.

**Table 1.** Data splits for concept drift study

| Data split | Dataset 1 | Dataset 2 |
|------------|-----------|-----------|
| No split   | 100 (CEAS2008) | 0 |
| Split1     | 90        | 10        |
| Split2     | 80        | 20        |
| Split3     | 70        | 30        |
| Split4     | 60        | 40        |
| Split5     | 50        | 50        |

The testing results, shown in Fig. 4 show an initial decrease in detection accuracy when the initial Dataset2 data is added to the test dataset, but there is an improvement in accuracy with continual re-training with increasing portions of data2 in the test dataset. The results also shows that the FPF performs well in CD.

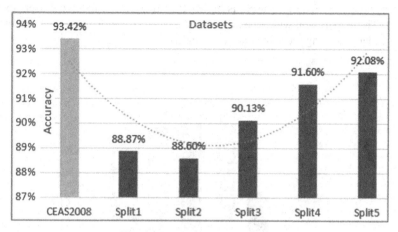

**Fig. 4.** Results with concept drift

## 4.1 Comparison of EWS and AWS Averaging Strategies

This section provides a comparison of the EWS and AWS averaging strategies. First, the four models described in Sect. 3.5 have been trained on various datasets, containing samples of sizes: 2k, 5k, 6k, 7k, and 10k. Then these four trained models were averaged with two EWS and WAS strategies. Finally, testing was performed using both the EWS and the AWS models is shown in Fig. 5.

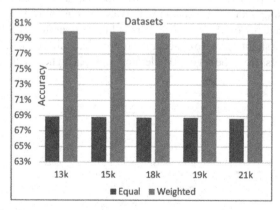

**Fig. 5.** Accuracy comparison with equal & weighted averaging

Results in Fig. 5 show that if models are trained on different datasets sizes, then models averaging with a weighted average gave better results. Further, we have identified that federated users benefit from the collective intelligence of all the users. To demonstrate this, phishing email detection accuracy studies have been conducted on four federated users and one non-federated user. Studies used dataset-H and the results in Fig. 6 show that the accuracy for federated users is higher than that of non-federated users except for user1. This experiment was done to compare the classification accuracy of federated and non-federated using the same dataset (dataset-H).

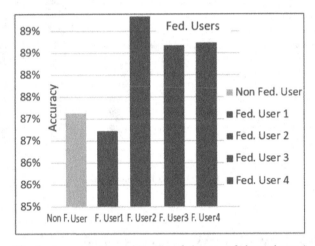

**Fig. 6.** Accuracy comparison (non-fed user vs federated users)

It is hypothesized, that the higher number of users in the federation will result in a higher collective intelligence for FL systems. As a result the accuracy of spam email detection should be higher. Accuracy studies were conducted for five individual federated users to demonstrate the claim. For the experimental purpose, a network of 5 and 10 federated users was prepared and accuracy was measured between these two sets of federated users using dataset-H (headers only data). Accuracy results in Fig. 7 depict that the system accuracy of 10 federated users is higher than that of 5 federated users. Although, there is a slight decrease of accuracy for one federated user, the accuracy of all other federated users (from the set of 10 federated users) is higher (with an improvement of more than 5%).

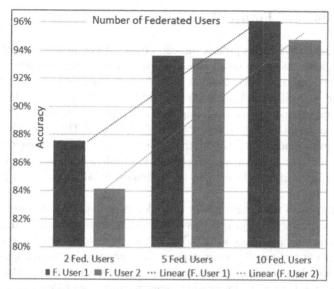

**Fig. 7.** Trendline showing accuracy increase with fed user count

## 4.2 Features Performance Comparison

Emails' header features have been used in ML previously. This section improves the spam email detection accuracy with the addition of NLP features to the header features. NLP features were extracted using TF/IDF vectorization technique. Accuracy was calculated using datasets containing NLP (dataset-NLP) features and non-NLP data (Headers only data dataset-H). Average accuracy was calculated for 5 federated users. Accuracy results in Fig. 8, show that accuracy with NLP data (dataset-NLP) is higher as compared with the accuracy of headers only data (dataset-H).

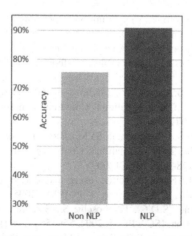

**Fig. 8.** Classification Accuracy - Email Header Vs Body Features

## 5   Conclusion and Future Work

The research conducted in this project was carried out on a limited number of federated users, and we have assumed that all federated users have the same number of features in their datasets. In future work, evaluation can be performed using a larger number of federated users with different features. Another future work is to develop other strategies for model averaging, including a weighting score of the class count in the training model. Other could be the weighting of True Positive (TF) classes. This research has shown that the addition of more features with the application of natural language processing has increased the accuracy. In this research word embedding implementation of TF/IDF was used. Future work should include an evaluation of additional word embedding techniques.

Although many of the techniques for spam detection have been developed, most of these techniques do not classify multi-class labelled data. Also, traditional ML techniques use strategies include that do not work in highly secure environments. These techniques are not used in our real-time/FL environments. Our proposed innovative approach uses NLP for FL and deep learning, which not only detects spam but also categorizes the spam. Our proposed technique trains and average the model for federated users without sharing the extracted features. We developed two strategies for model averaging in the FL environment. The proposed technique can train the model in four different ways using FL without sharing actual email data. Empirical evaluations have shown that with the proposed technique, accuracy is improved and federated user's data privacy is preserved. The developed technique can also cope with CD and multi-class data in real-time.

**Acknowledgment.** This research is an Industry Co-Funded Project, sponsored by Oceania Cyber Security and was conducted in the Internet Commerce Security Lab (ICSL).

## References

1. Lastdrager, E.: Achieving a consensual definition of phishing based on a systematic review of the literature, Crime Science (2014)
2. Drake, C.E., Oliver, J.J., Koontz, E.J.: Anatomy of a phishing email. In: CEAS (2004)
3. Kairouz, P., et al.: Advances and open problems in federated learning, arXiv preprint arXiv: 1912.04977 (2019)
4. Chirita, P.-A., Diederich, J., Nejdl, W.: Mailrank: using ranking for spam detection. In: Proceedings of the 14th ACM International Conference on Information and Knowledge Management, pp. 373–380 (2005)
5. Tseng, C.-Y., Chen, M.-S.: Incremental SVM model for spam detection on dynamic email social networks. In: 2009 International Conference on Computational Science and Engineering, vol. 4, pp. 128–135. IEEE (2009)
6. Sasaki, M., Shinnou, H.: Spam detection using text clustering. In: 2005 International Conference on Cyberworlds (CW 2005), pp. 4-pp. IEEE (2005)
7. Damiani, E., di Vimercati, S.D.C., Paraboschi, S., Samarati, P.: An open digest-based technique for spam detection. ISCA PDCS **2004**, 559–564 (2004)
8. Mohammad, R.M., Thabtah, F., McCluskey, L.: Phishing websites features. University of Huddersfield, School of Computing and Engineering (2015)

9. Qaroush, A., Khater, I.M., Washaha, M.: Identifying spam e-mail based-on statistical header features and sender behavior. In: Proceedings of the CUBE International Information Technology Conference, pp. 771–778 (2012)
10. Abhila, B., Koushika, M., Joseph, M.N., Dhanalakshmi, R.: Spam detection system using supervised ML. In: 2021 International Conference - ICSCAN, pp. 1–5 (2021)
11. Verma, R., Shashidhar, N.K., Hossain, N.: Automatic phishing email detection based on natural language processing techniques, Mar. 5 2015, uS Patent App. 14/015,524
12. Buber, E., Diri, B., Sahingoz, O.K.: Detecting phishing attacks from url by using NLP techniques. In: 2017 International Conference on Computer Science and Engineering (UBMK), pp. 337–342. IEEE (2017)
13. Seymour, J., Tully, P.: Weaponizing data science for social engineering: automated e2e spear phishing on twitter. Black Hat USA **37**, 1–39 (2016)
14. Mehta, S., et al.: Concept drift in streaming data classification: algorithms, platforms and issues. Procedia Comput. Sci. **122**, 804–811 (2017)
15. Gepperth, A., Hammer, B.: Incremental learning algorithms and applications. In: European Symposium on Artificial Neural Networks (ESANN) (2016)
16. Zliobaite, I.: Learning under concept drift: an overview (2010). arXiv preprint arXiv:1010. 4784
17. Trstenjak, B., Mikac, S., Donko, D.: Knn with tf-idf based framework for text categorization. Procedia Eng. **69**, 1356–1364 (2014)
18. Gopalakrishnan, R., Venkateswarlu, A.: Machine Learning for Mobile: Practical guide to building intelligent mobile applications powered by machine learning (2018)
19. Heaton, J., Polson, N.G., Witte, J.H.: Deep learning in finance, arXiv preprint arXiv:1602. 06561 (2016)
20. Roy, A., et al.: Systems and information engineering design symposium (SIEDS). IEEE **2018**, 129–134 (2018)
21. Hartmann, F.: Federated learning, línea (2018). https://orian.github.io/federated-learning/. Ultimo acceso. 15 Oct 2019

# SePass: Semantic Password Guessing Using k-nn Similarity Search in Word Embeddings

Maximilian Hünemörder[1]([✉])([iD]), Levin Schäfer[1], Nadine-Sarah Schüler[2], Michael Eichberg[3], and Peer Kröger[1]

[1] Christian-Albrechts Universität, Kiel, Germany
`mah@informatik.uni-kiel.de`
[2] Ludwig-Maximilians-Universität, Munich, Germany
[3] Federal Criminal Police Office, Wiesbaden, Germany

**Abstract.** Password guessing describes the process of finding a password for a secured system. Use cases include password recovery, IT forensics and measuring password strength. Commonly used tools for password guessing work with passwords leaks and use these lists for candidate generation based on handcrafted or inferred rules. These methods are often limited in their capability of producing entirely novel passwords, based on vocabulary not included in the given password lists. However, there are often semantic similarities between words and phrases of the given lists that are highly relevant for guessing the actual used passwords. In this paper, we propose *SePass*, a novel method that utilizes word embeddings to discover and exploit these semantic similarities. We compare SePass to a number of competitors and illustrate that our method not only is on par with these competitors, but also generates a significant higher amount of entirely novel password candidates. Using SePass in combination with existing methods, such as PCFG, improves the number of correctly guessed passwords considerably.

**Keywords:** Password guessing · Password cracking · Semantic word embeddings · Similarity search · Nearest neighbors · Law enforcement · Nlp

## 1 Introduction

Password-protected devices such as notebooks, tablets, smartphones or secure hard drives are ubiquitous and, thus, can be central to criminal investigations. In such cases, gaining access to these devices might lead to crucial evidence and may help preventing further crime.

Up until today, passwords are still the primary mechanism to protect a user's private information, even though additional measures, such as two-factor authentication, are steadily added. A huge benefit of passwords is that they do not involve additional devices or resources and are safe if the underlying passwords

---

M. Hünemörder and L. Schäfer—Contributed equally to this research.

© The Author(s), under exclusive license to Springer Nature Switzerland AG 2022
W. Chen et al. (Eds.): ADMA 2022, LNAI 13726, pp. 28–42, 2022.
https://doi.org/10.1007/978-3-031-22137-8_3

have enough entropy. In that case the possible search space is plainly too large to be attacked using brute force search, at least in any reasonable time frame.

Compared to application scenarios, such as internet forums or online accounts, mobile device users need to type in their password to unlock their device frequently and need to be able to remember them. Using password managers or similar tools is usually not practicable to unlock the devices themselves. This typically encourages users to utilize passwords which consist of or resemble real words and are usually relevant to their everyday life, their culture or social environment with little to no modification. Nevertheless, guessing passwords in this context remains a significant challenge.

The most common approach to password guessing is a deductive approach: it uses dictionaries based on previous leaks, e.g., the *rockyou* leak [4,5], potentially combined with some proven set of rules, e.g., provided by tools like hashcat [13], to derive password candidates. This is often enough to guess a certain amount of passwords but it is obviously bounded by the limits of the deductive model, i.e., by the dictionary and the rule set.

This limitation cannot be overcome by extending the model, e.g., by using a more general purpose dictionary or ontologies, which significantly enlarges the search space or performing a brute force attack, which in turn renounces a focused strategy to traverse the search space for generating promising candidates.

In contrast, a data-driven approach is more promising since it predicts candidates without being limited by predefined terms or rules. Recently, machine learning methods using statistical models (e.g. [17]) or deep learning (e.g. [9]) have reported promising results for password guessing in general.

However, these methods may be typically too generic in specific applications and, thus, fail to incorporate the hidden semantics of typical passwords found in leaks. Even though these methods may be able to guess passwords that are based on vocabulary not included in the training set (i.e., the leaked lists).

For example, when analyzing famous leaks, it becomes evident that one domain for passwords are the names of luxury brands. But, even if a leak already contains brand names such as *Armani* and *Chanel* none of the existing tools would propose a password based on *Burberry* because this term would be syntactically too different from the previous two passwords – though being an *obvious* candidate. Furthermore, these predictions can usually not be tried out in any practical time frame.

In order to address this shortcoming a method is needed that is able to extend the vocabulary used for the predictions, i.e., new terms not seen in the training set. We propose **SePass**, a method to generate passwords based on the vocabulary of an existing password list by semantically extending the given vocabulary using word embeddings. Focused leaks, most prominently the rockyou leak, often trainingshow semantic similarities between words and phrases. Our proposed method SePass uses pretrained word embeddings to suggest additional, semantically similar words. These plain words could potentially be the basis for passwords used by people belonging to the same peer-group. We refer to these words as *base words*.

Real passwords are built from such base words but usually follow certain rules of modification or have additional characters added, i.e., prefixes/postfixes to base words and/or combinations of multiple base words. By applying extracted password mangling rules to newly generated base words from our word embedding, we can extend the given list with additional passwords candidates exclusively found by SePass. We empirically show in our experiments that this will generate password candidates that are not produced by other methods. More importantly Sepass provides a new foundation on which existing or future methods can be built upon. Our experiments show, that SePass improves the prediction accuracy when combined with other existing methods. To summarize, the contributions of our paper are:

- We provide SePass, – to the best of our knowledge – the first method for password generation that unravels the hidden semantics in a password list by using word embeddings and, thus, is capable to semantically extend a given set of passwords.
- We present a working prototype implementation of a tool that addresses the creation of password candidates for people belonging to the same peer-group.
- We conduct an experimental study under realistic constraints comparing SePass with several state-of-the-art password generation methods.
- Our experiments show that using our proposed method as an augmentation to already existing password guessing methods, will improve both, the precision (number of correctly guessed passwords) and the effective time consumption (i.e. the number of guesses needed).

## 2   Related Work

Password guessing denotes the task of exactly matching an unknown string of characters used as a password for any kind of security system. Use cases include password recovery, IT forensics and measuring password strength. More generally, guessing a password is achieved by sequentially trying out password candidates until the correct one, then called a hit, is found. To be precise, passwords are generally not stored in clear text but rather as hash values. This requires that the true password must be recovered and is not readily available and therefore also limits the damage done by possible leaks. The used hash functions, such as *Sha512*, *PBKDF2* or scrypt, vary greatly in complexity but generally try to ensure that deriving the hash value from a given password always takes considerable time even on the most modern and specialized hardware. For the remainder of this paper, we only consider the basic problem of guessing the correct password. Hence, the concrete hash function is not relevant and is not further considered.

Password guessing methods differ by the way password candidates are generated. The most common methods for password guessing are brute force and dictionary attacks. A *brute force attack* consists of trying out all possible combinations of possible characters from a chosen alphabet to generate a password of a certain length. While brute forcing is the only method that guarantees a hit, it also evidently becomes unfeasible with increasing password length. *Dictionary attacks* on the other hand depend on lists of possible passwords, which are often times collected or designed by experts. Another common source of dictionaries are data breaches and passwords leaks, e.g. the *rockyou* leak consisting of more than 14 million password from the eponymous forum in 2009 [5].

While the methods mentioned above excel at generating vast amounts of password candidates they do not consider the plausibility of these passwords. In contrast, statistical password guessing methods utilize statistics based on existing password lists to focus on probable password candidates.

Assuming that human-created passwords are unlikely to be random combinations of characters, but rather follow a natural distribution stemming from the mother language of the user, password generation can be seen as a natural language processing (NLP) problem. Therefore statistical methods can be used to model the letter or character distribution of existing password lists and then sample new passwords. These kinds of methods aim for a high accuracy at a smaller amount of generated password candidates.

A method based on Markov modelling was introduced by Narayan et al. in 2005 [12]. The authors model the password distribution using a markovian assumption. Markov based NLP models predict which characters are likely to follow another character. Markov models intended for password guessing usually considers the last $n - 1$ characters, so called $n$-gram Markov model. Then they modify sampled passwords by applying predefined regular expressions, i.e. the mangling rules. Currently Hashcat and JTR include such markovian models as an additional attack mode.

Building on Narayan et al.'s method, Dürmuth et al. introduced OMEN [6], which specifically sorts the generated password candidates in order of decreasing probability, something the original method was not capable of. A more general improvement of markovian models using neural networks was introduced by Melchier et al. [10]. Weir et al. [17] introduce a method that learns word mangling rules from existing password lists based on probabilistic context-free grammars (PCFG), a method stemming from NLP. They learn template structures of passwords by finding common and frequent patterns in clear text password leaks. For example, 'L4D8S1' would describe all passwords consisting of 4 lowercase letters followed by 8 digits and a single special character.

A semantic extension of Weir et al.'s PCFG was introduced as Semantic Password Guesser by Veras et al. [14]. They combine PCFGs with Wordnet [11] to enhance their grammars with semantic meaning. Their structures then use overarching semantic categories of words, i.e. umbrella terms instead of defining characters and numbers. These can then be used to describe the string of characters that is supposed to be placed at a certain part of a generated password candidate. An example base structure would be '[sport][city][special]' and a password generated from this could be 'footballhamburg?'.

The semantic password generator is the most related approach to our method, but there are two major differences: First, they do not generate candidates based on words not present in the training data. Second, because their method is based on a hierarchical tree structure, they only consider a single context per base word.

For example, while the word "apple" would probably be categorized as a "fruit", it is also semantically similar to "tech companies".

In a further study from 2021 [15] the authors updated their method and investigated the semantic differences of commonly used password leaks. They found that semantic patterns found in some leaks correspond to the context of these leaks, i.e. the demographics of users of a forum or the general subject of the website the passwords were leaked from.

More recently, deep learning methods were introduced in order to depend less on strong assumptions about the word mangling rules that form passwords. These methods often use deep generative models and are trained on password lists to model the probabilistic space of passwords and can generate new passwords directly without applying rules.

An example of a deep generative model for password guessing is PasswordGAN [9]. This method uses a generative adversarial network, specifically a Wasserstein GAN [8] to generate large amounts of password candidates. In the course of their research, the authors found that the amount of candidates that need to be generated to reach similar or better results is significantly larger than those needed for statistical methods.

A review of other deep generative model architectures for password guessing was compiled by Biesner et al. in 2020 [2].

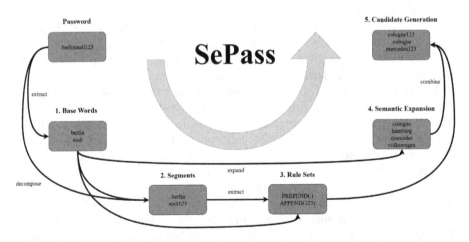

**Fig. 1.** Graphical illustration of the five steps of generating new password candidates.

## 3   Semantic Password Guessing

In this section, we describe our method and the procedures used to generate and sort a new candidate list for password guessing based on a well focused password list stemming from a specific peer group. The main focus of our method lies on the semantic context of the *base words* extracted from the given list. Deploying word embeddings, we derive new base words that are semantically similar. In this context, a base word is a substring of characters that is included in a password and has some kind of semantic meaning and cannot be broken down without losing its meaning. For example, the German city name "Berlin" would be considered a *base word* for passwords like "berlin123", "BeRl1naudi" as well as simply "Berlin".

### 3.1  Generation of New Password Candidates

Under the assumption that base words used by a specific clientele or distinct group of people are semantically similar, we use pretrained word embedding models to exploit these semantic similarities. These models enable us to find similar words to expand the given password list with previously unseen vocabulary. Word embeddings are a popular method from natural language processing and allow for words and other character strings to be mapped into a high dimensional vector space in order to be used in downstream tasks [1,16].

The goal is that semantically similar words are placed closely together according to some distance or similarity measure. For example, as euclidean distance is known for its adverse behavior in high dimensions, the cosine similarity is a popular choice when working with these high dimensional vectors. To obtain such a vector space, large-scale text corpora are processed. The resulting embedding is a vectorized representation of every single word in the training's corpora, where we can assume that semantically similar words are also similar in the vector space.

For the current version of *SePass* we use state of the art pretrained word embedding models from the *FastText* [3] toolkit. These models are available in 157 different languages [7] and are light-weight, extensive and publicly available[1]. We use the 10 most relevant European languages based on general usage and leaks that we analyzed. Those languages are: *English, German, French, Italian, Spanish, Portuguese, Turkish, Dutch, Finnish.*

In addition to finding new base words using word embeddings, we need to generate actual password candidates from these novel base words using a set of word mangling rules. These rules are simple functions that transform a base word in a step-by-step manner into a password candidate. Examples for such functions can be adding, removing or replacing certain characters as well as changing single characters to upper or lower case and much more[2]. The rule set [PREPEND(x), APPEND(123), LOWER(), REPLACE(s, $)] for example would transform the input word Password into xpa$$word123.

The following five steps, also summarized in Fig. 1, describe how the proposed method takes a list of known passwords and generates additional candidates and rule sets for each word embedding.

---

[1] https://fasttext.cc/docs/en/crawl-vectors.html.
[2] https://hashcat.net/wiki/doku.php?id=rule_based_attack.

*Step 1: Extraction of the Base Words* Using a given word list $P$ of known passwords, we extract base words from each password $p$, such that the list of base words for $p$ is of minimal length and covers as many of the password as possible. This is done in two steps using the vocabulary $V$ of a pretrained embedding model $e$. First, a decomposition into sub-words of $p$ belonging to $V$ is determined recursively. Hereby, the best decomposition is characterized by a minimum number of unused letters in $p$. In case of a tie, we prefer the solution with less base words. For example, for the password 'blueberry123a' the solution ['blueberry', '123'] wins against the solution ['blue', 'berry', '123'], each with a single unused character ('a') in the password. If no base words were detected for $p$, we additionally try to find non-obvious base words using the existing *rulegen* algorithm from PACK[3]. For example, to find the base words of passwords containing so called *leet speak*, i.e. replacements of characters with similar looking numbers such as 'passw0rd'[4].

*Step 2: Decomposition into Segments.* For every password $p$, for which base words were found in Step 1, the password is split into multiple segments. Each segment contains exactly one base word. With the exception of the last segment of $p$, segments contain only the unmatched letters to the left of the base word. For example, the password `berlin?audi123` would be split into the segments `berlin` and `?audi123`.

*Step 3: Extraction of a Rule Set.* Using existing methods from rulegen, based on all passwords in the source list $P$, a set of word mangling rules is derived such that all individual segments (from Step 2) can be created from the extracted base words (from Step 1). This is achieved by using the Levenshtein distance between the base words and the corresponding segments, e.g. Levenshtein distance (`bberlin`, `berlin`) = 1. We finally sort these rules by their occurrence frequency in $P$.

*Step 4: Semantic Expansion.* This fourth step is the cornerstone of our method and also where it deviates the most when compared to previous work. Using a pretrained word embedding – or embeddings if multiple languages or corpora are used – we collect for each base word in the source password list the $k$ most similar words in the vocabulary of the word embedding using a $k$-nearest neighbor query. Note, that $k$ is not a hyperparameter. Instead $k$ is calculated based on the number of password candidates that are intended to be generated.

Given, the number of pretrained models $|E|$, the intended number of password candidates $n$ overall, the intended number of password candidates for a single embedding $n_e = \frac{n}{|E|}$, the number of rules generated in Step 3 $|R|$, the hyper parameter *relevant ruleset ratio* $rr$, and the list of extracted base words $BW_{old}$, we first calculate the amount of base words we want to mangle

---

[3] https://github.com/iphelix/pack.
[4] https://github.com/hashcat/hashcat/blob/master/rules/unix-ninja-leetspeak.rule.

$$|BW| = \frac{n_e}{rr \cdot |R|} \qquad (1)$$

and then

$$k = \frac{|BW|}{|BW_{old}|} \qquad (2)$$

*Step 5: Generating New Password Candidates.* Finally, by applying every rule from $R$ to every word of the expanded base words list, we create the final list of new password candidates.

## 3.2 Sorting of the Password Candidates

Depending on the chosen initial word list and the parameters for the embedding, the newly generated list can grow in size considerably. This requires sorting the candidates based on the likeliness of being a real password – especially in cases where the time to guess a password is limited and does not allow trying out a large number of passwords. When executing the five steps described above, the candidate additionally is paired with a password score *pws*. The higher this score, the more suitable a candidate is considered to be. In accordance with the candidate being a combination of a base word and a rule, the password score is also made up of a word score *ws* and a rule score *rs* as shown in Eq. 3. The value of the rule score is simply determined by the relative occurrence of the specific rule in the total set of rules. The word score is calculated with the help of the embedding model. For every original base word we calculate how often a specific word $w$ is present in the k-neighbors by using the same methods mentioned above. The sum of all these distances of the base words BW to $w$ is used as word score for $w$. The formula for the calculation of the word score is shown in Eq. 4.

$$pws = ws \cdot rs \qquad (3)$$

$$ws(w) = \sum_{i=0}^{|BW_{old}|} \begin{cases} Cos\Theta(BW_i, w), & \text{if } w \in knn(BW_i) \\ 0, & \text{otherwise} \end{cases} \qquad (4)$$

## 4 Test Bed

In order to evaluate SePass and compare it with the current state-of-the-art, we performed a series of experiments. We primarily evaluate the use case of generating a list of novel password candidates from a relatively small training set, i.e., a highly focused leak, for example originating from a darknet or an extremist forum. As mentioned in Sect. 2, most other methods are based on learning candidates from large general password leaks and then testing the generated candidate lists on other smaller leaks. To capture the characteristics of this real application scenario, we opt for a different evaluation scheme, where we only use a small password list, which we split into a train and a test sets. We trained and evaluated all models on a compute server running Ubuntu 20.04.3 LTS with 62 GB of RAM and an AMD Ryzen 7 3700X 8-Core Processor.

### 4.1  Data Sets

We conducted all our experiments on two different datasets: a small real world list which is not yet widely available and is therefore only used for evaluation purposes and – for reproducibility purposes – we generated a second synthetic list which is a small excerpt of *rockyou* [5]. This synthetic list shares statistical similarities with the first list w.r.t. average length of the passwords, used languages and used rules. Both lists have the same length of entries, i.e., 66.490 passwords.

### 4.2  Compared Methods

We compared SePass to the following password prediction methods that offer publicly available code repositories and represent the different existing paradigms of password guessing.

**Hashcat Best64.** As a baseline we used hashcat with a basic rule set consisting of 64 word mangling rules that were created in a competition held by the community of hashcat[5]. These handcrafted rules are very simple instructions, such as appending single digits or letters, reversing the order of the password or replacing certain characters, for example, e with 3 or i with 1.

**OMEN.** In order to represent the various markovian methods we utilized the original implementation of OMEN[6]. OMEN is one of the best performing probabilistic password guessers, meaning it uses candidate occurrence frequencies to output the most likely passwords. It was written in C, making it extremely fast compared with its competitors.

**PCFG.** We picked Probabilistic Context-Free Grammars (PCFG) as a representative method based on statistical modeling. We used the pcfg cracker repository[7], which was developed by one of the authors of the original publication [17]. As the authors mention in the notes on their repository, the tool is actually aimed at a similar use case as ours.

**Semantic PCFG.** We chose this method because it is aimed at using semantic connections between words and, as such, follows a related concept to our approach. The authors have published their code on a git repository[8].

**PassGan.** We chose PassGan [9] as one of the most well-known deep learning approaches for password generation. While we could not find a code repository from the original authors, we used a re-implementation[9] which contains a pretrained version of PassGan.

---

[5] https://github.com/hashcat/hashcat/blob/master/rules/best64.rule.
[6] https://github.com/RUB-SysSec/OMEN/blob/master/README.md.
[7] https://github.com/lakiw/pcfg_cracker.
[8] https://github.com/vialab/semantic-guesser.
[9] https://github.com/brannondorsey/PassGAN.

## 4.3  Experimental Set-Up and Evaluation Metric

We evaluated the accuracy of the competitors by splitting both our real and synthetic password lists into a training and a test set. The test sets each contain a random sample of 20% of the full lists. We applied each method to the training sets and generated a password candidate list each. We then compared these lists to our test sets. For PassGan we did not train the model ourselves, but instead opted for the pretrained version that is included in the repository and was trained on *rockyou* [4], because our training sets would be magnitudes too small for PassGAN to be reasonably trained on. Still, this is a more than fair comparison, since both our training and test sets heavily overlap with the *rockyou* leak. We used the trained models to generate a list of 50 million password candidates to simulate a guessing attack on our test lists.

As usual in related work, our evaluation metric is the percentage of hits on the test set after $n$ guesses, called hits@$n$ which is defined as

$$\text{hits@}n = \frac{|P_m^{0...n} \cap P_{\text{test}}|}{|P_{\text{test}}|},$$

where $P_m$ denotes the set of password candidates generated by a single method $m$ and $P_{test}$ denotes the attacked test set. We report the results of the competitors for $n = 50$ million minus the number of duplicates in Table 1. In addition, we also report the hits@$n$ value of the competitors in Figs. 2a and 2b, which illustrate how quickly the corresponding methods may be able to successfully finish the attack.

**Table 1.** Prediction accuracy (hits@$n$) and number of unique passwords generated after duplicate removal.

| Method | Hits@n in % on synthetic data | Hits@n in % on real world data | # of unique candidates generated |
|---|---|---|---|
| Ours | 36.59 | 34.90 | $50 \cdot 10^6$ |
| hashcat-Best64 | 17.15 | | $3,199,660$ |
| OMEN | 32.35 | | $50 \cdot 10^6$ |
| PCFG | 36.52 | 39.39 | $50 \cdot 10^6$ |
| Semantic PCFG | 20.22 | | $24,903,549$ |
| PassGAN | 15.27 | | $24,761,815$ |
| Ours + PCFG | **44.25** | **45.33** | $50 \cdot 10^6$ |

## 5  Results and Discussion

In this section, we compare the results for the test bed and present further experiments performed to derive more insights into the strengths and weaknesses of the individual methods.

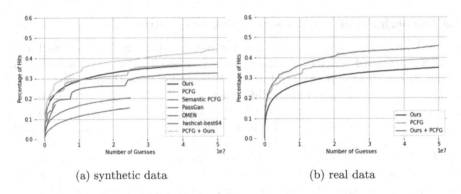

(a) synthetic data                    (b) real data

**Fig. 2.** Percentage of hits (hits@$n$) with increasing number of guesses $n$

## 5.1   Accuracy Results

For reproduction purposes, the implementation of the experiments on the synthetic data set is publicly available[10]. We decided to only use the methods for the real data set that performed best on the synthetic list. Table 1 displays the hits@$n$ results. In Figs. 2a and 2b the hits@$n$ are plotted as functions over $n$ guesses, i.e., the effective time consumptions.

*Hashcat Best64.* The hashcat Best64 rules seem to be a fitting baseline: while the method only produces a small number of unique candidates (e.g. 3.2 million on our synthetic data, i.e., the amount of passwords times 64 rules) a huge number of these are hits (17.5% correct guesses on the test set – see Table 1). Considering this method is based on applying fairly simple rules to mangle the base words from the training set, we can conclude that at least 17.5% of the passwords in our test set are rather trivially constructed. The graphs in Fig. 2a seem to indicate that our method, OMEN and PCFG are able to guess these trivial passwords at a faster rate then hashcat, while Semantic PCFG and PassGAN are slower, but do or will eventually surpass this threshold as well.

*PassGAN.* PassGan performs the worst of all methods tested on the synthetic data set, as seen in Fig. 2a. This is of particular interest since the pretrained PassGan model was expected to have an advantage on our synthetic data set since PassGan was trained on the rockyou leak and our synthetic data set consists of mostly passwords found in rockyou. This might be explained by the fact that PassGan and similar GPU based methods are designed to generate exorbitantly huge amounts of guesses. Therefore, as the authors state in the PassGAN paper, it might take a lot more guesses before it catches up with the other methods. Combined with the amount of duplicates this method produces, we have to concede that PassGAN does not fit our use case, which is extrapolating from a small, focused vocabulary. Another explanation might be, that it does not guess

---

[10] https://github.com/Knuust/SePass.

passwords that stem purely from rockyou and it might even be at a disadvantage since it was not trained on our specific data set. And lastly, we used a 3rd party reimplementation of PassGan, since no implementation from the authors is available, which might perform differently than originally presented. Therefore, as a consequence of the poor performance and not being suitable for our use case, we exclude PassGAN from experiments on the real dataset.

*Semantic PCFG.* The Semantic PCFG password guesser seems to be performing better than PassGAN, but not as well as the original PCFG. This is surprising since the semantic PCFG method is based upon the original PCFG method. We assume the reason for this performance is comparable to the problems found with PassGAN.

*SePass, OMEN and PCFG.* The best performing methods are OMEN, PCFG and our own method SePass. They all result in a similar percentage of hits at 50 million guesses, with our method SePass and PCFG coming out on top as seen in Table 1 and Fig. 2a. As mentioned earlier this was to be expected since all three methods aim at a similar use case of giving more weight to accuracy in less guesses rather than generating a large amount of password candidates in a short time. One advantage specifically concerning the implementation of PCFG and OMEN is that both actually come with a few a priori rules, similiar to the Best64 concept. For example, these rules include adding commonly used dates and keyboard walks (`qwerty`, etc.). These are applied additionally to enhance the base words and therefore lead to an enhanced performance for both OMEN and PCFG. PCFG and OMEN perform similar in Fig. 2a. This can also be explained by the fact that the PCFG implementation is based on the OMEN repository.

Considering these results, we only ran our method and the best other method, i.e., PCFG, on the real data set. The results can be found in Table 1 and Fig. 2b. It is evident, that when testing on this real world data set, PCFG outperforms SePass by a small percentage. To show that our method still provides additional benefit, we conducted an additional test, where we combined both lists proposed by SePass and PCFG.

*Combination.* We combined both lists by zipping them together, i.e. by taking the first element of each, then the second, etc., which results in a list with double the length. Then we cut this down to 50 million guesses in order to compare them to the other methods. We can observe that the combination does indeed perform even better than the individual methods on both the synthetic and the real dataset and are able to crack almost 50% of each test set. This is expected since our method adds the capability of using novel base words but can not generate the same amount of candidates using the mangling rules and the base words from the training set as PCFG does in 50 million guesses. This leads us to the conclusion that for a future version of our tool we should build upon the mangling rules of PCFG or other competitors and combine those with the novel base words found using our proposed approach.

## 5.2   Unseen Base Words

The main motivation behind our work was that existing methods, while very good at applying mangling rules to base words and creating passwords from existing lists, are generally not able to guess completely new base words without either using very specific handcrafted dictionaries or, at least partially, brute forcing passwords. We therefore investigated if and how well competitors find such new base words. Formally, given a vocabulary universe $V$, we are looking for base words $B_{new} \subseteq V$ consisting of all words that are included in the passwords from the test set $B_{test}$, but can not be found in the corresponding training set, $B_{train}$, i.e., $B_{new} = B_{test} - B_{train}$.

Firstly, in order to generate an extensive vocabulary universe, we collected the union of all vocabularies from the 10 language embedding models that we used in our method. This resulted in a set of exactly $12,953,300$ unique base words. This is about 7 million words less than expected because while each model has a vocabulary of 2 million words, often times languages overlap and use the same terms. Because the models were trained on very large internet corpora the vocabularies can also include artifacts, e.g. very long words or numbers and special characters that can include outliers and errors. In order to investigate only natural words for the following experiments on novel base words, we removed everything from these vocabularies that includes any digits or other special characters.

Next, we searched for each word found in our vocabulary universe $V$ in both the train and test set in both our password lists. We then subtracted the list of base words found in the train set from the ones found in the test set. This resulted in $13,428$ novel base words, i.e. a set of base words that are used only in the test set but cannot be found in the training set.

Afterwards, we checked how many of these test base words can be found by our method and PCFG. We therefore look at the set of hits for each method, i.e. the intersection between the list of password candidates and the test set. We then search for each base word in these two sets and build the intersection with the set of base words contained in $B_{test}$. We found that SePass found $2,439$ more novel base words than PCFG (which is almost 6 times more).

This demonstrates that SePass is able to extrapolate from the base words and significantly outperforms PCFG in this regard. On the other hand PCFG is also able to find a few novel base words. When taking a closer look at the new words that PCFG found, we can see that these often are random combinations of existing words or predetermined rules, for example `qwertyuiop` which is explicitly included in the PCFG repository as a keyboard walk. In order to validate the performance of SePass, we found the corresponding passwords these novel base words were used for. This resulted in a list of $5,296$ passwords, which consequently contain base words that are not included in the train set. The amount of passwords is lower than the amount of novel base words, since a password can include multiple base words.

We then calculated the percentage of novel passwords found by each method. The result is collected in a bar diagram in Fig. 3 for both the synthetic dataset

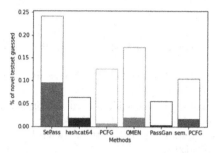
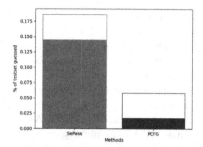

(a) % guessed on synthetic data          (b) % guessed on synthetic data

**Fig. 3.** Bar plots showing the percentages of passwords found in the synthetic (left) and real (right) test set, that include base words not found in the train set for each method. The filled part of each bar shows the percentage found exclusively by the corresponding method.

(a) and the real one (b). While the edge of each bar shows the percentage of passwords found, the filled areas represent the passwords this method found exclusively.

This means we see our expectations about SePass confirmed. Looking at the synthetic dataset, not only did SePass guess more of these novel passwords over-all, SePass also finds significantly more novel passwords than any other method. Additionally, while PCFG performed better on the real dataset overall, SePass guesses 6 times more exclusive novel passwords on this dataset as well. In general, we can see that our method performs similarly well to related methods and is able to guess a significant amount of unique passwords.

## 6   Conclusion

We introduced SePass, a novel password guessing algorithm. The foundation of SePass are word embeddings which are used to identify new base words given the vocabulary extracted from a list of passwords. After that, we applied the rules extracted from the passwords list to the found base words to generate password candidates that are semantically related to those found in the original passwords list. SePass compares favourably with the known methods used in this application field. It distinguishes itself from existing methods by being able to exclusively generate more passwords containing novel base words than any other method tested. We therefore conclude, that our tool, especially when used in combination with other methods like PCFG, can reach a high percentage of correctly guessed passwords, surpassing their individual scores.

# References

1. Almeida, F., Xexéo, G.: Word embeddings: A survey. CoRR abs/1901.09069 (2019). http://arxiv.org/abs/1901.09069
2. Biesner, D., Cvejoski, K., Georgiev, B., Sifa, R., Krupicka, E.: Generative deep learning techniques for password generation (2020)
3. Bojanowski, P., Grave, E., Joulin, A., Mikolov, T.: Enriching word vectors with subword information. Trans. Assoc. Comput. Linguist. **5**, 135–146 (2017)
4. Burns, W.J.: Common password list (rockyou.txt) (2019). https://www.kaggle.com/wjburns/common-password-list-rockyoutxt
5. Cubrilovic, N.: Rockyou hack: From bad to worse (2009). https://techcrunch.com/2009/12/14/rockyou-hack-security-myspace-facebook-passwords/
6. Dürmuth, M., Angelstorf, F., Castelluccia, C., Perito, D., Chaabane, A.: OMEN: faster password guessing using an ordered markov enumerator. In: Piessens, F., Caballero, J., Bielova, N. (eds.) ESSoS 2015. LNCS, vol. 8978, pp. 119–132. Springer, Cham (2015). https://doi.org/10.1007/978-3-319-15618-7_10
7. Grave, E., Bojanowski, P., Gupta, P., Joulin, A., Mikolov, T.: Learning word vectors for 157 languages. In: Proceedings of the International Conference on Language Resources and Evaluation (LREC 2018) (2018)
8. Gulrajani, I., Ahmed, F., Arjovsky, M., Dumoulin, V., Courville, A.C.: Improved training of wasserstein gans. CoRR abs/1704.00028 (2017). http://arxiv.org/abs/1704.00028
9. Hitaj, B., Gasti, P., Ateniese, G., Perez-Cruz, F.: PassGAN: a deep learning approach for password guessing. In: Deng, R.H., Gauthier-Umaña, V., Ochoa, M., Yung, M. (eds.) ACNS 2019. LNCS, vol. 11464, pp. 217–237. Springer, Cham (2019). https://doi.org/10.1007/978-3-030-21568-2_11
10. Melicher, W., et al.: Fast, lean, and accurate: Modeling password guessability using neural networks. In: Proceedings of the 25th USENIX Conference on Security Symposium, pp. 175–191. SEC'16, USENIX Association, USA (2016)
11. Miller, G.A.: WordNet: An electronic lexical database. MIT press (1998)
12. Narayanan, A., Shmatikov, V.: Fast dictionary attacks on passwords using time-space tradeoff. In: Proceedings of the 12th ACM Conference on Computer and Communications Security, CCS 2005, pp. 364–372. Association for Computing Machinery, New York (2005)
13. Steube, J.: hashcat (2002). https://hashcat.net/hashcat/
14. Veras, R., Collins, C., Thorpe, J.: On the semantic patterns of passwords and their security impact, January 2014
15. Veras, R., Collins, C., Thorpe, J.: A large-scale analysis of the semantic password model and linguistic patterns in passwords. ACM Trans. Priv. Secur. **24**(3), April 2021
16. Wang, S., Zhou, W., Jiang, C.: A survey of word embeddings based on deep learning. Computing **102**(3), 717–740 (2020)
17. Weir, M., Aggarwal, S., Medeiros, B.d., Glodek, B.: Password cracking using probabilistic context-free grammars. In: 2009 30th IEEE Symposium on Security and Privacy, pp. 391–405 (2009)

# DeMRC: Dynamically Enhanced Multi-hop Reading Comprehension Model for Low Data

Xiu Tang[1]([✉]), Yangchao Xu[2], Xuefeng Lu[1], Qiang He[2], Jun Fang[2], and Junjie Chen[1]

[1] Zhejiang University, Hangzhou, China
`tangxiu@zju.edu.cn`
[2] State Grid Shaoxing Power Supply Company, Shaoxing, China

**Abstract.** Multi-hop reading comprehension requires the aggregation of multiple evidence facts to answer complex natural language questions, and the answer should be avoided when there is no answer. Training a model that can handle such difficult tasks requires a large number of data sets to support, but the labeling of data sets is very expensive and time-consuming, so it is very important to explore reading comprehension models suitable for low data, and external data related to large-scale tasks. It will also effectively improve the performance of the model. This paper proposes a two-stage model with dynamically context-enhanced method for multi-hop reading comprehension tasks under low data called DeMRC. The first stage sentence filtering model filters the top k sentences that are strongly related to the question, and the second stage answer prediction model dynamically constructs the training set every time during training to expand the data set, and uses sentences selected by sentence filtering model as input to reduce the interference of irrelevant sentences to the model during inference. In addition, the self-training method is used to pseudo-label the external data and use it as an auxiliary data set to improve the performance of the model. We conducted experiments on the multi-hop reading comprehension data set of the Chinese "CAIL 2020" Judicial Artificial Intelligence Challenge Reading Comprehension Track and English cross-document level data set called HotpotQA, which are 3.5% and 21.3% higher than the powerful baseline model, showing the effectiveness of the method.

**Keywords:** Data augmentation · Machine reading comprehension · Self training · Multi-hot QA

## 1 Introduction

Machine reading comprehension tasks require machines to answer questions through a given context and can be used in areas such as search engines and intelligent assistants, to provide users with high-quality consulting services. With

W. Chen et al. (Eds.): ADMA 2022, LNAI 13726, pp. 43–57, 2022.
https://doi.org/10.1007/978-3-031-22137-8_4

advances in large-scale pre-trained language models, some machine reading comprehension models have shown significant performance improvements on single-hop machine reading comprehension data sets [1], but these models still lack the ability to reason across multiple sentences [2]. Recent proposals of multi-hop reading comprehension data sets such as WIKIHOP [3] and HotpotQA [4] require models to be able to reason across multiple disjunctive sentences or documents. There are a number of works based on it, all use pre-trained models [5,6] as feature extractors, then fine-tuned on a specific reading comprehension task. This approach requires a large amount of data driven in the training process. However, the process of labeling data in the real world is very time-consuming and laborious, and in some domains there are not enough samples for labeling.

Focusing on multi-hop reading comprehension tasks in the low-dataset case, this paper presents a dynamically context-enhanced multi-hop reading comprehension(DeMRC) approach on the Chinese CAIL 2020 reading comprehension data set[1] (see Fig. 1), and validates the method on HotpotQA data set, a cross-document English multi-hop reading comprehension data set.

In our model, the input is dynamically updated to generate different contextual statements so as to perform contextual enhancement during training, and the inference process directly uses the sentences filtered by the sentence filtering model. In addition, the prediction of the supporting sentences no longer uses the graph neural network to learn the correlation between sentences, but is based on the improved Transformer mechanism. The reading comprehension task proposed in CAIL 2019[2] is different from CAIL 2020 in that there is no supporting sentence prediction subtask. We increase the generalization ability of the model by learning external knowledge by using it as a external data set through self-training.

The contributions of this paper can be summarized as follows:

1. We propose a dynamically context-enhanced multi-hop reading comprehension model (DeMRC) for low data on Chinese CAIL 2020 reading comprehension data set. The labeled data is better utilized for dynamic contextual enhancement, and a sentence filtering model is designed to ensure the consistency in the training and inference process.
2. We use unlabeled data sets from the same domain as external data sets and generate pseudo-labels on DeMRC model using a self-training approach for data augmentation, enhancing the generalization ability of the model.
3. The method is validated on the English data set HotpotQA and compared with other classical models on this data set to demonstrate the effectiveness and generality.

---

[1] http://cail.cipsc.org.cn/instruction.html .
[2] http://cail.cipsc.org.cn:2019/ .

**文档:** [0] 本院经审理查明: 原、被告均系冀州市周村镇寇庄村第一队人, [1]1999年12月14日原告吴x0与周村镇寇家庄村民委员会签订土地承包合同书一份, [2]吴x0承包了寇家庄村土地8.19亩, [3]并由冀州市人民政府为原告吴x0颁发了土地承包经营权证书, [4]明确了地块座落、四至及亩数。[5]2001年原告吴x0将自己承包的土地转包给吴x22.1亩、陈x31.95亩、吴x61.98亩、吴x90.99亩、常玉171.47亩, [6]约定原告吴x0负担村里的费用, [7]种地户负担乡里的费用。[8]后又变更为村里及乡里的费用均由种地户负担。[9]2013年原告吴x0向种地户吴x2、陈x3、吴x6、吴x9、常玉17要求返还承包的土地, [10]常玉17将土地归还, [11]其余四人拒绝归还, [12]经村支部书记吴更海及小队长吴更新调解无效, [13]原告吴x0提起诉讼, [14]要求被告吴x2、陈x3、吴x6、吴x9返还耕地, [15]并返还2007年至2014年的粮补等补贴款3931.2元, [16]在审理过程中, [17]原告吴x0撤回了对被告吴x2的起诉, [18]本院予以准许。[19]被告陈x3、吴x6、吴x9以土地承包经营权证书上进行了变更, [20]与原告吴x0的转包关系变为转让关系为由抗辩, [21]认为已经享有了对原告吴x0诉争土地的承包经营权, [22]经调解, [23]原、被告坚持自己的主张, [24]调解无效。
**问题:** 被告目前起诉的人中谁承包的面积最大?
**答案:** 吴x6
**支撑句:** 5, 13, 14, 17

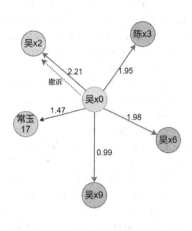

**Fig. 1.** A sample from CAIL 2020 reading comprehension data set. The input is a question and a legal decision document, the output is answer and supporting sentences.

## 2   Related Work

Early reading comprehension systems were small and had domain-specific limitations that did not allow for good applications, when the main approaches to address reading comprehension were rule-based or machine learning based which did not have good performance. With the superior performance of deep learning in contextual information acquisition and large benchmark data sets were proposed such as SQuAD [1], CNN & Daily Mail [7], many reading comprehension models were generated [5,8], and neural machine reading comprehension systems became the current research hotspots in academia and industry.

The proposed multi-hop reading comprehension data sets such as WIKIHOP [3] and HotpotQA [4] require that reading comprehension systems can perform answer inference across multiple sentences or documents, while HotpotQA [4] also expects models to provide supporting sentences that participate in answer inference to increase the interpretability of the model. Most approaches rely on graph neural networks to obtain the interrelationships between sentences [9,10]. Tu et al. [11] extended the entity graph to a heterogeneous graph by introducing document nodes and query nodes. However, C2Freader [12] demonstrated that graph structure is not necessary for multi-hop inference and that removing the entire graph structure does not have bad effects, and we did not use graph structures in the design of our model.

The original machine reading comprehension task assumes that the answer is always in the given context, however, this does not correspond to reality. 50,000 unanswerable questions were added to SQuAD 2.0, and the paper [13] sets an answer threshold to determine whether a question is answerable; tan et al. [14]

add a padding position to the original passage and reject to give an answer when the model predicted that position; SAE [9] use a multi-task learning approach to extract answers. Our model alse use multi-task learning approach by adding an answer type prediction subject and setting it as a "yes/no/unknown/span" four-class task.

Most current multi-hop reading comprehension tasks are supported by large amounts of data, and the low-data case has rarely been studied. CAIL 2020 reading comprehension track (See footnotee 1) presents a multi-hop reading comprehension task for the low-data case, where direct use of previous reading comprehension models does not yield good results. Data augmentation is a good alternative for the low-data case. However, current data augmentation methods in the text domain focus on text classification tasks. EDA [15] produces good results when the amount of text is low. Back-translation [16] based approaches tend to have good performance in multiple tasks, but require calls to API tools and the translation process is not efficient. In addition, textual mixed data augmentation [17], pre-training based contextual information enhancement [18,19], and text generation [20] have been used in classification tasks, but no paper has demonstrated significant results in reading comprehension tasks. Sliding window [21], as a means of data augmentation for reading comprehension tasks, does not guarantee that all supporting sentences are within the window, and it is not suitable for multi-hop cases. The paper [22] used pseudo-labeling of unlabeled data to expand the data set on the computer vision task with good results, and this paper extends it to the multi-hop reading comprehension domain by pseudo-labeling the supporting sentences to increase the data set.

## 3   Methodology

Our proposed dynamically context-enhanced model (DeMRC) under low data is a two-stage model. The first stage is a sentence filtering model to ensure consistency of DeMRC's input in the training and inference process. The second stage is an answer prediction model, which is trained with multi-task learning to complete the work of supporting sentence prediction as well as answer prediction. We use the supporting sentence superset as the input to DeMRC model, and there is a difference between the training and inference process. In the training process we propose a dynamically context-enhanced method to random generate it, the process does not require the involvement of the first-stage model. The sentences selected by the first-stage model are used as superset of the supporting sentences during inference to ensure consistency in the training and inference. Meanwhile, we use self-training method to introduce external data to expand the data set for optimization, and the overall flow of the model is shown in Fig. 2.

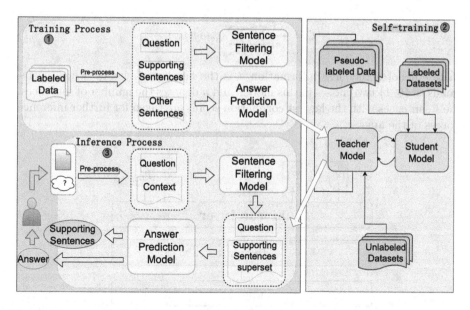

**Fig. 2.** Diagram of the proposed DeMRC model. We first train the sentence filtering model and the answer prediction model separately on the labeled data, followed by data augmentation using the self-training method of pseudo-labeling the unlabeled data, and finally the inference process of the model.

### 3.1 Sentence Filtering Model

The superset of supporting sentences in DeMRC is explicitly constructed from two parts, supporting sentences and other sentences. However, supporting sentences are unknown during model inference, which will lead to inconsistency in the input of the answer prediction model during training and inference. Meanwhile, the documents of CAIL 2020 reading comprehension data set are all legal judgment documents, considering that the legal documents are too long to be directly input to the model, and the interference of irrelevant document statements may increase the difficulty of model learning and reduce the model performance.

We designed a sentence filtering model (see Fig. 3) to reduce the interference information passed to the downstream answer prediction model by selecting the top $k$ most relevant sentences from the sentences of the legal decision document. For each data, we divide the document into sentences. For every sentence, we generate an input to feed through BERT [5] by concatenating "[CLS]" + question + "[SEP]" + sentence + "[SEP]". Then we use the vector [CLS] for each question/sentence pair as the global information. Then all vectors are stitched together and the linear layer is interacted for sentence features after first using the sigmoid function to obtain the probability distribution and then using the binary cross entropy for sequence labeling classification:

$$\mathcal{L} = -\frac{1}{n} \sum_{i=1}^{n} \sum_{c=1}^{t} y_{ic} log(p_{ic}) \qquad (1)$$

where $c$ denotes the class information, $t$ is the number of classes, which corresponds to $\{0, 1\}$ two classes in this model, and $n$ denotes the number of sentences. The $k$ sentences with the largest correlation scores are taken for further inference process of the answer.

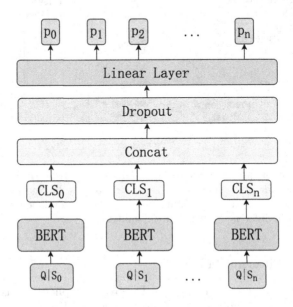

**Fig. 3.** Diagram of sentence filtering model.

## 3.2   Answer Prediction Model

The answer prediction model (see Fig. 4) is built based on the idea of multi-task learning and consists of three subtasks: answer category prediction, answer extraction, and supporting sentence prediction, which are learned as a 4-class classification, extraction, and 2-class classification task, respectively.

### 3.2.1   Input Layer

Since all sentences of the documents other than the supporting sentences are noisy for answering the questions, we propose a dynamically context-enhanced method to expand the data set. Specifically, the contexts of each document in the training phase are composed of the supporting sentences and a part of the sentences dynamically randomly selected from the remaining sentences. This dynamically expands the data set by obtaining a supporting sentence superset

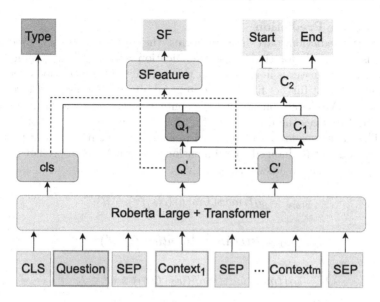

**Fig. 4.** Diagram of answer prediction model.

of different sentences each time. The sentences screened by the filtering model are directly used as the supporting sentence superset in the inference phase of the model to ensure the consistency to the answer prediction model during model training and inference. We use the superset as input.

### 3.2.2 Encoding Layer

The model takes the question and sentence splicing as input, where $Question$ denotes the question, $Context_i$ denotes the i-th sentence of input, then obtains the overall feature vector $cls$ and the features of the question $Q \in R^{m \times d}$, the sentence $C \in R^{l \times d}$, where $d$ is the output dimension of the pre-trained model, $m$ is the sentence length, and $l$ is the length sum of all sentences.

$$cls, Q, C = RoBERTa([CLS] + Question + ... + Context_i + ... + [SEP]) \quad (2)$$

Since the features encoded by the pre-trained model are more concerned with the connection between $Q$ and $C$, in order to explore their respective internal information, this paper recodes them using Transformer to obtain the recoded question feature $Q'$ and the context feature $C'$:

$$Q', C' = Transformer(Q, C) \quad (3)$$

### 3.2.3   Answer Prediction

The answer prediction part first performs the answer category prediction, directly using the $cls$ for the 4-class classification with a weight penalty $\tau$ for each class. If the answer is in the context then extracted from the document. The extraction task first further finds the contextual information using bi-directional attention (BiDAF-Attention) [8], get the contextual vector $C_1$ with the question features, and reduces the dimension of $Q'$ based on an attentive pooling mechanism. Then it is weighted and summed in each dimension of $C_1$ together with $cls$, and the output is changed to 2-dimensions using a MLP to predict the start and end position of the answer separately:

$$\hat{y}_{type} = softmax(Linear(cls)/\tau) \in R^{1 \times 4} \tag{4}$$

$$C_1 = BiDAF - Attention(C', Q') \tag{5}$$

$$Q_1 = softmax(w_1 \cdot (tanh(w_2 \cdot Q'))) \times Q' \in R^{1 \times d} \tag{6}$$

$$C_2 = Norm(w_3 \cdot cls + w_4 \cdot Q_1 + w_3 \cdot C_1) \in R^{l \ times d} \tag{7}$$

$$\hat{y}_{ans} = MLP(C_2) \in R^{l \ times 2} \tag{8}$$

where $w_1, w_2, w_3$ and $w_4$ are the weights, $tanh$ is the activation function, and $Norm$ is the normalization operation. Since the answers may have errors such as end position before start position, maximum value not in the sentence, exceeding the maximum answer length limit, etc., we take a candidate value in each dimension when taking the final answer, then match each start and end position pair, add the pairs that matche the answer to the candidate answer, and take the one with the maximum score as the final answer.

### 3.2.4   Supporting Sentence Prediction

Multi-hop inference requires reasoning across multiple sentences. The current mainstream practices all use graph neural networks to construct the connections between sentences or entities, but the graph construction process is complex and requires a lot of prior knowledge.

The paper [11] points out that both graph structure and adjacency matrix are task-related prior knowledge, while Transformer [23] itself is able to learn the relationship between sentences, so we use an improved Transformer based on the paper [24, 25] to recode the sentence vector to construct the relationship of them (see Fig. 5). The first improvement is to increase the dimension between $q$ and $k$, which can increase the representation of information when computing attention and alleviate the low-rank problem of original self-attention [24]. The second is that the individual heads of the original Transformer [23] are isolated from each other, and we use a parameter matrix to superimpose the information

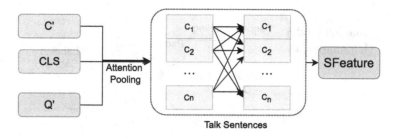

**Fig. 5.** Diagram of contextual features.

of each sentence obtained from the recoding downscaling, performing a fusion of features between sentences, so that the information can be better expressed.

$$SFeature = W \cdot AttentionPooling(C', cls, Q') \tag{9}$$

$$\hat{y}_{sf} = sigmoid(Linear(SFeature)) \tag{10}$$

The weight $W$ is initialized to the same constant matrix so that the initial phase of learning has the same attention for each sentence. Then the contextual features obtained by fusion are converted to the output $\hat{y}_{sf} \in R^{n \times 1}$ using a fully connected layer, those above a threshold value are judged as support sentences.

### 3.2.5  Loss Function

The loss function $L$ is composed of three loss functions for the answer, span, and supporting sentence prediction:

$$L = \alpha \cdot CE(\hat{y}_{type}, y_{type}) + \beta \cdot BCE(\hat{y}_{sf}, y_{sf}) + L_{ans} \tag{11}$$

$$L_{ans} = \frac{1}{2}(CE(\hat{y}_{ans}[:, 0], y_{start}) + CE(\hat{y}_{ans}[:, 1], y_{end})) \tag{12}$$

$CE$ denotes the cross-entropy loss, and $BCE$ denotes the cross-entropy loss. $L_{ans}$ denotes the loss function of the answer, the start and end position are calculated cross-entropy loss separately, take the average value and add to the whole loss calculation. Since the learning difficulty of each task is different, two weights $\alpha$ and $\beta$ are added to control the different subtasks.

### 3.3  Self-training Augmentation Based on External Data

Compared to labeled data, domain-related and task-related unlabeled data are much less difficult to obtain, and these data often contain knowledge that can help the model learn. Self-training augmentation is a method to expand the data set using external unlabeled data (see Fig. 6):

The inputs are the labeled data set $\mathcal{D}_1$ and the external data set $\mathcal{D}_2$ without support sentence labels, using $\mathcal{D}_1$ to train $n$ teacher models $\mathcal{T} = \{t_1, t_2, ..., t_i,$

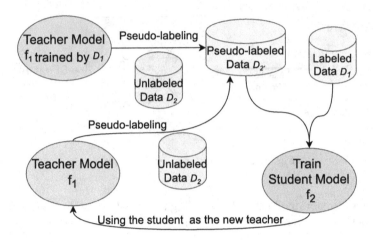

**Fig. 6.** Diagram of Self-training.

...,$t_n$}, and use the teacher model $\mathcal{T}$ to jointly generate the pseudo-labeled data set $\mathcal{D}_2'$ on $\mathcal{D}_2$. Then use $\mathcal{D}_1$ to train $n$ student models $\mathcal{S} = \{s_1, s_2, ..., s_i, ..., s_n\}$ with $\mathcal{D}_2'$, and return the student model $\mathcal{S}$ to the teacher $\mathcal{T}$, to iterate this process to generate new pseudo-labeled data to train students, enhancing the task information of the model by continuous iteration. The student uses the same size network as the teacher, and the teacher does not add noise to make the generated pseudo-labeled more quasi-group. However, the learning process of the student increases the learning difficulty of the student model by adding dropout to the model to add noise, encouraging the student to surpass the teacher.

## 4   Experiments

### 4.1   Data Set

CAIL 2020 (See footnotee 1) reading comprehension data set is a Chinese judicial domain data set with a training set of about 5100 samples, covering three domains: civil, criminal, and administrative. It has a validation set and a test set of about 1900 and 2600 samples. The external data set CJRC, proposed by CAIL 2019 (See footnote 2), contains two domains, criminal and civil, with 40,000 questions in the training set and about 5,000 questions each in the validation and test sets. HotpotQA [4] is an english document-level multi-hop reading comprehension data set containing about 110,000 question-answer pairs. Both the validation and test sets contain 7405 samples, each containing 10 unrelated documents. Since the usage scenario in this paper is low data, we only use about 10% of the randomly selected labeled data in the training, and remove the supporting sentence labels from the remaining as external data set.

Both data sets have two tasks: answer prediction and supporting sentence prediction. The models are evaluated based on F1 scores of the two tasks, while using joint F1 scores on these two tasks as the overall performance metrics.

**Table 1.** Results of different Model on the CAIL 2020 and HotpotQA, where "DeMRC" denotes the structure of our model, "−dl" denotes the removal of the dynamically context-enhanced mechanism, and "+sl" denotes the inclusion of the self-training.

| Data set | Model | $F1_{ans}$ | $F1_{sup}$ | $F1_{joint}$ |
|----------|-------|------------|------------|--------------|
| CAIL 2020 | Baseline | 0.759 | 0.64 | 0.516 |
| | DeMRC,-dl | 0.768 | 0.654 | 0.529 |
| | **DeMRC** | **0.76** | **0.671** | **0.536** |
| | **DeMRC,+sl** | **0.786** | 0.669 | **0.551** |
| HotpotQA | Baseline | 0.650 | 0.617 | 0.415 |
| | DeMRC,-dl | 0.668 | 0.828 | 0.563 |
| | **DeMRC** | **0.76** | **0.671** | **0.536** |
| | **DeMRC,+sl** | **0.786** | **0.669** | **0.551** |

### 4.2 Implementation Details

The experiments use the multi-task learning answer prediction model encoded by the BERT model as the baseline model (Baseline). Since the test set is closed, we divide the whole training set into 5 folds and experiment using cross-validation.

On CAIL 2020 reading comprehension data set, we use Chinese BERT [26] to encode the question sentence pairs in the sentence filtering stage, and uses the Chinese pre-trained RoBERTa [6] model in the answer prediction model, and HotpotQA data set uses BERT [5] base for encoding. Adam optimizer is used by default, and the warm_up strategy is used, the weight decay parameter weight delay is set to 0.01. To prevent overfitting, the training process is stopped early if the model does not improve for 5 epochs in the validation set.

## 5   Results

The final results are shown in Table 1. Compared with the baseline model, our model can bring 3.5% and 21.3% improvement, respectively, mainly in the supporting sentence prediction.

Since the maximum number of supporting sentences per sample in CAIL 2020 is 8, and there are at most 47 sentences per document, we set $k$ to 15 in the experiment to ensure that all support sentences are recalled, with an accuracy of 98.5%. In contrast, for HotpotQA, splicing the sentences of all documents would result in many sentences not being in the supporting sentences, so the final $F1_{joint}$ value is only 41.5%. The data set has only two documents per problem, and the correlation between the documents is weak, so we use a filtering model to filter out the candidate documents to reduce the interference from irrelevant paragraphs, which has the same structure as the sentence filtering model mentioned in Sect. 3.1, except that the input becomes each document and the $N$ most relevant documents to the problem are obtained. To be able to input all relevant documents into the answer prediction model, we take $N$

**Table 2.** Comparative experimental results of the HotpotQA data set.

| Model | $F1_{ans}$ | $F1_{sup}$ | $F1_{joint}$ |
|-------|-----------|-----------|-------------|
| DFGN | 0.568 | 0.775 | 0.467 |
| HGN | 0.539 | 0.534 | 0.440 |
| SAE | 0.626 | 0.820 | 0.538 |
| DeMRC | **0.672** | **0.844** | **0.577** |

**Table 3.** Results of contextual feature fusion. $C$ indicates no feature fusion; $CLS + C$ approach expands the $CLS$ to the same dimension as the contextual features, then adds them together; $Q + C$ uses bidirectional attention [8] to learn the interaction between questions and sentences.

| Fusion Method | $EM_{ans}$ | $F1_{ans}$ | $EM_{sup}$ | $F1_{sup}$ | $EM_{joint}$ | $F1_{joint}$ |
|---------------|-----------|-----------|-----------|-----------|-------------|-------------|
| $C$ | 0.677 | 0.775 | 0.480 | 0.658 | 0.377 | 0.539 |
| $CLS + C$ | 0.676 | 0.776 | 0.482 | 0.660 | **0.378** | **0.540** |
| $Q + C$ | 0.677 | 0.778 | **0.483** | **0.666** | 0.376 | 0.543 |
| $CLS + Q + C$ | **0.678** | **0.780** | 0.480 | **0.668** | 0.374 | **0.547** |

to be 3, and the filtering model achieves 100% accuracy. After that, our model without the dynamic filtering mechanism (DeMRC,-dl) improved by 14.8% over the benchmark.

After adding dynamic filtering, our model better learns the true combination of supporting sentences under different noise, and the $F1_{joint}$ improves by 0.7% and 1.4% on the two data sets, respectively. Self-training technique improved $F1_{joint}$ of the model by another 1.5% and 5.1%, respectively, and showed larger improvements in each subtask. The improvement of $F1_{joint}$ was more pronounced in the HotpotQA data set due to the larger number of external data sets. On the other hand, in order to demonstrate that our model is more suitable for low-data case, we compare it with other classical models of Hot-potQA, such as SAE [9]. We replace the training set in the official open source code with 10% of the data in the training set used, and keep the other parts of the original model unchanged.

Table 2 shows the results on HotpotQA. The DeMRC model proposed in this paper achieves a large improvement in all metrics with low data and surpasses the best performing SAE [9] model, with a 3.9% improvement in the $F1_{joint}$. Obviously, the methods of filtering irrelevant documents can effectively improve the performance of the supporting sentence subtask and simplify the training process, which is especially suitable for low-data case. For the extraction subtask in answer prediction model, how context is acquired is crucial. In this paper, we disentangle the acquisition of contextual features in 3.2.3, explore which feature fusion mechanisms have a greater impact on it, conduct experiments on the CAIL 2020 reading comprehension data set, and use the EM(Exact Match) [1] as an auxiliary indicator.

Table 3 shows the results of contextual feature fusion. The experiments were performed on a single-fold model. It can be seen that the improvement is weak after incorporating only $CLS$, because the linear layer gives different weights to the global features at each contextual location, but it does not bring substantial improvement; while the acquisition of the answer relies heavily on the question term, so the incorporation of the question vector into the original contextual features achieves a 0.4% improvement; Incorporating both question and global features into the context, $CLS + Q + C$, is more effective than adding the other two components separately.

**Table 4.** Experimental results of self-training, where $s_1$ denotes one iteration and $s_n$ denotes and multiple iterations. "All" indicates that all external data were used and "4000" indicates that 4000 data randomly and dynamically screened each time.

|       | Num  | $EM_{ans}$ | $F1_{ans}$ | $EM_{sup}$ | $F1_{sup}$ | $EM_{joint}$ | $F1_{joint}$ |
|-------|------|-----------|-----------|-----------|-----------|-------------|-------------|
| $s_n$ | All  | 0.667     | 0.773     | 0.454     | 0.663     | 0.357       | 0.541       |
|       | 4000 | 0.668     | 0.767     | 0.470     | 0.666     | 0.370       | 0.537       |
| $s_1$ | All  | 0.657     | 0.762     | 0.450     | 0.660     | 0.345       | 0.535       |
|       | 4000 | **0.683** | **0.786** | **0.481** | **0.669** | **0.376**   | **0.551**   |

We conducted self-training experiments. As shown in Table 4, the multiple iterations approach re-labeled the data from the 6th epoch onwards. The one-iteration approach learns the student model only once. When training with all pseudo-labeled data sets with the same number of training epochs, the multiple iterations achieved a significant advantage, but the results were much worse than one iteration after each dynamic screening of 4000 data, because the student model was not sufficiently trained to cause the subsequent teacher model to tag the unlabeled data with the wrong pseudo labeling. However, it is very time-consuming to fully train student with a large amount of data each time, so it is a good choice to randomly and dynamically filter some labeled data into the training set in one iteration, which not only saves time but also makes the model have better generalization ability.

## 6 Conclusion

In this paper, we take the multi-hop reading comprehension with low data as main research problem, and study the problem of model overfitting due to insufficient data from two perspectives: maximizing the utilization of the data set itself and augmentation using external knowledge. The input of training process is dynamically adjusted to allow the model to learn under different noisy interference, and the inference process uses a sentence filtering model to reduce the interference. We also expand the data set by pseudo-labeling external data to increase the generalization ability of the model. Experiments show the effectiveness of our model for multi-hop reading comprehension tasks with low data. We believe our model can be generalized to other natural language tasks, such as named entity recognition and recommend system.

**Acknowledgments.** This work is supported by State Grid Zhejiang Electric Power Co., Ltd. Science and Technology Project - Research and Application of Intelligent Operation and Inspection Technology Based on Natural Language Processing and Artificial Intelligence Technology.

The authors would like to thank AI+ High Performance Computing Center of ZJU-ICI.

# References

1. Rajpurkar, P., Zhang, J., Lopyrev, K., Liang, P.: Squad: 100, 000+ questions for machine comprehension of text. In: EMNLP, pp. 2383–2392 (2016)
2. Chen, J., Durrett, G.: Understanding dataset design choices for multi-hop reasoning. In: NAACL-HLT, pp. 4026–4032 (2019)
3. Welbl, J., Stenetorp, P., Riedel, S.: Constructing datasets for multihop reading comprehension across documents. Trans. Assoc. Comput. Linguistics **6**, 287–302 (2018)
4. Yang, Z., et al.: Hotpotqa: a dataset for diverse, explainable multi-hop question answering. In: EMNLP, pp. 2369–2380 (2018)
5. Devlin, J., Chang, M.-W., Lee, K., Toutanova, K.: Bert: Pre-training of deep bidirectional transformers for language understanding. arXiv preprint arXiv:1810.04805 (2018)
6. Liu, Y., et al.: Roberta: a robustly optimized bert pretraining approach. arXiv preprint arXiv:1907.11692 (2019)
7. Hermann, K.M., et al.: Teaching machines to read and comprehend. Adv. Neural. Inf. Process. Syst. **28**, 1693–1701 (2015)
8. Seo, M.J., Kembhavi, A., Farhadi, A., Hajishirzi, H.: Bidirectional attention flow for machine comprehension. In: ICLR (2017)
9. Tu, M., Huang, K., Wang, G., Huang, J., He, X., Zhou, B.: Select, answer and explain: Interpretable multi-hop reading comprehension over multiple documents. In: AAAI, pp. 9073–9080 (2020)
10. De Cao, N., Aziz, W., Titov, I.: Question answering by reasoning across documents with graph convolutional networks. arXiv preprint arXiv:1808.09920 (2018)
11. Tu, M., Wang, G., Huang, J., Tang, Y., He, X., Zhou, B.: Multi-hop reading comprehension across multiple documents by reasoning over heterogeneous graphs. In: ACL, pp. 2704–2713 (2019)
12. Shao, N., Cui, Y., Liu, T., Wang, S., Hu, G.: Is graph structure necessary for multi-hop reasoning? CoRR (2020)
13. Levy, O., Seo, M., Choi, E., Zettlemoyer, L.: Zero-shot relation extraction
14. Tan, C., Wei, F., Zhou, Q., Yang, N., Lv, W., Zhou, M.: I know there is no answer: Modeling answer validation for machine reading comprehension. In: Zhang, M., Ng, V., Zhao, D., Li, S., Zan, H. (eds.) NLPCC, pp. 85–97 (2018)
15. Wei, J., Zou, K.: Eda: Easy data augmentation techniques for boosting performance on text classification tasks. In: EMNLP-IJCNLP, pp. 6381–6387 (2019)

16. Xie, Q., Dai, Z., Hovy, E., Luong, M.-T., Le, Q.V.: Unsupervised data augmentation for consistency training. In: NeurIPS (2020)
17. Guo, H., Mao, Y., Zhang, R.: Augmenting data with mixup for sentence classification: An empirical study. arXiv preprint arXiv:1905.08941 (2019)
18. Kobayashi, S.: Contextual augmentation: Data augmentation by words with paradigmatic relations. In: NAACL-HLT, pp. 452–457 (2018)
19. Wu, X., Lv, S., Zang, L., Han, J., Hu, S.: Conditional BERT contextual augmentation. In: Rodrigues, J.M.F., Cardoso, P.J.S., Monteiro, J., Lam, R., Krzhizhanovskaya, V.V., Lees, M.H., Dongarra, J.J., Sloot, P.M.A. (eds.) ICCS 2019. LNCS, vol. 11539, pp. 84–95. Springer, Cham (2019). https://doi.org/10.1007/978-3-030-22747-0_7
20. Anaby-Tavor, A., et al.: Do not have enough data? deep learning to the rescue! In: AAAI, pp. 7383–7390 (2020)
21. Hewlett, D., Jones, L., Lacoste, A., Gur, I.: Accurate supervised and semisupervised machine reading for long documents. In: EMNLP, pp. 2011–2020 (2017)
22. Xie, Q., Luong, M.-T., Hovy, E., Le, Q.V.: Self-training with noisy student improves imagenet classification. In: CVPR, pp. 10687–10698 (2020)
23. Vaswani, A., et al.: Attention is all you need. In: NIPS, pp. 5998–6008 (2017)
24. Bhojanapalli, S., Yun, C., Rawat, A.S., Reddi, S.J., Kumar, S.: Low-rank bottleneck in multi-head attention models. In: ICML, pp. 864–873 (2020)
25. Shazeer, N., Lan, Z., Cheng, Y., Ding, N., Hou, L.: Talking-heads attention. arXiv preprint arXiv:2003.02436 (2020)
26. Cui, Y., Che, W., Liu, T., Qin, B., Yang, Z.: Pre-training with whole word masking for chinese BERT. IEEE ACM Trans. Audio Speech Lang. Process. **29**, 3504–3514 (2021)

# ESTD: Empathy Style Transformer with Discriminative Mechanism

Mingzhe Zhang[1], Lin Yue[2], and Miao Xu[1(✉)]

[1] The University of Queensland, Brisbane, QLD 4072, Australia
{mingzhe.zhang,miao.xu}@uq.edu.au
[2] The University of Newcastle, Callaghan, NSW 2308, Australia
Lin.Yue@newcastle.edu.au

**Abstract.** Language expressions without empathy can neither effectively convey the expresser's concern and goodwill, but also have a negative effect on the emotional and mental health of the recipients of the information. Compared to harsh or aggressive expressions, expressions with a high empathetic level can produce positive emotions. Unfortunately, non-empathetic expressions are generated daily without intention, causing negative feelings. Existing work has achieved certain success on style transfer, however, there are still limitations in language style selection. This paper addresses this challenge by using a corpus with multiple language styles. To this end, we employ ESTD to transfer a lower-empathetic expression to a higher-empathic expression. Experimental results on empathy style transfer task shows that our model outperforms some currently available baseline methods.

**Keywords:** Text style transfer · Transformer · Natural language processing · Empathy

## 1 Introduction

During the COVID-19 epidemic, nearly all the countries in the world have been in a state of physical isolation and blockade for a long time. This results in more communication happening via written expression on social media. Written expressions that lack empathy can easily create emotional stress for the person who receives such information. While long-term stress and negative emotions can seriously affect people's mental health. Empathy is a relatively broad concept, which refers specifically to a person's cognitive and emotional response to the experiences of others [5]. Showing empathy, either in written expression or oral expression plays a crucial role in relieving people from mental health issues; using the appropriate level of empathy in a conversation is an essential way of friendly and inclusive communication [24].

Usually, the definition of empathy is vague [9]. There are two main reasons for the lack of a precise human definition of empathy and related phenomena. First, as Shamasundar et al. [29] point out, empathy is highly related to the process of interpersonal interaction, which involves a wide range of information

© The Author(s), under exclusive license to Springer Nature Switzerland AG 2022
W. Chen et al. (Eds.): ADMA 2022, LNAI 13726, pp. 58–72, 2022.
https://doi.org/10.1007/978-3-031-22137-8_5

transfer in emotional, cognitive, and other domains. Therefore, the definition perspective is broad and multifaceted. Second, the specific degree of empathy is determined by the person's environment, experience, and state of mind, which means that empathy differs in each individual's perception.

Many previous works that focus on the text style transfer and text rewriting [7,13,33,39,40]. Notably, Sharma et al. first identified empathy style transfer as a major task [30]. Inspired by Sharma et al. [31], the criterion for discriminating the empathy level of a text proposed will be adopted in this work. Besides that, the main task is text style transfer. Figure 1 is a simple example showing that the rewritten sentence is highly semantically similar to the original one. Given an utterance, ( *"Stop it, right now!"*), we would like to transform it by converting it to ( *"It's okay to feel stuck. I'm here to help you."*). This makes the original sentence more compassionate while showing more understanding and encouragement. On the other hand, the rewritten sentence should show more empathy by understanding the message and experience of the text while ensuring that the rewritten sentence is semantically equivalent to the original one.

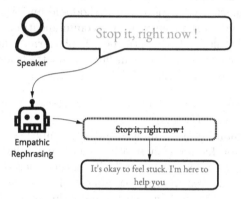

**Fig. 1.** Overview of the empathy rephrasing task. This task entails converting the original utterance that does not have an empathic or a low empathy level into a sentence with a high empathic expression. Given an utterance with low empathy, the task is to rewrite the original sentence to have a higher empathy level. Any samples in this paper were paraphrased to allow for anonymity [18].

As mentioned before, empathy is a complex and ambiguous concept. We must understand what has been conveyed in the original statement and how to make corresponding changes. Secondly, we must ensure that the rewritten sentences maintain the same semantic meaning as the original sentences. Meanwhile, common confusion needs to be considered from various aspects, such as language fluency and perplexity. In real situations, we can not rewrite each sentence similarly. Rewriting each low-sympathy sentence as ( *"I am sorry to hear that."*) cannot be applied in all cases. Finally, unlike traditional text style transfer, empathic style transfer is often much more complex than word-level substitutions. Traditional text style transfer tasks are usually based on the interchange of positive or negative words [4,13,33]. In brief, based on the original sentence, the original style of words is replaced using the opposite tag words. In this way, it can achieve the goal of style transfer. For example, to rewrite a negative sentence (e.g., *"It is bad."*) to a sentence with positive semantics, we only need to replace ( *"bad"*) to ( *"good"*). However, this approach cannot be applied to the empathic expression transformation. Take the example above; substituting a word can change the semantics of a sentence dramatically [11].

To address the above challenges, we proposed **ESTD**, a new empathy transformer with a discriminative mechanism. Empathy Transformer, as a generator, takes an utterance as input and generates an utterance with a higher empathy level. In addition, we discriminated the generated sentences and gave an empathy level to the output utterance by our discriminant model. We computed the cosine similarity between the generated sentences and the original sentences. The aim is to ensure that the generated sentences do not deviate from the original meaning.

Our experiments demonstrate that the proposed model outperforms existing baseline methods in the task of empathic transfer in terms of perplexity, empathy level, and cosine similarity. We believe our method and findings are a crucial step in establishing a friendly and inclusive online communication environment while furthering the development of a mental health platform. This work is one of the artificial intelligence's critical roles in human mental health [23]. The main contributions of this work include:

- We propose a transformer-based empathic expression converter with a discriminative mechanism (ESTD).
- ESTD is conceptually easy to understand and empirically powerful. ESTD improved the empathy level than other baseline models (+0.16 absolute improvement). It also reduces the perplexity of the generated text (1.15 absolute reductions) compared to other baseline models.
- ESTD is the first model to focus on the empathic expression style transfer task in the absence of contextual information. While training processes are via using supervised learning methods based on the parallel corpus.

## 2   Related Work

### 2.1   NLP for Online Mental Health Assistance

Overall, our work is highly relevant to existing research on online psychological help for NLP. These works are mainly applied in online psychological counseling, intelligent chatbots, and online psychological assistance platforms [1]. Relevant researchers have helped to establish a sound online communication platform to some extent by building assisted chat features, such as conversational agents and intelligent chatbots. Among them, AI with strong empathy may be particularly useful in mental health conversation applications [19].

In this work, we mainly focus on achieving empathic dialogue agents in mental health communication. While empathy is a crucial concept in mental health support [3]. We improve the empathic expression of models like those in conversational agents by using a transformer-based approach.

## 2.2  Text Style Transfer

Most existing work on textual style transfer has focused on some common linguistic styles. Such as Politeness Transfer [17], Sentiment Transfer [13,33], Formality Transfer [26], and Gender & Political Slant Transfer [27]. However, empathic dialogue rewriting has not received significant attention. Noteworthy, Sharma et al. first proposed a reinforcement learning-based model for rewriting empathic dialogue [30]. In real-world scenarios, upstream conversations cannot be supported in all cases. In highly active conversations, we cannot rely on historical information and translate it accordingly. Therefore, we will train the model based on a parallel corpus to accomplish the text style transfer task, aiming to make the model more adaptable.

## 2.3  Discriminatory Mechanism

To address the challenge of discerning levels of empathy, we incorporate a discriminative mechanism. This method is inspired by the GAN (Generative Adversarial Network) [8]. The discriminator network separates the candidates created by the generator from the actual data distribution. In contrast, the generative network learns to map from the latent space to the desired data distribution. The generative network's training objective is to increase the discriminator network's error rate. In our approach, the primary function of the discriminator is to perform empathy-level analysis on the generated sentences [8,15].

In our approach, we use the pre-trained discriminator. In the training stage of the generator, the improvement of the empathy level is used as one of the optimization goals.

## 3  Methodology

Given parallel samples of sentences $\mathcal{X}_1 = \{x_1^1, \ldots, x_n^1\}$ and $\mathcal{X}_2 = \{x_1^2, \ldots, x_n^2\}$ from original utterances and target utterances respectively. The goal of our task is to effectively generate samples with high levels of empathy, which is $\hat{\mathcal{X}}_1 = \{\hat{x}_1^1, \ldots, \hat{x}_m^1\}$ depends on $\mathcal{X}_1$.

We proposed an approach with two-step. The first step is pre-training the discriminator, a BERT-based discriminator model whose main task is to compute the empathy level of the sentences generated by the former while gradually transforming the sentences by the generator to approach the high empathy level. The goal of the discriminator is to compute the corresponding empathy scores $\mathcal{S}_{total}$ based on the input sentences. Secondly, we take the original utterance as input and generate the output $\hat{x}_i^1$ through the generator, which is the empathy transformer. Based on the output $\hat{x}_i^1$, we use the discriminator that has been pre-trained to calculate its corresponding empathy level $\mathcal{S}_{total}^i$. Our empathy transformer aims to efficiently convert the original sentences to higher empathy levels while maintaining semantic similarity by calculating the cosine similarity between original and rewritten sentences.

The model allows us to complete the transfer expression conversion without upstream history information, which significantly differs from previous work [30]. As mentioned before (Sect. 1), other similar text style transfer tasks focus more on addition and deletion [13]. However, we cannot achieve a transfer in empathy expression by simply adding and deleting.

### 3.1  Empathic Expression Calculation

In previous work, Sharma et al. developed a text-based framework for measuring empathic expression [31]. It contains three main communication mechanisms: Interpretations, Emotional Reactions, and Explorations. For each mechanism, there are three corresponding assessment scores:

- **0:** There is no expression.
- **1:** There is a relatively weak expression.
- **2:** Strong emotional expression.

**Table 1.** The training samples after filtering. The original utterance represents sentences with a low level of empathy, and the target utterance represents sentences with a certain level of empathy.

| Communication mechanisms | Weak expression | Strong expression |
|---|---|---|
| Interpretations | *It's really tough* | *I'm going through this too, and it's really, really bad* |
| Emotional reactions | *You can do it* | *I believe in you! You can do it!* |
| Explorations | *What's going on?* | *What happened? what can I do for you? Are you okay?* |

Table 1 shows some realistic sentences which illustrate the specific rank differences between weak expression and strong expression. There is a clear difference between the different levels of expression. Strong expressions are more likely to highlight the specific manifestations of the corresponding communication mechanisms than weak expressions.

**Interpretations.** Interpretation plays an essential linking role in perception, judgment, and communication [22]. Perfect dialogue and communication are generally based on clear and insightful interpretations [37].

**Emotional Reactions.** A person's emotional response plays an essential role in the development and maintenance of communication [14]. Emotional reactions mainly include expressions of concern, care, and deep feelings about the experiences or situations.

**Explorations.** Exploration implies a deeper exploration of the underlying meaning expressed by the utterance. In human communication, exploration usually refers to expressing curiosity and concern in an appropriate way [2].

## 3.2 ESTD Framework

This section will introduce the structure of our model and the objective functions of the empathy transformer and empathy level discriminator. We pre-trained the empathy rating model based on the empathy evaluation criteria proposed by Sharma et al. [31], and the corresponding corpus[1]. We train the ESTD based on the **Blended Skill Talk** dataset[2].

**Empathy Transformer.** Figure 2 shows the overall structure of our proposed method. We were given an input utterance $x_i^1$ with embedding and adding positional encoding. Aiming to allow Transformer [38] to retain information about the position of words by adding an encoding of the relative position of words in the sentence. The positional encode is represented by the sine and cosine formula [10].

$$\overrightarrow{p}_t^{(i)} = g(t)^{(i)} := \begin{cases} \sin(\omega_k \cdot t), & \text{if } i = 2k \\ \cos(\omega_k \cdot t), & \text{if } i = 2k + 1 \end{cases} \tag{1}$$

where $t$ is the desired position in the input sentence $x_i^1$, $\overrightarrow{p}_t^{(i)} \in \mathbb{R}^d$ is the corresponding encoding, and $d$ is the dimension. Meanwhile, $\omega_k = \frac{1}{10000^{2k/d}}$.

After the positional encoding, we generate the corresponding empathic expressions by a Transformer-based generator. As mentioned in Sect. 1, we accomplish our task objectives by fusing multiple loss functions. In the part of Empathy Transformer, we first calculate the CrossEntropyLoss[3] $\mathcal{L}_g$ of the output and target utterance. The loss function is defined as follows:

$$\mathcal{L}_g = -\omega_{x_i^2} \log \frac{\exp(x_{i,x_i^2}^1)}{\sum_{c=1}^{C} \exp(x_{i,c})} \cdot 1\{x_i^2 \neq ignore\_index\} \tag{2}$$

where $C$ is the number of classes. Second, we calculate the Cosine Similarity [34] $\Theta_g$ between the generated sentence and the original sentence by

$$\Theta_g = \frac{x_i^1 \cdot \hat{x}_i^1}{\|x_i^1\|\|\hat{x}_i^1\|}. \tag{3}$$

aiming to ensure that the semantics of generated sentences do not deviate from the original semantics.

---

[1] https://github.com/behavioral-data/Empathy-Mental-Health/tree/master/dataset.

[2] https://parl.ai/projects/bst/.

[3] https://pytorch.org/docs/stable/generated/torch.nn.CrossEntropyLoss.html.

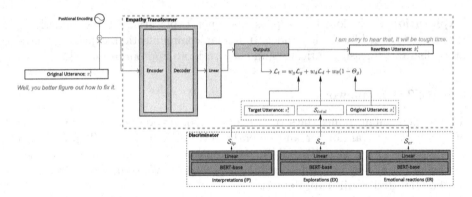

**Fig. 2.** The overall structure of **ESTD**. The model consists of two parts: an empathy transformer and a discriminator. The empathy transformer converts the input sentence $x_i^1$ into an output sentence $\hat{x}_i^1$ with a certain empathy level. The discriminator is pre-trained to judge the empathy level of the output sentence $\hat{x}_i^1$. Especially, the final loss function is a fusion of three loss functions, which are $\mathcal{L}_g$ $\mathcal{L}_d$, and $\Theta_g$ individually. The final generated sentence is semantically similar to the original sentence and has a certain level of empathy.

**Empathy Level Discriminator.** We fine-tuned the discriminant model based on BERT (*bert-base-cased*). Depending on the expression mechanism, we integrate three sub-models ($\mathrm{M}_{ip}$, $\mathrm{M}_{er}$ and $\mathrm{M}_{ex}$, individually) into the discriminator. For each sub-model, a linear layer with ReLU [20] is added to the based BERT model for classification, and each sub-model can be written as: $word\_emb(d, 768) - 768 - 3$.

The sentences generated by the empathy transformer are passed through a discriminator to obtain three empathy ratings, corresponding to different expression mechanisms [Interpretations (IP), Emotional Reactions (ER), and Explorations (EX)]. Each sub-model will have an output of the empathy level of the sentence ($\mathcal{S}_{ip}$, $\mathcal{S}_{er}$ and $\mathcal{S}_{ex}$). The final output value of the discriminator is obtained by linearly summing these three values. It can be expressed as

$$\mathcal{S}_{total} = \mathcal{S}_{ip} + \mathcal{S}_{er} + \mathcal{S}_{ex}. \tag{4}$$

Since we expect the rewritten sentences to tend to have the highest empathy level, which is equal to minimizing the difference between the maximum value ($\mathcal{S}_{max} = 6$) and the current value ($\mathcal{S}_{total}$). The goal of this stage can be expressed as

$$\arg\min_{\mathcal{S}_{total}} f(\mathcal{S}_{total}) = \{\mathcal{S}_{total} \mid f(\mathcal{S}_{total}) = \mathcal{S}_{max} - \mathcal{S}_{total}\}. \tag{5}$$

We use the Mean Squared Error (MSE) as the loss function of this part to achieve this goal. For ease of representation, here we set $\mathcal{S}_{total}$ equal to $\mathcal{S}_i$.

$$\mathcal{L}_d = \frac{1}{n} \sum_{i=1}^{n} (\mathcal{S}_i - \mathcal{S}_{max})^2. \tag{6}$$

Finally, we have discriminator loss objective ($\mathcal{L}_d$) given by Eq. 6. In the final training process, we only need to optimize these objectives simultaneously to accomplish the empathic expression transfer task. By this way, the final loss function of ESTD is

$$\mathcal{L}_t = w_g \mathcal{L}_g + w_d \mathcal{L}_d + w_\theta (1 - \Theta_g). \tag{7}$$

where $w_g$, $w_d$ and $w_\theta$ are the weights of each loss function. Here, we set $w_g = 0.7$, $w_d = 0.25$ and $w_\theta = 0.05$ individually.

## 4 Experiments and Results

In this section, we present details about the datasets and experiments (including the comparison with the baseline method and the final results of our method). The source code is available on GitHub[4].

### 4.1 Datasets

This section introduces the dataset used for the empathic expression transformation task. Also include some introduction to data pre-processing.

**Mental Health Subreddits.** The dataset was sourced from a sub-community of Reddit (reddit.com). Sharma et al. [32] performed in-domain pre-training on this publicly accessible dataset and annotated it as a subset of 10k interactions on empathy [31]. We counted the various data in the dataset. The number of dialogues corresponding to different scores in different expression mechanisms was calculated separately. Figure 3 shows the data visualization of statistics. We mainly use this dataset to pre-train the empathy rank classifier. To this end, we developed classifiers based on BERT (*bert-base-cased*)[5]. This model is

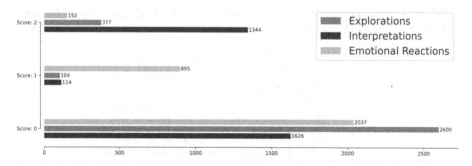

**Fig. 3.** Visualization of the distribution of empathy scores in **Mental Health Subreddits**, with three different colored bars corresponding to each of the three expression mechanisms. (Color figure online)

---

[4] https://github.com/masonzmz/ESTD.
[5] https://github.com/google-research/bert.

primarily for fine-tuning downstream tasks that use entire sentences (which may be masked) to make decisions, such as sequence classification, tag classification, or question and answer [6]. In this task, we use that to focus on different empathy scales, including three models (all with over 82% classification accuracy and over 81% F1 score, see Table 4), corresponding to Interpretations, Emotional Reactions, and Explorations, which will eventually focus on the pre-warm-up training of the discriminant model.

**Blended Skill Talk Dataset.** Smith et al. present a multi-task training dataset for various forms of multiple conversational skills [35], called Blended Skill Talk[6]. The dataset includes multitasking with the ConvAI2[7], Empathetic Dialogues[8] and Wizard of Wikipedia[9] datasets in their blend-debiased (topicifier) versions.

We constructed multiple parallel corpora for the multiple dialogue styles contained in this dataset. We need to reduce the bias due to the content of the dialogues since responses to upstream conversations in this dataset may have different meanings between different styles of responses. We first filtered the generated corpus for cosine similarity, aiming to filter out pairs that contain two utterances with too much difference. We train empathy expression transformation models based on this corpus.

In filtering the text data, we start by embedding the text using **Sentence-BERT**[10], which is a modification of the pre-trained BERT that uses siamese and ternary network structures to derive semantically meaningful sentence embeddings that can be compared by cosine similarity [28]. The final dimension of each sentence is **768**. We determine whether two sentences have the same meaning by calculating each pair's cosine similarity between the original and target utterance to have enough training data and enough similarity between the two utterances in the pair. Finally, we will choose those pairs with similarities greater than 0.5 as the experimental data. Table 2 shows some example sentences after filtering the original data.

**Table 2.** The training samples after filtering. The original utterance represents sentences with low level of empathy, and the target utterance represents sentences with certain level of empathy.

| Source dataset | Original utterance | Target utterance |
|---|---|---|
| Blended Skill Talk [35] | *I never learned much about graphic design but love art* | *I always wish i was good at design have tried, but i am pretty garbage, haha!* |
| | *I agree. Another thing we can do to fix the world we live in* | *Yes, that is all we can do. Keep trying to do better in life, and help others* |
| | *What else do you do for fun!!* | *Haha finally meet someone who is like me, game on, buddy* |

---

[6] https://parl.ai/.
[7] https://github.com/aliannejadi/ClariQ.
[8] https://github.com/facebookresearch/EmpatheticDialogues.
[9] https://parl.ai/projects/wizard_of_wikipedia/.
[10] https://www.sbert.net/docs/pretrained_models.html.

## 4.2   Baselines

We compare our systems against four baseline methods. Seq2Seq and Seq2Seq-Attn were evaluated under the same setting as the reference shows. Especially, we use the same dataset to train the GPT-2 fine-tuning and BART.

- **Seq2Seq** [36] Consists of an encoder and a decoder that converts a sequence to another sequence.
- **Seq2Seq-Attn** [38] Same as Seq2Seq, but with Attention mechanism. The attention mechanism is part of the neural network. It determines which source parts are more important.
- **GPT-2 Fine-Tuning** [25] Generative Pre-trained Transformer 2, a large text processing model.
- **BART** [12] is a autoencoder for pretraining sequence-to-sequence models.

## 4.3   Evaluation Metrics

Following the previous work [30], we use automatic metrics for the evaluation of our method. We mainly use the following metrics

- **Empathy changing.** It is mainly used to measure the degree of change in the empathy scale. We use the framework developed by Sharma et al. [31] to complete the measurement of this metric, where the value varies of empathy changing over the range of $[-6, 6]$.
- **Similarity.** Since one of the goals of our task is to ensure as much as possible that the meaning of the rewritten sentence does not deviate from the meaning of the original sentence. So we use Cosine Similarity as a measure for this metric.
- **Perplexity.** Following the previous work [16,30]. We used a pre-trained model to calculate this metric. The pre-trained model is GPT-2 language model.
- **BLEU.** We use the target utterance in the dataset as the ground truth while using the BLUE metric to compare with the output of the model [21].

## 4.4   Ablation Study

We conduct ablation studies on ESTD to empirically examine the contribution of its main mechanisms/components, including the use of Discriminator, the use of $\Theta_g$, and only training the generator without Discriminator and $\Theta_g$.

**Without Discriminator.** We analyze the specific differences in performance between the model using discriminators and not discriminators.

**Without $\Theta_g$.** We train the model without Cosine Similarity Loss. The loss function of the empathy transformer is changed in this experimental condition to $\mathcal{L}_t = \mathcal{L}_g + \mathcal{L}_d$.

**Without Discriminator and $\Theta_g$.** In the training process, we eliminate the discriminator and loss function. To investigate the model's performance without these two components, we train the model without both discriminator and $\Theta_g$.

## 4.5   Results

**Baseline Results.** The baseline experiments' results are shown in Table 3. By comparison, ESTD was found to have a significant advantage in terms of empathy level improvement (0.16 more than the next best approach, GPT-2 fine-tuning) while generating sentences with relatively low perplexity (1.15 less than the next best approach, GPT-2 fine-tuning). BART outperformed other baseline models, including ESTD, in terms of text similarity and BLEU score. However, its empathy changing was only −0.1712. While having relatively poor perplexity. The relatively low performance of ESTD in text similarity may be due to the bias of the training process toward the improvement of empathy scores. The inability to optimize each goal simultaneously is also a significant drawback of ESTD.

**Table 3.** Performance of ESTD and comparisons with other baseline methods on the set of automatic metrics. We can see that ESTD outperformed all other baseline models regarding empathy rating improvement. It also has the lowest perplexity level. However, it was inferior to BART in the utterance similarity and BLEU scores.

| Model | Empathy changing | Similarity | Perplexity | BLEU |
|---|---|---|---|---|
| ESTD | **0.2560** | 0.4387 | **9.5158** | 0.1046 |
| Seq2Seq | −0.3521 | 0.2078 | 15.9682 | 0.0047 |
| Seq2Seq-Attn | −0.3155 | 0.2547 | 14.3699 | 0.0051 |
| GPT-2 fine-tuning | 0.1006 | 0.4276 | 10.6703 | 0.0532 |
| BART | −0.1712 | **0.6550** | 13.9249 | **0.1392** |

Besides, Table 4 shows the classification accuracy and F1 score of our discriminator in different expression mechanisms.

**Table 4.** Accuracy and F1 score of discriminators in different expression mechanisms.

| Expression mechanisms | Accuracy | F1 score |
|---|---|---|
| Interpretations | 0.8467 | 0.8513 |
| Emotional reactions | 0.9160 | 0.9097 |
| Explorations | 0.8284 | 0.8120 |

**Ablation Results.** Table 5 reports results on ablated versions of ESTD. When only discriminator and $\Theta_g$ were used to compare with GPT-2 fine-tuning, the absolute increase of empathy change was +0.1554. Furthermore, although adding $\Theta_g$ can improve the model's overall performance to some extent, the improvement is not apparent. Besides, in the absence of the discriminator component, the output sentence of the model has the highest cosine similarity to the original sentence. But the empathy change is only 0.1222. Similarly, without the $\Theta_g$ component, ESTD only had an empathy change of 0.1912. Despite having the best BLEU score, 0.1125.

**Table 5.** The results of the ablation experiment. It can be seen that both discriminator and $\Theta_g$ play an essential role in the transformation of empathic expressions.

| Model | Empathy changing | Similarity | Perplexity | BLEU |
|---|---|---|---|---|
| ESTD | **0.2560** | 0.4387 | **9.5158** | 0.1046 |
| w/o Discriminator | 0.1222 | **0.4553** | 10.3686 | 0.0561 |
| w/o $\Theta_g$ | 0.1912 | 0.4301 | 9.6619 | **0.1125** |
| w/o Discriminator & $\Theta_g$ | −0.22954 | 0.4489 | 13.5586 | 0.1098 |

**Table 6.** The table shows the empathic rewriting of the sentences using different models. Where the $S_o$ is the empathy level of the original utterance and $S_r$ is the empathy level of the rewritten utterance.

| Original utterance | $S_o$ | Model | Rewritten Utterance | $S_r$ |
|---|---|---|---|---|
| Well, you better figure out how to fix it | 0 | ESTD | I'am sorry to hear that. It will be tough time | 2 |
| | | Seq2Seq | museums is is what not | 0 |
| | | Seq2Seq-Attn | Museums live any work through month chemo sushi in reason about cat, if hospital not | 1 |
| | | GPT-2 fine tuning | Well, you better figure out how to fix it. This is a change you made | 1 |
| | | BART | Stress over it. You cannot change anything at this point | 0 |
| Oh, just a nail? You are a nice person | 1 | ESTD | I'am sorry to hear that. How do you feel? | 4 |
| | | Seq2Seq | museums is of not believe never family sea | 0 |
| | | Seq2Seq-Attn | Cat have learning worry believe thailand sea | 0 |
| | | GPT-2 fine tuning | You are a nice person | 1 |
| | | BART | Oh, sorry | 0 |
| Some here would love to go | 2 | ESTD | I like to get my free time too. You can to me, lol | 3 |
| | | Seq2Seq | cat have start be sea chemo s | 0 |
| | | Seq2Seq-Attn | museums again family sea reason thailand most sea | 0 |
| | | GPT-2 fine tuning | Some here would love to go to blue moon resort | 0 |
| | | BART | Sounds pretty nice! Where are you going to? | 0 |

**Qualitative Examples.** We present examples of rewriting from ESTD and baseline methods in Table 6. ESTD works by transferring the original utterance to a new utterance with a higher empathy level while maintaining the original semantics. For example, given a sentence *"well, u better figure out how to fix it."*, the model can understand the underlying meaning in the original sentence and give a reasonable rewrite. The final rewritten sentence is *"I am sorry to hear that. it will be tough time."*.

## 5   Conclusion

Rewriting human-expressed sentences or dialogues through artificial intelligence may be an effective way to help provide inclusive expressions. We proposed a new method for converting non-empathetic or low-empathetic utterances to other utterances with appropriate levels of empathy. Our approach can help create a friendly and inclusive online environment by making human expressions more empathetic. Extensive experiments demonstrate that our model can effectively make sentences more empathetic, and the results outperform some existing baseline methods.

A potential problem due to the use of parallel corpora is the lack of rich content in the generated utterances. To solve this problem, a possible future work is to use unsupervised learning methods for this task.

## References

1. Althoff, T., Clark, K., Leskovec, J.: Large-scale analysis of counseling conversations: an application of natural language processing to mental health. Trans. Assoc. Comput. Linguist. **4**, 463–476 (2016)
2. Bohm, D., Senge, P.M., Nichol, L.: On dialogue. Routledge (2004)
3. Castonguay, L.G., Hill, C.E.: How and why are some therapists better than others?: Understanding therapist effects. JSTOR (2017)
4. Dai, N., Liang, J., Qiu, X., Huang, X.: Style transformer: unpaired text style transfer without disentangled latent representation. In: Proceedings of the 57th Annual Meeting of the Association for Computational Linguistics (2019)
5. Decety, J.: Empathy: From bench to bedside (2011)
6. Devlin, J., Chang, M.W., Lee, K., Toutanova, K.: Bert: pre-training of deep bidirectional transformers for language understanding. In: Proceedings of NAACL-HLT (2019)
7. Fu, Z., Tan, X., Peng, N., Zhao, D., Yan, R.: Style transfer in text: exploration and evaluation. In: AAAI (2018)
8. Goodfellow, I., et al.: Generative adversarial nets. In: Advances in Neural Information Processing Systems (2014)
9. Hall, J.A., Schwartz, R., Duong, F.: How do laypeople define empathy? J. Soc. Psychol. **161**(1), 5–24 (2021)
10. Kazemnejad, A.: Transformer architecture: The positional encoding. kazemnejad.com  (2019).  https://kazemnejad.com/blog/transformer_architecture_positional_encoding/

11. Kohut, H.: Introspection, empathy, and psychoanalysis an examination of the relationship between mode of observation and theory. J. Am. Psychoanal. Assoc. **7**(3), 459–483 (1959)

12. Lewis, M., et al.: Bart: denoising sequence-to-sequence pre-training for natural language generation, translation, and comprehension. In: Proceedings of the 58th Annual Meeting of the Association for Computational Linguistics (2020)

13. Li, J., Jia, R., He, H., Liang, P.: Delete, retrieve, generate: a simple approach to sentiment and style transfer. In: NAACL-HLT (2018)

14. Lovaglia, M.J., Houser, J.A.: Emotional reactions and status in groups. American Sociological Review, pp. 867–883 (1996)

15. Luc, P., Couprie, C., Chintala, S., Verbeek, J.: Semantic segmentation using adversarial networks. In: NIPS Workshop on Adversarial Training (2016)

16. Ma, X., Sap, M., Rashkin, H., Choi, Y.: Powertransformer: unsupervised controllable revision for biased language correction. In: EMNLP (2020)

17. Madaan, A., et al.: Politeness transfer: a tag and generate approach. In: Proceedings of the 58th Annual Meeting of the Association for Computational Linguistics (2020)

18. Matthews, T., et al.: Stories from survivors: privacy & security practices when coping with intimate partner abuse. In: CHI 2017, pp. 2189–2201 (2017)

19. Morris, R.R., Kouddous, K., Kshirsagar, R., Schueller, S.M.: Towards an artificially empathic conversational agent for mental health applications: system design and user perceptions. J. Med. Internet Res. **20**(6), e10148 (2018)

20. Nair, V., Hinton, G.E.: Rectified linear units improve restricted boltzmann machines. In: ICML (2010)

21. Papineni, K., Roukos, S., Ward, T., Zhu, W.J.: Bleu: a method for automatic evaluation of machine translation. In: ACL (2002)

22. Pedersen, R.: Empirical research on empathy in medicine-a critical review. Patient Educ. Couns. **76**(3), 307–322 (2009)

23. Pelau, C., Dabija, D.C., Ene, I.: What makes an ai device human-like? the role of interaction quality, empathy and perceived psychological anthropomorphic characteristics in the acceptance of artificial intelligence in the service industry. Comput. Hum. Behav. **122**, 106855 (2021)

24. Pfeil, U., Zaphiris, P.: Patterns of empathy in online communication. In: CHI (2007)

25. Radford, A., Wu, J., Child, R., Luan, D., Amodei, D., Sutskever, I.: Language models are unsupervised multitask learners (2019)

26. Rao, S., Tetreault, J.R.: Dear sir or madam, may i introduce the gyafc dataset: corpus, benchmarks and metrics for formality style transfer. In: NAACL-HLT (2018)

27. Reddy, S., Knight, K.: Obfuscating gender in social media writing. In: Proceedings of the First Workshop on NLP and Computational Social Science (2016)

28. Reimers, N., Gurevych, I.: Sentence-bert: sentence embeddings using siamese bert-networks. In: EMNLP-IJCNLP. ACL (2019)

29. Shamasundar, C.: Understanding empathy and related phenomena. Am. J. Psychother. **53**(2), 232–245 (1999)

30. Sharma, A., Lin, I.W., Miner, A.S., Atkins, D.C., Althoff, T.: Towards facilitating empathic conversations in online mental health support: a reinforcement learning approach. In: WWW (2021)

31. Sharma, A., Miner, A.S., Atkins, D.C., Althoff, T.: A computational approach to understanding empathy expressed in text-based mental health support. In: EMNLP (2020)

32. Sharma, E., De Choudhury, M.: Mental health support and its relationship to linguistic accommodation in online communities. In: Proceedings of the 2018 CHI Conference on Human Factors in Computing Systems (2018)

33. Shen, T., Lei, T., Barzilay, R., Jaakkola, T.: Style transfer from non-parallel text by cross-alignment. In: NeurIPS (2017)

34. Singhal, A., et al.: Modern information retrieval: a brief overview. IEEE Data Eng. Bull. **24**(4), 35–43 (2001)

35. Smith, E.M., Williamson, M., Shuster, K., Weston, J., Boureau, Y.L.: Can you put it all together: evaluating conversational agents' ability to blend skills. In: Proceedings of the 58th Annual Meeting of the Association for Computational Linguistics (2020)

36. Sutskever, I., Vinyals, O., Le, Q.V.: Sequence to sequence learning with neural networks. In: NeurIPS (2014)

37. Ten Have, P.: Doing conversation analysis. Sage (2007)

38. Vaswani, A., et al.: Attention is all you need. In: NeurIPS (2017)

39. Xu, W., Ritter, A., Dolan, W.B., Grishman, R., Cherry, C.: Paraphrasing for style. In: COLING (2012)

40. Yang, Z., Hu, Z., Dyer, C., Xing, E.P., Berg-Kirkpatrick, T.: Unsupervised text style transfer using language models as discriminators. In: NeurIPS (2018)

# Detection Method of User Behavior Transition on Computer

Yuki Ohkawa and Takafumi Nakanishi[(✉)]

Musashino University, 3-3-3, Tokyo 135-8181, Japan
g2250001@stu.musashino-u.ac.jp, tnakani@musashino-u.ac.jp

**Abstract.** IT general controls in a company are essential, and improper controls pose a significant risk to the company. One of those appropriate controls is the management and verification of daily logs. Companies utilize text-based logs (e.g., keystroke and application information) to monitor user behavior. However, some systems may only record PC screenshot image logs to prioritize stable operation. Therefore, we focus on PC screenshot image logs. Previous studies have attempted to analyze screenshot images as a classification problem or semi-supervised clustering. However, the number of tasks is indefinite, and the creation of training data and daily log checks is very costly. This paper proposes an efficient method for checking logs: detecting user action transitions using screenshot images. The proposed method detects user behavior transitions by grouping image and text features obtained from screenshot images based on their similarity without learning or labeling. We show that the proposed method can detect user behavior transitions with a reproduction rate of over 98% and reduces the total number of logs checked by auditors to about 1/4.

**Keywords:** Screenshot segmentation · Multi-modal features · Similarity · Audit system · User behavior analytics

## 1 Introduction

In recent years, risks such as information leakage due to inappropriate use of IT equipment by people have become important. In order to prevent such risks, IT general controls have become indispensable for companies. Inappropriate IT general controls are essential because they risk lowering corporate value and jeopardizing the company's existence. In general, information leakage and inappropriate use of equipment can be prevented to some extent by monitoring operations. Understanding user behavior on a PC is necessary, generally referred to as user behavior analytics (short as UBA).

There are two main types of operation logs: one is text logs (keystrokes, file access, active windows, running applications, Etc.). For example, UBA software typically uses text logs (application startup times, keystrokes, file operations, Etc.) to analyze business operations, improve employee productivity, and detect unauthorized operations. Another type of log is the screenshot image log on a PC. Image logs are more intuitive and easier to understand than text logs. Intuitive comprehensibility is an essential factor in ensuring

© The Author(s), under exclusive license to Springer Nature Switzerland AG 2022
W. Chen et al. (Eds.): ADMA 2022, LNAI 13726, pp. 73–85, 2022.
https://doi.org/10.1007/978-3-031-22137-8_6

reproducibility and traceability of operations. It is recommended that operation logs be checked daily as per PCI DSS [1]. Figure 1 shows an image of the flow in which auditors check employee operation log data accumulated daily.

Text logs can record user operations in detail, which can be analyzed to identify inappropriate operations and productivity. However, some companies may not want to install unnecessary software (applications to obtain operation text logs) to ensure the reliable operation of essential systems. Such companies often use RDP (Remote Desktop Protocol) or other means to record only screen operation logs. The goal of our study is to analyze user behavior limited to such situations where only images on the PC are available.

While screen operation logs are intuitive and easy to understand, the cost of checking them is high. Prior studies on screen screenshots have used CNNs for classification [2] or a combination of Active Learning and clustering for grouping [3]. It can be assumed that these methods can be used to perform daily log checks efficiently. However, information systems grow with the size of the business and the number of employees, and daily checking logs are very costly. Considering the modern business environment, it is unlikely that the type of work performed daily will remain the same. It would be inefficient to create and train data every time the work increases or decreases. To reduce the cost of daily checks, it would be practical to check logs by focusing on user behavior transitions instead of looking at all logs.

In this paper, we propose a detection method for user behavior transitions using image and text features obtained from screenshot image logs. It is taking advantage of the fact that operations on a PC change their display content when the operation target changes (e.g., opening a new window, scrolling, etc.). Focusing on user behavior is expected to effectively eliminate user stasis (reading information on the screen, taking a break, etc.), which are areas that do not need to be checked. Using this method shows that it can reduce the number of points to be reviewed by 70%, with a recall of at least 95% for the collected data range.

The contributions of this research are as follows.

- We propose a new method for image-based UBA.
- We propose a method for detecting user operation transitions based on the similarity of image and text features. We apply it to a new data collection, time-series PC screenshot image logs.
- We evaluate the detection results of user operation transitions based on the similarity of image and text features.
- We also discuss lessons learned from this study and directions for future research.

## 2   Related Work

### 2.1   Image Classification and Clustering

In [2], an image classification task, CNN is used to classify PC screenshot images. The method classifies models pre-trained on MIT's Places205CNN dataset into 14 class labels with a classification accuracy of 0.624. Our dataset is also similar in terms of data since it targets PC screenshots.

Another study [3] attempts semi-supervised clustering with image and text features, albeit on smartphone screenshot images. The labeling is being done in combination with active learning to evaluate performance. Image features are obtained with HOG (Histogram of Gradients) [4]. Text features are applied to text acquired by OCR and vectorized using GloVe's 300-dimensional word embedding using Wikipedia 2014 and Gigaword 5.

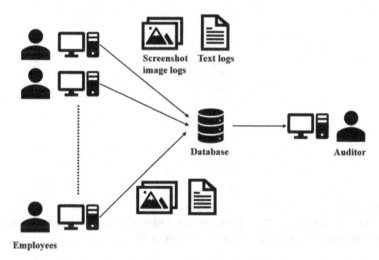

**Fig. 1.** An image of the flow in which the auditor checks employee operation log data stored daily. Based on the stored logs, the auditor works daily to check whether there is any irregularity or mishandling of the operations.

### 2.2   Search and Operation Automation

In [5], a system called Sikuli Search is proposed. It uses text describing images, image features extracted by SIFT [6] from screenshot images, and text read by OCR to construct a search system. This system makes it possible to search screenshot images and GUI elements (buttons, application icons, etc.) and is being studied to automate operations by combining it with other functions.

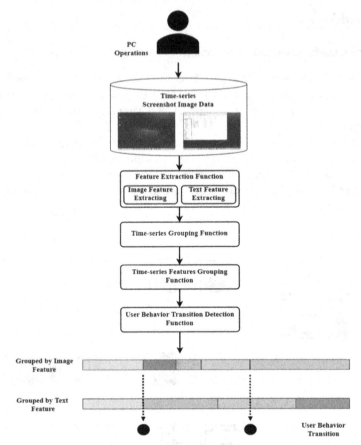

**Fig. 2.** The proposed method detects user behavior transition on a pc, using vectorized features images and text from pc screenshots. Image and text features are extracted from PC screenshots and grouped by similarity to detect user operation transitions.

## 2.3  User Behavior Analytics

In [7], foreground windows on a PC are considered user-operated windows. The types of open applications and operation times are output as text logs for time-series analysis and visualization.

Related studies use HOG or SIFT to extract image features and do not use pre-trained deep learning models. The number of classes is fixed or exploratory, and there is no mention of labeling or training costs. Also, user behavior transitions are treated as general information since they are output in text logs. Our research focuses not on class classification solutions but on image-based user behavior transitions that can be performed without costly labeling tasks or learning processes.

## 3   Detection Method of User Behavior Transition

### 3.1   Overview

Our proposed system extracts image and text features from pc screenshot images and groups them using time series and the same operation pattern. Grouped text and image features are used to determine user behavior transitions at pc operation. The proposed method is illustrated in Fig. 2. The proposed method consists of a feature extraction function, a time-series grouping function, a time-series feature grouping function, and a user behavior transition detection function.

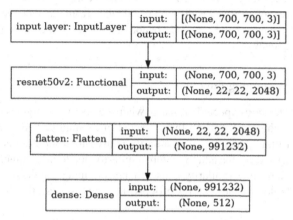

**Fig. 3.** Drawing of feature extractor; resized to 700 × 700 and input to the pre-trained model, the output is flattened and used as the image features.

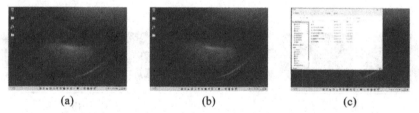

**Fig. 4.** It is assumed that (a) and (b) have no change and thus have high feature similarity, while (b) and (c) have low similarity due to changes in the image and text features caused by the Explorer is opened.

### 3.2   Feature Extraction

The feature extraction function consists of image feature extraction and text feature extraction. The image feature extraction realizes to use of a pre-trained feature extractor model, and the text extraction realizes to use of TF-IDF.

**Image Feature Extraction**

A pre-trained feature extractor is used to extract features from screenshot images. Using a pre-trained feature extractor avoids the cost of creating training data. This paper uses a ResNetV2 [8] feature extractor pre-trained on the Imagenet dataset [9]. Pre-processing of the screenshot images is resizing (700 × 700) and normalization. A drawing of the feature extractor is shown in Fig. 3.

**Text Feature Extraction**
The Tesseract OCR engine [10] extracts text from screenshots. The only preprocessing of the text excludes text less than 4 in length. Since the text output by OCR is not necessarily accurate, we do not exclude stop words, which is done in general natural language processing. Many texts were not outputting accurately as far as the human eye could see because exact accuracy is difficult to calculate from the data, we had prepared. TF-IDF to convert features the top 500 vocabulary words in the preprocessed text in order of word frequency.

### 3.3  Time-Series Grouping Function

The screen and the text are expected to change when the operation target is switched. This function detects such changes using the extracted image and text features and groups similar features. In this paper, cosine similarity is used to compute this similarity. As shown in Fig. 4, (a) and (b) have high feature similarity due to no change, while (b) and (c) are inferred to have low similarity due to changes in image and text features caused by opening the Explorer. The features grouped in a time series are called time-series features.

$$similarity(\mathbf{x_i}, \mathbf{x_j}) = \frac{\mathbf{x_i} \cdot \mathbf{x_j}}{|\mathbf{x_i}||\mathbf{x_j}|} \tag{1}$$

(a)                    (b)

**Fig. 5.** (a) and (b) are expected to be the same time-series features, although they are far apart in time series, so they are subject to grouping.

## 3.4    Time-Series Features Grouping Function

Since we expect the same time-series features to appear several times over time, we propose a function to detect and group the patterns. Calculate the average vector of each time series feature, calculate the cosine similarity, and group those with high similarity. As shown in Fig. 5, although (a) and (b) are far apart in time series, they are expected to be the same time series feature and thus should be included in the grouping.

## 3.5    User Behavior Transition Detection Function

This function determines user behavior transitions after grouping image and text features—the point at which the group switches is defined as the transition of user behavior. Since the group switching point is detected for each image and text feature value, the point where they overlap is considered the user behavior transition point. Figure 6 shows a sample illustration of determining user operation transitions.

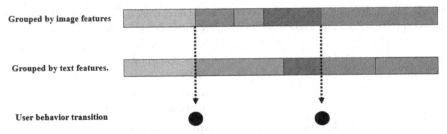

**Fig. 6.** This is a sample illustration of determining user operation transitions. The point at which the image and text feature groups are switched simultaneously is defined as the point of user operation transition.

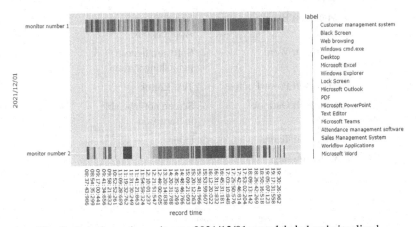

**Fig. 7.** Data for each monitor on 2021/12/01 were labeled and visualized.

# 4  Experiment

## 4.1  Our Dataset

We first looked into the publicly available data sets but found only text logs of operation logs and no image data sets. We collected five days of screenshot images recorded approximately once every 10 s on one employee's PC. The subject employee's primary job was customer management and software sales management. Since the employee was working on two monitors, the number of screenshot images was twice as large as when working on one monitor. Each screenshot image was labeled with an operational state to measure the accuracy of detecting user behavior transitions. Labels were given by one person with a visual check. Details of the labels are shown in Table 1. Figure 7 visualizes the labeled data in time series for each monitor on 2021/12/01. Figure 8 visualizes all labeled data. Table 2 shows the number of screenshot images per date, the number of user action transitions, and the average.

**Table 1.** List of classified label names. List of classified label names. The label was determined by looking at the before/after a relationship if multiple windows were open.

| | |
|---|---|
| Desktop | Sales Management System |
| Black Screen | Workflow Applications |
| Lock Screen | Windows cmd.exe |
| Attendance management software | Windows Task Manager |
| Customer management system | Windows Explorer |
| Microsoft Excel | Visual Basic for Applications |
| Microsoft Word | Software license publishing system |
| Microsoft PowerPoint | Text Editor |
| Microsoft Outlook | Screen not labelable in transition |
| Microsoft Teams | PDF |
| Web browsing | |

**Table 2.** Summarize the date and number of data collected, the number of user action transitions, and the data size. Operation transitions for which the label has changed shall be user action transitions. All image sizes are 1920 × 1080.

| Record date | Number of screenshot images | Number of user behavior transitions | Data size |
|---|---|---|---|
| 2021/12/01 | 8880 | 825 | 1038 MB |
| 2021/12/02 | 7802 | 618 | 1132 MB |
| 2021/12/03 | 9487 | 653 | 1233 MB |
| 2021/12/06 | 9257 | 620 | 1265 MB |
| 2021/12/07 | 9280 | 711 | 1352 MB |
| Average | 8941.2 | 685.4 | 1204 MB |

## 4.2  Experiment Results

We attempted to detect user behavior transitions with three features: image, text, and image and text features. The similarity thresholds for each function are shown in Table 3. Image feature extraction took an average of 1.5 s per image, and OCR processing took an average of 3 s per image. Detecting operational transitions took 16.6 s per day of data processing. Figure 9 (a) is an example of visualizing the results of grouping by image features. The number of groups created is 1391. Figure 9 (b) is an example of visualizing the results of grouping by text features. The number of groups created is 386. Table 4 shows the average accuracy, recall, and F-score of the user behavior transitions detection method.

**Table 3.** Similarity threshold for each function.

| | |
|---|---|
| Image time-series grouping function | 0.9 |
| Image time-series features grouping function | 0.9 |
| Text time-series grouping function | 0.6 |
| Text time-series features grouping function | 0.6 |

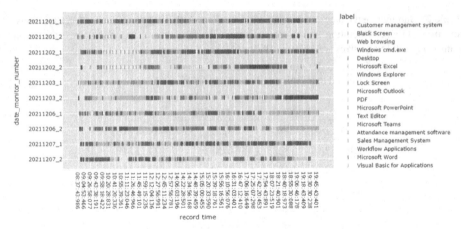

**Fig. 8.** Visualization of all labeled data.

**Table 4.** The user behavior transitions detection method's average accuracy, recall, and F-score.

|                          | Precision | Recall | F score |
|--------------------------|-----------|--------|---------|
| Only image features      | 0.302     | 0.986  | 0.461   |
| Only text features       | 0.403     | 0.885  | 0.550   |
| Image and text features  | 0.459     | 0.882  | 0.601   |

**Table 5.** The average number of transitions of detected operations when using each feature.

|                                        | The actual number of screenshots | Image features | Text features | Image and text features |
|----------------------------------------|----------------------------------|----------------|---------------|-------------------------|
| Number of user behavior transitions    | 8941                             | 2259           | 1501          | 1307                    |

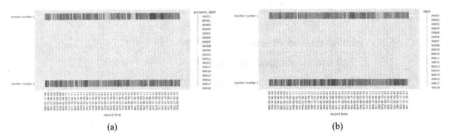

**Fig. 9.** (a) Example of visualizing the results of grouping by image features. The number of groups created is 1391. (b) Example of visualizing the results of grouping by text features. The number of groups created is 386.

### 4.3  Discussion

Image and text features are shown to be capable of detecting user behavior transitions in screenshots. This information on operational transitions will allow users to efficiently check the screen parts related to the screen changes. Table 5 shows the actual number of screens and the number of operations detected. Using each feature reduces the number of screenshot images that need to be checked.

**Data Generalization**

The data used in this experiment were 5-day records from a single user. Future work will include the generalization of the data. By incorporating data from different users, different work styles, tasks, operating systems, and software will be added to the data. Also, tasks should change fluidly, even for the same user.

**Active Window Detection**

Figure 10 (a) and (b) are examples of low image and text similarity. In (a) and (b), the same Explorer is opened, but the position of the Explorer and the text of the hidden desktop icon affect the similarity. As mentioned in previous studies, a way to reduce these effects is to detect active windows, as shown in (c). The features extracted by active

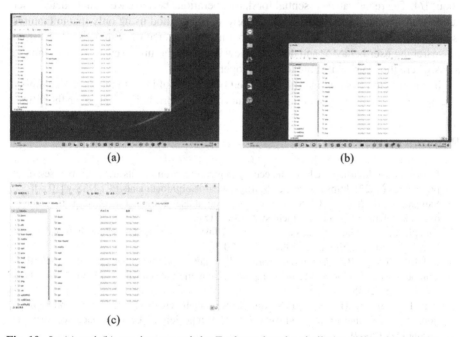

(a)                                      (b)

(c)

**Fig. 10.** In (a) and (b), we just moved the Explorer, but the similarity differs for both image and text features. The similarity of the image features reacts strongly to changes in the object's position, while the text features respond to changes in the displayed text information. One possible solution is to be able to detect windows, as in (c).

window detection are expected to be robust to position and will not need to deal with hidden text.

**User Behavior Search System and New User Behavior Detection**

In this experiment, user behavior transitions were detected, but the extracted features may be used for retrieval. The search function could support an efficient search of the target scene in the event of an incident or an incorrect operation.

In addition, the daily accumulation of features and the measurement of similarity may be used to detect new user behavior.

## 5  Conclusion

In this paper, we used to image and text features obtained from PC screenshot images to detect user operation transitions. It was found that the detection using only image features could detect with a recall rate of 98% or higher, reducing the overall confirmation cost to about 1/4. The recall rate is essential for daily operations because we want to detect user action transitions without omissions. Therefore, the method using only image features is optimal. The image and text features obtained from screenshot images can be used for searching user behavior and detecting new behavior. Future research can expand the areas where they can contribute to system operation.

In addition to researching utilization methods, we would like to increase the number of target users and the total amount of data and conduct correlation analysis in the future.

## References

1. Payment Card Industry (PCI): Data Security Standard: Requirements and security assessment procedures (2022). https://www.pcisecuritystandards.org/documents/PCI-DSS-v4_0.pdf
2. Sampat, A., Haskell, A.: CNN for task classification uses computer screenshots to integrate dynamic calendar/task management systems (2015)
3. Chiatti, A., et al.: Guess What's on my screen? Clustering smartphone screenshots with active learning. arXiv preprint arXiv:1901.02701 (2019)
4. Dalal, N., Triggs, B.: Histograms of oriented gradients for human detection. In: 2005 IEEE Computer Society Conference on Computer Vision and Pattern Recognition (CVPR 2005), vol. 1. IEEE (2005)
5. Yeh, T., Chang, T.-H., Miller, R.C.: Sikuli: using GUI screenshots for search and automation. In: Proceedings of the 22nd Annual ACM Symposium on User Interface Software and Technology (2009)
6. Lowe, D.G.: Object recognition from local scale-invariant features. In: Proceedings of the Seventh IEEE International Conference on Computer Vision, vol. 2. IEEE (1999)
7. Saito, R., Kuboyama, T., Yamakawa, Y., Yasuda, H.: Understanding user behavior through summarization of window transition logs. In: Kikuchi, S., Madaan, A., Sachdeva, S., Bhalla, S. (eds.) DNIS 2011. LNCS, vol. 7108, pp. 162–178. Springer, Heidelberg (2011). https://doi.org/10.1007/978-3-642-25731-5_14
8. He, K., Zhang, X., Ren, S., Sun, J.: Identity mappings in deep residual networks. In: Leibe, B., Matas, J., Sebe, N., Welling, M. (eds.) ECCV 2016. LNCS, vol. 9908, pp. 630–645. Springer, Cham (2016). https://doi.org/10.1007/978-3-319-46493-0_38

9. Deng, J., et al.: ImageNet: a large-scale hierarchical image database. In: 2009 IEEE Conference on Computer Vision and Pattern Recognition, pp. 248–255. IEEE (2009)
10. Smith, R.: An overview of the Tesseract OCR engine. In: Ninth International Conference on Document Analysis and Recognition (ICDAR 2007), pp. 629–633. IEEE (2007)

# Image, Multimedia and Time Series Data Mining

# Ensemble Image Super-Resolution CNNs for Small Data and Diverse Compressive Models

Yingnan Liu[✉] and Randy Clinton Paffenroth

Worcester Polytechnic Institute, Worcester, MA 01609, USA
{yliu18,rcpaffenroth}@wpi.edu

**Abstract.** Convolutional neural networks (CNN) have become some of the most powerful tools for image reconstruction problems thanks to the availability of very large data sets. Implementations of deep residual structures, adversarial generation networks and attention mechanisms have made great accomplishment. However, the good performance from complex and deep network architecture is not guaranteed when the training data set is small and not well preventative for the entire population. There are many real-world image reconstruction tasks where large and diverse training data is unavailable, such as problems in the physical sciences and engineering for which the data set generation process is complicated and large data sets are expensive to construct. For example, herein we discuss the application of deep-learning to challenging problems in material science. Inspired by compressive sensing and ensemble learning, we propose a method using ensemble image super-resolution CNNs in transform domains to overcome the challenges of small training data in image reconstruction problems. Ensemble methods provide a more robust approach when CNNs are trained with less representative data. Transform domains could support the CNNs with multiple sparse representations of the original image data which enrich the information so that the CNNs can be sufficiently trained even using small data sets. Particularly, we report here a successful application of CNN ensembles to the reconstruction of areal density maps of carbon nano-tube sheet materials. We show that applying the ensemble CNNs in transform domains can reveal finer details in the material texture and help to improve the quality control capabilities for carbon nano-tube sheet production with only a small collection of training data.

**Keywords:** Image reconstruction · Ensemble · Transform domain

## 1 Introduction

Convolutional neural networks (CNN) are popular techniques for high-resolution image reconstruction. The availability of large numbers of training images in the past few years has lead to the development of deep imaging neural network models. Following the breakthrough of single image super-resolution CNN (SRCNN)

© The Author(s), under exclusive license to Springer Nature Switzerland AG 2022
W. Chen et al. (Eds.): ADMA 2022, LNAI 13726, pp. 89–102, 2022.
https://doi.org/10.1007/978-3-031-22137-8_7

made by Dong et al. [1], numerous studies have been developed to improve the reconstruction quality. In particular, studies on deep residual neural networks, such as SRResNet [2], EDSR [3] and RCAN [4] provided benchmark results with the help of deep architectures and sufficient amounts of training data. However, in real-world applications, there are many cases where only a small collection of training data is available. For example, computerized tomography (CT) medical image reconstructions [5] and X-ray reconstructions are usually limited to the small training data from subject samples. Image super-resolution seeks to invert a *compressive model* which maps (theoretical) high-resolution images to low-resolution images, such as a camera or other imaging device. Such compressive models can be complex and diverse according to the application field. Previous studies focus on compressive models such as interpolation techniques using nearest neighbors and bicubic splines and Poisson noise, as they make comparisons with training data sets from benchmark database. As a result, some of the leading benchmark methods may not have stable performance in real-world applications.

Inspired by the classical approach of compressive sensing, transform domains are widely used for signal reconstruction [6]. Using bases in which signals can be sparsely represented, the high-resolution signal can be reconstructed with only a small number of nonzero coefficients in the sparse representation domain [6]. In particular, the connection between low-resolution images and high-resolution images can often be clearly represented in some transform domain where, for example, the spatial redundancy in the original image is reduced in the frequency domain. Fourier domain and higher-order wavelet sparsifying transform domains are both competitive choices for image processing problems [7]. Reducing memory requirements and computational cost compared with other types of Fourier transformations, the DCT is one of the most popular selections in the Fourier family as it considers only the real part of the expansion [8] and many algorithms were developed in the DCT domain before the recent breakthroughs in image super-resolution neural networks [9–11]. Wavelet transformations have also drawn substantial interest over many years, and have been adapted into image super-resolution neural networks to improve performance [12,13].

Reflecting on the success of both CNN and transform domains, we are inspired to extend current CNN based super-resolution techniques by leveraging appropriate sparse representations. We propose an ensemble CNN in multiple representation domains called EnsemNet. In particular, we demonstrate how our proposed technique improves the stability over strong base-line techniques for visual imagery with diverse compressive models, with a focus on the ability of our approach to function even in the presence of small data sets. While image processing of visual imagery is often done in the presence of large sets of training data, there are many important image processing problems that do not benefit from such large collections of training data. To that end, we also demonstrate the effectiveness of our techniques on an important problem in material manufacturing, namely the analysis of nano-scale material density maps arising from beta-particle transmission imaging.

## 1.1    Contribution

Herein we demonstrate how an ensemble model achieves a robust optimal result for various types of compressive models from small training data sets with the help of different transform domains. Bechmark algorithims SRResNet [2], EDSR [3] and RCAN [4] are utilized as basic image processors. We propose an ensemble CNN which assembles multiple basic image processors in different representation domains for an optimal combined result. We call this ensemble sparse model EnsemNet. Taking advantages from multiple representation domains, our model stands apart from previous algorithms by adapting different small training sets. The advantage of our model is demonstrated through the comparison among individual image super-resolution CNNs in single transform domains and the ensemble model. Our work is novel in three ways.

- First, our method provides a general solution for the image super-resolution problem on diverse and complex compressive models in real-world applications.
- Second, the optimal performance is stable over different selections of small training data sets.
- Third, based upon advantageous properties of sparsifying transform domains, our ensemble model combines the results from different domains to provide a robust solution from insufficient training data sets.

## 2    Foundational Work and Background

### 2.1    Sparse Representations

Inspired by the classical signal processing technique compressive sensing, we use sparse representations to enrich the training data. In compressive sensing theory, the compressed signal $y_s$ is a linear projection of the original signal $y_t$ [6]. With a certain sparse domain and the prior knowledge of the sensing matrix $R$, the system

$$y_s = Ry_t \tag{1}$$

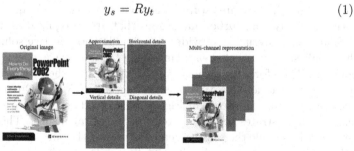

**Fig. 1.** Reprocessing wavelet representatives. A 2-level wavelet transformation gives four detail sub-matrices. The multi-channel representative is obtained by folding the detail sub-matrix.

can be solved with a small number of nonzero coefficients. One can represent the linear system as

$$y_t = \Psi\omega, R\Psi\omega = y_s, \tag{2}$$

where $\omega$ is the sparse representation of the original signal in basis $\Psi$. Problem (2) is computationally equivalent to a convex optimization problem. The sparse domain can be easily found by Fourier and wavelet transformations [14]. However, the prior knowledge of the sensing matrix is a restriction to the solutions. In contrast, single image super-resolution CNN (SRCNN) [1] solves the problem in a supervised fashion based on external example images without any requirement on prior knowledge. Moreover, studies on sparse representation [15,16] show that the signal sparsity can help to achieve improved results for image reconstruction neural networks. Taking the advantage of signal sparsity, we adapt image super-resolution CNNs to transform domains.

In addition to the limited information from the original space domain, the algorithms could extract more representing features from the sparse representations and therefore achieve an improved ensemble reconstruction. Specifically, we use the DCT Fourier domain and the db6 wavelet domain [17,18]. To prevent the algorithm from extracting features across different wavelet detail sub-matrices, a multi-channel wavelet representation is used, as illustrated in Fig. 1. In particular, four wavelet detail sub-matrices obtained from 2-level wavelet transformations have been folded into a 4-channel image representation. In this way, the convolutional kernel would not be applied across different detail sub-matrices which would degrade the algorithm effectiveness.

### 2.2 Miralon Areal Density Maps

In addition to our study of standard visual imagery, our proposed techniques is also demonstrated on the application of reconstructing beta transmission areal density maps of carbon nanotube material sheets called Miralon. Carbon nanotubes are seamless cylindrical hollow fibers, as shown in Fig. 2. The nature of its hexagonal pattern and the strong bond between carbon atoms provide carbon fiber materials impressive properties including strength, thermal and electrical conductivity, high-temperature resistance and so on. Miralon sheets are built with extremely long carbon nanotubes that are recognized as a state-of-the-art carbon fiber material. It provides sustainable and effective solutions to some of the toughest industry challenges involving the aerospace, energy, and electronics domains. Therefore, the quality control of Miralon sheets is crucial. To inspect the density variation in a Miralon sheet, a Mahlo QMS-12 Qualiscan Beta Transmission System[1], as shown in Fig. 3, is designed to generate the areal density map which illustrates the general texture of a Miralon sheet. The emitter releases beta particles while the sensing head moves over the surface. The number of particles which pass through the sheet and reach the receiver in each 20 ms window is converted into a compressive model. The areal density map shown in Fig. 4

---

[1] https://www.mahlo.com/en/products/process-control/details/traversing-quality-control-qualiscan-qms.html.

gives the texture of a Miralon areal density map. However, there is a lack of accuracy for fine details of the uneven distribution of material in the Miralon sheet. The zoom-in window in Fig. 4 shows a defective spot on the Miralon sheet. It is hard to observe and measure the shape and the area of the spot from the original areal density map.

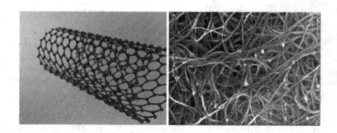

**Fig. 2.** Schematic structure of a carbon nanotube and the Miralon material under a microscope. The cylindrical hollow structured fiber gives Miralon material various properties, i.e. strong, lightweight and conductive. Electrons help the long fibers stick together naturally and form a tangled network.

As opposed to classic problems in image super resolution, this problem is a small sample problem. As the ground truth are generated from destruction tests, we have a limited number of samples to study. We need to ensure that the model is capable of capturing the compressive model from small training data sets. In addition, the unknown compressive model is complex. The behavior of beta particles, the spreading distribution from the emitter head, and the mathematical conversion in the equipment are all unknown. The measurement is clearly more complex than interpolation methods that are used in studies of regular visual imagery. With the help of the proposed algorithm, a high-resolution areal density map can be reconstructed, which successfully reveals finer patterns in the Miralon sheets.

## 3   Proposed Method

The ensemble algorithm, which we call EnsemNet, adapts the contents from the limited training data then decides the best way to combine the reconstructions from different domains to provide an optimal solution. The architecture is illustrated in Fig. 5. We parallel two algorithms on the space domain and the transform domain to generate two high-resolution reconstructions respectively. The algorithm then makes the ensemble in the original image domain. The ensemble method adds the two reconstructions element-wisely. Finally, with an extra convolutional layer for feature reconstruction and additional adjustment of the combination, an output is obtained. In this architecture, all the feature extraction convolutional layers use $3 \times 3$ kernels, and all the feature reconstruction convolutional layers use $1 \times 1$ kernels. The ensemble itself works on the combination

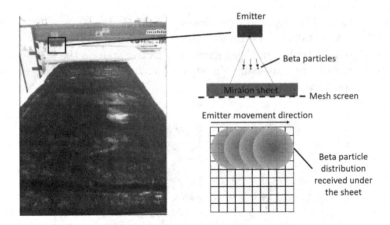

**Fig. 3.** Beta transmission equipment for measuring the Miralon sheet. As the sensing head moves over the surface, the emitter releases beta particles. The number of particles which reach the receiver in each 20 ms window is converted into a measurement.

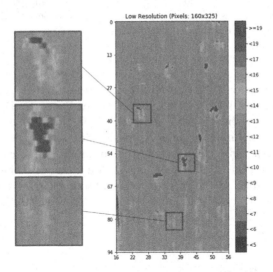

**Fig. 4.** Beta transmission areal density map of a Miralon sheet. The transmission sensor measures the material density with a low sampling rate. The original image is insufficient to identify variation and defects at the level needed.

and final feature reconstruction, it influences the training of individual algorithms by optimizing the ensemble based on different training sets and diverse compressive models. The element-wisely addition sufficiently provides a better result than simply using convolutional feature extraction from previous layers. Moreover, the performance of EnsemNet highly depends on the performances of its individual components. We try several candidates on different domains for algorithm 1 and algorithm 2 which are described in Fig. 5. The best performing

EnsemNet has better results than performing its individual components alone. For example, the EnsemNet with a SRResNet on the space domain as algorithm 1 and a EDSR on the wavelet domain as algorithm 2 should have improved result than SRResNet and EDSR. The best setting of EnsemNet will be found from experiments on regular visual imagery, and then applied to the Miralon sheet application.

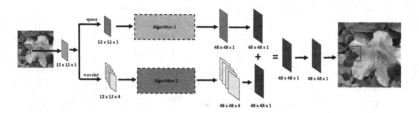

**Fig. 5.** The architecture for the ensemble method in both the space domain and the wavelet transform domain. Two algorithms are performed on the space representation and the multi-channel wavelet representation separately. Then the outcomes are combined by element-wise addition. Finally, the output is refined by an extra convolutional layer.

## 4  Experimental Results

In this section we show several experiments that were conducted to demonstrate the effectiveness of ensemble CNNs. We make comparisons among individual benchmark deep residual neural networks and ensemble CNNs over different combinations of individual algorithms. The efficiency of our best performing EnsemNet is indicated through experiments on multiple small data sets and different measurements.[2]

### 4.1  Training Details

Our target is to obtain an algorithm which performs stably well with small numbers of image signal for complex compressive models. The well-trained model should be able to accommodate both different small training data sets and diverse compressive models, then stably deliver a high-resolution image with improved reconstruction quality. The ground truths of high-resolution images are used as labels in the neural network. Mean square error (MSE) and Peak Signal-to-Noise Ratio (PSNR) are used as the metric. PSNR in decibels (dB) is defined as

$$PSNR = 10 \cdot \log_{10}\left(\frac{I^2}{MSE}\right), \tag{3}$$

---

[2] Code for all experiments can be found on github.com at https://github.com/innanliu426/EnsemNet.git.

where $I$ is the maximum pixel value of the data type [19], and

$$MSE = min\frac{1}{2N}\sum_{i=1}^{N} \|f(x_i) - x_i\|_2^2. \tag{4}$$

$f$ represents the operation of the neural network and $x_i$ represents the input. We use 8-bit images with $I = 255$. The experiments are performed in YCbCr format and the reconstruction is evaluated on Y channel.

We use standard benchmark image data sets as training and testing data. Training data are from the DIV2K data set [20]. Drawn from the 800 high-resolution images, we obtain four data sets with the size of 1, 2, 3 and 5 images respectively. Testing results are compared on data sets Set5 and Set14 which consist of natural scenes. In previous studies, cubic interpolation is often chosen to be the standard compressive model. We believe it is more faithful to consider more measurements. To present the complexity and diversity of compressive models in real world applications, four down-scaling procedures with super-resolution factors of 4 are carried out with different interpolation methods, Gaussian blur models and Gaussian pyramid degradation, as shown in Table 1. We use packages OpenCV and scikit-image in Python to implement the measurements. A $100 \times 100$ image would be reconstructed to $400 \times 400$. Each of the four down-scaling procedures consist of multiple measurements. Suppose we have a $400 \times 400$ high-resolution image. In down-scaling procedure No. 1, the image is compressed into $200 \times 200$ with linear interpolation first. Then, a Gaussian blur is added to smooth the image. Finally, the image is compressed into $100 \times 100$ with nearest interpolation.

To make full use of the limited training data, we extract small overlapping patches from the low-resolution image. The sizes for low-resolution patches are $12 \times 12$. Each patch is rotated in $90°$, $180°$ and $270°$ to enlarge the size of the training data set. The number of training patches are 100, 150, 200 and 250 for the four data sets respectively. Another validation set of 100 patches are randomly drawn from the DIV2K data set to train the models. We use the MSE loss function as our experiment suggests that other popular loss functions do not

Table 1. 4 compressive versions with multiple down-scaling measurements. Lanczos: Lanczos interpolation; Cubic: cubic interpolation; Nearest: nearest interpolation; Linear: linear interpolation; Gaussian: Gaussian Pyramid; GB: Gaussian Blur.

| Compressive versions | Down-scaling measurements |
|---|---|
| 1 | Linear x2 + GB + Nearest x2 |
| 2 | Lanczos x2 + Nearest x2 + GB |
| 3 | Gaussian x4 + GB |
| 4 | Gaussian x2 + GB + Gaussian x2 + GB |

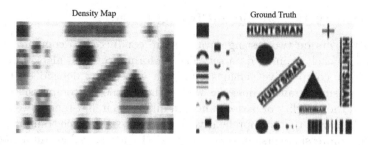

**Fig. 6.** Areal density maps and the ground truth for the patterned shim-stock.

help to train models well with the sparse representatives in transform domains. Other specific settings include Adadelta optimizer [21], 32 batch-size and 1000 maximum epochs. We use a 3.20 GHz Intel core i7-8700 CPU and 64 GB memory to run the implementations.

For areal density maps, a ground truth condition was designed using patterned shim-stock of uniform areal density. As shown in Fig. 6, geometric figures of different dimensions and orientations were laser-cut into the patterned shim-stock. Then, using the same parameters used for Miralon sheets, an areal density map was generated. Due to the cost of laser-cutting, we have only 2 training shim-stock areal density maps. The size of the low-resolution areal density map is $50 \times 140$ pixels. The high-resolution ground truth of the geometric figures can be easily rescaled. We rescale it into $200 \times 560$ pixels for training models with super-resolution factors equal to 4.

SRResNet [2], EDSR [3] and RCAN [4] are implemented as basic models. For individual algorithms, we apply EDSR on DCT domain, both SRResNet and EDSR on the space domain and RCAN on the Wavelet domain. Applying the architecture in Fig. 5, three combinations of ensemble algorithms are implemented on both the space domain and the Wavelet domain, as illustrated in Table 2. Packages SciPy and PyWavelets in Python are used to implement the transformations. The shape of the multi-channel wavelet representation is adjusted with zero-padding.

**Table 2.** Implemented models and the corresponding domains for the three settings of EnsemNet.

| Names | Algorithm 1 | Algorithm 2 |
|---|---|---|
| EnsemNet1 | EDSR - Wavelet domain | EDSR - DCT domain |
| EnsemNet2 | RCAN - Wavelet domain | SRResNet - space domain |
| EnsemNet3 | EDSR - Wavelet domain | SRResNet - space domain |

## 4.2   Reconstruction Quality on Testing Images

Four experiments on two testing data sets are conducted to compare the performances for each training data set. The performance of all algorithms varies from different compressive versions and data sets. They depend on the randomly drawn small training image as well as the 'unknown' compressive models. To compare the stability of performances in different circumstances, we evaluate the algorithms with their comprehensive performances. For each training data set with each compression version, the testing PSNR (in dB) values from the seven algorithms are ranked from the highest to the lowest. Then an average ranking for each algorithm is obtained over the four training data sets. Table 3 shows the ranking result obtained from the testing PSNR values from Table 4. For example, the testing PSNR values from ENsemNet3 for compressive version 1 on Set5, rank 2, 1, 1 and 3 for training data sets of size 100, 150, 200 and 250 respectively. Therefore, the overall average ranking for ENsemNet3 on compressive version 1 is 1.75. The rankings is also reflected in Fig. 7. For individual algorithms, EDSR on the original space domain has the most competing results on different small training sets. In two of the eight cases, SRResNet slightly outperforms EDSR. Moreover, individual algorithms have frequent gradient explosion while training with the extreme small data sets, which demonstrate the instability of performance in circumstances of this study. For ensemble algorithms, we find EnsemNet3, which combines a EDSR in the wavelet domain and a SRResNet in the space domain, has the most stable performance. This ensemble algorithm delivers a robust outstanding result in all the cases from different small training data sets on diverse and complex compressive models.

## 4.3   Application of Miralon Areal Density Maps

We implement the version of EnsemNet3 to the application of beta transmission areal density maps for Miralon sheets. The measurement parameters of the beta transmission Mahlo system is assumed to be fixed. If the setting of

**Table 3.** Average rankings for testing PSNR (in dB) values. For each training data set with each compression version, the seven algorithms are ranked from the highest PSNR to the lowest PSNR. An average ranking for each algorithm is then obtained with the performances from all the four training data sets.

| Models | Compressive version 1 | | Compressive version 2 | | Compressive version 3 | | Compressive version 4 | |
|---|---|---|---|---|---|---|---|---|
| | Set5 | Set14 | Set5 | Set14 | Set5 | Set14 | Set5 | Set14 |
| EDSR-DCT | 7 | 7 | 5.75 | 5.75 | 6.5 | 6.5 | 7 | 6.75 |
| EDSR | 2.5 | 2 | 2.25 | 2.75 | 3 | 3.25 | 2 | 2.25 |
| SRResNet | 2.25 | 3.25 | 3.75 | 3.75 | 3.75 | 3.75 | 2.25 | 2.25 |
| RCAN-WVT | 5 | 5 | 5 | 4.5 | 5.25 | 4.75 | 5.25 | 5.25 |
| EnsemNet1 | 5.75 | 5.75 | 5.25 | 5.25 | 5.25 | 5 | 5.75 | 6 |
| EnsemNet2 | 3.5 | 3.25 | 4 | 4 | 2.5 | 2.25 | 4 | 3.5 |
| EnsemNet3 | 1.75 | 1.5 | 2 | 2 | 1.75 | 2.25 | 1.75 | 2 |

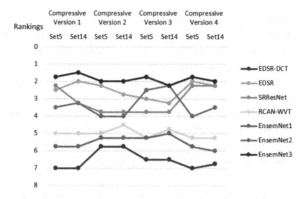

**Fig. 7.** Average rankings of testing PSNR from the four training sets over the four down-scale procedures and the two testing data sets.

**Table 4.** Testing PSNR (in dB) with different models for different training data sets. The performance of all algorithms varies from different compressive versions and data sets. Table 3 shows the summary of these results. This table is provided for completeness as one can see the averages in Table 3 demonstrate that EnsemNet3 has the most stable performance on different training data sets and diverse compressive models. (Data Size: # of training patches, CV: Compressive Versions)

| Data Size | CV | Test Set | EDSR-DCT | EDSR | SRRes-Net | RCAN-WVT | Ensem-Net1 | Ensem-Net2 | Ensem-Net3 |
|---|---|---|---|---|---|---|---|---|---|
| 100 | 1 | Set5 | 22.474 | 22.788 | 22.752 | 22.742 | 22.681 | 22.831 | 22.816 |
| | | Set14 | 20.975 | 21.157 | 21.149 | 21.126 | 21.114 | 21.198 | 21.190 |
| | 2 | Set5 | 22.345 | 23.105 | 22.073 | 22.976 | 22.631 | 22.926 | 22.978 |
| | | Set14 | 20.869 | 21.169 | 20.633 | 21.188 | 21.009 | 21.116 | 21.207 |
| | 3 | Set5 | 21.759 | 21.103 | 23.226 | 19.279 | 22.269 | 24.118 | 23.335 |
| | | Set14 | 20.976 | 20.375 | 22.099 | 20.558 | 21.541 | 22.437 | 22.043 |
| | 4 | Set5 | 21.753 | 23.988 | 24.147 | 23.479 | 22.778 | 23.610 | 24.162 |
| | | Set14 | 20.586 | 22.234 | 22.331 | 21.881 | 21.368 | 21.940 | 22.317 |
| 150 | 1 | Set5 | 22.597 | 22.863 | 22.870 | 22.785 | 22.643 | 22.772 | 22.870 |
| | | Set14 | 21.086 | 21.225 | 21.211 | 21.172 | 21.096 | 21.168 | 21.222 |
| | 2 | Set5 | 22.142 | 23.015 | 22.625 | 22.600 | 22.477 | 22.736 | 22.783 |
| | | Set14 | 20.725 | 21.158 | 20.936 | 20.982 | 20.901 | 21.043 | 21.023 |
| | 3 | Set5 | 23.987 | 25.680 | 24.879 | 24.669 | 24.745 | 25.072 | 25.337 |
| | | Set14 | 22.256 | 23.352 | 22.976 | 22.627 | 22.724 | 23.107 | 23.180 |
| | 4 | Set5 | 24.476 | 26.078 | 25.610 | 25.096 | 24.869 | 25.422 | 25.486 |
| | | Set14 | 22.608 | 23.693 | 23.396 | 23.025 | 22.882 | 23.335 | 23.352 |
| 200 | 1 | Set5 | 22.520 | 22.869 | 22.849 | 22.620 | 22.700 | 22.839 | 23.041 |
| | | Set14 | 21.042 | 21.249 | 21.233 | 21.071 | 21.118 | 21.201 | 21.373 |
| | 2 | Set5 | 22.951 | 22.822 | 23.166 | 22.661 | 22.858 | 22.851 | 23.194 |
| | | Set14 | 21.150 | 20.995 | 21.188 | 20.956 | 21.096 | 20.966 | 21.199 |
| | 3 | Set5 | 24.808 | 25.462 | 25.221 | 25.557 | 25.180 | 25.571 | 25.578 |
| | | Set14 | 22.718 | 23.134 | 22.968 | 23.223 | 22.994 | 23.270 | 23.223 |
| | 4 | Set5 | 24.864 | 25.959 | 25.811 | 25.515 | 25.392 | 25.755 | 25.842 |
| | | Set14 | 22.798 | 23.615 | 23.416 | 23.295 | 22.793 | 23.432 | 23.456 |
| 250 | 1 | Set5 | 22.656 | 22.952 | 22.969 | 22.830 | 22.703 | 22.914 | 22.936 |
| | | Set14 | 21.104 | 21.304 | 21.297 | 21.206 | 21.165 | 21.297 | 21.323 |
| | 2 | Set5 | 22.934 | 23.459 | 23.440 | 23.212 | 23.045 | 23.261 | 23.410 |
| | | Set14 | 21.167 | 21.464 | 21.489 | 21.304 | 21.237 | 21.321 | 21.436 |
| | 3 | Set5 | 23.991 | 25.917 | 25.583 | 25.214 | 25.055 | 25.323 | 25.656 |
| | | Set14 | 22.351 | 23.507 | 23.378 | 23.083 | 22.955 | 23.261 | 23.442 |
| | 4 | Set5 | 23.617 | 24.805 | 24.895 | 24.136 | 24.327 | 24.703 | 25.088 |
| | | Set14 | 21.988 | 22.843 | 22.939 | 22.426 | 22.491 | 22.845 | 23.062 |

the beta transmission sensor is adjusted, i.e. sensor speed, viewable range, etc., then the model needs to be trained for the new setting before applying to the manufacturing batches. First, we train the algorithm on the shim stock data set. Then, we apply the model on the areal density map of a recently manufactured Miralon sheet. Figure 8 shows the reconstruction result from EnsemNet3. Additional details about defects and areal density variation are revealed in the high-resolution reconstructed density maps. Taking the zoomed section from the sheet in Fig. 8 as examples, the shape and the area of defective spots can be discovered more precisely from the super-resolution reconstruction. This indicates that our method is practical for the application.

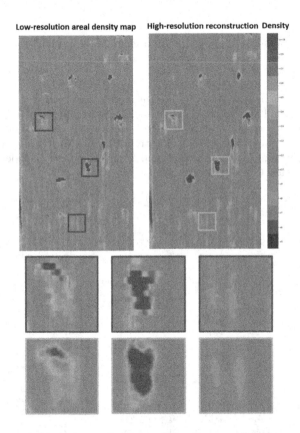

**Fig. 8.** Reconstructed beta transmission areal density maps by the proposed algorithm for Miralon sheets from the production line. Details for uneven density distribution are recovered.

## 5    Conclusion

In this work, EnsemNet is proposed to overcome the challenge of small training data for the image super-resolution problem. Instead of suffering unstable performance on limited data and complex 'unknown' compressive models, our proposed ensemble method leverages the advantage from wavelet sparsifying transform domains. A general solution is provided for various types of applications. The advantages of using EnsemNet is illustrated through experiments on both regular testing images and the Miralon density maps for different complex compressive models. Overall, EnsemNet provides a more robust and efficient solution for the image super-resolution problem with small training data sets and diverse compressive models.

# References

1. Dong, C., Loy, C.C., He, K., Tang, X.: Image super-resolution using deep convolutional networks. IEEE Trans. Pattern Anal. Mach. Intell. **38**(2), 295–307 (2016)
2. Ledig, C., et al.: Photo-realistic single image super-resolution using a generative adversarial network. In: Proceedings of the IEEE Computer Society Conference on Computer Vision and Pattern Recognition, pp. 4681–4690 (2017)
3. Lim, B., Son, S., Kim, H., Nah, S.: Enhanced deep residual networks for single image super-resolution. In: IEEE Computer Society Conference on Computer Vision and Pattern Recognition Workshops, pp. 136–144 (2017)
4. Zhang, Y., Li, K., Li, K., Wang, L., Zhong, B., Fu, Y.: Image super-resolution using very deep residual channel attention networks. In: Ferrari, V., Hebert, M., Sminchisescu, C., Weiss, Y. (eds.) ECCV 2018. LNCS, vol. 11211, pp. 294–310. Springer, Cham (2018). https://doi.org/10.1007/978-3-030-01234-2_18
5. Herman, G.T., Davidi, R.: Image reconstruction from a small number of projections. Inverse Prob. **24**(4), 045011 (2008)
6. Candès, E.J., Romberg, J., Tao, T.: Robust uncertainty principles: exact signal reconstruction from highly incomplete frequency information. IEEE Trans. Inf. Theory **52**(2), 489–509 (2006)
7. Strang, G.: Wavelet transforms versus Fourier transforms. Bull. Amer. Math. Soc. **28**(2), 288–305 (1993)
8. Park, S.C., Park, M.K., Kang, M.G.: Super-resolution image reconstruction: a technical overview. IEEE Signal Process. Mag. **20**(5), 21–36 (2003)
9. Rhee, S., Kang, M.G.: Discrete cosine transform based regularized high-resolution image reconstruction algorithm. Opt. Engvol. **38**, 1348–1356 (1999)
10. Marichal, X., Ma, W.Y., Zhang, H.J.: Blur determination in the compressed domain using DCT information. Proc. Int. Conf. Image Process. **2**, 386–390 (1999)
11. Demirel, H., Anbarjafari, G.: Image resolution enhancement by using discrete and stationary wavelet decomposition. IEEE Trans. Image Process. **20**(5), 1458–1460 (2011)
12. Guo, T., Seyed Mousavi, H., Huu Vu, T., Monga, V.: Deep wavelet prediction for image super-resolution. In: Proceedings of the IEEE Computer Society Conference on Computer Vision and Pattern Recognition Workshops, pp. 104–113 (2017)
13. Huang, H., He, R., Sun, Z., Tan, T.: Wavelet-SRNet: a wavelet-based CNN for multi-scale face super resolution. In: IEEE International Conference on Computer Vision, pp. 1689–1697 (2017)
14. Welland, G.: Beyond Wavelets. Academic Press, San Diego, CA (2003)
15. Yang, J., Wright, J., Huang, T.S.: Image super-resolution via sparse representation. IEEE Trans. Image Process. **19**(11), 2861–2873 (2010)

16. Gu, S., Zuo, W., Xie, Q., Meng, D.: Convolutional sparse coding for image super-resolution. In: IEEE International Conference Computer Vision, pp. 1823–1831 (2015)
17. Besar, R., Eswaran, C., Sahib, S., Simpson, R.J.: On the choice of the wavelets for ECG data compression. In: Proceedings IEEE International Conference on Acoustics, Speech, and Signal Processing, vol. 6, pp. 3614–3617 (2000)
18. Bairagi, V.K., Sapkal, A.M.: Selection of wavelets for medical image compression. In: International Conference on Advances in Computing, Control, and Telecommunication Technologies, pp. 678–680 (2009)
19. Huynh-Thu, Q., Ghanbari, M.: The accuracy of PSNR in predicting video quality for different video scenes and frame rates. Telecommun. Syst. **49**, 35–48 (2012)
20. Timofte, R., et al.: Ntire 2017 challenge on single image super-resolution: methods and results, In: Proceedings of the IEEE Computer Society Conference on Computer Vision and Pattern Recognition Workshops, pp. 114–125 (2017)
21. Swastika, W., Ariyanto, M.F., Setiawan, H., Irawan, P.L.T.: Appropriate CNN architecture and optimizer for vehicle type classification system on the toll road. J. Phys. Conf. Ser. **1196**, 012044 (2019)

# Optimizing MobileNetV2 Architecture Using Split Input and Layer Replications for 3D Face Recognition Task

Phattharaphon Romphet⬤, Supasit Kajkamhaeng⬤,
and Chantana Chantrapornchai$^{(\boxtimes)}$⬤

Kasetsart University, Bangkok, Thailand
{phattharaphon.ro,supasit.k}@ku.th, fengcnc@ku.ac.th

**Abstract.** Facial recognition is one of the problems that has been focused for a long time. In this paper, we consider the 3D face data set, and explore its facial recognition task considering automatic architecture finding. We present the approach to customize MobileNet architecture and automatically find a good architecture variant for the 3D face recognition task. The main concept is based on the split input and lengthening the network by the layer replication. The evaluation is done by using the dataset generated by the GAN model with style transfer to augment the makeup faces. The results show that the found modified model from our automatic finding approach yields the more cost-effective model, i.e., with a 0.005% increase in size compared to baseline 3D Mobilenet and 0.01% compared to a simple Mobilenet while the found model has 12% more accuracy compared to the 3D MobileNetV2 and 11% compared to the traditional MobileNetV2.

**Keywords:** 3D Face recognition · MobileNetV2 · Deep learning · Convolution Neural Network

## 1 Introduction

Nowadays, common machine learning models have been used in an everyday life, such as face recognition, handwriting recognition, etc. These are basic tasks for many applications. There is still continuous development of the machine learning models for these tasks to increase the ability to predict various situations and purposes.

For this research, we consider the enhancement of the face recognition task. Current face recognition models such as FaceNet [8], rely on 2D images and usually work on limited constraints such as lights and face poses.

To overcome these challenges, 3D facial recognition systems have been developed. By adding such depth information, 3D face recognition has a high level of

Department of Computer Engineering, Faculty of Engineering, Kasetsart University Bangkok, Thailand.

W. Chen et al. (Eds.): ADMA 2022, LNAI 13726, pp. 103–115, 2022.
https://doi.org/10.1007/978-3-031-22137-8_8

accuracy and reliability, being more robust to face variation due to the different factors [10].

As technology development grows, many small chips can be placed inside a small device. For example, mobile phones are equipped with a 3D camera, making it possible to extract more features from images. However, using the small models on the edge device is challenging. The model needs to be optimized for accuracy while considering the model size constraint. In this research, we are interested in finding a small deep learning model with high accuracy. In particular, we focus on the 3D face recognition task.

A normal face-based recognition system consists of an input unit, preprocessing unit and face detection unit, and the last one is a recognition unit. An input unit refers to the 2D or 3D camera. The preprocessing unit will preprocess the capture frame, such as detecting faces, cropping and aligning the facial area to eliminate the irrelevance of information and improve the feature extraction. Then, the recognition unit performs the recognition task at last.

A neural network search has been a trend for finding the efficient neural network structure. For example, Neural Architecture Search (NAS) [14] has been proposed to automatically tune deep neural networks, but existing search algorithms usually have expensive computational resources. In this research, we find a proper strategy for searching for suitable models while considering the accuracy and model size for the 3D face recognition task.

Our strategy is as follows: first, we consider the state-of-the-art models of image recognition task, namely MobilenetV2 [7], ResNet [3] and VGG [9] to be baseline models. We first examine the performance of these models for 2D face recognition compared to the 3D face recognition. Then, we consider the MobileNetV2 as a baseline for 3D tasks and attempt to modify it to increase the performance while considering the small model size.

## 2   Backgrounds

This section highlights the literature reviews and necessary basic models used in the paper.

### 2.1   Related Works

**Deep Learning Method.** In 2017, the deep learning method was used in 3D face recognition. Donghyun [4] proposed a deep convolution neural network using VGG convolutional network and 3D augmentation techniques for a 3D face recognition task. In this research, the VGG network was used to extract the feature map from 3d facial cloud points and the augmentation techniques in the research are Pose variation, Random patch and Expression generation. This research generated the transformation matrices for the 3D point cloud. The random patch is the augmentation that puts eight $18 \times 18$ size patches on the 2D depth map to prevent overfitting to specific regions of the face. The expression generation augmentation creates the expression in the input to make more data.

In 2018, Ying Cai [1] proposed 3 methods for 3D face recognition: using a fast 3D scan preprocessing method then, using a fast Principal Component Analysis (PCA) pose correction and performing nose-tip refining on the raw 3D points cloud to be in the correct position. Then, the 3D scan is projected to a range image and normalized in scale with only three facial landmarks that can make the 3D face recognition easier to apply in real-world scenarios. In addition, this research combined multiple data augmentation techniques such as rotation in 3D space, shearing, zooming and resolution augmentation. The deep learning models were built from 4 deep neural networks. This model can achieve a higher accuracy. In 2019, Zheng, Siming, et al. [13] proposed the 3D texture-based face recognition system using fine-tuned deep residual networks. This system contains four main parts: the first is 3D face detection, the second is face alignment, and third is the facial feature extraction and recognition. In face detection and face alignment, the Dlib tool was used to detect and align faces. Dlib can extract 68 key points of the face in real-time to obtain the position and posture of the face. The histogram of orientation gradient (HOG) was used to extract the feature of the 3D faces. The main idea of the HOG algorithm is to describe the texture of the detected face by the gradient or distribution of edge direction which can represent the local texture. The deep learning model in this work was based on the ResNet model. In their experiments, four pooling layers with adaptive average pooling have been reconstructed using the new architecture.

## 2.2 Convolutional Neural Network (CNN)

CNN is usually used for image recognition tasks. It takes image pixels as inputs, and the output is a vector of classes. Each layer processes the portion of the inputs (not all) by using some filters. This simulates humans' vision as it looks at sub-areas and tries to visual features in the sub-areas. For an image problem, the features are such as lines, curves, patterns, and textures. Looking into this small area is done by filters, a mathematical matrix to calculate convolutions. Each sub-feature of the image is applied to the filter for the following computation in the next layer. The early layers in the CNN process image feature directly while the later layers gather the features into abstracted features, called high-level features. The high-level features are used for classification at last. There are state of the art models that are based on CNN, considered in this paper, MobilenetV2, VGG, ResNet and GAN models.

**MobilenetV2.** It is a convolutional neural network architecture that performs well on mobile devices. It is designed based on an inverted residual structure where the residual connections are between the bottleneck layers. The intermediate expansion layer uses lightweight depthwise convolutions to filter features as a source of non-linearity (the image was shown in Fig. 1). As a whole, the architecture of MobileNetV2 contains the initial fully convolution layer with 32 filters, followed by 19 residual bottleneck layers.

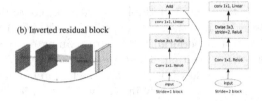

(b) Inverted residual block

**Fig. 1.** The structure of the Inverted residual and the infrastructure of the Mobilenet architecture.

**VGG.** is a convolution neural net (CNN) architecture which won the ILSVR (Imagenet) competition in 2014. It is considered to be one of the most excellent vision model architectures to date. The most unique thing about VGG16 is that instead of having a large number of hyper-parameter they focused on having convolution layers of a $3 \times 3$ filter with a stride 1 with the same padding and max pooling layer of $2 \times 2$ with stride 2. It utlizes this arrangement of convolution and max pooling layers consistently throughout the whole architecture. In the last part, it has 2 fully connected layers followed by softmax for output classification. This network has 16 weighted layers which contain about 138 million parameters.

**ResNet.** Residual Network (ResNet) is a specific type of neural network that was introduced in 2015 by Kaiming He, Xiangyu Zhang, Shaoqing Ren, and Jian Sun in their paper "Deep Residual Learning for Image Recognition". The ResNet models were extremely successful since it won 1st place in many competitions such as ILSVRC 2015, COCO 2015 etc. It replaced VGG-16 layers in Faster R-CNN with ResNet-101. The accuracy improvement of 28% was obtained. The fundamental of ResNet architecture is Residual connection (shown in Fig. 2).

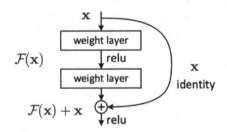

**Fig. 2.** Structure of the Residual network.

**GAN.** A generative adversarial network (GAN) is a machine learning model in which two neural networks compete with each other to become more accurate in their predictions. The network learns to generate from a training distribution through a 2-player game Generator and Discriminator. The generator's task is to generate the images and the Discriminator's task is to identify/score generated images from the generator. During the training process, the generator tries to fool the discriminator by the real-looking images, while the discriminator tries to censor the images. In this work, we use a style transfer called CPM (Color-Pattern Makeup Transfer) [6] based on BeautyGAN this model can transfer the makeup feature to another image (shown in Fig. 3)

**Fig. 3.** Result from CPM model (left image is makeup style, right image is the result after applied makeup style transfer.)

## 3    Methodology

We present the whole methodology of experiments, starting from data gathering, preprocessing, and then model refinement process.

### 3.1    Data Gathering

The facial dataset in this research given by the Institute of Computing Technology, Chinese Academy of Sciences, and University of Chinese contains 403,067 pairs of face images of 1,208 people.

### 3.2    Preprocessing

For the face recognition model, the size of the face image is resized to 224 × 224 and the face in the image is detected and aligned using MTCNN.

**Transforming 2D to 3D Data.** In this research, to add more local data we added the 2D faces of Thai people into the data set. The 2D dataset is converted the 3D images using PRNet [2]. PRNet is the Position Map Regression Network to map the image input to 3D cloud-point position. Thus, PRNet was used to generate Depth images from 2D face images (shown in Fig. 4)

**Image Augmentation.** The random horizontal flipping was done to increase the variety of the data and improve the training performance.

**Fig. 4.** Result from PRnet.

### 3.3 Model Overview

In this section, we explain the models that were used in our research. We first focus on the input layer of the deep learning model.

**The RGB Input Layer.** The RGB input is the normal input layer for normal image data. The image data contains 3 color channels; thus, we have to create a CNN layers with 3-input channels.

**The RGBD Input Layer.** Since we consider the depth of the face, the RGB images and the depth images are concatenated. The inputs of the model become 4 channels: RGB and depth channels. The images are shown in Fig. 5.

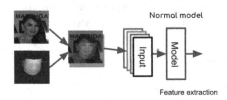

**Fig. 5.** RGBD input.

**The RGB+D Input Layer.** Another way is to consider the depth channel separately. RGB and depth are considered separately as inputs to the models, called the RGB+D input layer. Figure 6 shows this example.

**Extended MobilNetV2 Structure.** Our research focuses on MobileNetV2 model. The architecture of this model is based on an inverted residual structure. An Inverted Residual Block follows a narrow-wide-narrow structure. In some layers, e.g. 3rd, 5th, 8th, 13th and 15th layers, the dimension of the input and output are the same. Hence, we can duplicate at these layers. For example, in the third invert residual block, the input size and output size are [24, 56, 56], we can duplicate the 3rd layers as in Fig. 7.

In addition, using this technique, we can transfer the weights from the old model to the modified model according to the order of the referenced layer. For

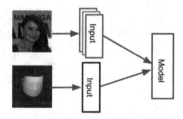

**Fig. 6.** RGB+D input.

example, if we duplicate the 3rd layer, we can transfer the weights from other layers except the 3rd layer to reduce the training time. From the experiments, using this technique with early stopping can reduce the training time to 70% compared to not using the weight transfer.

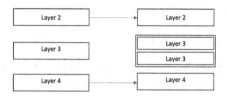

**Fig. 7.** Example of extended structure.

## 3.4   Metrics

The metric used in this evaluation is accuracy calculated by the number the correct predictions divided by the number of total images. In addition, we use the CPM model to create the augmented makeup images to evaluate the model accuracy (an example of augmenting style was shown in Fig. 8). The model evaluated using the augmented makeup images yields the good results. This shows the more possibility of using in the real situation when people wear makeup in their everyday lives.

$$Accuracy = \frac{Correct}{Total} \times 100 \tag{1}$$

$$Score = \frac{Accuracy_0 + \sum_{i=1}^{n} Accuracy_i}{n} \tag{2}$$

for i is augment number i and $Accuracy_0$ represent Accuracy on evaluate dataset.

**Fig. 8.** Examples of augmented style.

## 3.5   Training Configuration

Cross entropy loss [12] is used as a loss function and decoder. SGD is used for optimization. We assume the batch size is 16, the learning rate is 0.0001, and the total number of training epochs is 50. In addition, the early stopping and learning rate scheduler is used to reduce the training time.

The experiments were run on the Kasetsart university AI Server with 2 NVIDIA v100 GPUs, and 256 GB memory.

## 3.6   Automatic Model Finding

In this section, we explain how we find the best modified MobileNetV2 model. Our modified model can be extended in any of 5 Invert Residual layers [3rd, 5th, 8th, 13th, 15th], for example, if we replicate the third and fifth layers, the configuration will be [3, 5] (same as [5,3]) or we double third and fifth layers, the configuration will be [3, 3, 5].

The following summarizes our approach.

1. Train model with a single layer replication e.g. [3], [5], [8], [13], or [15]. Note that [3] means the replication of 3rd layer.
2. Use the best accuracy for the single-layer replication. For example [15], the model is best at the single replication for the 15th layer.
   (a) Add 15 to the configuration list, for example ([15,3], [15,5], ... ,[15,15] ).
   (b) Create the models from the above configuration list.
   (c) Transfer model weights from model([15]) to the new model.
3. Train the models.
4. Consider the model with the best accuracy. For example, model no.35, where the model configuration is ([15,13]).
   (a) Add 13 to the configuration list, for example ([15, 13, 3], [15, 13, 5], ... , [15, 13, 15] )
   (b) Again, create the above models and transfer the model weights from previous model([15, 13]) to the new model.

The exploration process continues until the accuracy does not improve more than the given threshold. In our experiments, we use 0.0001.

By using weight transferring with early stopping the training time was reduced to 73% (e.g. training epoch reduce from 36 epoch to 23 epoch). The result shown in Table 1.

**Table 1.** Training time between standard training and our strategy

| Model | Epoch | Time elapsed |
|---|---|---|
| Normal 2D | 36 | 2555 min |
| Our strategy 3D | 23 | 1685 min |

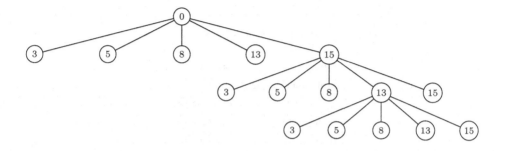

## 4    Experimental Results

The experiments are considered on the following cases. 1) We compare the performance of 2D face model and 3D face models on various baseline models. 2) We compare how the depth should be added as input between RGBD and RGB+D on the baseline models. 3) We consider MobileNetV2 considering RGB+D applying automatic layer replications.

### 4.1    Comparison Between 2D and 3D Face Recognition Models

We first show that 3D face recognition can perform better than 2D face recognition although they both use the same initial data.

The difference between the 2D and 3D face recognition models in this part is the depth of the first layer. In the 2D face recognition model, the depth of the first layer is 3 to handle RGB data and in the 3D face recognition model, the depth is 4 to handle the depth information which is added from the 2D model (shown in Fig. 9).

From Table 2, the results show that 3D face recognition models perform slightly better than 2D face recognition models.

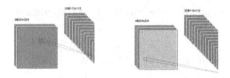

**Fig. 9.** RGB input and RGBD input.

**Table 2.** 2D and 3D Face recognition models.

| Model | Loss | Acc |
|---|---|---|
| MobilenetV2 2D | 0.78 | 82.25% |
| MobilenetV2 3D | 0.71 | 83.88% |
| Resnet 2D | 0.63 | 88.43% |
| Resnet 3D | 0.59 | 89.40% |
| VGG 2D | 0.4 | 91.81% |
| VGG 3D | 0.43 | 91.02% |

## 4.2 Comparison Between RGBD and RGB+D Face Recognition Models

We modify the input layer from RGBD to RGB+D in order to extract more features from the dataset.

**Table 3.** Comparison between RGBD and RGB+D for 3D faces.

| Model | Loss | Best accuracy |
|---|---|---|
| MobilenetV2 3D | 0.71 | 83.88% |
| SpMobilenetV2 3D | 0.33 | 92.26% |
| Resnet 3D | 0.59 | 89.40% |
| SpResnet 3D | 0.26 | 94.03% |
| VGG 3D | 0.43 | 91.02% |
| SpVGG 3D | 0.97 | 78.95% |

From Table 3, the results show that the performance increases significantly in MobilenetV2 and ResNet. In MobileNetV2, the accuracy increases around 10%, in ResNet model, the accuracy increases 5%.

However, in VGG, the accuracy drops 13%. In overall, we conclude that the RBB+D input layer can achieve better accuracy.

### 4.3   Comparison Between Baseline MobileNetV2 and RGB+D MobileNetV2 with Layer Replication

After performing the above layer replication exploration, we found the best model with configuration ([15, 15]) which has a triple fifteenth invert of the residual layer (the model was shown in Fig. 10). The exploration algorithm stopped at 3 layers of exploration when the accuracy is not further improved. In this experiment, we also compare our derived model with EfficientNet [11]. EfficientNet that we used is Efficientnet B0 has the parameter around 5 million parameters while ours is around 3 million.

**Table 4.** Comparison among baseline models.

| Model | Eval dataset | Aug 1 | Aug 2 | Aug 3 | Aug 4 | Avg |
|---|---|---|---|---|---|---|
| MobilenetV2 | 82.25% | 78.72% | 75.63% | 77.84% | 77.22% | 78.33% |
| ResNet50 | 88.43% | 58.41% | 54.20% | 58.74% | 56.88% | 63.33% |
| VGG | 91.81% | 74.76% | 70.34% | 73.81% | 71.53% | 76.45% |
| SpMobilenetV2 | 92.26% | 69.55% | 63.87% | 69.65% | 67.15% | 72.50% |
| SpResNet50 | 94.03% | 75.70% | 69.28% | 75.83% | 72.53% | 77.47% |
| SpVGG | 78.95% | 67.59% | 61.69% | 67.52% | 69.32% | 69.01% |
| SpMobilenetV2 [15] | 93.86% | 89.47% | 87.33% | 89.41% | 88.52% | 89.72% |
| SpMobilenetV2 [15, 15] | **94.37%** | **89.20%** | **87.60%** | **89.10%** | **88.84%** | **89.82%** |
| EfficientNet | 84.72% | 76.10% | 74.10% | 75.93% | 75.73% | 77.34% |
| EfficientNet 3D | 85.46% | 77.69% | 74.41% | 77.07% | 75.88% | 85.46% |
| Sp EfficientNet | 76.48% | 66.13% | 59.90% | 62.86% | 76.48% | 65.65% |

From Table 4, the results show that the explored model can achieve the best accuracy on the evaluation dataset compared with other models. In addition, in the augment dataset our explored model also yields higher accuracy.

### 4.4   Comparison Between Our Baseline Model and EffiencientNet on CelebA Dataset

**Table 5.** RGB+D MobileNetV2 with layer replication and EfficientNet.

| Model | Accuracy | Parameter |
|---|---|---|
| SpMobilenetV2 | 22.89% | 3249024 |
| SpEfficienctnet | 8.95% | 5545100 |

In this experiment, we use the CelebA face dataset [5] which has 10,177 identities. In this part, we keep the class which has the identities images less than 30 images to reduce the training time. After the pre-processing, there are 800 classes (identities) for training. The results are shown in Table 5. The results shows that our model has better accuracy utilizing a small number of image datasets.

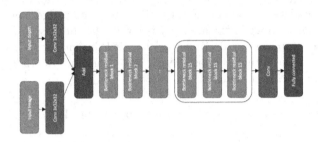

**Fig. 10.** Architecture of RGB+D MobileNetV2 with layer replication

## 5    Conclusion and Future Work

We explore the variety of optimizing the CNN for the 3D face recognition task. The optimization approaches replace the input layer by using the separate RGB input and depth to extract more features and replicating more inverted residual blocks. We also propose the exploration scheme to modify MobileNetV2 layers with prior knowledge transfer which can reduce training time by around 70%. The experiments show that achieved model can yield better accuracy about 8–12% higher than that of the baseline in the evaluation dataset and 10% higher in the augmenting makeup dataset.

In the future, we will use another model extraction to segment the model into small blocks, for example, ResNet or VGG to find a better model architecture.

**Acknowledgment.** We are grateful to Asst. Prof. Dr Putchong Uthayopas facilitated the usage of the HPC cluster for this project. The work is supported in part by PMU-B scholarship, TRF-RSA Grant, and Faculty of Engineering, Kasetsart University graduate innovation scholarship.

## References

1. Cai, Y., Lei, Y., Yang, M., You, Z., Shan, S.: A fast and robust 3D face recognition approach based on deeply learned face representation. Neurocomputing **363**, 375–397 (2019)
2. Feng, Y., Wu, F., Shao, X., Wang, Y., Zhou, X.: Joint 3D face reconstruction and dense alignment with position map regression network. In: Ferrari, V., Hebert, M., Sminchisescu, C., Weiss, Y. (eds.) Computer Vision – ECCV 2018. LNCS, vol. 11218, pp. 557–574. Springer, Cham (2018). https://doi.org/10.1007/978-3-030-01264-9_33

3. He, K., Zhang, X., Ren, S., Sun, J.: Deep residual learning for image recognition. In: Proceedings of the IEEE Conference on Computer Vision and Pattern Recognition, pp. 770–778 (2016)
4. Kim, D., Hernandez, M., Choi, J., Medioni, G.: Deep 3D face identification. In: 2017 IEEE International Joint Conference on Biometrics (IJCB), pp. 133–142. IEEE (2017)
5. Liu, Z., Luo, P., Wang, X., Tang, X.: Deep learning face attributes in the wild. In: Proceedings of International Conference on Computer Vision (ICCV), December 2015
6. Nguyen, T., Tran, A.T., Hoai, M.: Lipstick ain't enough: beyond color matching for in-the-wild makeup transfer. In: Proceedings of the IEEE/CVF Conference on Computer Vision and Pattern Recognition, pp. 13305–13314 (2021)
7. Sandler, M., Howard, A., Zhu, M., Zhmoginov, A., Chen, L.C.: MobileNetV 2: inverted residuals and linear bottlenecks. In: Proceedings of the IEEE Conference on Computer Vision and Pattern Recognition, pp. 4510–4520 (2018)
8. Schroff, F., Kalenichenko, D., Philbin, J.: FaceNet: a unified embedding for face recognition and clustering. In: Proceedings of the IEEE Conference on Computer Vision and Pattern Recognition, pp. 815–823 (2015)
9. Simonyan, K., Zisserman, A.: Very deep convolutional networks for large-scale image recognition. arXiv preprint arXiv:1409.1556 (2014)
10. Soltana, W.B., Huang, D., Ardabilian, M., Chen, L., Amar, C.B.: Comparison of 2D/3D features and their adaptive score level fusion for 3D face recognition. In: 3D Data Processing, Visualization and Transmission (2010)
11. Tan, M., Le, Q.: EfficientNet: rethinking model scaling for convolutional neural networks. In: International Conference on Machine Learning, pp. 6105–6114. PMLR (2019)
12. Zhang, Z., Sabuncu, M.R.: Generalized cross entropy loss for training deep neural networks with noisy labels. In: 32nd Conference on Neural Information Processing Systems (NeurIPS) (2018)
13. Zheng, S., Rahmat, R.W.O., Khalid, F., Nasharuddin, N.A.: 3D texture-based face recognition system using fine-tuned deep residual networks. PeerJ Comput. Sci. 5, e236 (2019)
14. Zoph, B., Le, Q.V.: Neural architecture search with reinforcement learning. arXiv preprint arXiv:1611.01578 (2016)

# GANs for Automatic Generation of Data Plots

João Tomás Caldeira$^{(\boxtimes)}$ and Cláudia Antunes

Instituto Superior Técnico, Universidade de Lisboa, Lisbon, Portugal
joao.tomas.brazao.caldeira@tecnico.ulisboa.pt

**Abstract.** In the last years, online learning saw a surge in relevance, along with interest in the automation of the learning process. Variety and amount of practice exercises are of particular importance, since they limit the amount of practice a student can have. In this paper, we propose the use of *generative adversarial networks* (GANs) to produce scatter plots, for students' training on data profiling tasks in data-driven courses. Our results show that progressively grown GANs (ProGANs) generate scatter plots with little tuning and display an adequate level of randomness, diversity, and quality. These properties show promise in allowing for diversity of exercises created from the generated plots.

**Keywords:** Generative adversarial networks · MOOCs · Learning resources

## 1 Introduction

Online learning, and Massive Open Online Courses (MOOCs) in particular, have seen a recent rapid increase in usage, bringing about a set of new opportunities to enhance education [17]. Among its major challenges is the ability to automatically personalize the learning process, in particular by recommending the most adequate learning resources for each student. The abundance of practice exercises is one of the keys to the success of this personalization, by allowing students to practice as many times as needed, over different questions along time. Approaches for the automatic production of practice exercises are scarce, and, to our knowledge, there has been no proposal to address the generation of exercises around the analysis of data charts (such as scatter plots, histograms, box plots, etc. used in data profiling tasks), as often explored in subjects like data science.

Exploring the advances on the area of image classification and generation brought by deep learning research, in this paper we propose a methodology to train *generative adversarial networks* (GANs) for automatically generating scatter plots.

The rest of the paper is organized as follows: next (Sect. 2), we overview how GANs work and explain their fundamentals; after this, in Sect. 3, we briefly

W. Chen et al. (Eds.): ADMA 2022, LNAI 13726, pp. 116–125, 2022.
https://doi.org/10.1007/978-3-031-22137-8_9

discuss the progress done on the automatic generation of questions, from textual to visual-based ones. In Sect. 4, we present the methodology to generate the charts, followed by its validation in Sect. 5. The paper concludes in Sect. 6 with a critical analysis of the results obtained and some guidelines for future work.

## 2  Generative Adversarial Networks

Generative modeling has traditionally been addressed by approaches based on maximum likelihood estimation [2]. Though images are very high-dimensional, it is common that they are supported by low-dimensional manifolds. Such is the case also for the models trying to learn the distribution underlying some set of training images, which output high-dimensional samples (images) but are supported by low-dimensional manifolds [3]. Consequently, it is very likely that the model and real data distributions are disjoint, which means that the Kullback-Leibler divergence (KLD) between them is undefined. Since maximum likelihood estimation is equivalent to minimizing the KLD between the model and real distributions, it becomes very hard to train models under the former. Variational autoencoders [11] address this issue by adding noise terms, which "stretch" the distribution over the space and cause the distributions not to be disjoint. However, it is well known that noise tends to decrease the quality of generated images, making it an inappropriate solution to this problem. Furthermore, a theoretical analysis on different measures of distance between distribution illustrates how the KLD favors placing a lot of probability mass in non-data regions, which would cause the model to generate atypical (low-fidelity) samples [23].

*Generative Adversarial Networks* (GANs) are a framework proposed by Goodfellow et al. [7] in 2014, in which maximum likelihood estimation is replaced with an adversarial training regime that approximately minimizes the Jensen-Shannon divergence (JSD). In particular, two models engage in a minimax game, where one of the models creates samples (for simplicity, images) that the other model tries to recognize as being fake. One such model is the *generator* $G$, which takes as input some noise vector $z \sim \mathbb{P}_z$ where $\mathbb{P}_z$ is a well known distribution (e.g., the standard uniform distribution) and yields a "fake" sample $G(z)$. The *discriminator* model $D$, given some input $x$, outputs a prediction $D(x)$ for whether $x$ is real (i.e., belonging to the real distribution $\mathbb{P}_r$) or fake (i.e., generated by $G$). We can then write the GAN framework as

$$\min_G \max_D V(D, G) = \mathbb{E}_{x \sim \mathbb{P}_r}[log(D(x))] + \mathbb{E}_{z \sim \mathbb{P}_z}[log(1 - D(G(z)))]. \quad (1)$$

For image generation tasks, $G$ and $D$ are usually convolutional neural networks due to their ability to deal with this kind of data. Indeed, since the introduction of the deep convolutional GAN architecture [20] in 2015, GANs used in image generation tasks almost always use convolutional networks [9,14,27].

The JSD had been proposed to place a lot of probability mass under one or only a few modes of the data [23], which makes sense given that one of the biggest issues with GANs is *mode collapse*, whereby they only produce samples from a small subset of the modes of the data. Furthermore, since the supports of $\mathbb{P}_r$ and $\mathbb{P}_\theta$ (the model distribution) are disjoint or lie low dimensional manifolds, training under the JSD is unstable [2].

The *Wasserstein GAN* [3] (WGAN) was introduced in 2017 and proposed the use of the Earth Mover distance (EMD) (also known as Wasserstein-1 distance) as an alternative to the JSD. The EMD is given by the following objective function:

$$W(\mathbb{P}_r, \mathbb{P}_\theta) = \inf_{\gamma \in \prod(\mathbb{P}_r, \mathbb{P}_\theta)} \mathbb{E}_{(x,y) \sim \gamma} \left[ \|x - y\| \right], \tag{2}$$

where $\mathbb{P}_\theta$ is the distribution represented by a model with parameters $\theta$. However, the infimum in Eq. 2 is computationally intractable. From the Kantorovich-Rubinstein [24] duality, we have:

$$W(\mathbb{P}_r, \mathbb{P}_\theta) = \sup_{\|f\|_L \leq 1} \mathbb{E}_{x \sim \mathbb{P}_r} \left[ f(x) \right] - \mathbb{E}_{x \sim \mathbb{P}_\theta} \left[ f(x) \right], \tag{3}$$

where the supremum is over all 1-Lipschitz $f$. Instead of a discriminator $D$ that predicts whether a sample is real or fake, the WGAN has a *critic C* that outputs a score of "realness" of a given sample. We can therefore write Eq. 3 in the previously introduced GAN notation

$$W(\mathbb{P}_r, \mathbb{P}_\theta) = \sup_{\|C\|_L \leq 1} \mathbb{E}_{x \sim \mathbb{P}_r} \left[ C(x) \right] - \mathbb{E}_{z \sim \mathbb{P}_z} \left[ C(G(z)) \right]. \tag{4}$$

Intuitively, this objective function tries to maximize the interval between scores for real images $x$ and generated images $G(z)$.

WGAN uses weight clipping on the critic (e.g., to the [-0.01, 0.01] range) as a means to enforce the 1-Lipschitz constraint. However, they note that weight clipping is a terrible way to enforce this constraint [3].

Follow-up work introduced a WGAN with a gradient penalty (WGAN-GP) [8] as an alternative to weight clipping, and obtained better results in many relevant tasks. The WGAN loss with the gradient penalty term is

$$WGP(\mathbb{P}_r, \mathbb{P}_\theta) = W(\mathbb{P}_r, \mathbb{P}_\theta) + \lambda \mathbb{E}_{\hat{x} \sim \mathbb{P}_{\hat{x}}} \left[ (\|\nabla_{\hat{x}} C(\hat{x})\|_2 - 1)^2 \right], \tag{5}$$

where $\hat{x} = \epsilon x + (1 - \epsilon)G(z)$, $z \sim \mathbb{P}_z$, $\epsilon \sim U[0, 1]$, $W$ is the WGAN loss (Eq. 4), and $WGP$ is the newly introduced Wasserstein loss with gradient penalty. The gradient penalty can be seen as penalizing gradients whose norm becomes larger than 1, thus enforcing the 1-Lipschitz constraint.

Finally, *progressively grown GANs* (ProGAN) [10] introduced a new training method for GANs which increased training speed and stabilized it. As the name implies, both the generator and the discriminator (or critic, in the case of WGAN) are trained progressively: starting from a low resolution of $4 \times 4$, both models are trained until reasonable convergence, and then another layer (with resolution $8 \times 8$) is added, doubling the resolution. This is done over and over, with impressive results having been attained up to a $1024 \times 1024$ resolution.

The motivation behind growing the networks progressively is that they are first allowed to learn coarse information about the image distribution, and then pay attention to increasingly finer details. Furthermore, new layers are introduced smoothly into each network via a scaling mechanism which starts by not letting them contribute much to the output at first, but then increasing their weight as the model trains. This is so that new layers don't greatly disturb the already well-trained lower resolution layers.

## 3   Related Work

Massive Open Online Courses (MOOCs) have been around for a few years under other names [17], but *The New York Times* called 2012 "The Year of the MOOC" [19]. It saw platforms such as Coursera, Udacity, and edX become mainstream, and associate with education institutions such as universities, which will remain the focus of the discussion herein.

The growth of these platforms was accompanied by that of online learning in general: *blended learning*, which combines face-to-face with online education, has seen an increase in popularity, particularly in higher education [6]. Furthermore, as of 2013, around 70% of education institutions in the U.S.A. claimed that online learning was fundamental to their long-term strategy [1]. Finally, the COVID-19 pandemic has forced a widespread temporary adoption of full online learning [5].

All these factors continuously contribute to the adoption of MOOCs by education institutions, which now face a different paradigm in providing courses. While a discussion of the challenges of online learning is beyond the scope of this document, it is relevant that the informatization of learning systems introduces the opportunity to leverage automated tools to improve the quality of online courses.

Indeed, along the years, several have been the challenges faced by the development of MOOCs, either in their creation or in supporting their operation. Among these, *personalization* has been one of the most researched [15], aiming for creating a learning experience as unique as possible for each student. However, personalization demands the existence of a diversity of learning resources that are difficult to create manually, in particular evaluations items, like textual questions or more complex exercises. As noted by Kurdi et al. [12], the need of providing a "continuous supply" of these resources led to a new research field in *automatic question generation.*

Unlike other fields such as educational data mining and learning analytics, automatic question generation hasn't received much attention, and its results have been seldom deployed in applications [13]. In the last few years however, advances in NLP (natural language processing) techniques brought new and more powerful tools that have been explored in this context. Their usage in the automatic generation of sentences has been considerably successful, allowing for the creation of different types of textual questions, such as fill-in-the-blank, word formation, multiple choice, and error correction questions [18,25].

Despite the promising application of NLP in this context, in formal domains, such as maths, physics, and engineering in general, questions are more than sentences - often they involve the existence of diagrams, charts and even more complex images. The work by Singhal and Henz [22] is a good example of this, where the goal was to produce geometry questions, or the work by Moura Santos on the automatic generation of Markov chains [21]. More recently, GANs have been applied to image generation, mostly for training purposes in the medical domain to help students recognize real images [4, 26].

These examples show the diversity of situations we may face in generating visual-based questions. Indeed, these can vary from simple diagrams to real images. While the former are simple enough to be accomplished by parameterized solutions, the latter are nowadays possibly generated through GANs. However, to our knowledge, the generation of data charts has not been addressed before, and it is our main goal.

In particular, we will attempt to train GANs that are capable of producing scatter plots that display an adequate level of randomness, diversity, and quality. In the 2D scenario, a scatter plot is a type of plot that uses dots to represent values for two variables (i.e., a 2-tuple) in a given dataset. This would allow for the creation of an arbitrary amount of scatter plots that exhibit specific features to varying degrees, which bears direct relevance to the task of automatically generating learning resources.

## 4    Methodology

Data charts however, have a particularity: they depend on an external element - the data being plotted. Indeed, synthetic data generation has been the usual approach to generate examples for training purposes, either for machine learning algorithms or for humans. However, it has long been shown that these datasets are not good enough to train those algorithms, since they are rough approximations to real data. And if they are not good enough to train algorithms, they shouldn't be accepted to train humans. Furthermore, the synthetic data generation has to be parametrized, which limits the kinds of datasets generated, defeating the main purpose of generating practice exercises - to expose students to some diversity of situations. Now, if datasets become too similar, the answer for a given question becomes fixed, and consequently students learn it by heart, instead of analysing the data to conclude the answer.

Taking advantage of the advances on the field of image generation, we propose a methodology to train GANs for generating data charts. The generated charts will then be used to support the production of visual-based questions. By applying GANs, we expect to gain variability, due to their freedom on creation and ability to capture the common features among the set of training images, making free variations of them. Indeed, the more different the charts used for training are, the more distinct we expect the produced charts to be.

Let's consider a *data chart* to be composed by a set of elements, namely *axes* (two or three), *titles*, *legend*, *plot area* and *gridlines*. Scatter plots, bar charts, pie charts and boxplots are examples of these kinds of data charts.

It is important to note that those elements are not atomic. An *axis* is composed by a set of *ticks*, usually equidistant, the corresponding *tick labels*, its *title* and *unit label*. *Titles* by their side are strings of characters, representing meaningful sentences or isolated words; likewise for the *legend*, which includes some meaningful identifiers and their corresponding visual elements. *Plot areas* are rectangles or parallelepipeds, depending on the number of existing axes, which contain all the data points. *Gridlines* are lines that are parallel to each axes, and that cross them in some specific ticks. Following these definitions, it is clear that data charts do not correspond to random images, but to heavily constrained ones, which should put additional challenges to the use of GANs. The proposed methodology is illustrated in Fig. 1.

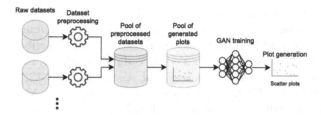

**Fig. 1.** Tool pipeline.

The **raw datasets** correspond to the data to be used as a "seed" to generate the synthetic data charts. It should consist of a collection of datasets, as diverse and large as possible. From these datasets, we will generate the set of data charts to be used as training set for the GAN, which is accomplished in the first step in the methodology - the **dataset preprocessing**. The importance of this step lies in the simplification of the learning process, choosing the the type of data charts to generate, and their required elements.

In this step, we separate different types of data (numerical real *versus* categorical, for example), since the resulting data charts follow different constraints. While real values spread continuously in the plot area, categorical ones are confined to specific places in the area. Now, if we use both kinds of charts to train the GAN, it is not able to distinguish between them, possibly generating continuous distributions for categorical variables. As such, we keep only continuous numerical variables for each dataset.

In terms of visual elements, we also choose to use the more basic ones, avoiding semantically demanding elements like titles, legends, and labels. Additionally, all data shall be scaled to a common range, so that the tick values are fixed and can thus be learned by the GAN. We thus keep only the axes, ticks, and tick values (with all variables being scalled to the $[0, 100]$ range) on each scatter plot.

After choosing the data and elements to use, it is then time to produce the corresponding data charts. In order to maximize the variety of data charts to train the GAN, we discard all variables whose correlation coefficient is larger than 0.99 for each dataset considered. In this manner, by ignoring pairs of highly correlated variables, we increase the variability of the charts used for training, avoiding any undesired bias into the model. From the selected variables we then created the data charts, following a common format. The image resolution of 128 × 128 (with 3 channels for RGB) is chosen in advance and fixed for all training charts.

In sum, we propose to train a GAN that can generate scatter plots with enough variability and fidelity to be used as tools in the automatic generation of learning resources. In order to qualitatively assess image diversity and quality, we rely on the assessment of domain experts: teachers in data science.

## 5   Results

We collected 14 datasets from the ones available in the UCI ML repository, in order to validate our proposal. From them, we selected all numerical variables, avoided titles and gridlines, and followed the constraints described before: *correlation threshold* set to 0.99 and scaled all variables to the [0, 100] range. From these, we produced 11,552 two-dimensional scatter plots as 128 × 128 images (with 3 channels for RGB), which were used to train a ProGAN. Examples of real plots used in training can be seen in Fig. 2.

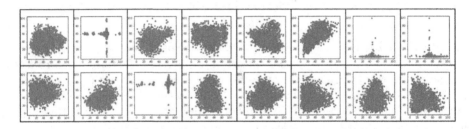

**Fig. 2.** Original scatter plots (with ticks and tick labels)

In these examples, we are able to see some of the diversity shown. Most of the charts show heavy dense data, but four of them almost only occupy two orthogonal lines. Then we identify different levels of correlation between the variables, either positive and negative, or entirely nonexistent.

Figure 3 shows the synthetic scatter plots generated by our ProGAN, trained for 3 d over the original scatter plots described, but without ticks and tick labels.

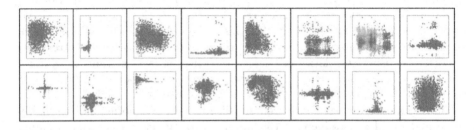

**Fig. 3.** Scatter plots without ticks, generated by the ProGAN

The results are quite interesting, presenting a diversity similar to the original ones, with different levels of correlation and different patterns of dispersion along the axis. From the sixteen plots shown, only the $7^{th}$ one in the first row doesn't pass a human validation.

The plots in Fig. 4 were produced by a ProGAN which was trained on plots with both ticks and tick labels for 3 days and 6 h.

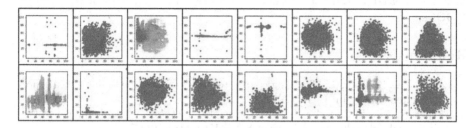

**Fig. 4.** Intermediate scatter plots with ticks and tick labels, generated by a ProGAN (training not complete)

Again, the plots show impressive results, being able to generate perfect axes with their ticks and tick labels. Indeed, it is now clear that beside being able to generate images where all data elements are represented inside the plot area, axes may also be learnt perfectly, when the data is previously scaled.

It is important to mention that the existence of more plots with blurred data points (3 in 16), is due to unfinished training of the GAN.

All the generated artificial samples result from passing noise vectors (sampled from the standard normal distribution) through two different ProGANs. Due to computational infrastructure limitations, we were not able to make use of the original accompanying code in the ProGAN paper [10]. Instead, we used a PyTorch implementation of the same model developed by Facebook Research[1]. We trained the models on an NVIDIA Tesla T4 using all the default hyperparameters, which match those in the paper and its original implementation.

## 6  Conclusion

The need of online learning personalization and continuous supply of learning resources demands the automatic generation of learning resources; questions in particular. Despite the promising advances in textual questions generation through NLP tools, the results on visual-based questions are far from satisfactory.

In this paper, we propose to train GANs on data charts created from publicly available datasets to generate synthetic ones. We obtained good quality samples

---

[1] The code can be found at a GitHub repository.

with no hyperparameter tuning and a modest training time of about three days on an NVIDIA Tesla T4.

Interesting future work includes conditioning the GAN on features of the data (cf. CGAN [16]) so that the user may have explicit control over several features of the generated plots. Doing so could allow the user to generate plots with certain general properties (e.g., data point dispersion, correlation, relative location in the plot) while maintaining sample diversity. The successful implementation of such a model could thus be a useful tool for educators in data-driven subject MOOCs.

# References

1. Allen, I.E., Seaman, J.: Changing course: ten years of tracking online education in the United States. In: ERIC (2013)
2. Arjovsky, M., Bottou, L.: Towards principled methods for training generative adversarial networks. arXiv preprint arXiv:1701.04862 (2017)
3. Arjovsky, M., Chintala, S., Bottou, L.: Wasserstein generative adversarial networks. In: International Conference on Machine Learning, pp. 214–223. PMLR (2017)
4. Chuquicusma, M.J.M., Hussein, S., Burt, J., Bagci, U.: How to fool radiologists with generative adversarial networks? a visual turing test for lung cancer diagnosis. In: 2018 IEEE 15th International Symposium on Biomedical Imaging (ISBI 2018), pp. 240–244 (2018). https://doi.org/10.1109/ISBI.2018.8363564
5. Dhawan, S.: Online learning: a panacea in the time of COVID-19 crisis. J. Educ. Technol. Syst. **49**(1), 5–22 (2020)
6. Drysdale, J.S., Graham, C.R., Spring, K.J., Halverson, L.R.: An analysis of research trends in dissertations and theses studying blended learning. Internet High. Educ. **17**, 90–100 (2013)
7. Goodfellow, I.J., et al.: Generative adversarial networks (2014). http://arxiv.org/abs/1406.2661, arxiv:1406.2661
8. Gulrajani, I., Ahmed, F., Arjovsky, M., Dumoulin, V., Courville, A.C.: Improved training of Wasserstein gans. In: Guyon, I., et al. (eds.) Advances in Neural Information Processing Systems 30: Annual Conference on Neural Information Processing Systems 2017(December), pp. 4–9, 2017. Long Beach, CA, USA, pp. 5767–5777 (2017). https://proceedings.neurips.cc/paper/2017/hash/892c3b1c6dccd52936e27cbd0ff683d6-Abstract.html
9. Isola, P., Zhu, J.Y., Zhou, T., Efros, A.A.: Image-to-image translation with conditional adversarial networks (2016). http://arxiv.org/abs/1611.07004, cite arxiv:1611.07004Comment: Website: https://phillipi.github.io/pix2pix/, CVPR 2017
10. Karras, T., Aila, T., Laine, S., Lehtinen, J.: Progressive growing of gans for improved quality, stability, and variation. In: 6th International Conference on Learning Representations, ICLR 2018, Vancouver, BC, Canada, 30 April - 3 May 2018, Conference Track Proceedings. OpenReview.net (2018). https://openreview.net/forum?id=Hk99zCeAb
11. Kingma, D.P., Welling, M.: Auto-encoding variational bayes (2013). https://doi.org/10.48550/ARXIV.1312.6114, https://arxiv.org/abs/1312.6114
12. Kurdi, G., Leo, J., Parsia, B., Sattler, U., Al-Emari, S.: A systematic review of automatic question generation for educational purposes. Int. J. Artif. Intell. Educ. **30**(1), 121–204 (2019). https://doi.org/10.1007/s40593-019-00186-y

13. Le, N.T., Kojiri, T., Pinkwart, N.: Automatic question generation for educational applications - the state of art. Adv. Intell. Syst. Comput. **282**, 325–338 (2014). https://doi.org/10.1007/978-3-319-06569-4_24

14. Ledig, C., et al.: Photo-realistic single image super-resolution using a generative adversarial network. In: CVPR, pp. 105–114. IEEE Computer Society (2017)

15. Mayer, R.E.: Thirty years of research on online learning. Appl. Cogn. Psychol. **33**(2), 152–159 (2019). https://doi.org/10.1002/acp.3482, https://onlinelibrary. wiley.com/doi/abs/10.1002/acp.3482

16. Mirza, M., Osindero, S.: Conditional generative adversarial nets (2014). http:// arxiv.org/abs/1411.1784, cite arxiv:1411.1784

17. Moe, R.: The brief & expansive history (and future) of the MOOC: why two divergent models share the same name. Curr. Issues Emerg. elearn. **2**(1), 2 (2015)

18. Nwafor, C.A., Onyenwe, I.E.: An automated multiple-choice question generation using natural language processing techniques. CoRR abs/2103.14757 (2021). https://arxiv.org/abs/2103.14757

19. Pappano, L.: The New York times: the year of the MOOC. https://www.nytimes. com/2012/11/04/education/edlife/massive-open-online-courses-are-multiplying-at-a-rapid-pace.html. Accessed 09 Jan 2022

20. Radford, A., Metz, L., Chintala, S.: Unsupervised representation learning with deep convolutional generative adversarial networks (2015). http://arxiv.org/abs/ 1511.06434, cite arxiv:1511.06434Comment: Under review as a conference paper at ICLR 2016

21. Santos, A.M., Ribeiro, P.: Assessment in an online mathematics course. In: EMOOCs-WIP, pp. 22–27 (2017)

22. Singhal, R., Henz, M.: Automated generation of region based geometric questions. In: 2014 IEEE 26th International Conference on Tools with Artificial Intelligence, pp. 838–845. IEEE (2014)

23. Theis, L., Oord, A.v.d., Bethge, M.: A note on the evaluation of generative models. arXiv preprint arXiv:1511.01844 (2015)

24. Villani, C.: Optimal Transport: Old and New, vol. 338. Springer, Cham (2009). https://doi.org/10.1007/978-3-540-71050-9

25. Wang, Y., Okamura, K.: Automatic generation of e-learning contents based on deep learning and natural language processing techniques. In: Barolli, L., Okada, Y., Amato, F. (eds.) Advances in Internet, Data and Web Technologies, pp. 311–322. Springer International Publishing, Cham (2020). https://doi.org/10.1007/978-3-030-39746-3_33

26. You, A., Kim, J.K., Ryu, I.H., Yoo, T.K.: Application of generative adversarial networks (GAN) for ophthalmology image domains: a survey. Eye Vis. **9**(1), 6 (2022). https://doi.org/10.1186/s40662-022-00277-3

27. Zhu, J.Y., Park, T., Isola, P., Efros, A.A.: Unpaired image-to-image translation using cycle-consistent adversarial networks. CoRR abs/1703.10593 (2017)

# An Explainable Approach to Semantic Link Mining in Multi-sourced Dynamic Data

Zhe Wang[✉], Yi Sun, Hong Wu, and Kewen Wang

Griffith University, Brisbane, Australia
{zhe.wang,k.wang}@griffith.edu.au, {yi.sun,hong.wu2}@griffithuni.edu.au

**Abstract.** It is challenging to mine semantic links among multi-sourced data. Knowledge graphs can capture the semantics of data to support implicit links (cross-data sources) to be inferred through reasoning, known as the link prediction task. However, existing link prediction approaches are limited in their adaptability to data changes and cannot provide explanations for the predictions. In this work, we introduce a framework for semantic link mining through knowledge graphs and rule-based link prediction. In particular, rules representing higher-order patterns in the data are automatically mined and updated according to the dynamics of the data. We present a practical use case and a system for the semantic link mining of aviation data from multiple online sources. Besides, we evaluate our system by comparing it with several link prediction models to demonstrate the effectiveness of our approach in both static and dynamic link prediction and explanation.

## 1 Introduction

It is challenging to establish semantic connections among data from multiple sources to provide uniform access [5]. Knowledge graphs (KGs) offer a promising semantic link mining solution by extracting objects across data sources as entities and describing semantic links between them. A KG is usually modelled as a set of triples ($subject$, relation, $object$). For instance, (737-800, isAircraftOf, $Air France$) states that entity 737-800 is linked to the entity $Air France$ with the relation isAircraftOf. Implicit semantic links across data sources can be discovered via reasoning, known as the *link prediction* task. The objective of link prediction is to validate whether a pair of entities are connected via a given relation. For example, a link prediction task (737-800, isAircraftOf, ?) asks which airlines have 737-800 in their fleet. A rich body of research has been dedicated to accurate link prediction in large KGs, including approaches based on embedding models [1,7,15,17] and those based on rules [8,10,11].

While existing link prediction models show strong performance on standard benchmarks, they are still inadequate in mining semantic links in real-life applications. First, real-life data are often highly incomplete, especially cross-links between data sources. Existing models often require a significant amount of initial links to train. Also, data from many sources are dynamic due to various changes, which poses significant challenges to the current link prediction models. Existing models are developed for static data and thus cannot be directly applied or conveniently adapted when the data change. In particular, embedding-based models have difficulty when new entities and links are dynamically added. Moreover, in real-life applications, it is often necessary to explain the predicted links so that human experts can approve or disapprove them. Link prediction models adopting a black-box approach suffer from the lack of explainability; that is, they cannot provide human-comprehensible explanations to the predicted links.

Rules provide a natural and yet explainable approach to link prediction in knowledge graphs as rules are easily comprehensible to humans. Also, rules reveal higher-order patterns in the data, and they can be applied directly to new entities, such as new flights and persons. Consider a logic rule

$$0.7 \; : \; \mathsf{isAircraftOf}(X, Y) \leftarrow \mathsf{isAircraftOf}(X, A), \mathsf{hasAlliance}(Y, A).$$

If the triple $(737\text{-}800, \mathsf{isAircraftOf}, \mathsf{AirFrance})$ is in the knowledge graph and Qantas is an alliance partner of Air France, then there is a good chance that Qantas has aeroplanes 737–800, that is, $(737\text{-}800, \mathsf{isAircraftOf}, \mathsf{Qantas})$. Recent advancement in rule-based methods also shows promising accuracy in link prediction on commonly used benchmarks [12]. Yet we are unaware of any attempts to combine KGs with rule mining and reasoning for mining semantic links over multi-sourced data, and neither has rule-based link prediction for dynamic data been explored.

In this paper, we introduce a framework for semantic link mining through rule mining and reasoning in KGs. The core of our framework is an incremental rule mining method over dynamic data, which can refine the mined rules when data changes. We demonstrate our framework through the integration of aviation data by constructing a KG with the guidance of an ontology. Finally, we evaluate our rule-based link prediction method in comparison with commonly used models and demonstrate its benefits, especially in link prediction under a dynamic setting. The main contributions in the paper are as follows.

- We introduce a framework for semantic link mining, which supports rule-based link prediction and explanation. In particular, rules are mined automatically from the integrated data and rule reasoning are used for link prediction.
- We present a concrete KG contracted by integrating aviation-related data from online resources and demonstrate a use case of querying such data, including predicted links and explanations.
- We present experimental evaluations of our rule-based link prediction by comparing it with embedding-based models and inductive neural models. The evaluation shows the competitiveness of our method, besides its clear and unique advantage of explainability.

Source code for our system, the dataset, and experiment setting details are all available at https://drive.google.com/file/d/1-jeJ1ahLaZAkKZvGgKVzhy j4QWp_yyTY/view?usp=sharing.

## 2  Related Work

In this section, we summarise major existing research works closely related to the paper.

### 2.1  Knowledge Graph Link Prediction

KGs have shown to be a promising approach for interlinking data and capturing semantic connections. Link prediction is the task of predicting missing links between entities in KGs and is one of the major reasoning tasks studied for KGs. Existing approaches include embedding-based [1,7,15,17] and rule-based [8,10,11] ones. The embedding-based methods encode entities and relations in KGs as low-dimensional latent representations, such as vectors and matrices, called *embeddings* [16]. While most existing link prediction methods are embedding-based and show high accuracy on standard benchmarks, their major limitations are the lack of adaptability and explainability. In particular, the embedding-based models are trained for KGs with fixed entities and relations and cannot effectively handle changes. Once a KG is updated, embedding-based models need to be retrained, which is often highly time and resource-consuming. To address such a limitation, inductive neural models [13] are proposed to handle new (called unseen) entities introduced after the models are trained by encoding subgraph patterns instead of individual entities. Nevertheless, neither embedding-based nor inductive neural models themselves can provide human comprehensible explanations to the predictions made.

On the other hand, rule-based approaches [8,10,11] mine logic rules from KGs and apply rule-based reasoning for link prediction. Rules and rule-based reasoning have the clear advantage of being comprehensible to humans, and as rules capture higher-order knowledge, they can be directly applied to previously unseen entities in the KG. Recent advancement in rule mining [10,11] allows significant amounts of high-quality rules to be mined from large-scale KGs, and their performance in link prediction is comparable with major embedding-based models, often with significantly better efficiency. Yet, to the best knowledge, logic rule mining and reasoning have not been applied in semantic link mining, and existing rule learners have not been evaluated on real-life datasets that involve data dynamics. In this work, we present a rule-based approach for link prediction over dynamic integrated data.

### 2.2  Semantic Data Integration

*Semantic data integration* [5] refers to the integration of multi-sourced and heterogeneous data by establishing links between the entities across the original datasets to provide a single interface to process the data. KGs offer a promising solution to semantic data integration [6] by representing entities of interest as vertices and their semantic links as edges in a graph, which supports a single interface to access and explore the data. KGs have been increasingly adopted in the industry to integrate business-related data, provide a virtual schema to browse and query data, and use reasoning to answer user queries [4,9]. Also, ontology-based data access (OBDA) approaches [3] have been developed to integrate relational data from multiple sources. These approaches often require manually developed knowledge bases such as ontologies to capture the domain knowledge and to provide reasoning capability, and they focus on the query answering task.

Our framework is different in the following aspects. First, while we use a manually developed ontology to guide the auto-construction of the KG, the reasoning is not only based on the ontology. Instead, we use automated rule mining to extract high-level knowledge as logic rules for reasoning, which makes it possible to capture the high-order effects of data changes. Also, our framework allows users to query both explicit facts and implicit links through link prediction. Moreover, by combining rule mining and reasoning in link prediction, our framework provides adaptability and explainability.

## 3 Preliminaries

A knowledge graph (KG) consists of a set of *entities* $\mathcal{E}$ as its vertices and its edges are directed and labelled with a set of *relations* $\mathcal{P}$. An entity $e \in \mathcal{E}$ is an object such as a place, a person, etc., and link between two entities is a *triple* $(e, p, e')$, which means that the entity $e \in \mathcal{E}$ is related to another entity $e' \in \mathcal{E}$ via the relation $p \in \mathcal{P}$. Following the convention in knowledge representation, we can also denote such a triple as a fact $p(e, e')$.

A *probabilistic rule* (or simply a *rule*) $r$ is of the form

$$\alpha : H \leftarrow B_1, B_2, \ldots, B_n,$$

where $H$ and each $B_i$ $(1 \leq i \leq n)$ is of the form $p(X, Y)$ with $p \in \mathcal{P}$ being a relation and $X, Y$ being variables, and $\alpha$ is a number between 0 and 1. Intuitively, the rule $r$ reads that if $B_1, B_2, \ldots, B_n$ hold, then $H$ holds with a confidence of $\alpha$. $H$ is the *head* of the rule and the set of atoms $\{B_1, B_2, \ldots, B_n\}$ is the *body* of the rule.

The *confidence degree* of a rule is usually defined as the number of instances that make the body valid, divided by those instances that make both the body and the head valid. In particular, for a rule $r$ with its head being $p(X, Y)$, $\#(e, e') : head(r, e, e')$ is the number of pairs of entities $e, e' \in \mathcal{E}$ making the head of $r$ valid, i.e., fact $p(e, e')$ is in the KG. Similarly, $\#(e, e') : body(r, e, e')$ is the number of pairs of entities making the body of $r$ valid, i.e., there is a way of substituting variables in the body of $r$ with entities in $\mathcal{E}$ such that (i) all the facts in the body of $r$ (after substitution) hold in the KG, and (ii) $X$ and $Y$ are substituted with $e$ and $e'$ respectively. And $\#(e, e') : head(r, e, e') \wedge body(r, e, e')$

is the number of pairs of entities making both the head and body of $r$ valid at the same time. The *confidence degree* of a rule $r$ is defined as follows:

$$\frac{\#(e, e') : head(r, e, e') \wedge body(r, e, e')}{\#(e, e') : body(r, e, e')}$$

The *link prediction* task is to predict, given the subject (or object) $e \in \mathcal{E}$ and the relation $p \in \mathcal{P}$ in a triple $(e, p, ?)$ (resp., $(?, p, e)$), the missing object (resp., subject). Unlike embedding-based approaches that rank the possible entities $e'$ via scoring functions, a rule-based approach tries to derive plausible facts $p(e, e')$ by applying the mined rules to the existing facts in the KG. For a rule $r$ with its head $H = p(X, Y)$, the derived facts are $p(e, e')$ for $e, e' \in \mathcal{E}$ such that there exist a way to substituting the variables in the rule so that $X$ and $Y$ are replaced with $e$ and $e'$, respectively, and all the facts obtained from the body of $r$ occur in the KG. The ranking of the derived fact is obtained from the confidence of the rules deriving it.

## 4    Our Approach

In this section, we describe our approach for semantic link mining with rule-based link prediction.

### 4.1    Our Framework

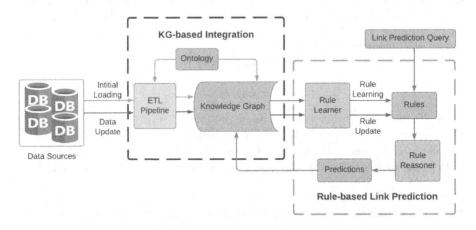

**Fig. 1.** Our framework for semantic link mining and link prediction.

Figure 1 shows an overview of our framework. There are two major components in our approach, one is the KG-based integration module, and the other is the rule-based link prediction module. In the KG-based integration module, we integrate heterogeneous data from multiple sources by mapping them into a KG. As a result, the KG provides a uniform way to access information from multiple data sources and also connects the data for querying. To guide the data mapping, we use an ontology manually developed for the related domains.

The links between data sources in the KG are still incomplete due to the data sizes and their complex semantic connections. Hence, in the rule-based link prediction module, we mine logic rules to capture higher-order knowledge stored in the KG, which can be considered as metadata. The rules are automatically mined using a rule learner. Such rules can be applied to infer new links with the confidence degrees of the rules propagated to the inferred links. The confidence degrees can be used to rank the inferred links for humans to decide whether such links should be added to the KG. When new data are added to the system, they can be integrated through the mapping to expand the existing KG. Rules can be updated accordingly, and the inferred links can be re-evaluated, or new links can be inferred.

## 4.2   KG-Based Integration

Unlike conventional data integration approaches that design an extended schema to cater to all the data attributes, our approach uses a KG, which does not require a rigid schema, to integrate the data by treating objects of interest such as aircraft, airports, flights, events and organisations as entities. Entities are linked by relations manually selected from the data sources.

The mapping of data objects from the sources to entities and relations in the KG is based on the Extract, Transform and Load (ETL) pipeline approach, which has been widely applied in semantic data applications. Major challenges in developing such an ETL pipeline include identifying important data objects as entities of interest, discovering correlations between entities from different sources, and establishing meaningful relations between entities while maintaining the integrity and quality of all data.

To address such challenges, our approach adopts ontology-based data access (OBDA) techniques to guide the ETL process. The ontology provides a virtual schema for integrating multiple data sources to build our final KG. The instances of each class in the ontology schema become the entities in the KG. Similarly, object properties between classes in the ontology schema correspond to the relations and data properties become entity attributes in the KG. With the emerging requirements for FAIR data, an ontology also provides a foundation for integrating data based on the linked open data principles by enabling the development of common vocabularies.

## 4.3   Rule-Based Link Prediction

The semantically integrated data can be used to infer missing links through link prediction. We are interested in cross-source link prediction, for example, the link prediction task (737-800, isAircraftOf, ?), which requires aircraft data and airline data to be combined. In what follows, we describe our rule-based link prediction module.

To achieve this, we first mine rules across data sources by exploring the paths in the integrated data. For example, to mine the following rule for predicting links about the relation isAircraftOf:

$$\text{isAircraftOf}(X, Y) \leftarrow \text{isAircraftOf}(X, A), \text{hasAlliance}(A, Y),$$

we explore paths where aircraft entities $X$ (e.g., from FAA) are connected to some airline entities $A$ (e.g., from openflights.org) through relation isAircraftOf, which in turn are connected with some other airline entities $Y$ via relation hasAlliance. Hence, we essentially want to extract a sequence of relations like (Aircraft, hasAlliance). Formally, a *relation path* is a sequence of relations $(p_1, p_2, \ldots, p_n)$ in the KG.

Path-based rule mining has been used in [10], where paths are examined locally independent of other paths and thus lack a global view of potentially more plausible paths. Unlike [10], we use a vector representation of paths to efficiently evaluate large numbers of paths together before assessing the rules corresponding to individual paths.

To efficiently assess whether a relation path has a significant number of instances in the data, instead of retrieving all its instances using a SPARQL query, which is relatively inefficient, we use a lightweight check for rapid evaluation as below. For each relation $p \in \mathcal{P}$, let $dom(p) = \{e \in \mathcal{E} \mid p(e, e') \text{ in the KG}\}$ and $ran(p) = \{e \in \mathcal{E} \mid p(e', e) \text{ in the KG}\}$. Intuitively, for a relation path $(p_1, p_2, \ldots, p_n)$ that forms the body of a rule

$$p(X_0, X_n) \leftarrow p_1(X_0, X_1), p_2(X_1, X_2), \ldots, p_n(X_{n-1}, X_n),$$

there should be a significant number of entities exist in the following sets

- $dom(p) \cap dom(p_1)$ (in the place of $X_0$),
- $ran(p) \cap ran(p_n)$ (in the place of $X_n$), and
- $ran(p_i) \cap dom(p_{i+1})$ (in the place of $X_i$ for $1 \leq i < n$).

We use one-hot encodings for $dom(p)$ and $ran(p)$ as a computational method to obtain statistics on entity distributions over relations. Let $\mathbf{p}^{dom}$ (resp., $\mathbf{p}^{ran}$) be a vector of length $|\mathcal{E}|$, such that the scalar at position $i$ is $\frac{1}{|dom(p)|}$ (resp., $\frac{1}{|ran(p)|}$) if $e_i \in dom(p)$ (resp., $e_i \in ran(p)$), and 0 otherwise, for $1 \leq i \leq |\mathcal{E}|$. We use the following scoring function to filter relation paths.

$$f_{path}(r) = sim(\mathbf{p}_1^{dom}, \mathbf{p}^{dom}) + sim(\mathbf{p}_n^{ran}, \mathbf{p}^{ran}) +$$
$$sim(\mathbf{p}_1^{ran}, \mathbf{p}_2^{dom}) + \ldots + sim(\mathbf{p}_{n-1}^{ran}, \mathbf{p}_n^{dom}),$$

where $sim(\cdot, \cdot)$ is defined by the Frobenius norm, i.e. $sim(\mathbf{v}_1, \mathbf{v}_2) = \exp(-\|\mathbf{v}_1 - \mathbf{v}_2\|)_F$. Paths with high scores form candidate rules, which are validated through their confidence degrees.

When the data change, if the change involves only a small amount of data, the mined rules may still be valid. As rules are naturally inductive, they can be applied directly to new entities introduced. For example, if a new triple is added (*DeltaAirlines*, hasAlliance, *AirFrance*) with a new entity *DeltaAirlines*, supposing a triple (737-800, isAircraftOf, *AirFrance*) is in the data, the aforementioned rule can be directly applied to the new entity and derive a new link (737-800, isAircraftOf, *DeltaAirlines*).

If a considerable amount of data are involved in the change, the mined rules should be updated to reflect the higher-order impact of such a change. Our one-hot encodings can be utilised to update mined rules. In particular, when new data is added, or some existing data is deleted, such changes are reflected in the KG through our integration pipeline. Then, the one-hot encodings $\mathbf{p}^{\mathrm{dom}}$ and $\mathbf{p}^{\mathrm{ran}}$ can be efficiently updated for each relation $p$ involved in the changes. Then, the following changes will be made to update the mined rules. For each relation paths $(p_1, p_2, \ldots, p_n)$ where a significant change occur to the one-hot encodings of a relation $p_i$ $(1 \le i \le n)$,

- if the path corresponds to previously mined rules then it is reassessed using the scoring function;
- if the path had a high score but the confidence of its corresponding rule fell slightly below the threshold, then the path may be reassessed; and
- if the path can be obtained from the above paths by replacing relation $p_{i-1}$ or $p_{i+1}$, then the path may be reassessed.

These operations will update rules and their confidence degrees based on the data changes, allowing higher-order changes in the data to be reflected in the rule model.

For link prediction such as (737-800, isAircraftOf, ?), our rule parser translates the query and all the related rules into SPARQL queries to retrieve entities from the data, as the subject or objection in the place of the question mark. The translated SPARQL query of the first rule above would be as follows.

```
SELECT  DISTINCT ?Y
WHERE {
        ?X              rdf:type                ai4dm:Aircraft .
        ?A              rdf:type                ai4dm:Airline .
        ?X              airgraph:isAircraftOf   ?A .
        ?Y              rdf:type                ai4dm:Airline .
        ?A              airgraph:hasAlliance    ?Y .
        FILTER( ?X = '737-800' )
}
```

The retrieved answers are attached with the same confidence degrees as the rules used to infer them and are ranked by their confidence degrees. Our system will display the answers ordered by their confidence degrees together with the rules as an explanation of the predictions. These answers can be evaluated by human experts and added to the KG if their validity is confirmed.

## 5   An Application Case

In this section, we present a practical use case by integrating aviation-related data from four major sources: Federal Aviation Administration (FAA)[1], ourairports.com[2], openflights.org[3], and DBpedia[4]. FAA is a transport agency in the USA that aggregates aircraft and airport data around the world. Ourairports.com is a free online portal for users to explore world airports. It comprises information on airports and runways. Openflights.org is an online service for logging and sharing flight data, which provides information on flight routes, origin, and destination airport information. DBpedia is a large KG, which contains structured general knowledge extracted from Wikipedia. It describes general concepts and entities related to airlines, airports, and aircraft.

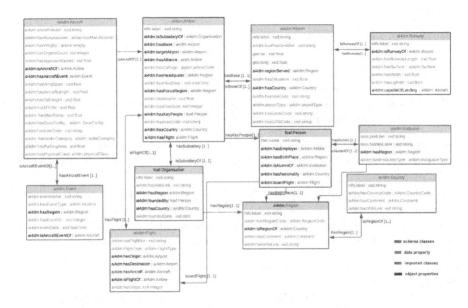

**Fig. 2.** Key classes and relations in our ontology expressed in an ER model.

The data from the above sources contain complementary information. For example, there are a rich collection of aircraft data in the FAA database, which is not contained in the other data sources. Hence, integrating the data from these sources allows us to build connections between airports and aircraft. To express such semantic links are cross data sources, relations are extracted from the ontology. For example, the relation isAircraftOf connects entities about aircraft in the

[1] https://www.faa.gov/data_research/.
[2] https://ourairports.com/.
[3] https://openflights.org/data.html.
[4] https://www.dbpedia.org/.

FAA database with those about airlines in the openflights.org database, and the relation targetAirport connects entities in ourairports.com and openflights.org.

Our ontology provides a collection of key classes and their relations (a.k.a. object properties), including 32 classes and 82 relations. Some key classes and relations in the ontology schema are shown in Fig. 2.

Initially, a total of 59K entities and 77K triples are extracted through the ETL process. The distribution of the entities in several major classes is shown in Table 1. We refer to this KG as *AirGraph*. Note that as a proof of concept and for the ease of data cleaning, the KG contains only a fragment of all the information from the original data sources, while the framework allows further data to be added.

**Table 1.** Entities distributions.

| Classes | Entities | Classes | Entities | Classes | Entities |
|---------|----------|---------|----------|---------|----------|
| Flight | 42491 | Aircraft | 521 | Institution | 36 |
| Airport | 9131 | Country | 246 | Airline | 13 |
| Region | 3958 | Person | 54 | Organisation | 10 |
| Runway | 3171 | Event | 38 | | |

## 6 Evaluation

We have conducted two sets of experiments to evaluate the performance of our rule-based link prediction module in both static and dynamic settings. For the static setting, we used two commonly used benchmarks and a static version of AirGraph and used an incremental version of AirGraph for the dynamic setting. Statistics of the four datasets are shown in Table 2. In particular, FB15K237 [14] and WN18RR [7] are widely used benchmarking datasets obtained from respectively Freebase and WordNet.

**Table 2.** Statistics of datasets.

| KG | #Entity | #Relation | #Triple |
|----|---------|-----------|---------|
| FB15K237 | 14541 | 237 | 310116 |
| WN18RR | 40943 | 11 | 89969 |
| AirGraph | 21685 | 23 | 59619 |

For rule mining, we used both our own rule learner and an efficient rule learner AnyBURL [10]. AnyBURL is an anytime rule learner that can be configured with a very short training time, hence can be efficiently retrained for online prediction over dynamic data. We are particularly interested in the time efficiency of rule-based link prediction, whether our module can be used for online processing where predictions need to be made within a matter of minutes. We also evaluate the accuracy of our rule-based link prediction in comparison with some baselines.

We compare our module with embedding-based models TransE [1], DisMult [17], ComplEX [15], and ConvE [7], which cover a range of geometric, matrix factorization, and deep learning models [12]. We also compare with an inductive neural model GraIL [13], which is particularly designed for dynamic setting where new entities are introduced after the model is trained. GraIL is inspired by the inductive capability of rules and is based on graph neural networks (GNN), but it cannot generate explicit rules.

Our experiments are designed to validate the following statements.

- In the static setting, the accuracy of our rule-based link prediction module is superior than the compared embedding-based models.
- The time efficiency of our module is superior than the embedding-based models, and significantly better than GraIL.
- In the dynamic setting, the accuracy of our module is still compared to the inductive neural models, with the clear advantage of explainability.

The experiments were conducted on a machine with 2 CPU, 16G RAM, and 1 GeForce GTX1605 GPU.

## 6.1   Static Link Prediction

In the first set of experiments, we evaluate the accuracy of our module in link prediction on FB15K237 and WN18RR. The predictions are measured using filtered Mean Reciprocal Rank (MRR), Hits@1, and Hits@10, where a higher value indicates better performance. These quality measurements are commonly used in the literature. In particular, each prediction is ranked by its confidence score, and MRR is the average number of the reciprocal ranks of correctly predicted entities. Hits@1 and Hits@10 are the proportions of correctly predicted entities that are ranked number one and among the top ten, respectively. Table 3 shows the results, with those for embedding-based models for FB15K237 and WN18RR obtained from [2], and those for AnyBURL are from [10]. The best results are highlighted in bold, and the second best ones are underlined.

**Table 3.** Performance of link prediction on FB15K237 and WN18RR.

| Models | FB15K237 | | | WN18RR | | |
|---|---|---|---|---|---|---|
| | MRR | Hits@1 | Hits@10 | MRR | Hits@1 | Hits@10 |
| RESCAL | <u>35.6</u> | <u>26.3</u> | **54.1** | 46.7 | 43.9 | 51.7 |
| TransE | 31.3 | 22.1 | 49.7 | 22.8 | 5.3 | 52.0 |
| DistMult | 34.3 | 25.0 | 53.1 | 45.2 | 41.3 | 53.0 |
| ComplEx | 34.8 | 25.3 | <u>53.6</u> | <u>47.5</u> | 43.8 | <u>54.7</u> |
| ConvE | 33.9 | 24.8 | 52.1 | 44.2 | 41.1 | 50.4 |
| AnyBURL | 31.0 | 23.3 | 48.6 | 47.0 | <u>44.1</u> | **55.2** |
| Ours | **33.7** | **26.7** | 52.0 | **48.3** | **44.8** | 54.1 |

In the next set of experiments, we evaluate both the accuracy and time efficiency of the models on the statistic version of AirGraph, which includes a core fragment of the data we collected. We separate the dataset into approximately 70% train, 10% validate, and 20% test. Our module mines rules from and applies rules on the train data to predict links in the test. For AnyBURL, as it is an anytime rule learner, we set a time limit according to the time our rule learner took (11 min). Table 4 summarises the results, and the time is in minutes.

**Table 4.** Performance of link prediction on static AirGraph.

| Models | Time | MRR | Hits@1 | Hits@10 |
|--------|------|-----|--------|---------|
| TransE | 15.03 | 0.10 | 0.07 | 0.17 |
| DistMult | 16.67 | 0.10 | 0.07 | 0.16 |
| ComplEX | 34.09 | 0.12 | 0.09 | 0.19 |
| ConvE | 16.18 | 0.14 | 0.10 | 0.20 |
| GraIL | 378.20 | 0.48 | 0.37 | 0.68 |
| AnyBURL | 11.00 | 0.10 | 0.11 | 0.17 |
| Ours | 10.52 | 0.22 | 0.18 | 0.30 |

Compared to well-established KGs, our AirGraph is highly incomplete regarding the inter-source links, which affects the overall prediction accuracy. Our module demonstrates superior accuracy and time efficiency compared to the evaluated embedding-based models. The accuracy of our rule learner is also better than AnyBURL configured with the same rule mining time as ours. The accuracy of GraIL is outstanding, likely due to its powerful GNN-based model, yet it is also extremely time-consuming to train. GraIL took over 5 h for training and prediction, which is 30 times more than ours. Please note that our goal is not to compete with deep neural models like GraIL on the prediction accuracy, as the advantages of our rule-based module are in its explainability and time efficiency. For decision-critical applications where explainability is desired and for time-critical applications, our rule-based module would be more suitable.

Our rule learner extracted a totally 373 rules with lengths up to 3, and we show some examples of the rules and their confidence.

$0.36$ : $\text{isAircraftOf}(X, Y) \leftarrow \text{hasAircraft}(A, X), \text{isFlightOf}(A, Y)$.

$0.39$ : $\text{isAircraftOf}(X, Y) \leftarrow \text{hasAircraft}(A, X), \text{hasOrigin}(A, B), \text{hasBase}(Y, B)$.

$0.50$ : $\text{hasAlliance}(X, Y) \leftarrow \text{isSubsidaryOf}(X, A), \text{foundedBy}(A, B),$
$$\text{hasEmployer}(B, Y).$$

$0.50$ : $\text{isSubsidaryOf}(X, Y) \leftarrow \text{hasEmployer}(A, X), \text{foundedBy}(Y, A)$.

$0.60$ : $\text{isRegionOf}(X, Y) \leftarrow \text{hasHeadquarter}(A, X), \text{hasCountry}(A, Y)$.

$0.67$ : $\text{hasNationality}(X, Y) \leftarrow \text{hasEmployer}(X, A), \text{hasFocusRegion}(A, B),$
$$\text{isRegionOf}(Y, B).$$

The rules are intuitive, for example, the first rule says an aircraft $X$ is likely to be in the fleet of an airline $Y$ if a flight $A$ from the airline $Y$ uses the aircraft. The final rule tells a person $X$ has a nationality $Y$ if $X$ is employed by an organisation $A$ that has a focus region $B$ where country $Y$ is part of.

## 6.2   Dynamic Link Prediction

As discussed before, data changes often pose significant challenges in link prediction. Hence, in the second set of experiments, we compare our module with GraIL in a dynamic setting. In particular, we consider the scenario where data arrive and are integrated into AirGraph in batches.

We separate the initial train data that consist of around 25% of the triples used in the previous section. Based on a common assumption of inductive neural models, the initial train data contain all the relations, while only new entities (i.e., unseen entities in [13]) will be introduced later. All the models are trained once on the initial train data. Then, new data arrive in 4 batches, each consisting of roughly the same number of triples as in the initial train data. Each batch is divided into approximately 70% train, 10% validate, and 20% test, similar to the data division as in [13]. When a new batch arrives, say batch $n$ ($n = 1, 2, 3, 4$), all the models make predictions with access to train and validate in the current and previous batches, i.e., batches $i$ for all $1 \leq i \leq n$. Such a setting is based on the following considerations: (1) retraining the models may not be possible for online prediction, (2) GraIL and our rule-based module can apply the trained models on new data to make predictions about unseen entities.

Moreover, as AnyBURL is the only model that can be retained within a minute, we include a version of it with retraining on each batch (marked with

**Table 5.** Performance of link prediction on dynamic AirGraph.

| Batch | Models | Time | MRR | Hits@1 | Hits@10 |
|---|---|---|---|---|---|
| 1 | GraIL | 28.00 | 0.078 | 0.044 | 0.076 |
|   | AnyBURL | 0.03 | 0.009 | 0.010 | 0.019 |
|   | AnyBURL* | 0.83 | 0.039 | 0.043 | 0.068 |
|   | Ours | 0.58 | 0.156 | 0.131 | 0.199 |
| 2 | GraIL | 57.00 | 0.193 | 0.133 | 0.232 |
|   | AnyBURL | 0.05 | 0.025 | 0.028 | 0.041 |
|   | AnyBURL* | 0.83 | 0.061 | 0.067 | 0.101 |
|   | Ours | 1.17 | 0.197 | 0.167 | 0.257 |
| 3 | GraIL | 124.00 | 0.280 | 0.197 | 0.402 |
|   | AnyBURL | 0.07 | 0.040 | 0.045 | 0.073 |
|   | AnyBURL* | 0.83 | 0.078 | 0.087 | 0.133 |
|   | Ours | 1.89 | 0.220 | 0.188 | 0.288 |
| 4 | GraIL | 220.00 | 0.329 | 0.235 | 0.496 |
|   | AnyBURL | 0.08 | 0.059 | 0.064 | 0.099 |
|   | AnyBURL* | 0.83 | 0.099 | 0.108 | 0.167 |
|   | Ours | 2.83 | 0.225 | 0.191 | 0.296 |

an ∗). Our rule learner took 2.53 min to train, and AnyBURL was given 10 min for the initial training and 50 s for each retraining. The results are shown in Table 5. The times are for link prediction only, except for AnyBURL∗ whose times include retraining. Times are in minutes.

From Table 5, our rule-based module outperforms GraIL and AnyBURL on the first two batches, which shows our path-based rule mining method is capable of mining quality rules from medium-sized data. As more batches arrive, the accuracy of all the models increases as more data becomes accessible. The accuracy of GraIL increases rapidly and outperforms ours in the final two batches. Yet GraIL also took significant amount of time for link prediction even after it is trained. For the first batch with relatively small amount of data, it already took 28 min, making it unsuitable for online predictions. For time-critical applications, our rule-based module would have a clear advantage in its time efficiency. It is worth noting that the accuracy of our rule learner outperforms AnyBURL∗, which is allowed to retrain on each batch. This shows the effectiveness of our dynamic rule update method, because unlike AnyBURL∗ that explores paths from scratch each time when it is retrained, our rule update method reassesses and refines previously seen paths.

# 7  Conclusion

In this paper, we have introduced a semantic link-mining framework using knowledge graphs and rules. In particular, semantic connections between data from different sources are captured through links in knowledge graphs. Rules are automatically extracted from knowledge graphs and applied in prediction to discover missing links. A major benefit of our rule-based link prediction approach is that rules involved in the reasoning process can be used to provide human-comprehensible explanations for the predictions. Another advantage of our framework is link prediction over dynamic data by incrementally refining the mined rules based on data changes.

We demonstrate our framework's usefulness through an application case and experimental evaluation. In particular, we developed a prototype system based on the semantic integration of aviation data from four online sources. It demonstrates the use case of link prediction and explanation in an aviation information system. Moreover, we conducted experimental evaluations on our rule-based link prediction in static and dynamic settings. The evaluation shows that our approach has a good balance in time efficiency and prediction accuracy, both of which are desired for real-life online prediction.

This paper can be extended with the following research directions: First, it is worth further exploring incremental learning to boost the time efficiency of the system and reduce its response time. The gain in time efficiency may also allow a more accurate inference approach in an online setting. Moreover, the framework can be extended for event handling and predictions by adopting data stream processing techniques and frameworks.

**Acknowledgements.** This work was partially supported by the Office of National Intelligence of Australia under the AI for Decision Making Initiative.

# References

1. Bordes, A., Usunier, N., García-Durán, A., Weston, J., Yakhnenko, O.: Translating embeddings for modeling multi-relational data. In: Proceedings of NeurIPS-13, pp. 2787–2795 (2013)
2. Broscheit, S., Ruffinelli, D., Kochsiek, A., Betz, P., Gemulla, R.: LibKGE - a knowledge graph embedding library for reproducible research. In: Proceedings of EMNLP-20: System Demonstrations, pp. 165–174 (2020)
3. Calvanese, D., et al.: Ontop: answering SPARQL queries over relational databases. Semantic Web **8**(3), 471–487 (2017)
4. Chaves-Fraga, D., Priyatna, F., Cimmino, A., Toledo, J., Ruckhaus, E., Corcho, Ó.: GTFS-Madrid-Bench: a benchmark for virtual knowledge graph access in the transport domain. J. Web Semant. **65**, 100596 (2020)
5. Cheatham, M., Pesquita, C.: Semantic data integration. In: Zomaya, A.Y., Sakr, S. (eds.) Handbook of Big Data Technologies, pp. 263–305. Springer, Cham (2017). https://doi.org/10.1007/978-3-319-49340-4_8
6. Cudré-Mauroux, P.: Leveraging knowledge graphs for big data integration: the XI pipeline. Semantic Web **11**(1), 13–17 (2020)
7. Dettmers, T., Minervini, P., Stenetorp, P., Riedel, S.: Convolutional 2D knowledge graph embeddings. In: McIlraith, S.A., Weinberger, K.Q. (eds.) Proceedings of AAAI-18, pp. 1811–1818. AAAI Press (2018)
8. Galárraga, L., Teflioudi, C., Hose, K., Suchanek, F.M.: Fast rule mining in ontological knowledge bases with AMIE+. VLDB J. **24**(6), 707–730 (2015)
9. Kalaycı, G., et al.: Semantic integration of bosch manufacturing data using virtual knowledge graphs. In: Pan, J., et al. (eds.) ISWC 2020. LNCS, vol. 12507, pp. 464–481. Springer, Cham (2020). https://doi.org/10.1007/978-3-030-62466-8_29
10. Meilicke, C., Chekol, M.W., Ruffinelli, D., Stuckenschmidt, H.: Anytime bottom-up rule learning for knowledge graph completion. In: Proceedings of IJCAI2019, pp. 3137–3143. ijcai.org (2019)
11. Omran, P.G., Wang, K., Wang, Z.: An embedding-based approach to rule learning in knowledge graphs. IEEE Trans. Knowl. Data Eng. **33**(4), 1348–1359 (2021)
12. Rossi, A., Barbosa, D., Firmani, D., Matinata, A., Merialdo, P.: Knowledge graph embedding for link prediction: a comparative analysis. ACM Trans. Knowl. Discov. Data **15**(2), 14:1–14:49 (2021)
13. Teru, K.K., Denis, E., Hamilton, W.: Inductive relation prediction by subgraph reasoning. In: Proceedings of ICML2020. Proceedings of Machine Learning Research,,vol. 119, pp. 9448–9457. PMLR (2020)
14. Toutanova, K., Chen, D., Pantel, P., Poon, H., Choudhury, P., Gamon, M.: Representing text for joint embedding of text and knowledge bases. In: Proceedings of EMNLP-2015, pp. 1499–1509. The Association for Computational Linguistics (2015)
15. Trouillon, T., Welbl, J., Riedel, S., Gaussier, É., Bouchard, G.: Complex embeddings for simple link prediction. In: Proceedings of ICML-16. JMLR Workshop and Conference Proceedings, vol. 48, pp. 2071–2080. JMLR.org (2016)

16. Wang, Q., Mao, Z., Wang, B., Guo, L.: Knowledge graph embedding: A survey of approaches and applications. IEEE Trans. Knowl. Data Eng. **29**(12), 2724–2743 (2017)
17. Yang, B., Yih, W., He, X., Gao, J., Deng, L.: Embedding entities and relations for learning and inference in knowledge bases. In: Proceedings of ICLR-2015 (2015)

# Information Mining from Images of Pipeline Based on Knowledge Representation and Reasoning

Raogao Mei[1], Tiexin Wang[1(✉)] (iD), Shenpeng Qian[1], Huihui Zhang[2], and Xinhua Yan[3]

[1] College of Computer Science and Technology, Nanjing University of Aeronautics and Astronautics, 29#, Jiangjun Road, Jiangning District, Nanjing 211106, China
{gavinnell,tiexin.wang,qianshenpeng}@nuaa.edu.cn
[2] Weifang University, Weifang 261061, China
huihui@wfu.edu.cn
[3] Nanjing DENET System Technology Co. Ltd., Nanjing, China

**Abstract.** Urban drainage pipeline plays a crucial role in the construction of modern cities, and Closed Circuit Television (CCTV) is one of the most commonly used technologies for pipeline inspection. However, CCTV is only in charge of reflecting the pipeline's status with videos and images. Identifying and evaluating pipeline defects still require professional knowledge and the participation of experienced practitioners. Therefore, intelligent approaches designed to improve the effectiveness and efficiency of pipeline inspection are deadly expected in practice. To address this issue, this paper presents a knowledge-driven approach with a prototype software tool. More specifically, one domain ontology is defined to formalize the required knowledge of pipeline defects identification, e.g., the types and classifications of pipeline defects. Furthermore, a set of reasoning rules for deducing pipeline defects and relevant defects parameters are designed to work with the proposed domain ontology. To verify the validity of our proposed domain ontology and reasoning rules, we conducted one industrial case study based on original images of pipeline defects provided by Nanjing BeiKong Enterprises Water Group Co., Ltd. Results show that defects in the selected pipeline images can be inferred correctly, which indicates that our proposed method can assist the automatic identification of pipeline defects.

**Keywords:** Pipeline defects identification · Knowledge model · Domain ontology · Semantic reasoning

## 1 Introduction

With the continuous increase of the urban population and the continuous development of the urban scale, the scale of urban underground pipelines has also continued to expand [1]. Urban drainage pipeline is one of the most important infrastructures in urban underground network management, just like the blood vessels and meridians, which is closely related to the normal operation of the city [2]. Due to the fact that the urbanization development in the early stage has not been fully integrated with the actual situation, it has led to various problems such as imperfect urban drainage systems, and backward pipeline maintenance, resulting in the frequent occurrence of drainage pipeline accidents in later use [1, 3].

At present, Closed Circuit Television (CCTV) inspection has been utilized to detect underground drainage pipelines, which has a wide range of applications [4]. The internal structure of the pipeline and its condition are recorded in real-time, and the inspection videos are uploaded to the external monitor. Then, the relevant inspector can check the cause and location of the defects according to the images formed deep inside the pipeline. Although CCTV inspection can intuitively reflect the condition of the pipeline through images or videos, the defect identification of these pipeline images or videos still requires the participation of experienced personnel, and such manual interpretation of inspection images or videos is time-consuming and labor-intensive [4, 5]. In addition, different inspectors may have different understandings of the defects, resulting in the discriminative results may be subjective and error-prone [5]. Therefore, transforming the process of manual defect identification into computer processing is particularly important.

However, pipeline defects identification (PDI) is a knowledge-intensive task, which affects by various types of information such as inspection data, safety specifications, pipeline regulations, and practitioners' experience, and this kind of knowledge mainly exists in unstructured forms. Therefore, knowledge formalization is to play a significant role in the field of PDI, which can promote knowledge sharing and reuse [6].

Ontology, a knowledge representation technique, can be used to represent knowledge of a specific domain because of its explicit and rich semantics [7]. By defining concepts and various relationships among concepts, it can convert regulatory knowledge of textual documents into a specific and understandable format, which has been widely used in knowledge representation due to its benefits on knowledge management [8]. Since the process of manual interpretation for PDI is time-consuming and error-prone, a knowledge model combined with semantic reasoning is developed to formalize PDI knowledge in this paper. Consequently, images of pipelines can be transformed as instances of the constructed ontology model for defects reasoning based on the pre-defined reasoning rules. Specifically, the defect type, the defect grade, the defect score and other relevant parameters of pipeline images can be inferred with the constructed ontology model.

Overall, the two main contributions of this paper are as follows:

1) Integrating pipeline regulations, safety specifications, and practitioners' experience, we construct the domain ontology of PDI to formalize this unstructured knowledge into a machine-readable format, which promotes domain knowledge sharing, reuse, and expansion.
2) Combined with semantic reasoning, we define a series of reasoning rules to link the pipeline defect images from the industry with the constructed ontology model to realize the defect identification from pipeline images.

This paper is organized as follows. The related work is given in Sect. 2. A general overview of the PDI ontology construction is illustrated in Sect. 3. Section 4 conducts a case study demonstrating the validity of the constructed ontology model. Some concluding remarks and future work are given in Sect. 5.

## 2  Related Work

### 2.1  Pipeline Defects Identification

In recent years, the rapid growth of computing power has promoted the development of deep learning technology, and one of its important applications is object detection [12]. In the field of PDI, there are various deep learning algorithms applied for defects detection, such as faster R-CNN [9], single-shot detector [10], You Only Look Once (YOLO) [11], etc. Yin et al. [12] used CCTV video as the input data and YOLOv3 deep learning algorithm as the object detector to extract the defect feature information of pipeline images in the video for model training, and finally to detect the pipeline defects information in the video automatically. Cheng et al. [13] proposed a method of pipeline defects inspection based on a region-based convolutional neural network (faster R-CNN), using the collected 3000 CCTV images to train the detection model, and the trained model can be used to automatically identify pipeline defects images. Kumar et al. [14] evaluated the accuracy and speed of three object detection models based on Single Shot multi-box Detector (SSD), YOLOv3, and faster R-CNN deep learning algorithms by identifying root intrusion and sediment defects in pipeline CCTV images.

The above-mentioned deep learning based solutions for identifying pipeline defects are efficient, but they still have limitations. First of all, the automatic identification of pipeline defects by deep learning models is un-interpretable and cannot quantify the severity of the detected pipeline defects. In other words, deep learning models can only identify the types of pipeline defects, but not the level of pipeline defects. In addition, a pipeline image may contain two or more types of defects, and the above deep learning models cannot identify multiple defects simultaneously.

### 2.2  Ontology for Knowledge Formalization

Ontology, based on the ability to formalize domain knowledge with explicit and rich semantics, has recently played a crucial role in various areas, such as the semantic web, risk identification, and knowledge management [15]. For instance, considering that hazard identification in the process of metro construction is a knowledge-intensive task, Wu et al. [7] used a domain ontology (SRI-Onto) to facilitate safety knowledge management. In stark contrast, Lu et al. [16] formalized metro accident knowledge into domain ontology for knowledge retrieval and reasoning. Domain ontology can not only link different knowledge information together for convenient sharing and reuse but also support consistency checking and semantic reasoning. Zhong et al. [17] realized the formalization and reasoning of risk knowledge by combining the construction process of the knowledge reuse method with ontology and semantic reasoning.

Overall, the aforementioned studies formalized the domain knowledge into ontology with different research purposes, which indicates that ontology can represent knowledge formally and explicitly with its knowledge interoperability and reasoning ability. In terms of no existing ontology applied in the field of PDI, our proposed ontology model of PDI will further promote ontology applications.

# 3 PDI Ontology Construction

The purpose of constructing ontology is to transform domain knowledge into a machine-readable format for computer processing, which is convenient for knowledge sharing and reuse [15]. Moreover, concepts and their semantic relationships in PDI can be intuitively represented in the form of classes and properties of the ontology [18]. Protégé is an open-source ontology building tool with an intuitive user interface and numerous plug-ins that enables developers to create and edit domain ontologies [19]. The process of constructing PDI ontology is based on the latest version of Protégé 5.0 and is explained in the following subsections.

## 3.1 Knowledge Resource

The PDI knowledge was mainly obtained from the textual regulatory document < <Technical Regulations for Testing and Evaluation of Urban Drainage Pipelines (CJJ 181–2012) > >, which includes pictures, tables, and relevant pipeline evaluation formulas. Part of the knowledge was obtained from the published academic articles in the digital library. The remaining small part of knowledge was pulled from the public internet platform such as Baidu Baike, Baidu library, and Wikipedia, as a supplement to the above knowledge. In addition, some details involved in technical regulations are filtered out through knowledge preprocessing, mainly by manual method relying on the relevant practitioners' experience.

## 3.2 Ontology Development for PDI

The most common modeling method for domain ontology is the seven-step method [21] proposed by Stanford University. Combining the above seven-step method, we simplify it into four procedures to construct the ontology model based on the ontology editing tool Protégé, which is more suitable for our ontology construction process. These four procedures are: 1) Determine the meta-ontology model; 2) Determine the knowledge hierarchy; 3) Define the object and data properties of the class and create instances; 4) Consistency checking.

**Determine the Meta-ontology Model.** This step is to determine the scope of the PDI ontology that we construct. Since there is no existing ontology in the field of PDI, we need to construct the domain ontology from scratch. Through the learning of <<Technical Regulations for Testing and Evaluation of Urban Drainage Pipelines (CJJ 181–2012)>>, we focus the meta-ontology on the scope of pipeline evaluation and pipeline defects, which is the core content of PDI.

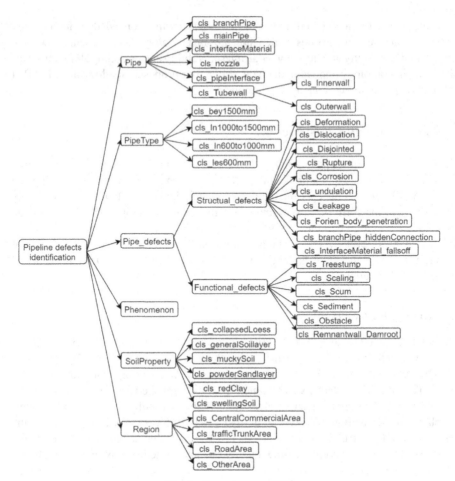

**Fig. 1.** Core classes of PDI

**Determine the Class Hierarchy.** We need to list the important terminologies of PDI, namely, the core concepts in the domain of pipeline evaluation and pipeline defects. In terms of pipeline defects, important terms include structural defects, functional defects, defect grade, defect score, defect influence parameters, etc.

After identifying the important terminologies in the domain of PDI, we need to divide the hierarchy of knowledge, which plays a key role in knowledge representation. There are a total of 6 top-level classes: Pipe, PipeType, Pipe_defects, Phenomenon, SoilProperty, and Region, among them, including 40 subclasses, as shown in Fig. 1.

**Define the Object and Data Properties of the Classes and Create Instances.**
The class hierarchy cannot completely present sufficient information on PDI. Based on the core concepts, we defined the object properties between classes as their relations and the data properties as properties of classes. There are a total of 27 essential object

and data properties defined in PDI ontology, including Domain and Range information, which are presented in Table 1 and Table 2.

**Table 1.** Part of the defined object properties

| Object Property | Domain | Range | Characteristic |
|---|---|---|---|
| pty_hasPhenomenon | Pipe | Phenomenon | Functional |
| pty_hasDefects | Pipe | Pipe_defects | Functional |
| pty_hasSoilproperty | Pipe | Soilproperty | Functional |
| pty_locatedIn | Pipe | Region | Functional |
| pty_Belongto | Pipe | PipeType | Functional |
| pty_cause | Phenomenon | Pipe_defects | Inverse Functional |
| pty_causeBy | Pipe_defects | Phenomenon | Inverse Functional |

**Table 2.** Part of the defined data properties

| Data Property | Domain | Range | Characteristic |
|---|---|---|---|
| pty_grade | Pipe | int | Functional |
| pty_score | Pipe | double | Functional |
| pty_diameter | Pipe | double | Functional |
| pty_disjointDistance | cls_pipeInterface | double | Functional |
| pty_thicknessofTubewall | cls_Tubewall | double | Functional |

After determining the class hierarchy, the object and data properties are added to the corresponding classes. The instantiation of the class is followed, that is, the creation of instances, which plays a key role in coding SWRL rules and is essential for reaching knowledge sharing and semantic interoperability [15]. Some of the instances are listed in Table 3.

**Table 3.** Part of the instances

| Class | Instance |
|---|---|
| Phenomenon | ind_Finecracks, ind_Les60ofCirCover, ind_slightshedding |
| Region | ind_CentralCommercialArea, ind_trafficTrunkArea, ind_RoadArea |
| Soilproperty | ind_redClay, ind_collapsedLoess, ind_muckySoil, ind_littlesoilSqueeze |
| PipeType | ind_bey1500mm, ind_In1000to1500mm, ind_Les600mm |
| Pipe defects | ind_liehen, ind_liekou, ind_posui, ind_tanta |

### 3.3  Reasoning Rules for PDI

PDI is a knowledge-intensive task, which concerns a large number of conceptual definitions and detailed descriptions. In this paper, SWRL rules are used to formalize the various definitions and descriptions into corresponding reasoning rules owing to their powerful deductive reasoning abilities based on OWL concepts [22].

Each SWRL rule consists of an antecedent and consequent part, both of which are formed by the conjunction of atoms. When the antecedent part is satisfied, the consequent part will be triggered to execute reasoning [22]. In brief, the SWRL rules can be interpreted as: if all the atoms in the premises are true, then the conclusion is true. Table 4 shows part of SWRL rules corresponding to the regulations.

**Table 4.** Part of the SWRL rules corresponding to the regulations

| No | Regulations | SWRL rules |
|----|-------------|------------|
| 1 | If the deformation range of the pipe is 15%–25% of the pipe diameter, the corresponding defect grade is 3, and the defect score is 5 | Pipe(?p) ^ pty_deformArea(?p, ?area) ^ pty_diameter(?p, ?d) ^ swrlb:multiply(?d1, ?d, 0.15) ^ swrlb:greaterThan(?area, ?d1) ^ swrlb:multiply(?d2, ?d, 0.25)^ swrlb:lessThanOrEqual(?area, ?d2) -> pty_hasDefects(?p, ind_yanzhongBX) ^ pty_gradeBX(?p, 3) ^ pty_scoreBX(?p, 5.0) |
| 2 | If the area where the pipeline is located is a central business area with Class A civil construction projects attached, then the regional importance parameter K of the pipeline is 10 | Pipe(?p) ^ pty_locatedIn(?p, ind_CentralCommercialArea) ->pty_has_K(?p, 10) |
| 3 | If the soil quality of the soil layer where the pipeline is located is red clay, the soil quality influence parameter T of the pipeline is 8 | Pipe(?p) ^ pty_hasSoilproperty(?p, ind_redClay) ->pty_has_T(?p, 8) |
| 4 | If the diameter of the pipeline is greater than 1000 mm and less than or equal to 1500 mm, then the pipeline importance parameter E of the pipeline is 6 | Pipe(?p) ^ pty_diameter(?p, ?d) ^ swrlb:greaterThan(?d, 1000) ^ swrlb:lessThanOrEqual(?d, 1500) -> pty_Belongto(?p, ind_In1000to1500mm) ^ pty_has_E(?p, 6) |

Those starting with "pty_" represent object or data properties of a class, such as *pty_locatedIn* and *pty_hasSoilproperty* are object properties between classes, and *pty_diameter*, *pty_deformArea* are data properties, and those starting with "ind_" represent an instance of a class, such as *ind_CentralCommercialArea* is the instance of class *Region*.

The SWRL specifications contain some built-in functions or predicates, which can be used to calculate addition, subtraction, multiplication, and division, and can also reflect the magnitude relationship through value comparison, such as the predicate swrlb:

*greaterThan( ?d, 1000)* is used to represent whether or not the variable "?d" is greater than 1000 and if this relationship holds, it will return true.

## 4   Case Study

To validate the validity of the PDI ontology model, we selected some pipeline images with common defect types which contain the original labels of defect type and relevant attribute information, such as pipe diameter, the phenomenon description of the image defects, and defect parameters, provided by Nanjing Beikong Enterprises Water Group Co., Ltd. The relevant attribute information is saved in an XLSX file, which will be mapped into the constructed ontology model to generate corresponding image instances. Then, the pre-defined SWRL rules will be executed in an inference engine to recognize the image defect types and the relevant defect parameters. Finally, the reasoning results are compared with the original defect type labels and defect parameters of the selected images to validate the validity of the ontology model. The above process is shown in Fig. 2.

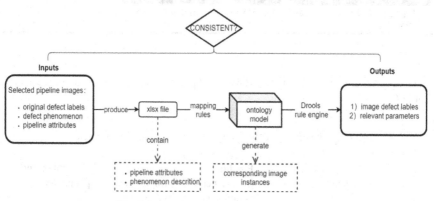

**Fig. 2.** The process of ontology validation

### 4.1   Selected Pipeline Images with Common Defect Types

The image datasets provided by Nanjing Beikong Enterprises Water Group Co., Ltd. are categorized as two groupes, namely structural defects and functional defects. Among all defects, there are 10 structural defects and 6 functional defects, with a total of 16 sub-categories. Functional defects can be repaired manually in the process of pipeline inspection, such as residual walls, dam roots, obstacles, etc. After discussing with the pipeline experts, in our study, four most common structural defects (i.e., Rupture, Deformation, Dislocation and Disjointed), are determined as the main source of our selection.

Because one image cannot reflect the overall defect condition of the pipeline, we select a set of defect images from different perspectives of the same pipeline, which are shown in Table 5.

**Table 5.** The selected pipeline images

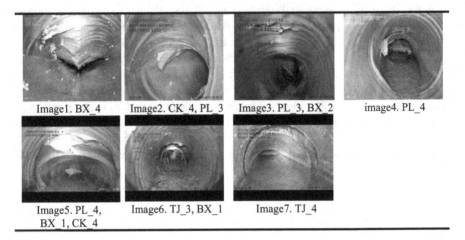

| Image1. BX_4 | Image2. CK_4, PL_3 | Image3. PL_3, BX_2 | image4. PL_4 |

| Image5. PL_4, BX_1, CK_4 | Image6. TJ_3, BX_1 | Image7. TJ_4 |

The original defect labels and relevant defect parameters of selected images are listed in Table 6.

**Table 6.** The original defect labels and relevant defect parameters

| Defect images | Original defect labels | Pipeline importance parameters E | Regional importance parameters | Soil importance parameters T |
|---|---|---|---|---|
| Image1 | BX_4 | 3 | 6 | 8 |
| Image2 | CK_4, PL_3 | 3 | 6 | 8 |
| Image3 | PL_3, BX_2 | 3 | 6 | 8 |
| Image4 | PL_4 | 3 | 6 | 8 |
| Image5 | PL_4, BX_1, CK_4 | 3 | 6 | 8 |
| Image6 | TJ_3, BX_1 | 3 | 6 | 8 |
| Image7 | TJ_4 | 3 | 6 | 8 |

The capital letter abbreviations: "PL", "BX", "CK" and "TJ" represent Rupture, Deformation, Dislocation, and Disjointed respectively and PL_4 means the grade of Rupture is 4. Since the images are derived from the same pipeline, the value of Pipeline Importance Parameters E, Regional Importance Parameters K, and Soil Importance Parameter T are all the same.

## 4.2 The Attribute Information of Pipeline Images

Since Protégé 5.0 provides a plug-in for mapping entities stored in the spreadsheet to the ontology model, the attribute information will be saved in the XLSX form. In addition, the keywords describing the pipeline defect images in the XLSX file should correspond to the concepts (classes, properties, and instances) defined above in the constructed ontology model, so as to facilitate the mapping of the attribute information to the ontology model. The attribute information is shown in Table 7.

**Table 7.** The attribute information of defect images

| Class | Instance | Diameter(mm) | DefromArea (mm) | LateralDeviofNozzle (mm) | ThicknessofTubewall (mm) | DisjointDistance (mm) |
|-------|----------|--------------|-----------------|--------------------------|--------------------------|-----------------------|
| Pipe | Image1 | 1000 | 764 | 0 | 30 | 0 |
| | Image2 | 1000 | 0 | 68 | 30 | 0 |
| | Image3 | 1000 | 124 | 0 | 30 | 0 |
| | Image4 | 1000 | 0 | 0 | 30 | 0 |
| | Image5 | 1000 | 36 | 73 | 30 | 0 |
| | Image6 | 1000 | 47 | 0 | 30 | 39 |
| | Image7 | 1000 | 0 | 0 | 30 | 62 |

Images 1 to 7 come from different parts of the same pipeline, and the pipeline is located in the area of the main road, and the corresponding soil property is red clay, so the values of "soil"s and |"traffic" in columns 3 and 4 of the XLSX file are "ind_traffic" and "ind_redClay" respectively, which are not shown due to the limited space of the table.

## 4.3 Mapping Rules for Images Instantiation in PDI Ontology

Two mapping rules are created for image instantiation in PDI ontology, which is saved in the JSON format and shown as follows.

**Mapping rules 1:**
```
{
  "Collections":[
    {
      "sheetName":"Attribute information",
      "startColumn":"B",
      "endColumn":"B",
      "startRow":"2",
      "endRow":"8",
      "comment":"",
      "rule":"Individual: @B*\nTypes: @A2\n Facts: pty_hasSoilproperty @C*,
      pty_locatedIn    @D*,    pty_diameter    @F*,    pty_deformArea
      @G*,\npty_haslateralDeviofNozzle @H*, pty_thicknessOfTubewall @I*,
      pty_disjointDistance @J*",
      "active":true
    }
  ]
}
```
**Mapping rules 2:**
```
{
  "Collections":[
    {
      "sheetName":"Attribute information",
      "startColumn":"B",
      "endColumn":"B",
      "startRow":"3","
      endRow":"6",
      "comment":"",
      "rule":"Individual:@B*\nTypes: @A2\n Facts: pty_hasPhenomenon @E*",
      "active":true
    }
  ]
}
```

After executing the above mapping rules, the mapping results are shown in Fig. 3. Image instances from 1 to 7 correspond to the selected 7 defect images. The right half shows the mapping results of object and data properties, which correspond to the attribute information in the above XLSX file.

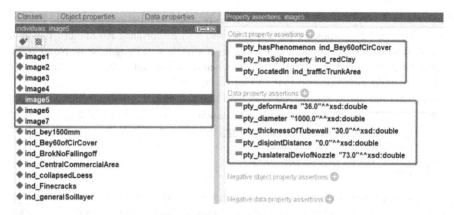

**Fig. 3.** The mapping results

## 4.4   Knowledge Reasoning

The pre-defined SWRL rules can be executed in an inference engine named Drools, which consists of a fact base, a rule base, and an execution engine [22]. Drools can be used to conduct knowledge reasoning in Protégé after mapping the attribute information into instances in the ontology model. After executing the pre-defined SWRL rules, the defect type, defect grade, defect score, and relevant defect parameters are automatically inferred, which is shown in Fig. 4.

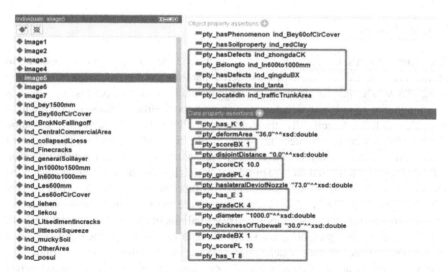

**Fig. 4.** The reasoning results

Take image5 as an example, the reasoning results show that image5 has three defect types: *ind_zhongdaCK*, *ind_qingduBX*, *ind_tanta*, and the corresponding defect grades are *pty_gradeCK*, *pty_gradeBX*, *pty_gradePL*, with the value of 4, 1 and 4, respectively.

The *pty_has_E*, *pty_has_K*, and *pty_has_T* correspond to Pipeline Importance Parameters E, Regional Importance Parameters K, and Soil Importance Parameters T, which are consistent with the original defect parameters. For the rest images, same patterns are observed, i.e., the reasoning results are all the same as original defect labels and we do not discuss the details here due to the space limitation.

Based on the above results, we conclude that the defect types and relevant defect parameters can be inferred correctly by pre-defined SWRL rules and it is straightforward to verify the validity of the proposed ontology model.

### 4.5  Discussion

Although the constructed ontology model combined with semantic reasoning can infer defect types and relevant defect parameters automatically, there are still some open issues.

First, the attribute XLSX file is created manually, which is not intelligent, and how to automatically extract the attribute information from pipeline images and map it into the ontology model needs further research. Second, only images of common defect types in different parts of the same pipeline are selected to verify the constructed ontology model. More complex scenes involving various defect types in different pipelines should also be considered. Third, the SWRL rules encoded in this paper are a finite set, mainly encoding the definitions of different types of pipeline defects, and more safety regulations should be transformed into corresponding SWRL rules into satisfy the requirement of PDI in various scenes.

## 5  Conclusion

This paper proposes an ontology model of PDI combined with semantic reasoning to automatically infer the defect types and defect parameters of the pipeline, formalizing the PDI knowledge into the machine-readable format for computer processing, which is more intelligent and objective in comparison with the traditional manual identification of pipeline defects. Specifically, the concepts of PDI knowledge and their semantic relationships can be represented intuitively in the form of classes and properties of ontology, supporting consistency checking and reasoning. Then, the attribute information of defect images is imported into the ontology model through the mapping rules, and the defect types and relevant defect parameters of the pipeline are inferred based on the pre-defined SWRL rules. Finally, the reasoning results are compared with the original labels of defect types and defect parameters to verify the validity of the ontology model. Overall, the constructed ontology model incorporates the knowledge of pipeline regulations and the experience of technicians, which facilitates knowledge sharing and reuse. Importing the attribute information of images into the ontology model can recognize the defect types and the relevant defect parameters, avoiding the ambiguity of PDI caused by artificial subjectivity, and saving human resources. Since the attribute file of defect images is still created manually, we intend to combine deep learning techniques with the ontology model in future research. The deep learning model can automatically extract attribute information of defect images after training and then import it into the ontology model for knowledge reasoning. Relying on the development of deep learning and image recognition technologies, an automatic defect identification of pipeline images could be realized.

**Acknowledgement.** This work was partially supported by the Shandong Provincial Natural Science Foundation (No. ZR2021MF026) and the technical service project No. 1015-KFA21833. The authors would like to thank Nanjing BeiKong Enterprises Water Group Co., Ltd and the staff for their cooperation and assistance in providing industrial datasets.

# References

1. Wang, J., et al.: Current status, existent problems, and coping strategy of urban drainage pipeline network in China. Environ. Sci. Pollut. Res. **28**(32), 43035–43049 (2021)
2. Pikaar, I., Sharma, K.R., Hu, S., Gernjak, W., Keller, J., Yuan, Z.: Reducing sewer corrosion through integrated urban water management. Science **345**(6198), 812–814 (2014)
3. Bai, D.: Application and development of detection technology for urban drainage pipeline. World Build Mater. **40**(4), 83–86 (2019)
4. Halfawy, M.R., Hengmeechai, J.: Optical flow techniques for estimation of camera motion parameters in sewer closed circuit television inspection videos. Autom. Constr. **38**, 39–45 (2014)
5. Pan, G., Zheng, Y., Guo, S., Lv, Y.: Automatic sewer pipe defect semantic segmentation based on improved U-Net. Autom. Constr. **119**, 103383 (2020)
6. Cortés, B.J., et al.: Formalization of gene regulation knowledge using ontologies and gene ontology causal activity models. Biochim. Biophys. Acta (BBA)-Gene Regul. Mech. **1864**(11–12), 194766 (2021)
7. Xing, X., Zhong, B., Luo, H., Li, H., Wu, H.: Ontology for safety risk identification in metro construction. Comput. Ind. **109,** 14–30 (2019)
8. Zhong, B., Wu, H., Li, H., Sepasgozar, S., Luo, H., He, L.: A scientometric analysis and critical review of construction related ontolog y research. Autom. Constr. **101,** 17–31 (2019)
9. Girshick, R., Fast R-CNN. In: Proceedings of the IEEE International Conference on Computer Vision, pp. 1440–1448 (2015)
10. Liu, W., Anguelov, D., Erhan, D., Szegedy, C., Reed, S., Fu, C.-Y.: SSD: single shot multibox detector. In: Leibe, B., Matas, J., Sebe, N., Welling, M. (eds.) ECCV 2016. LNCS, vol. 9905, pp. 21–37. Springer, Cham (2016). https://doi.org/10.1007/978-3-319-46448-0_2
11. Redmon, J., Divvala, S., Girshick, R., Farhadi, A.: You only look once: Unified, real-time object detection. In: Proceedings of the IEEE Conference on Computer Vision and Pattern Recognition, pp. 779–788 (2016)
12. Yin, X., Chen, Y., Bouferguene, A., Zaman, H., Al-Hussein, M., Kurach, L.: A deep learning-based framework for an automated defect detection system for sewer pipes. Autom. Constr. **109**, 102967 (2020)
13. Cheng, J.C.P., Wang, M.: Automated detection of sewer pipe defects in closed-circuit television images using deep learning techniques. Autom. Constr. **95**, 155–171 (2018)
14. Kumar, S.S., Wang, M., Abraham, D.M., Jahanshahi, M.R., Iseley, T., Cheng, J.C.: Deep learning– based automated detection of sewer defects in CCTV videos. J. Comput. Civil Eng. **34**(1), 04019047 (2020)
15. Zhang, S., Boukamp, F., Teizer, J.: Ontology-based semantic modeling of construction safety knowledge: towards automated safety planning for job hazard analysis (JHA). Autom. Constr. **52**, 29–41 (2015)

16. Wu, H., Zhong, B., Medjdoub, B., Xing, X., Jiao, L.: An ontological metro accident case retrieval using CBR and NLP. Appl. Sci. **10**(15), 5298 (2020)
17. Lu, Y., Li, Q., Zhou, Z., Deng, Y.: Ontology-based knowledge modeling for automated construction safety checking. Saf. Sci. **79**, 11–18 (2015)
18. Zhong, B., Li, Y.: An ontological and semantic approach for the construction risk inferring and application. J. Intell. Rob. Syst. **79**(3), 449–463 (2015)
19. Noy, N.F., et al.: Protégé-2000: an open-source ontology-development and knowledge-acquisition environment. In: AMIA... Annual Symposium Proceedings. AMIA Symposium, pp. 953–953 (2003)
20. Zhong, B.T., Ding, L.Y., Luo, H.B., Zhou, Y., Hu, Y., Hu, H.: Ontology-based semantic modeling of regulation constraint for automated construction quality compliance checking. Autom. Constr. **28**, 58–70 (2012)
21. Noy, N.F., McGuinness, D.L., Ontology development 101: a guide to creating your first ontology. Technical report SMI-2001-0880 (2001). Stanford Medical Informatics, Stanford University, Palo Alto, CA, USA
22. Wu, H., Zhong, B., Li, H., Love, P., Pan, X., Zhao, N.: Combining computer vision with semantic reasoning for on-site safety management in construction. J. Build. Eng. **42**, 103036 (2021)

# Binary Gravitational Subspace Search for Outlier Detection in High Dimensional Data Streams

Imen Souiden[1](✉), Zaki Brahmi[2], and Mohamed Nazih Omri[1]

[1] MARS Laboratory, University of Sousse, Sousse, Tunisia
imen.sui@gmail.com
[2] Taibah University, Medina, Saudi Arabia

**Abstract.** In recent years, technology has continued to rapidly evolve, resulting in the generation of high-dimensional data streams. Combining the streaming scenario and high dimensionality is a particularly complex task specifically for outlier detection. This is due to the data stream's unique properties, such as restricted space and time, and concept drift, in addition to the influence of the curse of dimensionality in high-dimensional space. Typically, interesting knowledge including outliers resides in low-dimensional subspaces of the full feature space. Finding these subspaces is considered an NP-Hard problem and requires careful attention, especially in the context of data streams. To address these issues, we proposed BGSSA (Binary Gravitational Subspace Search Algorithm), a novel metaheuristic-based subspace search method for outliers in high dimensional data streams. The idea behind is to adapt the binary GSA algorithm by producing the top best solutions instead of a single one in the original method to find, for each streaming window, relevant subspaces composed of independent features, where outlier detection will be performed. The relevance of a subspace is evaluated by the contrast measure. Experiments on real and synthetic datasets confirm the feasibility of our solution as well as its performance improvement in comparison with the main approaches studied in the literature.

**Keywords:** Outlier detection · Data streams · High dimensional data · Subspace

## 1 Introduction

Outlier detection has received special attention in many research areas due to its importance. It seeks irregular patterns, showing high deviation in contrast to regular ones in data sets. This paper focuses on the problem of outlier detection in high-dimensional data streams which are being used in a variety of fields, including network intrusion detection [20], fault detection and prevention [24], and so on. The data tend to be infinite, evolving, and arriving continuously at a rapid rate with a large number of features. Examples include sensors in industrial

© The Author(s), under exclusive license to Springer Nature Switzerland AG 2022
W. Chen et al. (Eds.): ADMA 2022, LNAI 13726, pp. 157–169, 2022.
https://doi.org/10.1007/978-3-031-22137-8_12

settings, online financial transactions, etc. In this situation, detecting outliers becomes more difficult. It must cope with data streams non-stationary in which the underlying distribution changes over time [18]. In addition to respecting resource constraints while providing good accuracy. Further, as the irrelevant features mislead the mining process and increase the computing burden, it is essential to reduce the feature space by keeping only interesting ones. One silent characteristic of high dimensional data is that outliers are often embedded in low dimensional subsets (subspaces) [1]. Motivated by this fact, we opt for the outlier detection in subspaces instead of the full feature space.

In a d-dimensional space, there exist $2^d - 1$ possible subspaces growing exponentially with increasing dimensionality [16]. Thus, searching for relevant subspaces for outlier detection is an NP-hard problem [5]. Due to the combinatorial explosion, an exhaustive search through subspaces is not a scalable strategy, especially in the data stream setting. Therefore, it is essential to provide a scalable and appropriate method that, while accommodating data stream limitations, finds subspaces containing relevant information. This relevance may be dependent or independent of the mining task, i.e. outlier detection in our context. In the first case, it is limited to and closely tied to an outlier criterion. In the second case, it only considers the features and the relation between them. The obtained subspaces often contain interesting information useful for a multitude of tasks. Dependency estimators like mutual information [7] and contrast [6] are some measures of relevance in this context. Different works have been proposed based on these measures [6,9,10]. Yet, the literature in the context of the data stream is still scarce. A multivariate dependency estimator that meets various desirable properties for data streams, like efficiency and robustness, was recently proposed in [4]. So, we decided to use the suggested contrast. As a search strategy, we used the Binary Gravitational Search Algorithm (Binary GSA) [13]. It is a metaheuristic-based approach that provides good results with low computational cost. Hence, it can be very effective in solving this problem. Binary GSA utilizes the concept of the law of gravity to find the near-optimal solution. To our knowledge, we are the first to use GSA in this context. We proposed an adapted version of Binary GSA over windowed data streams. We adjusted the algorithm to output, in every window, the set of optimal solutions represented by subspaces instead of a single one. As well, to keep track of subspace relevance, we customized the initial population generation by providing a portion of the best subspaces from the previous window (except the initial window). To process the stream and respond to its changing nature, we used a non-overlapping sliding window model. Overall, our main contributions are: 1) Formalizing the subspace search problem as a mono-objective optimization function. 2) Proposing an adapted version of Binary GSA producing the best solutions set instead of a single one. 3)Proposing BGSSA approach (Binary Gravitational Subspace Search Algorithm) to find high contrast subspaces using the adapted Binary GSA. 4) Using BGSSA in outlier detection problem. 5) Testing BGSSA on real and synthetic data sets. All of this is applied on high-dimensional data streams.

The rest of this paper is structured as follows: Sect. 2 presents the main related works. Section 3 introduces the problem formulation. Section 4 explains our methodology and the proposed algorithms. Section 5 examines our proposed solutions experimentally. Section 6 concludes the work and gives some outlooks.

## 2   Related Work

Recently, numerous studies have been conducted for outlier detection in high dimensional data streams [5,14,23,24]. They are broadly categorized into full and subspace-based methods. The former assumes the same relevance of all features for outlier detection. Thus, the algorithms may be biased by irrelevant features downgrading their outlierness estimation. The latter addresses this issue by limiting the use of the original feature set to subsets. Feature selection and extraction [17,20,21] are powerful in reducing the dimensionality to only one subspace and are commonly used as a pre-processing step. Nonetheless, they cannot deal with outliers masked into different subsets of attributes. Multiple subspace-based methods overcome this problem by finding outliers in subspaces obtained in a deterministic or random way. Random methods operate on randomly generated subspaces [8,11,14] and avoid the costly subspace search strategy. The random subspace generation may include many irrelevant features into subspaces while omitting relevant ones notably when irrelevant features are the dominants.

In contrast to random methods, deterministic ones sacrifice speed for accuracy. They search for relevant subspaces satisfying a search criterion and then compute outlier scores in those subspaces. In sparse subspace-based methods, low-density subspaces are identified, indicating the presence of outliers. They usually rely on density estimation-based measures to characterize the sparseness of subspaces, such as KNN in ABSAD-SW [24] and the sparsity coefficient in SPOT [23]. Sparse subspace-based methods are task-specific since they find subspaces specific to outlier detection. Relative subspace methods seek subspaces composed of meaningful attributes and thus they are general and used for a multitude of tasks. HPC-StreamMiner [19] and SGMRD [5] have been proposed in this context. They seek subspaces composed of independent features using the concept of contrast representing the deviation between the marginal and conditional density distributions. In [19], the high contrast subspaces are progressively searched with time and their ranking is incrementally updated to keep track of data changes. SGMRD leveraged a Hill climbing greedy heuristic with novel multivariate quality estimators. As well, it proposed a monitoring technique based on bandit theory to update the results of subspace search over time. It tries to find optimal subspaces for each dimension instead of the full feature space. Random subspace methods lack accuracy while the deterministic one suffers from the resource burden. Consequently, it's crucial to develop a method that provides a compromise between efficiency and effectiveness which is the goal of our approach. We proposed an adapted BGSA algorithm over windowed data streams to find the high contrast subspaces from the original feature space, where outlier detection will be performed. Our subspace search strategy is deterministic, and we look for relative and general subspaces.

# 3   Problem Formulation

Given an incoming infinite sequence of high dimensional stream items: $DS = \langle X_1, t_1 \rangle, \langle X_2, t_2 \rangle, \langle X_3, t_3 \rangle, ..., \langle X_\infty, t_\infty \rangle$. Each data object $X_i = (x_i^1, ..., x_i^d)$ is a d-dimensional vector. $d$ is the dimensionality of the full data space represented by the feature set $F = \{f_1, ..., f_d\}$. $i$ is the number of data points arriving at time $t$. $x_i^d$ is the value of the feature $d$ for data object $i$. To deal with the data stream, we used a non-overlapping sliding window structure $W$ of size $ws$ due to its generality and efficiency. The stream is split into fixed-size windows and the algorithm always operates on the latest window. At any time point $t$, we only consider the latest $ws$ data in the window $W[t - ws + 1, t]$.

In high dimensional settings, outliers are often embedded in lower dimensional subspaces. Therefore, we opt to detect outliers in subspaces instead of the full-feature space. A subspace $S \subseteq F$ is a non-empty subset of the full data space with dimensionality $d' < d$. The objective is to decide for each data point $X_n \in DS$ if it is an outlier or not at any time $t$. The scoring of $X_n$ should be done before $X_{n+1}$ and within the available memory. Our general aim is a two-step processing: Subspace search and outlier scoring. In the subspace search, we seek to find the best set of relevant subspaces denoted by $RS$. The relevance of subspaces $S \in F$ is assessed using the contrast denoted by $C(S)$. A high contrast projection $rs \subseteq RS$ is a selection of dimensions showing a data distribution with a high dependency. This dependency leads to clear clustered structures vs individual outliers. To compute the contrast $C(S)$ of a subspace $S$, we adopted the MCDE dependency estimator [4]. It treats the attributes in $F$ as random variables. MCDE quantifies the contrast of an attribute set as the average discrepancy between marginal and conditional distributions estimated via a statistical test $\tau$ and approximated via $M$ Monte Carlo simulations (see Ref. [4]). In our case, we instantiated the contrast with a two-sample Kolmogorov-Smirnov (KSP) test [15]. It is non-parametric and widely used in testing the equality of two continuous one-dimensional probability distributions. An index structure $I$ containing the ordered values per feature is employed to facilitate the subspace slicing operation used for conditional distribution estimation.

The maximum number of subspaces is set to $Q$, a user-defined parameter. The number of features in each subspace is in $[2, \sqrt{d}]$ range. In real-data sets, the implicit dimensionality usually does not grow much faster than $\sqrt{d}$ with increasing dimensionality $d$ [2]. As well, the minimum number of features is set to 2, since in one-dimensional subspace the notion of dependence cannot exist. Furthermore, $RS$ must contain unique subspaces. A relevant subspace $rs_i$ must differ from subspace $rs_j$ for all subspaces in $RS$. It is formally expressed as follows:

$$RS = argmax_{S_i \in F} C(S_i)$$

Subject to

$$|RS| = Q \text{ and } |rs| = [2, \sqrt{d}] \forall rs \in RS$$

$$rs_i \neq rs_j \forall rs_i, rs_j \in RS$$

(1)

Once the relevant subspaces are identified, the points are scored using an outlier detection algorithm in each such subspace. In our case, we used LOF algorithm

[3]. Then, the scores of each point $sc(X_n)_{rs}$ in various subspaces $rs \subseteq RS$ are computed and averaged to provide a unified score $sc(X_n)$ of each point. $X_n$ will be labeled as outlier if its score $sc(X_n)_{rs}$ is lower than a predefined threshold $\beta$.

# 4 Binary Gravitational Subspace Search for Outlier Detection in High Dimensional Data Streams

Our solution finds relevant subspaces, where outlier detection will be performed while coping with NP-hardness of subspace search. We adapted the binary GSA algorithm to find subspaces with high contrast. Further, we used the popular density-based LOF outlier detector [3] to detect outliers within these subspaces. As a processing unit, we employed a non-overlapping sliding window structure.

## 4.1 Subspace Search with Adapted Binary GSA

**Binary GSA:** GSA is a successful swarm-based metaheuristic algorithm inspired by Newton's law of gravity [12]. In GSA, a swarm is a collection of physical particles (search agents), each with a mass representing the solution's fitness value. The fitness is the performance of the solutions (contrast measure in our case). The agents attract each other based on the gravitational force. This force pushes lighter objects (lower mass) towards heavier objects (higher mass). The heavier objects representing better solutions will move slower than the lighter ones. GSA was designed to search spaces of real-valued vectors. The search space in binary optimization problems is modeled as a hypercube, with an agent's location updated by altering one or more bits of its position vector. GSA in binary format was proposed first by [13]. A transfer function bounded within interval [0,1] is needed to convert a continuous algorithm into its binary version.

**Adapted BGSA for Subspace Search:** In our context, a solution consists of a subspace $S$. All solutions are represented by binary strings with fixed and equal length $d$, where $d$ is the number features $F$. Each bit in the individual will take 0 or 1 indicating whether or not its corresponding feature is selected for a particular subspace. To find the set of relevant subspaces $RS$, BGSA is applied with a slight modification in the output. The regular BGSA returns a single best solution. Yet, in our approach, we return the set of best solutions represented by subspaces. To find this set, the algorithm first initializes a population. In first window, this population consists of $N$ randomly generated candidate solutions. For the subsequent windows, it will be customized by adding a portion $ts$ of the best subspaces $subspace_w$ obtained in the previous window. The inclusion of this set permits enhancing and refining the subspace search with time. Thereafter, at each iteration $it$ of BGSA, we assess the different combinations of subspaces basing on their fitness. For each solution, the fitness is evaluated by calculating its corresponding contrast value using the MCDE estimator. Thereafter, the whole population is sorted in ascending order of fitness. Subsequently, the specific parameters related to the BGSA, such as the acceleration, velocity, and position of each agent are updated. At each iteration, the $topQ$ subspaces in each population are stored in a temporary list. In the end, the $topQ$ subspaces with high fitness among all subspaces are returned as the best subspaces. Figure 1 illustrates the flowchart of the proposed adapted Binary GSA, where our modifications compared to the original Binary GSA are highlighted in gray.

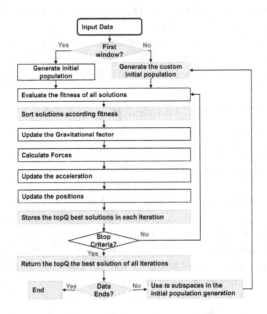

**Fig. 1.** Flow chart of Adapted BGSA

## 4.2  Solution Overview

In this section, we will give an overview of our approach basing on the algorithm pseudo-code illustrated in Algorithm 1. The algorithm takes as input a stream $DS$ with n-d-dimensional features and the global parameters for the different components. At the beginning, "Subspaces" and "outliers" sets are empty. "Subspaces" is the best subspace base and "outliers" is the outlier base. The first phase is the algorithm initialization, where the first window $W_1$ will be created and treated. First, the index structure $I$ used for contrast calculation will be created (Line 10). Then, the adapted binary GSA will be applied for the first time to find the $topQ$ subspaces (Line 15). The subspaces found will be used for outlier detection (Line 16). Thereafter, this window will be cleared and a new window will be created and treated. At this time, the new window is employed and a model update is performed to deal with possible changes in the data distribution. The model update consists of updating the index structure according to the new inputs. Further, a portion $ts$ of the obtained subspaces in the previous

**Algorithm 1. BGSSA**

1: **Data**: DS: data stream, Q: top subspaces, ts: the size of subspaces added in the initial population, ws: Window size, binarygsa_params: Binary GSA related parameters, contrast_params: Contrast calculation related parameters
2: **Result**: set of outliers and top Q subspaces
3: **Begin**
4: Subspaces← ∅
5: Outliers← ∅
6: $i$←1
7: **while** (DS.hasData() ) **do**
8:      $W_i$ ← $createwindow(ws)$
9:      **if** ($i$=1) **then**
10:          $I$ ← $CreateIndexstructure(W_1)$
11:          $Subspacesw_{i-1}$ ← ∅
12:      **else**
13:          $I$ ← $UpdateIndexstructure(W_i)$
14:      **end if**
15:      $Subspaces_i$ ← $AdapatedBGSA(W_i, binarygsa\_param, contrast\_params, Subspacesw_{i-1})$
16:      $Outliers_i$ ← $OutlierDetection(Subspaces_i, )$
17:      $Clearwindow(W_i)$
18:      $i$←$i$+1
19: **end while**
20: **END**

window will be added to GSA's initial population to ensure the continuity of the process. In the first window, this set is empty (Line 11). The relevance of these subspaces will be re-evaluated and hence they will be discarded if they are not relevant. These steps will be repeated for every window until the stream ends.

## 5   Experimental Study and Results Analysis

### 5.1   Experimentation Setting

The performance of our method is evaluated through experiments on synthetic and real datasets described in Table 1. KDDcup99[1] and Activity[2] are frequently used real data with outlier ground truth. We selected data similar to [14]. For the synthetic data sets, we used the datasets generated in [5]; Synth10, Synth20, and Synth50 as they satisfy our requirements. Furthermore, we compared our solution with LOF [3] adapted to the stream setting (LOF-Stream), SGRMD [5], and xStream [8] algorithms. These approaches were used with the configuration recommended by their authors. Except for approaches using LOF, such as ours, LOF-stream, and SGRMD, we repeat the calculation with parameter $K=\{1, 5, 10, 20, 50, 100\}$ and report the best result in terms of effectiveness. We average every result from 5 runs. To assess the effectiveness, we used the AUC and average precision (AP). AUC describes how well outliers are ranked relatively to inliers. AP indicates whether the model can correctly identify inliers. To measure efficiency, we used the processing time in seconds. We performed additional comparisons in terms of outlier scoring and subspace searching time

---

[1] http://kdd.ics.uci.edu/databases/kddcup99/kddcup99.html.
[2] https://archive.ics.uci.edu/ml/datasets/PAMAP2+Physical+Activity+Monitoring.

for our algorithm and SGMRD. We also implemented them in python to effectively assess the time. As for xStream and LOF-stream, we used the PySAD framework implementation [22]. All the experiments were conducted on a laptop Intel (R) Core(TM) i7 CPU @2.80 GHz, 8 GB RAM, Windows 10 (Professional).

**Table 1.** Datasets

| Dataset | Dimension | Instances | Outlier percentage |
|---------|-----------|-----------|--------------------|
| *Real datasets* | | | |
| Kddcup99 | 38 | 25000 | 7.12 |
| Activity | 51 | 22253 | 10 |
| *Synthetics datasets* | | | |
| Synth10 | 10 | 10000 | 0.86 |
| Synth20 | 20 | 10000 | 0.88 |
| Synth50 | 50 | 10000 | 0.81 |

## 5.2   Results and Analysis

During the experimentation, we relied on different scenarios to evaluate the performance of our approach. First, we assessed the impact of the parameters on the performance. We set the number of BGSA iterations *nbiter* to 10 S, after obtaining the best parameter setting giving a trade-off between efficiency and effectiveness, we compared our approach with the chosen competitors.

**Results of the Proposed Approach:** In this experiment, we assessed the efficiency and effectiveness of our approach under different parameter settings. First, for every dataset, we fixed the window size to 1000, population size *pop* and *topQ* subspaces to $d/2$. Additionally, we set the dependency estimators' parameters to the value recommended in [5]. We only varied $K$ to obtain the value giving the best effectiveness. In the next scenario, we used the best $K$ value for each dataset and varied *pop* and *topQ* according to these combinations:{ *(topQ = d, pop = d/2), (topQ = d/2, pop = d), (topQ = pop = d)*}. We set the upper bound for both parameters to the size of the original space as they should not exceed it. Further, we set the lower bound to half the search space.

*Efficiency Results:* From the performed evaluation, we noticed that our processing time increases with dimensionality, population size *pop*, *topQ* subspaces, and $K$ value. This is evident since the increase in these values implies more search space. The subspace search and scoring time make up the approach's processing time. Figure 2 describes the subspace search time of the three synthetic datasets. They are sufficient to reveal the impact of these parameters on the processing time. We can see that; the dimensionality has the biggest impact on the time consumption comparing the other parameters followed by the population size. Yet, the topQ parameter has less impact. As for the scoring time, it increases with increasing $K$ values, and it is also affected by the *topQ* subspaces where outliers will be detected since the latter vary with the number of dimensionalities.

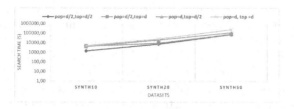

**Fig. 2.** Our approach search time with different parameter settings

*Effectiveness Results:* Figure 3 resumes the AUC of the different datasets with varying dimensionality, *pop* and *topQ* values in the best K setting. As we can see, the ability of the algorithm to correctly detect outliers decreases with increasing dimensionality and this is an artifact of the curse of dimensionality for the three synthetic datasets. Yet, it showed the best performance with Activity dataset having 50 dimensions and the worst performance with KDDcup99 dataset having 38 dimensions. Concerning the AP, our algorithm performs well across all datasets, with values between 0,96 and 0,99. As for the *pop* and *topQ* parameters, we noticed that they do not have a big impact on the results as the AUC and AP vary slightly from case to case. According to the obtained results, the scenario where *pop* and *topQ* are equal to *d/2* gives a trade-off between efficiency and effectiveness. In this setting, we have the lowest processing time and an acceptable AUC. For this reason, we choose this configuration to compare our algorithm with other competitors.

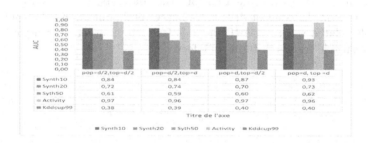

**Fig. 3.** AUC of our approach with different parameter settings

**Comparison with the Competitors:** We first compared our algorithm with the closely related approach SGMRD. It finds the optimal subspace set composed of the best subspace per dimension. The subspace relevance is assessed using the deviation between the marginal and conditional distribution of every dimension compared to the other features in the subspace. Contrary, in our case the subspace relevance is the average deviation of random features in the subspace. Both approaches use statistical tests to test this deviation. In our experiments, we chose the Kolmogorov-Smirnov test. SGRMD adopts different monitoring strategies to obtain the set of subspaces in every window slide. We chose the MPT strategy recommended by the authors [5]. We first compared the subspace search strategies. As shown in Fig. 4, our strategy is faster than the one used in SGRMD, although it performs the search only when the window slides for specific features. This difference is also clear in the scoring time since we select the top $d/2$ subspaces instead of $d$ subspaces in SGRMD. All of this has led to a decrease in the whole processing time compared to SGMRD. As well, SGMRD requires a further monitoring and decision time corresponding to the time required for subspace update and selection. As for the effectiveness, we noticed comparable performance. Yet, we outperform SGMRD in the data with the lowest dimensionality (synth10) and Activity dataset. On his side, SGMRD slightly outperforms our approach in the other datasets. The increase ranges between 0,01% and 0,03% for the AUC. Considering the other approaches, they outperform our approach in terms of processing time (see Fig. 5). Yet, this comes with a noticeable performance degradation in terms of effectiveness. This is typical, especially for the random subspace-based method xStream which avoid the costly subspace search strategies. All of this is illustrated in Fig. 6. According to the performed experimentation, we can say that our approach presents good results in terms of both efficiency and effectiveness. The use of BGSA permits exploring the feature space efficiently. Yet, it requires critical parameter tuning especially *pop, nbiter,* and *topQ* which has a great impact on the performance. During the experimentation, we noticed that the combination *(pop = d/2 and topQ = d/2)* gives the best compromise between time and accuracy. However, this needs to be validated for other datasets. Furthermore, we need to focus more on the concept drift handling. Overall, our approach outperforms the closely related work SGRMD, as it is faster by 2 orders of magnitude and gives comparable results in terms of effectiveness.

**Fig. 4.** SGRMD vs our approach search time

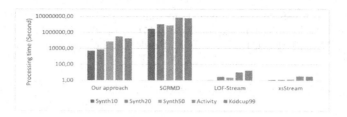

**Fig. 5.** Processing time of the approaches

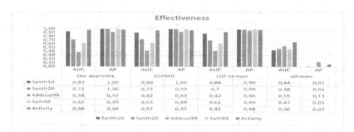

**Fig. 6.** Effectiveness of the approaches

## 6    Conclusion and Future Works

The availability of high-dimensional data streams from various domains as well as the great importance of outlier detection, demands the development of tools capable of detecting outliers. In this work, we have proposed an approach for outlier detection within relevant subspaces obtained by using an adapted binary GSA algorithm. Different from typical general subspace search methods [5, 6, 19], where subspace search is performed separably from the downstream task. Our approach keeps the notion of generality in the subspaces definition and couples the subspace search with outlier detection in every window. Therefore, it still be used with any task by replacing the outlier detection algorithm. We have compared our solution to SGRMD, xStream, and LOF-stream in terms of efficiency and effectiveness. Results revealed that our solution has good performance and outperforms its competitors in terms of effectiveness. In the future, we intend to improve the current algorithm by using other stream processing structures, and different statistical tests, as well as handling the concept drift more efficiently. Furthermore, proposing a new approach based on deep learning that will take into account the various questions that we have dealt with in this study.

## References

1. Aggarwal, C.C.: High-dimensional outlier detection: the subspace method. In: Outlier Analysis, pp. 149–184. Springer, Cham (2017). https://doi.org/10.1007/978-3-319-47578-3_5

2. Aggarwal, C.C., Sathe, S.: Theoretical foundations and algorithms for outlier ensembles. ACM SIGKDD Explor. Newsl. **17**(1), 24–47 (2015)
3. Breunig, M.M., Kriegel, H.P., Ng, R.T., Sander, J.: Lof: identifying density-based local outliers. In: Proceedings of the 2000 ACM SIGMOD International Conference on Management of Data, pp. 93–104 (2000)
4. Fouché, E., Böhm, K.: Monte carlo dependency estimation. In: Proceedings of the 31st International Conference on Scientific and Statistical Database Management, pp. 13–24 (2019)
5. Fouché, E., Kalinke, F., Böhm, K.: Efficient subspace search in data streams. Inf. Syst. **97**, 101705 (2021)
6. Keller, F., Muller, E., Bohm, K.: HiCS: high contrast subspaces for density-based outlier ranking. In: 2012 IEEE 28th International Conference on Data Engineering, pp. 1037–1048. IEEE (2012)
7. Kraskov, A., Stögbauer, H., Grassberger, P.: Estimating mutual informationn. Phys. Rev. E **69**(6), 066138 (2004)
8. Manzoor, E., Lamba, H., Akoglu, L.: xStream : outlier Dete x 'ion in feature-evolving data streams. In: Proceedings of the 24th ACM SIGKDD International Conference on Knowledge Discovery & Data Mining, pp. 1963–1972 (2018)
9. Nguyen, H.V., Müller, E., Böhm, K.: 4s: Scalable subspace search scheme overcoming traditional apriori processing. In: 2013 IEEE International Conference on Big Data, pp. 359–367. IEEE (2013)
10. Nguyen, H.V., Müller, E., Vreeken, J., Keller, F., Böhm, K.: CMI: an information-theoretic contrast measure for enhancing subspace cluster and outlier detection. In: Proceedings of the 2013 SIAM International Conference on Data Mining, pp. 198–206. SIAM (2013)
11. Pevnỳ, T.: Loda: Lightweight on-line detector of anomalies. Mach. Learn. **102**(2), 275–304 (2016)
12. Rashedi, E., Nezamabadi-Pour, H., Saryazdi, S.: GSA: a gravitational search algorithm. Inf. Sci. **179**(13), 2232–2248 (2009)
13. Rashedi, E., Nezamabadi-Pour, H., Saryazdi, S.: BGSA: binary gravitational search algorithm. Nat. Comput. **9**(3), 727–745 (2010)
14. Sathe, S., Aggarwal, C.C.: Subspace outlier detection in linear time with randomized hashing. In: 2016 IEEE 16th International Conference on Data Mining, pp. 459–468. IEEE (2016)
15. Siegel, S., Castellan Jr., N.: Jr. Nonparametric Statistics for the Behavioral Sciences, 2nd edn. Mcgraw-Hill Book Company, New York (1988)
16. Souiden, I., Omri, M.N., Brahmi, Z.: A survey of outlier detection in high dimensional data streams. Comput. Sci. Rev. **44**, 100463 (2022)
17. Su, S., Sun, Y., Gao, X., Qiu, J., Tian, Z.: A correlation-change based feature selection method for IoT equipment anomaly detection. Appl. Sci. **9**(3), 437 (2019)
18. Tran, L., Fan, L., Shahabi, C.: Outlier detection in non-stationary data streams. In: Proceedings of the 31st International Conference on Scientific and Statistical Database Management, pp. 25–36 (2019)
19. Vanea, A., Emmanuel, M., Keller, F., Klemens, B.: Instant selection of high contrast projections in multi-dimensional data streams. In: Proceedings of the Workshop on Instant Interactive Data Mining (IID 2012) in Conjunction with ECML PKDD (2012)
20. Wen, J., et al.: On-line anomaly detection with high accuracy. IEEE/ACM Trans. Netw. **26**(3), 1222–1235 (2018)
21. Xue, L., Chen, Y., Luo, M., Peng, Z., Liu, J.: An anomaly detection framework for time-evolving attributed networks. Neurocomputing **407**, 39–49 (2020)

22. Yilmaz, S.F., Kozat, S.S.: Pysad: a streaming anomaly detection framework in python. arXiv preprint arXiv:2009.02572 (2020)
23. Zhang, J., Gao, Q., Wang, H., Liu, Q., Xu, K.: Detecting projected outliers in high-dimensional data streams. In: Bhowmick, S.S., Küng, J., Wagner, R. (eds.) DEXA 2009. LNCS, vol. 5690, pp. 629–644. Springer, Heidelberg (2009). https:// doi.org/10.1007/978-3-642-03573-9_53
24. Zhang, L., Lin, J., Karim, R.: Sliding window-based fault detection from high-dimensional data streams. IEEE Trans. Syst. Man Cybern Syst. **47**(2), 289–303 (2017)

# Classification, Clustering
# and Recommendation

# Signal Classification Using Smooth Coefficients of Multiple Wavelets to Achieve High Accuracy from Compressed Representation of Signal

Paul Grant[✉] and Md. Zahidul Islam

School of Computing and Mathematics, Charles Sturt University,
Bathurst, NSW 2795, Australia
{pgrant,zislam}@csu.edu.au

**Abstract.** Classification of time series signals has become an important construct and has many practical applications. With existing classifiers, we may be able to classify signals accurately; however, that accuracy may decline if using a reduced number of attributes. Transforming the data and then undertaking a dimensionality reduction may improve the quality of the data analysis, decrease the time required for classification and simplify models. We propose an approach, which chooses suitable wavelets to transform the data, then combines the output from these transformations to construct a dataset by applying ensemble classifiers. We demonstrate this on different data sets across different classifiers and use different evaluation methods. Our experimental results demonstrate the effectiveness of the proposed technique, compared to the approaches that use either raw signal data or a single wavelet transform.

**Keywords:** Signal classification · Energy distribution · Wavelets · Ensembles

## 1 Introduction

Classification is a methodology that determines categories within a collection of data, which then allows the analysis of large data sets. Using decision tree induction for classification has become a common approach in machine learning. Decision trees attempt to allot symbolic decisions to new samples and provide a visual representation of the derived rule set. Automatic rule induction systems for inducing classification rules have already proved valuable as tools for assisting in knowledge acquisition for expert systems [1]. If a labelled dataset is implemented to train algorithms that classify the data, this is considered an instance of supervised learning.

In most research, a signal is related to the main topic of interest. It is common to apply wavelet transforms to that signal. Wavelets may be used to reduce

W. Chen et al. (Eds.): ADMA 2022, LNAI 13726, pp. 173–186, 2022.
https://doi.org/10.1007/978-3-031-22137-8_13

noise[1] in time series data, with the aim to provide better classification performance which may be achieved after conducting a wavelet transform on the original data, also compression using wavelets speeds up the classification process [2]. Wavelet coefficients in preference to raw data has been previously mentioned [3], where a single wavelet filter was used. Using wavelet transformed data to select various frequency levels within the signal, enabling reduction or elimination of specific inherent frequencies (*and noise*), to then undertake classification has proved successful [4]. Multiple Wavelets have been previously used [5,6] and applied to noise reduction as well as Image and ECG data compression. However finding a close to optimal wavelet filter (*or filters*) to use which provides a close to the best representation of the underlying data, is not a trivial task.

In this paper a method is presented which first determines a manner to select suitable wavelets, then implements and combines these selected wavelets to not only transform but reduce the dimension of the data by using subset of the available wavelet coefficients and yet still maintains or improve upon classification accuracy when using decision trees as classifiers. This paper built upon and results from previous work [5] where an approach using multi-wavelet decomposition to construct a set of attributes composed of the full set of wavelet coefficients derived from different wavelet filters to enable noisy signal classification.

From the wavelet transforms the smooth components resulting from multiple wavelets are utilised to provide the attributes used for the classification process. This is demonstrated by Classification across three different sets of signals[2], where classifiers are applied to both raw and transformed signals. It is shown that accuracy may be improved or even maintained with reduced elements in the attribute space by using wavelets, and even further enhanced by multiple wavelet transformed data combined with decision tree ensembles.

## 2   Wavelets

Wavelets are linear transforms, see Definition 1, that can be used to segment the data into separate non overlapping frequency bandwidths. They have advantages where the signal has discontinuities and sharp spikes. Wavelets have been applied in various areas such as image compression, turbulence, human vision, radar and digital signal processing [7]. A wavelet transform is the representation of a function by wavelet coefficients.

**Definition 1.** *Linear Transform*
   *A linear transformation is a transformation $T : R^n \to R^m$ satisfying*

$$T(u + v) = T(u) + T(v)$$
$$T(cu) = cT(u)$$

*for all vectors $u, v$ in $R^n$ and all scalars $c$.*

---

[1] Noise here includes missing or misclassification of values as well as other induced random fluctuations in the data.

[2] We designate these signals as raw or unmodified Data.

By sampling a wavelet discretely then applying that filter to the raw data, the result is a Discrete Wavelet Transform, (**DWT**). The DWT allows analysis *(decompose)* of a time series, segmented into coefficients **W**, from which its possible to synthesise *(reconstruct)* the original series [8]. The DWT in principle provides more information than the time series raw data points in classification because the DWT locates where the signal energies are concentrated in the frequency domain [3]. Following is an overview of the DWT and the Multiple Discrete Wavelet transform (**MDWT**), adapted from [5].

## 2.1 DWT

Let a sequence $X_1, X_2, \ldots, X_n$ represent a time series of $n$ elements, denoted as $\{X_t : t = 1, \ldots, n\}$ where $n = 2^J : J \in \mathbb{Z}^+$, $X_t \in \mathbb{R}$, the discrete wavelet transform is a linear transform which decomposes $X_t$ into $J$ levels giving $n$ DWT coefficients; the wavelet coefficients are obtained by premultiplying $X$ by $\mathcal{W}$.

$$\mathbf{W} = \mathcal{W}X \tag{1}$$

- **W** is a vector of DWT coefficients ($j$th component is $W_j$)
- $\mathcal{W}$ is $n \times n$ orthonormal transform matrix; i.e.,
  $\mathcal{W}^T \mathcal{W} = I_n$, where $I_n$ is $n \times n$ identity matrix
- inverse of $\mathcal{W}$ is its transpose, $\implies \mathcal{W}\mathcal{W}^T = I_n$
  $\therefore \mathcal{W}^T \mathbf{W} = \mathcal{W}^T \mathcal{W}X = X$

**W** is partitioned into $J + 1$ subvectors

$$\mathbf{W} = [\mathbf{W}_1, \mathbf{W}_2, \ldots, \mathbf{W}_j, \ldots, \mathbf{W}_J, \mathbf{V}_J] \tag{2}$$

- $\mathbf{W}_j$ has $n/2^j$ elements[3]
- $\mathbf{V}_J$ has one element[4]

conversely the synthesis equation for the DWT is:

$$X = \mathcal{W}^T \mathbf{W} = [\mathcal{W}_1^T, \mathcal{W}_2^T, \ldots, \mathcal{W}_J^T, \mathcal{V}_J^T] \begin{bmatrix} \mathbf{W}_1 \\ \mathbf{W}_2 \\ \vdots \\ \mathbf{W}_J \\ \mathbf{V}_J \end{bmatrix} \tag{3}$$

Equation 3 leads to additive decomposition which expresses $X$ as the sum of $J + 1$ vectors, each of which is associated with a particular scale $\mathcal{T}_j$

$$X = \sum_{j=1}^{J} \mathcal{W}_j^T \mathbf{W}_j + \mathcal{V}_J^T \mathbf{V}_J \equiv \sum_{j=1}^{J} \mathcal{D}_j + \mathcal{S}_J \tag{4}$$

---

[3] note: $\sum_{j=1}^{J} \frac{n}{2^j} = \frac{n}{2} + \frac{n}{4} + \cdots + 2 + 1 = 2^J - 1 = n - 1$.
[4] Decomposing $X_t : n = 2^J$, to level $J_0 : 1 \leq J_0 \leq J$ then $\mathbf{V}_{J_0}$ has $n/2^{J_0}$ elements.

- $\mathcal{D}_j \equiv \mathcal{W}_j^T \mathbf{W}_j$ is portion of synthesis due to scale $\mathcal{T}_j$, called the $j$th 'detail'
- $\mathcal{S}_J \equiv \mathcal{V}_J^T \mathbf{V}_J$ is a vector called the 'smooth' of the $J$th order [8].

*Remark 1.* If the DWT is used to decompose $X_t$ to only the first level, then from Eq. 4 the transformed signal consists of $\frac{n}{2}$ detail coefficients and $\frac{n}{2}$ smooth coefficients. Similarly, decomposition to second level, $\frac{3n}{4}$ detail coefficients and $\frac{n}{4}$ smooth coefficients. *(provided signal length n, is a factor of 2 or 4 respectively.)*

## 2.2  MDWT

To construct a Multiple Discrete Wavelet Transform from a time series $\{X_t : t = 1, \ldots, n\}$ use the DWT to deconstruct the signal to level $J_0 : J_0 \in \mathbb{Z}^+$, $X_t \in \mathbb{R}$, choose $N$ different DWT filters and apply to $X_t$ sequentially. From Eq. 4 this results in:

$$\sum_{i=1}^{N} \left[ \sum_{j=1}^{J_0} \mathcal{D}_{ij} + \mathcal{S}_{iJ_0} \right] \tag{5}$$

giving a sequence of vectors, where each $\text{DWT}_i$ is $\mathcal{D}_{i1}, \mathcal{D}_{i2}, \ldots, \mathcal{D}_{iJ_0}, \mathcal{S}_{iJ_0}$ *(which consists of the wavelet coefficients resulting from level Jo decomposition)*. If we choose only the smooth coefficients $s_{i,k} \in \mathcal{S}_{iJ_0} : 1 \le k \le \frac{n}{2^{J_0}}$ then

$$MDWT = s_{1,1}, s_{1,2}, \ldots, s_{1,\frac{n}{2^{J_0}}}, s_{2,1}, s_{2,2}, \ldots, s_{2,\frac{n}{2^{J_0}}}, \ldots, s_{N,1}, s_{N,2}, \ldots, s_{N,\frac{n}{2^{J_0}}} \tag{6}$$

*Remark 2.* This has $\frac{N}{2^{J_0}}$ times as many elements as in $X_t$.

## 2.3  Energy Distribution

Define the energy within a signal $X_t$ as the squared norm $||X||^2$, see Definition 2, then it is possible to derive the energy distribution in the signal via a normalised partial energy sequence; **NPES** [8].

For a signal $\{X_t : t = 1, \ldots, n\}$ if we reorder by squared magnitude such that,

$$|x_{(1)}|^2 \ge |x_{(2)}|^2 \ge \cdots \ge |x_{(n)}|^2.$$

This enables us to compute the NPES[5], $: n \ge M$

$$C_M \equiv \frac{\sum_{j=1}^{M} |x_{(j)}|^2}{\sum_{j=1}^{n} |x_{(j)}|^2} = \frac{\text{energy in largest M terms}}{\text{total energy in signal}} \tag{7}$$

and similarly for a NPES of wavelet[6] coefficients.

---

[5] Which permits construction of a plot of cumulative energy% in the signal (or representation of), against the number of data points, see Fig. 2.

[6] As the DWT is an orthonormal transform, the energy in the transform *(consisting of all J + 1 subvectors)* equates to the energy in the signal.

**Definition 2.** *Vector Norm*

*Given an n-dimensional vector $X = x_1, x_2, \ldots, x_n$*

*The vector norm $||X||^p$ for $p = 1, 2, \ldots$ is defined as*

$$||X||^p \equiv \left\{ \sum_i |x_i|^p \right\}^{1/p}$$

## 3   Proposed Technique

Presented here is a method, "Multi-Wavelet Compression Signal Classification", (**MWCSC**) which utilises wavelet transforms of the signal data, First we introduce the methodology of the main concepts followed by the advantages provided, then a succinct overview of the steps required to implement this technique.

- From the dataset used, consider the different classes within the signal data. Using wavelet transforms of these classes, map the distribution of the original signal energy/information against the corresponding representation provided by the wavelet coefficients. Use the NPES, see Sect. 2.3, to select a group of suitable wavelet filters to transform the data. NPES is an existing methodology that enables us to choose wavelets that may better represent the distribution of energy in the signal. The aim is to use the NPES to provide a sparse representation of the original data, i.e. many coefficients have low values.

- From multiple wavelets, construct a Multiple Discrete Wavelet transform, (MDWT, see Sect. 2.2). The rationale for using multiple wavelets in the transform is, to include wavelets that are either symmetric, have short support, provide higher accuracy and are orthogonal. No single wavelet may provide all such properties simultaneously [9].

The MDWT provides the wavelet coefficients used to form the attributes on which the classification methods derive their rule set from. For a transform consisting of a single wavelet, one would obtain the same number of data points (*here wavelet coefficients*) as provided in the original signal. Here only the smooth wavelet coefficients $S_J$, see Eq. 4, are chosen. This enables reduction in the number of coefficients used to form the attributes for classification.

- At the first level of wavelet decomposition, the MDWT consisting of $N$ wavelets (*smooth coefficients only at $J = 1$*), then from Eq. 5 we would have $\frac{N}{2}$ times the number of original data points, which could be considered as attributes. Similarly at the second level of decomposition, MDWT consisting of $N$ wavelets contains $\frac{N}{4}$ time the number of original data points.

Similarly by reducing the overall number of attributes we may speed up the classification process [2]. Our method enables us reduce the overall number of attributes yet still maintain or even improve classification accuracy. For evaluation of classification we utilise a different method for each dataset chosen, to highlight the results of this approach regardless of the manner used for grading the results.

## 3.1   Advantages

Our methodology does require some additional computation, however

- The NPES identifies suitable wavelets to use,
- By using only smooth coefficients from the wavelet transform, it is possible reduce the number of attributes required for the classification process
- Using multiple wavelets increases accuracy.

While the construction of the MDWT is not a complicated procedure, the results obtained by using only the smooth coefficients highlight that the extra computation is worthwhile, providing us with similar levels of accuracy yet reducing number of required attributes, *(a form of data reduction or compression)*.

We are providing a set of attributes for the classifiers where a considerable amount of energy in the signal is then represented by a smaller of number of components. The classification methods combined with MDWT tend to return smaller sets of decision rules (or smaller less complex trees) to arrive at their final rule set [5]. Using our MDWT compared to a single wavelet transform, see Sects. 4.2 to 4.4, the gain in classification accuracy when used with single decision tree methods is apparent and even more so when used with ensemble classification methods, The software used in construction and application is freely available on-line : R [10], the R package WMTSA [11] and WEKA [12].

## 3.2   Steps in the Proposed Technique: MWCSC

**Step 1** *Wavelet Selection*.
    From a set of Time series

$$\{X_{t_i} : t = 1, \ldots, n, \; i = 1, \ldots, K\}$$

which has a distinct number of classes, compare the energy distribution of each of the signal classes as represented by varied wavelet transforms, using the NPES described in Sect. 2.3.

**Step 2** *Discrete Wavelet Transforms*
    For each single time series $X_{t_i}$ take the DWT of the signal using a different wavelet filter (as determined by the NPES in **Step 1**) for each of the transforms and extract the wavelet smooth coefficients $S_{J_0}$. Use the same decomposition level for each DWT.

**Step 3** *DataSet Construction via MDWT*
    **3a.** Construct a new data series, *(MDWT, see* Sect. 2.1) placing each of the individual vectors of the wavelet smooth coefficients (resulting from each DWT, with level of decomposition $= J_0$) in a continuous sequence, one after each other, see Eq. 6.

    **3b.** For each each of these MDWT, stack each transformed signal, to form a data array or matrix as depicted in Table 1.

**Table 1.** Array of transformed data as developed by MDWT

$\text{MDWT}(X_{t_1}):$ $s_{1,1_1},$ $s_{1,2_1},$ $\dots,$ $s_{1,\frac{n}{2^{J_0}}_1},$ $s_{2,1_1},$ $\dots,$ $s_{2,\frac{n}{2^{J_0}}_1},$ $\dots,$ $s_{N,1_1},$ $\dots,$ $s_{N,\frac{n}{2^{J_0}}_1}$

$\text{MDWT}(X_{t_2}):$ $s_{1,1_2},$ $s_{1,2_2},$ $\dots,$ $s_{1,\frac{n}{2^{J_0}}_2},$ $s_{2,1_2},$ $\dots,$ $s_{2,\frac{n}{2^{J_0}}_2},$ $\dots,$ $s_{N,1_2},$ $\dots,$ $s_{N,\frac{n}{2^{J_0}}_2}$

$\vdots$

$\text{MDWT}(X_{t_K}):$ $s_{1,1_K},$ $s_{1,2_K},$ $\dots,$ $s_{1,\frac{n}{2^{J_0}}_K},$ $s_{2,1_K},$ $\dots,$ $s_{2,\frac{n}{2^{J_0}}_K},$ $\dots,$ $s_{N,1_K},$ $\dots,$ $s_{N,\frac{n}{2^{J_0}}_K}$

**Step 4 *Build a Classifier***

Using the new data as generated in previous steps, apply ensemble classifiers.[7] to the transformed data.

# 4 Experimental Results

The data used was sourced from the UCR time series archive [13]. Here we simply chose three data sets, each of different length and number of records, see Table 2. These data sets exhibit widely different levels of smoothness[8] when compared to each other. We also use 5 Tree based classifiers from WEKA.

## 4.1 Classification Methods Used

We utilise the following tree based classifiers[9].

- **J48** a decision tree is an extension of ID3. Some additional features of J48; accounting for missing values, decision trees pruning, continuous attribute value ranges and derivation of rules [14].
- **Random Forest\*** Class for constructing a forest of random trees. [15].
- **ForestPA\*** Decision forest algorithm Forest PA, using bootstrap samples and penalised attributes [16].
- **SysFor\*** Decision forest algorithm SysFor, a systematically developed forest of multiple decision trees [17].
- **SimpleCart** Classification and Regression Tree, Class implementing minimal cost-complexity pruning [18].

---

[7] We also applied single Decision tree classifiers to the MDWT data as a baseline to compare with ensemble classifiers.

[8] Smoothness defined here as: standard deviation of the of first differences of a time series elements. i.e. standard deviation of $(X_S) : X_S = x_1 - x_2, x_2 - x_3, \dots, x_{n-1} - x_n$.

[9] For the Ensemble Classifiers\* throughout our experiment we set number of trees used to 100, no fine tuning of parameters was undertaken.

**Table 2.** Dataset descriptions

| Data name | Test set | Training set | No.of classes | Length |
|-----------|----------|--------------|---------------|--------|
| ArrowHead | 35 | 176 | 3 | 251 |
| Mallet | 2345 | 55 | 8 | 1024 |
| FordA | 1320 | 3601 | 2 | 500 |

## 4.2 Arrowhead Data

The first dataset consists of profiles of Arrowheads where we utilise the profiles as time series. The dataset has three classes, see Fig. 1, we combined the test and training sets together to give 211 records This enabled us to use *Ten-fold Cross-Validation* with WEKA for evaluation.

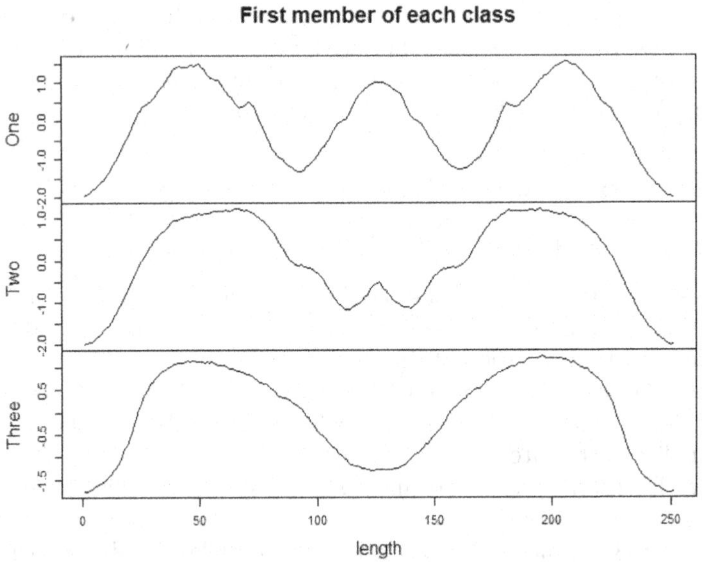

**Fig. 1.** Profile of each Arrowhead class

Following the **MWSCS** procedure, initially we applied various wavelet filters to each of the arrowhead classes to plot the NPES. Given the 251 data points in length, the wavelet transform deconstructing the signal to the first level, provided us with 125 detail and 125 smooth coefficients but also an "Extra" class of coefficients, how this class is calculated and implemented in the transform, by the chosen software is fully described in [8, Chap 4.11].

These additional coefficients are simply untransformed data. For this dataset as the "Extra" coefficients contain considerable signal energy/information, they are combined with the smooth coefficients in constructing our new data. For

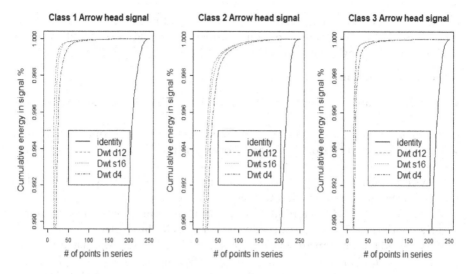

**Fig. 2.** NPES of each Arrowhead Class using Wavelet Transforms

each transformed signal, the Extra coefficients contain an average of 1.41% of signal energy per transformed signal at level 1 and 3.45% at level 2.

The NPES of samples taken from each Arrowhead class highlight suitable wavelet filters to represent the energy in a smaller number of data points as shown in Fig. 2. Little difference is apparent between the wavelet filters shown, however the filters s16 and d12 represent the energy slightly more efficiently than d4. Hence we use these wavelet filters to transform our signal data and build our transformed dataset, this however provides no compression in the signal representation as we will still have same number of wavelet coefficients as data points in the original signal.

By selecting only the smooth coefficients[10] at the respective level we are able to reduce the number of attributes within our data and maintain classification accuracy as shown in Table 3. Here classification accuracy is defined as the number of correctly Classified Instances with respect to the dataset's Class labels which are initially provided within the original data.[11]

### 4.2.1 Arrowhead Results
Table 3 demonstrates the benefits of the wavelet transforms, s16 wavelet filter performs slightly better than d4[12] for the ensemble classifiers, The use of the smooth coefficients resulting from s16 (*and when combined with d12*), maintain accuracy while using reduced numbers of attributes.

---

[10] In this instance we also include the Extra coefficients as they represent considerable signal energy.

[11] Using Accuracy % here is a suitable metric, as the dataset is reasonably balanced i.e. Class 1 has 81 records, Class 2 and 3 have 65 records each.

[12] As indicated by the NPES, Fig. 2.

Using the smooth wavelet coefficients and including the Extra coefficients, where these Extra coefficients contain considerable signal energy, still provide considerable accuracy, especially when using the ensemble classifiers.

**Table 3 Description:**

- **Classifier** The first column is the list of tree based classifiers used, see Sect. 4.1
- **Raw data** results from classifying original data, being 251 units in length
- **Wavelet d4** results from data transformed using Wavelet filter d4, signal decomposed to four levels, 248 units in length
- **Wavelet s16** results from data transformed using using Wavelet filter s16, signal decomposed to four levels, 248 units in length
- **Wavelet s16 $S_1$ + Extra** results from data transformed using s16 using only the smooth level 1 and Extra coefficients, 126 units in length
- **Wavelet s16 d12 combined $S_2$** results from data transformed using s16 and d12 filters, combining the smooth level 2 coefficients from both transforms to form a new data series , $62 + 62 = 124$ units in length. A MDWT, see Sect. 2.2
- **Wavelet s16 d12 combined $S_2$ + Extras** results from data transformed using s16 and d12 wavelet filters, combining the smooth level 2 and Extra coefficients from both transforms to form a new data series , $64 + 64 = 128$ units in length, again a MDWT.

**Table 3.** Classification of arrowhead data, *cross validation 10-fold*

| Classifier | Classification accuracy% | | | | | |
|---|---|---|---|---|---|---|
| | Raw data | Wavelet d4 4 levels | Wavelet s16 4 levels | Wavelet s16 $S_1$ + Extra | Waves s16 d12 combined $S_2$ | Waves s16 d12 comb. $S_2$ + Extra |
| *J48* | 75.35 | 76.03 | 74.41 | 79.14 | 77.72 | 72.72 |
| *Rforest* | 86.25 | 84.36 | 86.25 | 90.05 | 89.1 | 89.57 |
| *ForestPA* | 80.09 | 79.14 | 81.51 | 79.62 | 82.46 | 83.88 |
| *SysFor* | 82.46 | 81.52 | 81.99 | 83.41 | 84.36 | 84.36 |
| *SimpleCart* | 73.46 | 73.93 | 72.51 | 70.14 | 70.61 | 70.61 |

### 4.3  Mallat Data

From the UCR dataset, Mallat Curve data, 8 distinct classes of 1024 units in length, see Fig. 3. We combined the data, both testing and training sets to have a larger set containing 2400 records, *eight classes of 300 records each.*

Following the **MWCSC** procedure as with the Arrowhead data, using the NPES to determine suitable wavelets to represent the signal energy, we apply the classifiers to the wavelet transformed data. This time we use 20% of data for training and 80% for testing, (the actual records in these sets were chosen by WEKA).

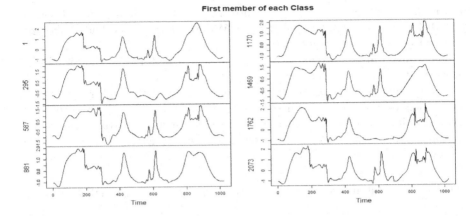

**Fig. 3.** Mallat data

### 4.3.1. Mallat Data, Results

Table 4 highlights results from the Classifiers. The original data is 1024 unit long, hence the full wavelet transform also 1024 units long. Using the smooth wavelet coefficients from $S_1$ results 512 units in length, similarly $S_2$ results in 256 units and $S_3$, 128 units.

Using the various levels of decomposition, increasing the effective compression of the signal transform we note that accuracy is maintained, especially with the ensemble classifiers. However when we combine the two $S_3$ levels (a MDWT) from different wavelets (as determined by the NPES), then the ensemble classifiers maintain considerable accuracy, given the reduced transformed signal size, $S_3 \times 2 = 128 + 128 = 256$ units. A small accuracy gain over the single wavelet transforms using $S_2$ or $S_3$.

**Table 4.** Classification of Mallat data, 20% training, 80% testing

| Classifier | Classification accuracy% | | | | | |
|---|---|---|---|---|---|---|
| | Raw data | Wavelet s16 10 levels | Wavelet s16 $S_1$ | Wavelet s16 $S_2$ | Wavelet s16 $S_3$ | Waves s16 d8 combined S3 |
| *J48* | 98.88 | 94.48 | 97.29 | 95.99 | 95.67 | 94.74 |
| *Rforest* | 97.18 | 98.85 | 98.33 | 98.01 | 98.07 | 98.25 |
| *ForestPA* | 96.61 | 97.34 | 97.39 | 97.23 | 97.81 | 97.86 |
| *SysFor* | 95.52 | 95.93 | 96.15 | 94.63 | 95.15 | 95.36 |
| *SimpleCart* | 95.57 | 95.0 | 96.3 | 95.93 | 94.17 | 94.53 |

### 4.4  Ford Data

From the UCR dataset, Ford data, 2 distinct classes of 500 units in length. here the classification problem is to diagnose whether a specific symption exists or not in an automotive subsystem. We use the original training and test sets as provided in the UCI data and follow the **MWCSC**. The test set, had 681 records in one class and 629 in the other.

NPES determined wavelet filters d16 and s20 as suitable. Results in Table 5 shows that accuracy is maintained even at higher levels of compression,(or attribute reduction). Using the smooth components at level 1, $S_1$ results in 250 units, using $S_2$ provides 125 units, $S_3$ gives 62 units and $S_3$ + Extra coefficient gives 63 units in length. Very little gain is evident in this last transform as only a minor levels of signal energy is contained within the Extra coefficients at the $3^{rd}$ level.[13]

**Table 5.** Classification of Ford data, 3601 training records, 1320 testing records

| Classifier | Classification accuracy% | | | | | |
|---|---|---|---|---|---|---|
| | Raw data | Wavelet d16 S1 | Wavelet d16 S2 | Wavelet d16 S3 | Waves d16 s20 combined S3 | Waves d16 s20 comb. S3 + Extra |
| *J48* | 56.13 | 58.03 | 56.44 | 52.42 | 58.71 | 58.18 |
| *Rforest* | 73.25 | 72.95 | 71.37 | 73.41 | 74.2 | 75.53 |
| *ForestPA* | 74.24 | 75.91 | 74.17 | 74.69 | 74.09 | 74.92 |
| *SysFor* | 62.27 | 59.17 | 59.59 | 63.79 | 64.47 | 60.38 |
| *SimpleCart* | 58.18 | 58.11 | 56.37 | 59.02 | 59.92 | 59.15 |

*Remark 3.* For the single tree method J48, It would appear that our choice of specific wavelets within the MDWT might not be crucial as no evidence of consistent additional gain between the two MDWT variations. However the choice of additional attributes (as provided by the MDWT) to train upon would seem to provide some minor gain over using a single Wavelet transform.

# 5    Conclusion

It has been shown previously that wavelets may be used for signal compression with little loss of accuracy [2]. Our MWCSC methodology utilizing the NPES and MDWT provides us with a method to determine suitable wavelets as well as add additional information to the attribute space. This enables us to use transformed data sets with smaller dimensions than the original data yet still provide similar or enhanced accuracy. From the NPES graphic (Fig. 2), we note that wavelet d4 is not as efficient at energy representation as wavelet s16, for the arrowhead dataset. This is similarly represented in Table 3 by comparing classification results, across the various classifiers of raw data as well as the data transformed by the d4 or the s16 wavelet filters.

Construction of the MDWT from suitable wavelets enhances the accuracy while offering an effective data reduction or compressed representation of the

---

[13] The average energy/information provided by the Extra coefficients at level $S_3$, per transformed signal is only 0.5% hence little if at all any gain in accuracy is achieved by including the Extra coefficients at this level.

signal to which the classifiers may be applied. Inclusion of "Extra" coefficients into the construction of the MDWT, where such coefficients contain considerable signal energy, adds additional accuracy for little extra computation, as they are an included class in the transform calculation where $\{X_t : t = 1, \ldots, n\}$, $n \neq 2^J : J \in \mathbb{Z}^+$.

# References

1. Clark, P., Niblett, T.: Induction in noisy domains. In: Progress in Machine Learning (from the Proceedings of the 2nd European Working Session on Learning), 11–30, Bled, Sigma Press, Yugoslavia (1987)
2. Li, D., Bissyand'e, T., Klein, J., Le Traon, Y.: Time series classification with discrete wavelet transformed data: insights from an empirical study. In: The 28th International Conference on Software Engineering and Knowledge Engineering (2016)
3. Fong, S.: Using hierarchical time series clustering algorithm and wavelet classifier for biometric voice classification. J. Biomed. Biotechnol. **2012**, Article ID 215019 (2012)
4. Grimaldi, M., Kokaram, A., Cunningham, P.: Classifying music by genre using wavelet packet transform and a round-robin ensemble. Comp Science Dept.; Electronic and Electrical Engineering Dept, Trinity College Dublin, Ireland (2002)
5. Grant, P., Islam, M.Z.: A novel approach for noisy signal classification through the use of multiple wavelets and ensembles of classifiers. In: Li, J., Wang, S., Qin, S., Li, X., Wang, S. (eds.) ADMA 2019. LNCS (LNAI), vol. 11888, pp. 195–203. Springer, Cham (2019). https://doi.org/10.1007/978-3-030-35231-8_14
6. Moazami-Goudarzi, M., Moradi, Md. H., Abbasabadi, S.: High performance method for electrocardiogram compression using two dimensional multiwavelet transform. In: IEEE 7th Workshop on Multimedia Signal Processing (2006)
7. Graps, A.: An Introduction to wavelets. IEEE Comput. Sci. Eng. Sum. **2**(2), 50–61 (1995)
8. Percival, D.B., Walden, A.T.: Wavelet Methods for Time Series Analysis, Cambridge University Press, New York (2000)
9. Nason, G.P.: Wavelet Methods in Statistics with R. Springer, New York. (2008). https://doi.org/10.1007/978-0-387-75961-6
10. R Core Team: R: A language and environment for statistical computing. R Foundation for Statistical Computing, Vienna, Austria (2013). https://www.r-project.org
11. Constantine, W., Percival, D.: WMTSA: Wavelet Methods for Time Series Analysis. R package version 2.0-3 (2017)
12. Frank, E., Hall, M.A., Witten I.H.: The WEKA Workbench. Online Appendix for "Data Mining: Practical Machine Learning Tools and Techniques, 4th edn. Morgan Kaufmann, New York (2016)
13. Chen, Y., et al.: The UCR time series classification archive. October 2018. https://www.cs.ucr.edu/~eamonn/time_series_data_2018
14. Quinlan R.: C4.5: Programs for Machine Learning. Morgan Kaufmann Publishers, San Mateo (1993)
15. Breiman, L.: Mach. Learn. **45**, 5–32 (2001). https://doi.org/10.1023/A:1010933404324

16. Adnan, M.N., Islam, M.Z.: Forest PA: constructing a decision forest by penalizing attributes used in previous trees. Expert Syst. Appl. **89**, 389–403 (2017)

17. Islam, Md. Z., Giggins, H.: Knowledge discovery through SysFor: a systematically developed forest of multiple decision trees, In: Proceedings of the Ninth Australasian Data Mining Conference (AusDM 2011)

18. Breiman, L., Friedman, J.H., Olshen, R.A., Stone, C.J.: Classification and regression trees. Cytometry **8**, 534–535 (1987)

# On Reducing the Bias of Random Forest

Md. Nasim Adnan$^{(\boxtimes)}$

Department of Computer Science and Engineering, Jashore University of Science and
Technology, Jashore 7408, Bangladesh
`nasim.adnan@just.edu.bd`

**Abstract.** Random Forest is one of the most popular decision forest
building algorithms that uses decision trees as the base classifier. Decision trees for Random Forest are formed from the records of a training
data set. This makes the decision trees almost equally biased towards the
training data set. In reality, testing data set can be significantly different
from the training data set. Thus, to reduce the bias of decision trees
and hence of Random Forest, we introduce a random weight for each of
the decision trees. We present experimental results on four widely used
data sets from the UCI Machine Learning Repository. The experimental results indicate that the proposed technique can reduce the bias of
Random Forest to become less sensitive to noisy data.

**Keywords:** Bias · Decision tree · Random Forest

## 1 Introduction

At the moment, our "Digital Universe" is experiencing an unprecedented growth.
More data was generated in the last two years than in the entire human history
before that [1]. Nowadays, sophisticated computer hardware technologies enable
us to store the generated data. This huge pool of stored data can be regarded as
a valuable resource if they can be analyzed effectively and automatically. Data
mining is collection of automated tasks to identify valid, novel, potentially useful
and ultimately understandable patterns in data [2]. Classification and clustering
are two widely used data mining tasks that are applied for knowledge discovery
and pattern understanding.

Conventionally, classification is tasked to generate a model (commonly known
as the classifier) that maps a set of non-class attributes to a predefined class
attribute from an existing data set [3]. In this paper, we consider a data set
$D$ as a two-dimensional table where rows are records $R = R_1, R_2, ..., R_n$ and
columns are attributes $A = A_1, A_2, ..., A_m, C$. We also consider that a data set
can have two types of attributes; numerical (such as Salary) and categorical (such
as Designation). Among all attributes, one categorical attribute is chosen to be
the class attribute ($C$) and the rest are considered to be non-class attributes.

There are different types of classifiers including Decision Trees [4,5], Bayesian
Classifiers [6], Artificial Neural Networks [7,8], and Support Vector Machines [9].
Among these classifiers, decision trees are quite popular as they can be easily
interpreted into more reasonable logic rules to help infer valuable knowledge [10].

W. Chen et al. (Eds.): ADMA 2022, LNAI 13726, pp. 187–195, 2022.
https://doi.org/10.1007/978-3-031-22137-8_14

There are many decision tree induction algorithms such as CART [4], C4.5 [11,12], SLIQ [13], SPRINT [14] and ComboSplit [15]. Most of these algorithms follow the structure of Hunt's algorithm [3]. According to Hunt's algorithm, a decision tree is induced in a recursive manner from the training data set, e.g. the data set where all records are labeled with class values. The induction process starts by selecting every non-class attributes to divide the training data set $D$ into a disjoint set of horizontal segments/partitions [11,12,16]. If the non-class attribute $A_i$ is categorical with $k$ different domain values i.e. $A_i = a_{i1}, a_{i2}, ..., a_{ik}$ (domain values of an attribute are the set of all possible values for the attribute) then $D$ is divided (split) into $k$ segments $D_1, D_2, ..., D_k$, where all records of a segment $D_j$ have the same value $a_{ij}$, and the records belonging to different segments have different values [11,12,17]. However, if the splitting attribute is numerical $A_i = [l, u]$ ($l$ is the lower limit and $u$ is the upper limit of the domain values of $A_i$) then the data set is typically divided into two segments; $D_1$, and $D_2$. All records in segment $D_1$ have values of $A_i$ "lower than or equal to" a splitting point $p$ and the records in the other segment $D_2$ have values higher than the splitting point $p$, where $l \leq p < u$ [11,12,18].

The reason behind this splitting is to create a comparatively purer class distribution in the succeeding partitions/segments than the class distribution within $D$. Therefore, for a numerical non-class attribute, all possible split points (i.e. all values between $l$ and $u$ present in the data set) are used to find out the split point that gives the best distribution of class values. Finally, the splitting attribute that gives the purest class distribution among all splitting attributes is selected as the test attribute. The process of selecting the test attribute continues recursively in each subsequent data segment $D_i$ until either every partition gets the purest class distribution or a stopping criterion is satisfied. By "purest class distribution" we mean the presence of a single class value $c_i \in C$ for all records.

In recent years, ensembles of classifiers have been studied rigorously by the research community in exploration for more accurate classification models [19–21]. One of the most interesting finding of these ongoing research is that ensemble methods work better with unstable classifiers such as a decision tree [3,22,23]. Decision forest is an ensemble of decision trees where an individual decision tree acts as the base classifier and the classification is performed by taking a vote based on predictions made by each decision tree of the decision forest [3,24].

Though decision trees are unstable yet they are entirely formed from the training data set. This phenomenon makes every decision tree extremely biased towards the training data set and consequently enables each decision tree to have remarkable classification performance on the examples of the training data set. In reality, testing data sets can be significantly different from the training data set. Thus, a classification model that is highly biased towards the training data set may not perform well on testing data sets. In literature, Geurts et al. [25] proposed "Extremely Randomized Trees" to make the forest more general and receptive to the testing data set. In [25], the authors proposed to select a random number of attributes (between 1 to the total number of attributes) and then a random cut point for each of the selected attributes i.e. independently from the target training data set. Among these attributes with their random cut points, the attribute having the highest test value is selected as the splitting attribute. In this way, the bias towards the training data set is desensitized. One significant problem of the Extremely Randomized Trees algorithm [25] is that it can be applied only on those data sets that have all-numerical attributes. This makes the application domain of the Extremely Randomized Trees algorithm [25] considerably smaller.

Random Forest [26] is a popular state-of-the-art decision forest building algorithm that is essentially a combination of Bagging [27] and Random Subspace [28] algorithms. Bagging generates new training data set $D_i$ iteratively where the records of $D_i$ are chosen randomly from the original training data set $D$. $D_i$ contains the same number of records as in $D$. Thus, some records from $D$ can be chosen multiple times and some records may not be chosen at all. This approach of generating a new training data set is known as bootstrap sampling. On an average, 63.2% of the original records are present in a bootstrap sample and the rest 36.8% are repeated [29,30]. The Random Subspace algorithm is then applied on each bootstrap sample $D_i(i = 1, 2, ..., k)$ in order to generate $T$ number of decision trees for the forest.

The Random Subspace algorithm randomly draws a subspace $f$ from the entire attribute space $m$ in order to determine the test attribute. Attributes in $f$ can either be drawn at the node level or at the decision tree (in short, tree) level. When drawn at the node level, attributes in $f$ may differ from one node to another in a tree; however when drawn at the tree level, attributes in $f$ remain the same for a tree. In the simplest form of Random Forest, attributes in $f$ are selected randomly at the node level and the size of $f$ is chosen to be $|f| = int(\log_2 |m|)+1$ [26]. The Random Forest algorithm can be applied on every type of data sets such as all-numerical, all-categorical and mixed (both numerical and categorical) data sets. Even though the Random Forest algorithm generates trees from the bootstrap samples $(D_i)$, the trees may still show almost equal biasness towards the training data set $(D)$ as the records of $D_i$s are necessarily the subset of that of $D$. Hence, in order to make Random Forest less biased towards training data sets and more robust on significantly different/noisy testing data sets, we propose the following technique.

## 2   The Proposed Technique

The proposed technique is very straightforward. We first generate trees using the Random Forest algorithm. Then we assign weights for each of the trees. In order to assign a weight for a tree, we first generate a random number from a uniform distribution in the interval of [0.00, 1.00]. For example, if the number 0.25 is generated randomly from the uniform distribution [0.00, 1.00], we assign 0.25 as the weight of the tree.

We use these weights to classify a new record of a testing data set whose class value is unknown. Original Random Forest generally uses Majority Voting [22, 26] for computing the classification result. In Majority Voting, each base learner (tree) determines the class value of a new record. Then the class value from the highest number of base learners (trees) is determined to be the classification result of the ensemble (Random Forest).

In the proposed technique, in order to obtain the classification result for the Random Forest, we use the associated weight of the trees. For example, if a tree with weight 0.25 predicts a class value $c_1$ for a new record, we consider the weight of the tree as the value of the vote for $c_1$. If another tree with weight 0.35 predicts $c_1$, then the total vote-value for $c_1$ will be $(0.25 + 0.35) = 0.60$. In this way, the class value with the highest vote-value is determined to be the classification result.

## 3    Experimental Results

We analyze the impact of the modified Majority Voting on reducing the bias of Random Forest. In this process, we select four widely used data sets that are publicly available from the UCI Machine Learning Repository [31]. The data sets used in the experimentation are listed in Table 1. We generate 100 trees for every decision forest since the number is considered to be large enough to ensure convergence of the ensemble effect [22, 32]. All the results reported in this paper are obtained using 10-fold Cross Validation (10-CV) [16, 33] for every data set. The best results are distinguished through bold-face.

**Table 1.** Description of the Data Sets

| Data set name | Non class attributes | Records | Distinct class values |
| --- | --- | --- | --- |
| Car Evaluation | 6 | 1728 | 4 |
| Glass Identification | 9 | 214 | 6 |
| Pima Indians Diabetes | 8 | 768 | 2 |
| Statlog Vehicle | 18 | 846 | 4 |

Classification accuracy is one of the most important performance indicators of any classifier [34, 35]. In this paper, we inject 0%, 10%, 20% and 30% noise in all the data sets used and analyze the impact on classification accuracy for both the original Random Forest (O_RF) and the Random Forest with Random-Weight Trees (RF_RWT). Next, we report the average classification accuracy for all data sets with all noise levels in Figs. 1, 2, 3, and 4.

To be more conclusive in comparison between O_RF and RF_RWT, we put aggregated classification accuracy for all noise levels in Table 2. From Table 2, it is evident that the introduction of random weighs on trees helps Random Forest to reduce the bias and become less sensitive to noisy data.

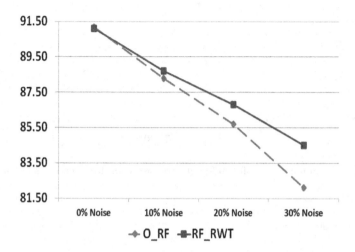

**Fig. 1.** Classification accuracy for different noise levels on Car Evaluation data set

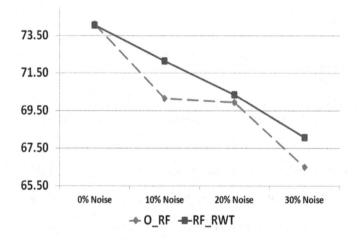

**Fig. 2.** Classification accuracy for different noise levels on Glass Identification data set

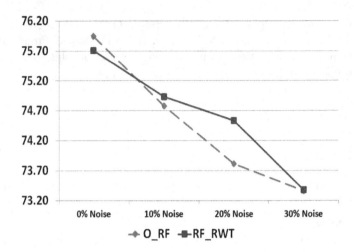

**Fig. 3.** Classification accuracy for different noise levels on Pima Indians Diabetes data set

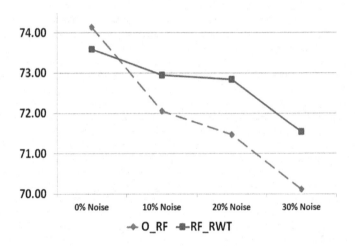

**Fig. 4.** Classification accuracy for different noise levels on Statlog Vehicle data set

**Table 2.** Aggregated classification accuracy for different noise levels

| Noise Level | O_RF | RF_RWT |
|---|---|---|
| 0% Noise | **78.85%** | 78.62% |
| 10% Noise | 76.32% | **77.18%** |
| 20% Noise | 75.23% | **76.14%** |
| 30% Noise | 73.02% | **74.38%** |

# 4    Conclusion

In this paper, we have proposed a new technique to reduce the bias of Random Forest to become less sensitive to noisy data. The results reported in this paper show great potential of the proposed technique. In future, we intend to extend our work by including more decision forest algorithms such as Bagging [27], Random Subspace [28], Forest CERN [34], Forest PA [35] and BDF [36]. We shall also include more data sets in experimental paradigm.

# References

1. Big Data Stats for the Big Future Ahead. https://hostingtribunal.com/blog/big-data-stats/
2. Fayyad, U., Piatetsky-Shapiro, G., Smyth, P.: From data mining to knowledge discovery in databases. AI Mag. **17**(3), 37–53 (1996)
3. Tan, P.N., Steinbach, M., Kumar, V.: Introduction to Data Mining, vol. 12. Pearson Education (2011)
4. Breiman, L., Friedman, J.H., Olshen, R.A., Stone, C.J.: Classification and Regression Trees. Wadsworth International Group (2017)
5. Quinlan, J.R.: Induction of decision trees. Mach. Learn. **1**(1), 81–106 (1986)
6. Abramson, N., Braverman, D., Sebestyen, G.: Pattern Recognition and Machine Learning, vol. 9. Springer, Heidelberg (1963)
7. Jain, A.K., Mao, J., Mohiuddin, K.M.: Artificial neural networks: a tutorial. Computer **29**(3), 31–44 (1996)
8. Zhang, G.P.: Neural networks for classification: a survey. IEEE Trans. Syst. Man Cybern. Part C Appl. Rev. **30**(4), 451–462 (2000)
9. Burges, C.J.C.: A tutorial on support vector machines for pattern recognition. Data Min. Knowl. Discov. **2**(2), 121–167 (1998)
10. Murthy, S.K.: Automatic construction of decision trees from data: a multidisciplinary survey. Data Min. Knowl. Disc. **2**(4), 345–389 (1998)
11. Quinlan, J.R.: C4.5 - Programs for Machine Learning. Morgan Kaufmann Publishers, San Mateo (1993)
12. Quinlan, J.R.: Improved use of continuous attributes in C4.5. J. Artif. Intell. Res. **4**, 77–90 (1996)
13. Mehta, M., Agrawal, R., Rissanen, J.: SLIQ: a fast scalable classifier for data mining. In: Apers, P., Bouzeghoub, M., Gardarin, G. (eds.) EDBT 1996. LNCS, vol. 1057, pp. 18–32. Springer, Heidelberg (1996). https://doi.org/10.1007/BFb0014141
14. Srivastava, A., Singh, V., Han, E.-H., Kumar, V.: An Efficient, Scalable, Parallel Classifier for Data Mining, pp. 544–555 (1996). http://www.Cs.Umn.Edu/~Kumar/Papers.Html
15. Adnan, Md.N., Islam, Md.Z.: ComboSplit: combining various splitting criteria for building a single decision tree. In: International Conference on Artificial Intelligence and Pattern Recognition, AIPR 2014, Held at the 3rd World Congress on Computing and Information Technology, WCIT, pp. 1–8 (2014)

16. Adnan, Md.N.: Decision tree and decision forest algorithms: on improving accuracy, efficiency and knowledge discovery. Ph.D. thesis, School of Computing and Mathematics, Charles Sturt University, Bathurst, Australia (2017)
17. Adnan, Md.N., Islam, Md.Z., Akbar, Md.M.: On improving the prediction accuracy of a decision tree using genetic algorithm. In: Gan, G., Li, B., Li, X., Wang, S. (eds.) ADMA 2018. LNCS (LNAI), vol. 11323, pp. 80–94. Springer, Cham (2018). https://doi.org/10.1007/978-3-030-05090-0_7
18. Adnan, Md.N., Islam, Md.Z., Kwan, P.W.H.: Extended space decision tree. In: Wang, X., Pedrycz, W., Chan, P., He, Q. (eds.) ICMLC 2014. CCIS, vol. 481, pp. 219–230. Springer, Heidelberg (2014). https://doi.org/10.1007/978-3-662-45652-1_23
19. Adnan, Md.N., Islam, Md.Z.: A comprehensive method for attribute space extension for Random Forest. In: 2014 17th International Conference on Computer and Information Technology, ICCIT 2014, pp. 25–29 (2003)
20. Adnan, Md.N., Islam, Md.Z.: Complement random forest. In: Conferences in Research and Practice in Information Technology Series, vol. 168, pp. 89–97 (2015)
21. Adnan, Md.N., Islam, Md.Z.: Improving the random forest algorithm by randomly varying the size of the bootstrap samples for low dimensional data sets. In: 23rd European Symposium on Artificial Neural Networks, Computational Intelligence and Machine Learning, ESANN 2015 - Proceedings, pp. 391–396 (2015)
22. Polikar, R.: Ensemble based systems in decision making. IEEE Circuits Syst. Mag. 6(3), 21–44 (2006)
23. Adnan, Md.N., Islam, Md.Z.: Effects of dynamic subspacing in random forest. In: Cong, G., Peng, W.-C., Zhang, W.E., Li, C., Sun, A. (eds.) ADMA 2017. LNCS (LNAI), vol. 10604, pp. 303–312. Springer, Cham (2017). https://doi.org/10.1007/978-3-319-69179-4_21
24. Adnan, Md.N., Islam, Md.Z.: Optimizing the number of trees in a decision forest to discover a subforest with high ensemble accuracy using a genetic algorithm. Knowl.-Based Syst. 110, 86–97 (2016)
25. Geurts, P., Ernst, D., Wehenkel, L.: Extremely randomized trees. Mach. Learn. 63(1), 3–42 (2006)
26. Breiman, L.: Random forests. Mach. Learn. 45(1), 5–32 (2001)
27. Breiman, L.: Bagging predictors. Mach. Learn. 24(2), 123–140 (1996)
28. Ho, T.K.: The random subspace method for constructing decision forests. IEEE Trans. Pattern Anal. Mach. Intell. 20(8), 832–844 (1998)
29. Han, J., Kamber, M., Pei, J.: Concepts and Techniques: Data Mining. Morgan Kaufmann Publishers (2012)
30. Adnan, Md.N., Islam, Md.Z.: One-vs-all binarization technique in the context of random forest. In: 23rd European Symposium on Artificial Neural Networks, Computational Intelligence and Machine Learning, ESANN 2015 - Proceedings, pp. 385–390 (2015)
31. Lichman, M.: UCI Machine Learning Repository (2013). http://archive.ics.uci.edu/ml. http://archive.ics.uci.edu/ml/datasets.html
32. Adnan, Md.N., Islam, Md.Z.: ForEx++: a new framework for knowledge discovery from decision forests. Australas. J. Inf. Syst. 21, 1–20 (2017)
33. Arlot, S., Celisse, A.: A survey of cross-validation procedures for model selection. Stat. Surv. 4, 40–79 (2010)

34. Adnan, Md.N., Islam, Md.Z.: Forest CERN: a new decision forest building technique. In: Bailey, J., Khan, L., Washio, T., Dobbie, G., Huang, J.Z., Wang, R. (eds.) PAKDD 2016. LNCS (LNAI), vol. 9651, pp. 304–315. Springer, Cham (2016). https://doi.org/10.1007/978-3-319-31753-3_25

35. Adnan, Md.N., Islam, Md.Z.: Forest PA: constructing a decision forest by penalizing attributes used in previous trees. Expert Syst. Appl. **89**, 389–403 (2017)

36. Adnan, Md.N., Ip, R.H.L., Bewong, M., Islam, Md.Z.: BDF: a new decision forest algorithm. Inf. Sci. **569**, 687–705 (2021)

# A Collaborative Filtering Recommendation Method with Integrated User Profiles

Chenlei Liu[1], Huanghui Yuan[1], Yuhua Xu[1], Zixuan Wang[1], and Zhixin Sun[1,2(✉)]

[1] Post Big Data Technology and Application Engineering Research Center of Jiangsu Province, Post Big Data Technology and Application Engineering Research Center of Jiangsu Province, Nanjing University of Posts and Telecommunications, Nanjing 210003, Jiangsu, China
sunzx@njupt.edu.cn

[2] Broadband Wireless Communication Technology Engineering Research Center of the Ministry of Education, Nanjing University of Posts and Telecommunications, Nanjing 210003, Jiangsu, China

**Abstract.** In the article recommendation, text information as the main body of the recommendation is rich in semantic content. Especially for content-based recommendation methods, whether an accurate and concise feature representation can be extracted from existing text information is the key to the effective recommendation. Since the long-term use of content-based recommendation methods to generate personalized result sets can make the recommendation variety too homogeneous, the collaborative filtering recommendation method compensates for the above problem by finding other preferred articles of similar users for the recommendation. In this paper, we propose a collaborative filtering recommendation method that incorporates user profiles. This method designs a user portrait labeling system for the article recommendation scenario. Moreover, it uses relevant text processing techniques to extract multi-dimensional user features, which can alleviate the cold start and matrix sparsity problems when performing collaborative filtering recommendations. Finally, we tested our scheme with the MIND Data Set and analyzed the advantages of our scheme.

**Keywords:** Recommendation system · Collaborative filtering · User profiles · Article recommendation

## 1 Introduction

The recommendation system reduces the user's product exploration time by providing personalized interests based on user preferences and past behavior patterns, thus greatly improving the user experience. Recommendation systems are usually classified into seven types: collaborative filtering based, knowledge based, content based, demographics-based, context aware based, and hybrid based [1]. The collaborative filtering-based method has been widely applied in recommendation systems that can produce recommendations based on past interactions

W. Chen et al. (Eds.): ADMA 2022, LNAI 13726, pp. 196–207, 2022.
https://doi.org/10.1007/978-3-031-22137-8_15

between users and items. In article recommendation systems, the general practice of user-based collaborative filtering is to obtain a collection of articles with user behavior records, generate a user-article feature matrix, find similar users of a specific user, and recall other users' preferences. User profiling effectively describes user characteristics and is increasingly used in collaborative filtering-based recommendation systems. It can accurately abstract the essential characteristics of users through the labeling of multidimensional information of users. A good user profile model represents the user characteristics needed in a recommendation scenario with the most accurate dimensions to improve the effectiveness of recommendations. Therefore, it has gradually become one of the core tasks of the recommendation system and has received widespread attention.

In the innovation of the user profile construction method in recommendation systems, [10] proposed a user profile construction method based on the stacking model. [5] proposed a text feature extraction method for constructing user profiles. In this method, a new topic model algorithm, LDA-RCC, is designed for the matrix sparsity problem that the traditional LDA topic model is easy to produce. [2] proposed a multidimensional user profile construction method based on text clustering, automatically constructing the profile and reflecting users' interesting topics. In [6], scholars exploited external criteria to calculate the similarity between items and users. [3] proposed a graphical deep collaborative filtering (GraphDCF) algorithm for providing personalized mutual fund recommendations. Scholars can model different latent relationships among customers with similar shopping habits in this scheme. [8] attempted to extract an appropriate number of negative cases from missing cells in a user-item interaction matrix instead of considering all missing values as negative cases to solve the One-Class Collaborative Filtering problem. [4] introduced a personalized news recommendation framework that would improve the accuracy of news article recommendations. However, these studies do not consider both cold-start and user sparsity simultaneously. In this paper, we propose a collaborative filtering recommendation method that incorporates user profiles. We construct user profiles from multiple dimensions and rely on the basic information features of users to address the cold start problem. Moreover, the topic model performs deeper feature extraction of the user, and the text content is fully utilized to enrich the user profile features, which avoids the matrix sparsity problem commonly found in collaborative filtering. Finally, the users are reasonably grouped by using the multidimensional feature similarity calculation method and the improved clustering algorithm to alleviate the problem of excessive computation caused by searching the whole user set.

## 2    Proposed Method

The specific methodological process of the proposed model includes user profile labeling system, user profile construction and similarity calculation, user clustering, and collaborative filtering to generate result sets.

## 2.1   User Profile Labeling System

The user profile is the labeling of user-related information. The premise of the user profile is to establish a profile labeling system. User profile labels can generally be divided into static and dynamic information labels from the attributes. Static information tags refer to the basic attribute information of users, including gender, age, and marital status. In order to further alleviate the cold start problem of new users, the system will collect the user's interest tags during the registration of new users to further enrich the initial user profile.

Dynamic information tags refer to the feature tags mined through users' behaviors. Based on the scenario of article recommendation, the dynamic information tags are divided into two dimensions: preferred article information and reading habits. Preferred articles are obtained based on relevant user behaviors, including browsing, retweeting, commenting, and like records; reading habits include reading frequency and reading time. Reading habit is chosen as a dimension of user profile because having the same reading habit may reflect their same resting habits and other implied information, and similar users tend to pay more attention to the same content. The structure of a user profile labeling system is shown in Fig. 1.

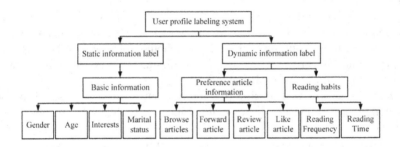

**Fig. 1.** User profile labeling system structure

## 2.2   User Profile Construction and Similarity Calculation

A comprehensive user profile model can more accurately describe users' characteristics and interest preferences. This method takes the articles on user-generated behaviors as the basic information of the portrait. It proposes a multi-feature fusion user portrait model, which is represented explicitly as $P = \{B, T, R, HT\}$. The first dimension $B = \{Ge, Ag, Ma, Ho\}$ represents the basic information characteristics of the user, which are the user's gender, age, marital status, and interests. The second dimension $T$ is a collection of articles based on user-generated behaviors to extract user topic distribution features. The third and fourth dimensions, $R$ and $HR$, represent users' reading frequency and reading time in one day, respectively. The construction methods of the proposed model are as follows.

(1) Similarity calculation of basic information of users. The basic information of users in this method includes gender, age, marital status, and initial interests, because this information is helpful for the accuracy of personalized recommendations. However, it is not easy to measure the user similarity based on these dimensions directly, so in this method, the user label information of each dimension is pre-normalized and constructed. Finally, we characterize the basic information of users in vectorized form.

Firstly, the gender dimension is only male and female. Age can be divided into different age groups. Marital status can be divided into two categories: married status and unmarried status. Finally, the interest dimension is collected when new users register and can be multi-selected.

In order to facilitate the similarity calculation of basic user information, the features of each dimension are represented using a binary sequence. The gender feature is set as a flag bit. If it is male, it is set to 1. Otherwise, it is set to 0; the age group is set to 6 flag bits, in which age group it is set to 1 and the rest to 0; the marital status is set to two flag bits, one of them is set to 1 and the other to 0; the interests are assigned to 8 flag bits, the interest position is 1, and the rest is set to 0. The final binary representation of the user's basic information is obtained. Then the binary Jaccard similarity calculation method is used to calculate the similarity of users and is shown in Eq. 1.

$$\text{BSim}\,(U_i, U_j) = \frac{\sum_{i=1}^{l} (B_i \wedge B_j)}{\sum_{i=1}^{l} (B_i \vee B_j)} \tag{1}$$

The binary $B_i$ and $B_j$ are the representations of the user $U_i$ and $U_j$ respectively. The binary $l$ represents the sequence length. The final similarity result is in the interval $[0, 1]$. And the similarity matrix $BS$ is obtained by calculating the similarity of basic information among all users in turn.

(2) Similarity calculation of users based on topic model

A user-article feature matrix is usually constructed in the traditional collaborative filtering method, with 1 if the user acts on the article and 0 if not. Then the binary sequence similarity between each user is calculated for comparison. However, this approach does not use the feature information of the article to characterize the user and calculate the similarity but only represents the user's interest sequence in the form of 0, 1. The similar results obtained in this way are often not comprehensive and accurate. Therefore, a topic model is used to characterize users and compare their similarities. The flowchart is shown in Fig. 2.

Step 1: Collect the data set $D_u = \{d_{u,1}, d_{u,2}...d_{u,n}\}$ of articles in which user $U_i$ have generated operational behaviors in the last week, where $d_{u,k}$ represents the $k$-th article in which users have recently generated behaviors.

Step 2: Subject the articles used as the training corpus to text pre-processing. The specific process includes designing, stopping words, and Chinese word separation.

Step 3:

– Use the articles in the corpus for LDA topic model training.

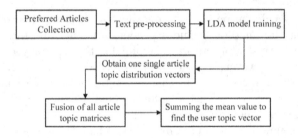

**Fig. 2.** User topic feature extraction

- Use the total number of classified articles in the article corpus $m$ as the number of topics parameter.
- Obtain the topic model of the training set after the training.

Step 4:

- Input the articles in the article collection $D_u$ as the validation set into the trained topic model in Step 3.
- Obtain a topic distribution vector $T_d = (p_{d,i}, p_{d,i}, ...p_{d,k}..., p_{d,m})$ for each article, where $p_{d,k}$ denotes the probability that article $d$ belongs to topic $k$.
- Use this vector to characterize each article.

Step 5: Get the user's dimensional topic portrait matrix from the topic distribution of all articles in the set. $SS$ denotes the common feature of topic features of all articles in the set.

Step 6: Apply the time decay mechanism to weight each article, as shown in Eq. 2. Then sum up the topic vector of each article to find the mean value to obtain the topic-based user feature vector, calculated as shown in Eq. 3, where $n$ means there are $n$ preferred articles.

$$TW = e^{-\lambda(t-t_0)} \tag{2}$$

$$F_u = \frac{1}{n}\sum_{i=1}^{n} T_i * TW_i \tag{3}$$

(3) User reading frequency calculation As a form of dynamic user behavior, reading frequency can effectively reflect the user's reading habits. Therefore, using reading frequency as a one-dimensional user profile can effectively differentiate users and help personalized recommendations. The calculation method of reading frequency $R$ is shown in Eq. 4.

$$R = \frac{1}{k}\sum_{l=1}^{k} R_l \tag{4}$$

where $k$ denotes the total number of days with reading behavior, and $R_l$ denotes the number of articles read by users on the $l$-th day. This formula can calculate the average number of reading by users.

(4) Users' reading time statistics As a kind of behavioral information, reading time can also reflect users' reading habits. The time of day is divided in this method to facilitate the measurement of a person's reading time distribution, as shown in Table 1. In order to obtain the vector of reading time distribution, we obtain the set of articles $D_u = \{d_{u,1}, d_{u,2}...d_{u,n}\}$ in which users generate reading behaviors and classify the articles according to the time division in Table 1. Then we can calculate the reading quantity $RN$ of each time and the vector of reading time distribution based on the distribution of reading quantity. The formula is shown in Eq. 5.

$$RT = \frac{RN}{\sum_{i=1}^{k} RN_i} \tag{5}$$

**Table 1.** Time division table

| Time | Label |
|------|-------|
| 00:00:00–04:59:59 | Midnight |
| 05:00:00–07:59:59 | Early morning |
| 08:00:00–10:59:59 | Morning |
| 11:00:00–12:59:59 | Noon |
| 13:00:00–17:59:59 | Afternoon |
| 18:00:00–21:59:59 | Evening |
| 22:00:00–23:59:59 | Night |

(5) Multi-dimensional similarity calculation For a user profile $P = \{B, T, R, HR\}$ with four-dimensional features, the basic information $B$'s similarity matrix $BS$ can be obtained from Eq. 1. And for the remaining three-dimensional features $PT = \{T, R, HR\}$, we can calculate the common similarity directly by using the cosine similarity calculation method. For example, for users $U_i$ and $U_j$, the user profile $PT_i$ and $PT_j$ can be expressed as vectors. The calculation method is shown in Eq. 6.

$$\text{Sim}\,(PT_i, PT_j) = \frac{PT_i \cdot PT_j}{\|PT_i\| \cdot \|PT_j\|} \tag{6}$$

We can get the $n \times n$-dimensional user similarity matrix $PTS$ and add the basic information similarity matrix $BS$ and $PTS$ of the same $n \times n$-dimensional to get the end-user similarity matrix $US$.

## 2.3   User Clustering

Since the article data is constantly updated and the volume of data will become larger, the amount of computation will also increase rapidly. The general clustering method clusters an object after it belongs to one class, but a user can have many different features and belong to multiple classes for user clustering. Therefore, in this method, an ordered clustering method is used to cluster users, which results in the same user existing in different user groups. The flowchart is shown in Fig. 3.

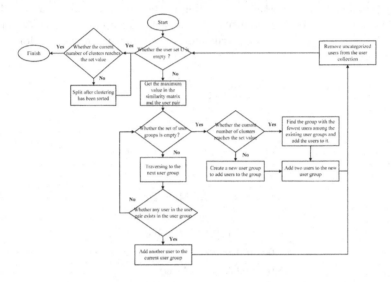

**Fig. 3.** User clustering

Step 1: Set the initial user class capacity $C$ as 0, the set of users to be clustered as $U$, and the number of clusters as $m$.

Step 2: Determine whether the set of users $U$ is empty. If it is empty, end the main clustering step and go to step 6. If it is not empty, get the user similarity matrix $US$ calculated in the previous subsection, get the most similar pair of users $(u_i, u_j)$ and similarity value $Max$. It is then setting the corresponding position of the original matrix to 0.

Step 3: Determine whether the current is the first time clustering. If it is the first time, skip to step 4. If it is not the first time clustering, then skip to step 5.

Step 4: Determine whether the total number of clusters is equal to the set number of clusters $m$. If not, creating a new class, adding the current two users to the new class and deleting both from the user set $U$. If the number of clusters has been reached, selecting the one containing the least number of users from all existing clusters to add new users. When it finishes, skip to step 2.

Step 5: Iterate through the categories $c_i$ in the set $C$. If a user $u_i$ or $u_j$ belonging to the category, add another user to the category group directly. Then add the user to the group, delete it in the set $U$, and jump out of the traversal loop to step 2. After traversing all categories, it is found that neither user belongs to either category, then skip to step 4.

Step 6: After all users have been traversed determine whether the current number of clusters reaches the set m value. If it reaches it, end directly; if not, go to step 7.

Step 7: Sort the users by the decreasing number of users in the user class, traverse the sorted user group for group partitioning. After partition, back to step 6 to judge until the m value is reached, end the clustering process, and get the clustering matrix $CM$, where indicates whether the $i$-th group contains the user information of the $j$-th user, 1 is included, 0 is not included.

Since the pair of users with the highest similarity is selected each time in the clustering process, there are cases where the same user appears in different groups simultaneously. In addition, since the similarity value is decreasing, for a particular user in a group, his right neighbor is more similar to him than his left neighbor, i.e., the similarity gap between neighboring users in the whole group is decreasing, which provides convenience for finding nearest neighbor users in subsequent collaborative filtering.

## 2.4 Collaborative Filtering

The candidate set generation for the collaborative filtering method can be performed based on the user clustering matrix obtained in the previous subsection, as shown in Fig. 4.

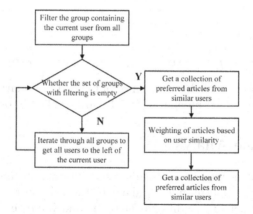

**Fig. 4.** Collaborative filtering result generation

Step 1: Obtain the user clustering information matrix $CM$. Then filtering the user groups containing user $U_i$ from the clustered user groups for the recommended users $U_i$.

Step 2: Iterate through the group of the user $U_i$ and finding all users on the left side to obtain the set of similar users.

Step 3: Obtain the set of preferences $SU = \{SU_1, SU_2...SU_n\}$ of these users, where $SU_n$ denotes the set of preferred articles of the $n$-th similar user.

Step 4: The articles in the obtained set $SU$ are weighted using the previously obtained user similarity matrix $SM$ for the current recommended user $U_i$, as shown in Eq. 7. The articles are sorted in descending order based on the weights.

$$SU_n = SM\,(U_i, U_n) * SU_n \tag{7}$$

Step 5: Set the article hotness threshold. The specific method is to first calculate the current time to the previous week of the article heat score, heat weight distribution as shown in Table 2. Here the reading as the main weighting behavior is to avoid recommending to the user the current more popular and more people have read the article. The calculation method is shown in Eq. 8. After calculating the hotness score of each article in the recent week, the first 10% of articles with higher hotness are not recommended, and the hotness value of the first 10% is the hotness threshold.

**Table 2.** Heat weight distribution

| Operations | Symbols | Weights |
|---|---|---|
| Read | $OP_{re}$ | 0.4 |
| Comment | $OP_{co}$ | 0.2 |
| Favor | $OP_{fa}$ | 0.2 |
| Forward | $OP_{fo}$ | 0.2 |

$$P = OP_{re} * RE + OP_{fa} * FA + OP_{co} * CO + OP_{fo} * TR \tag{8}$$

Step 6: According to the hotness threshold and the similarity ranking result obtained in Step 4, a fixed number of articles are recalled in order.

## 3    Performance Analysis

This method combines traditional collaborative filtering recommendation methods with user profile information to alleviate the cold start and matrix sparsity problems prevalent in traditional methods and improve the recommendation effectiveness. In order to ensure the feasibility of this method, this section conducts a experimental design and analyzes the compared results with the recommendation method proposed in [4]. The experimental environment and implementation of the experimental method are presented below.

### 3.1    Experimental Method

This experiment still uses the news article recommendation dataset **MIND** provided by Microsoft [9]. This information dimension is temporarily ignored in the experiment due to the difficulty of obtaining user demographic information.

Only the user's behavior records and article information data are used for the experimental analysis.

The LDA model provided by the **Gensim** library in Python [7] is used to train the topic model in the experiments. For the feature information of user reading habit dimension, the reading time in the user behavior record needs to be parsed to get the feature vector of user reading frequency and reading time point. The final obtained user portrait feature vectors are compared for similarity and clustered and combined with collaborative filtering for the recommendation.

When performing collaborative filtering, we need to determine how many users are selected as nearest neighbors to get the associated preference articles, so we need to set the number of nearest neighbors. In this experiment, the number of clusters is set to 11, the recall set is set to 35, and the number of nearest neighbors is set to 20, 25, 30, 35, 40, 45, and 50, respectively.

### 3.2 Experimental Result

In this experiment, several experiments were conducted for different numbers of recall sets to obtain the experimental results of the recommended scheme and the comparison scheme proposed in this paper, to calculate the precision and recall of the results, and finally to evaluate the effectiveness of the method with the calculated F-measure and AUC values. The number of articles specifically recalled in the experiment was set between 20 and 50, and a total of seven experiments were conducted. The comparison of the F-measure and AUC values obtained from the experiments is shown in Fig. 5 and 6.

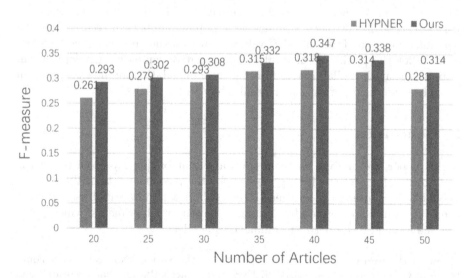

**Fig. 5.** F-measure comparison chart

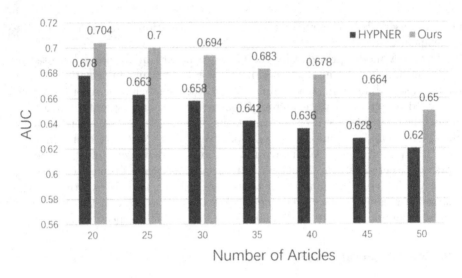

**Fig. 6.** AUC comparison chart

By analyzing the above figures, we can find that the collaborative filtering recommendation method proposed in this paper is better than the collaborative filtering recommendation method in [4] in terms of F-measure and AUC values for different numbers of recalls. And the experiments show that the F-measure value, which integrates the recommendation accuracy and recall rate, will start to fall down after the number of recalls increases to a certain upper limit, and this method falls down more slowly compared with the comparison method, so this method has stronger stability for different numbers of recalls. And the experimental results from the AUC values show that the present method is more accurate for the prediction of positive and negative samples. Therefore, it can be proved that the collaborative filtering recommendation method proposed in this subsection incorporating user profiles has better results.

## 4    Conclusion

In this paper, we analyze the cold start and matrix sparsity problems of the traditional collaborative filtering method for the recommendation. We propose a collaborative filtering recommendation method with integrated user profiles. Then the results are verified by designing simulation experiments, which prove that the method has a better recommendation effect.

**Acknowledgements.** Our work was supported by the National Natural Science Foundation of China under Grant No. 61972208, and Jiangsu Postgraduate Research and Innovation Plan under Grant No. KYCX20_0761.

# References

1. Anwar, T., Uma, V.: Comparative study of recommender system approaches and movie recommendation using collaborative filtering. Int. J. Syst. Assur. Eng. Manag. **12**(3), 426–436 (2021). https://doi.org/10.1007/s13198-021-01087-x
2. de Campos, L.M., Fernández-Luna, J.M., Huete, J.F., Redondo-Expósito, L.: Automatic construction of multi-faceted user profiles using text clustering and its application to expert recommendation and filtering problems. Knowl.-Based Syst. **190**, 105337 (2020)
3. Chou, Y.C., Chen, C.T., Huang, S.H.: Modeling behavior sequence for personalized fund recommendation with graphical deep collaborative filtering. Expert Syst. Appl. **192**, 116311 (2022). https://doi.org/10.1016/j.eswa.2021.116311. https://www.sciencedirect.com/science/article/pii/S0957417421016122
4. Darvishy, A., Ibrahim, H., Sidi, F., Mustapha, A.: HYPNER: a hybrid approach for personalized news recommendation. IEEE Access **8**, 46877–46894 (2020). https://doi.org/10.1109/ACCESS.2020.2978505
5. Ding, Z., Yan, C., Liu, C., Ji, J., Liu, Y.: Short text processing for analyzing user portraits: a dynamic combination. In: Farkaš, I., Masulli, P., Wermter, S. (eds.) ICANN 2020. LNCS, vol. 12397, pp. 733–745. Springer, Cham (2020). https://doi.org/10.1007/978-3-030-61616-8_59
6. Pirasteh, P., Bouguelia, M.-R., Santosh, K.C.: Personalized recommendation: an enhanced hybrid collaborative filtering. Adv. Comput. Intell. **1**(4), 1–8 (2021). https://doi.org/10.1007/s43674-021-00001-z
7. Řehůřek, R., Sojka, P.: Software framework for topic modelling with large corpora. In: Proceedings of the LREC 2010 Workshop on New Challenges for NLP Frameworks, Valletta, Malta, pp. 45–50. ELRA (2010). http://is.muni.cz/publication/884893/en
8. Song, G.J., Song, H.S.: Algorithm for generating negative cases for collaborative filtering recommender. Expert Syst. **n/a**(n/a), e12986. https://doi.org/10.1111/exsy.12986. https://onlinelibrary.wiley.com/doi/abs/10.1111/exsy.12986
9. Wu, F., et al.: Mind: a large-scale dataset for news recommendation. In: ACL 2020 (2020). https://www.microsoft.com/en-us/research/publication/mind-a-large-scale-dataset-for-news-recommendation/
10. Wu, Y., Yu, P.: User portrait technology based on stacking mode. In: 2020 IEEE International Conference on Dependable, Autonomic and Secure Computing, International Conference on Pervasive Intelligence and Computing, International Conference on Cloud and Big Data Computing, International Conference on Cyber Science and Technology Congress (DASC/PiCom/CBDCom/CyberSciTech), pp. 245–250. IEEE (2020)

# A Quality Metric for K-Means Clustering Based on Centroid Locations

Manoj Thulasidas[(✉)] [iD]

School of Computing and Information Systems, Singapore Management University,
80 Stamford Road, Singapore 178902, Singapore
manojt@smu.edu.sg

**Abstract.** K-Means clustering algorithm does not offer a clear methodology to determine the appropriate number of clusters; it does not have a built-in mechanism for K selection. In this paper, we present a new metric for clustering quality and describe its use for K selection. The proposed metric, based on the locations of the centroids, as well as the desired properties of the clusters, is developed in two stages. In the initial stage, we take into account the full covariance matrix of the clustering variables, thereby making it mathematically similar to a reduced $\chi^2$. We then extend it to account for how well the clustering results comply with the underlying assumptions of the K-Means algorithm (namely, balanced clusters in terms of variance and membership), and define our final metric ($\mathcal{M}_C$). We demonstrate, using synthetic and real data sets, how well our metric performs in determining the right number of clusters to form. We also present detailed comparisons with existing quality indexes for automatic determination of the number of clusters.

**Keywords:** K-Means clustering · Quality metrics · K selection problem · Number of clusters

## 1 Introduction

K-Means clustering [15] is conceptually simple and easily explained and understood. Practically, however, one of the difficulties that we face in using the algorithm is that we cannot clearly and objectively articulate why one clustering output is better than another one for a given data set. We lack a quality measure. Because of the lack of a quality measure, we face difficulties when it comes to selecting the optimal number of clusters to form.

In this paper, we propose a new quality metric that can be easily computed during (or after) K-Means clustering and argue from basic principles that it accurately captures the validity of the clustering run. We will study its performance in determining the optimal number of clusters to form on a wide range of synthetic data as well as some real data sets. We will demonstrate that it compares favourably against the current metrics, several of which are reviewed in [4].

W. Chen et al. (Eds.): ADMA 2022, LNAI 13726, pp. 208–222, 2022.
https://doi.org/10.1007/978-3-031-22137-8_16

## 2   Related Work

We have several quality indexes and statistics in the literature, which are frequently used to automatically determine the right number of clusters $(K)$. The ones we will consider in the article for comparison are:

- Variance Ratio Criterion [5]: **VRC**
- Akaike Information Criterion [3]: **AIC**
- Bayesian Information Criterion [24]: **BIC**
- Silhouette Width [23]: **Sil. Wid.**
- Gap Statistic [26]: **Gap**
- Evaluation Function [20]: **$f(K)$**

In addition to these "classic" quality indexes, we have several other candidates, some of which are algorithms specifically designed to determine the right $K$ automatically. A recent study [10] introduces the Projected Gaussian (PG-Means) method, which performs a K-Means clustering for all $K$s in the range of interest and projects both the data and model to a linear subspace. It then looks for a good fit between the model and data using the Kolmogorov-Smirnov (KS) test. PG-Means runs with ten sets of random starting seeds, which our studies indicate may be too small to ensure convergence.

X-Means [19], originally developed to address the scalability issue of K-Means, also helps determine the right $K$. An extension [16] of X-Means is found in the literature, designed to automatically determine $K$ through progressive iterations and merging of clusters based on a BIC stopping rule. This method, however, does not give an index, which is needed for other purposes such as feature selection.

G-Means [14] is a method to repeatedly perform K-Means with increasing $K$ until statical tests show that the resulting clusters are Gaussian within a specified confidence level. This method again does not provide a quality metric. Other attempts to determine $K$ include a visual assessment of clustering tendency [18], again with no overall quality metric.

A recent comparative study [13] argues that relying on any single internal metric or index is unwise, while noting that the WB index [28] (based on sum of squares similar to **VRC**) seems to perform best. Our index, also loosely based on sum of squares, seems to work well both in synthetic and real data sets.

One of the more recent studies that define quality metrics or indexes is a probabilistic approach [6] on external validation of fuzzy clustering, where one data point may belong to multiple clusters. Our approach also uses within-standard deviations, and applies only to K-Means clustering, which is distinctly non-fuzzy. Another approach [12] introduces a cluster-level similarity index called the centroid index, focusing on the overall clustering output to quantify the clustering quality. An external quality measure that can apply to many different clustering algorithms, it is not directly comparable to our internal metric focusing on K-Means. Lastly, in a paper proposal [27], a new separation measure, (termed "dual center") is developed, based on which a validity index is proposed for fuzzy clustering. It is not, however, employed for $K$ selection.

## 3   New Quality Metrics

To develop the metric proposed in this article, we will start from the standard $(z)$ scores of the centroid locations and combine them into a metric. We will then generalize it using the full covariance matrix of the clustering variables (grouped by cluster) to define a reduced $\chi^2$ metric. At the second stage, we will extend the $\chi^2$ metric to incorporate extra information about how well the clusters conform to the implicit assumptions in the K-Means algorithm and come up with the proposed metric, $\mathcal{M}_C$.

Given the centroids $(\vec{\mu}_k)$ and the population mean $(\vec{\mu})$, we can compute the significance of the difference between them for each variable as,

$$z_{k_j} = \frac{\delta_{k_j}}{\sigma_{k_j}^{(c)}} = \frac{\delta_{k_j}}{\frac{\sigma_{k_j}}{\sqrt{n_k}}} = \frac{\sqrt{n_k}(\mu_j - \mu_{k_j})}{\sigma_{k_j}} \tag{1}$$

where $\delta$ stands for the difference and $\sigma^{(c)}$ for the within-cluster standard deviation. Since the $k^{th}$ cluster has $n_k$ members, the standard error is $\sigma^{(c)}$ divided by $\sqrt{n_k}$. In order to interpret the squared sum as a weighted average, we divide it by the number of observations $n$, so that each term in the sum has a weight of $n_k/n$, the fraction of the observations belonging to the cluster, and call it our quality **Score**.

$$\begin{aligned} \textbf{Score} &= \frac{1}{K(p-1)} \sum_{k=1}^{K} \frac{|\vec{z}_k|^2}{n} \\ &= \frac{1}{nK(p-1)} \sum_{k=1}^{K} n_k \sum_{j=1}^{p} \left( \frac{\mu_j - \mu_{k_j}}{\sigma_{k_j}} \right)^2 \end{aligned} \tag{2}$$

### 3.1   Reduced $\chi_R^2$ Metric

The generalized version of the distance to be used in the presence of correlations is the Mahalanobis Distance [17], $D_M(\vec{\mu}_k, \vec{\mu})$ corresponding to the $K$ cluster centroids. The square of each one (denoted by $D_M^2(\vec{\mu}_k, \vec{\mu})$) is a random variable which follows a $\chi^2$ distribution with a parameter (or degrees of freedom, **DoF**) $p-1$, where $p$ is the number of clustering variables. We can combine these Mahalanobis distances in quadrature using the same weightage as in the definition of **Score**.

$$\begin{aligned} \chi_R^2 &= \frac{1}{K(p-1)} \sum_{k=1}^{K} \frac{D_M^2(\vec{\mu}_k, \vec{\mu})}{n} \\ &= \frac{1}{nK(p-1)} \sum_{k=1}^{K} n_k (\vec{\mu}_k - \vec{\mu}) \Sigma_k^{-1} (\vec{\mu}_k - \vec{\mu})^\mathsf{T} \end{aligned} \tag{3}$$

The sum of the squares of the $K$ Mahalanobis distances, being the sum of $K$ random variables, each with a $\chi^2$ (of **DoF** $= p - 1$) distribution, is another $\chi^2$ random variable of **DoF** $= K(p-1)$.

Since $K(p-1)$ is actually the number of degrees of freedom, $\chi_R^2$ can be thought of as the reduced $\chi^2$ per cluster, but with an extra (constant) scaling factor of $n$. This scaling, being constant, does not impact the usage of $\chi_R^2$ in determining the right $K$. We call this reduced and *scaled* $\chi_R^2$ our "Reduced $\chi_R^2$ Metric."

### 3.2 Implicit Assumptions in K-Means Algorithm

The K-Means algorithm works best when the data set has spherical clusters of roughly equal sizes. The clusters are expected to be similar in terms of membership, density and variance. If this assumption is violated, the K-Means algorithm is likely to give unreliable results. Furthermore, if one cluster has significantly smaller variance or number of members, it tends to "scavenge" observations belonging to other clusters. This is because the cluster boundaries are perpendicular bisectors and the statistical fluctuations in the observations always favor the tighter or smaller cluster. The soft requirement of balanced clusters in terms of membership and variance forms an implicit assumption in the algorithm.

### 3.3 Covariant Metric ($\mathcal{M}_C$)

Since the metric is a reduced $\chi^2$, it may be possible extend it to include components that quantify these assumptions in the K-Means algorithm. We will show how the cluster membership (or frequency) and the cluster standard deviation are compared against their expected or ideal values, and a standard score for each is generated, to be combined with $\chi_R^2$. We will call the extended metric the Covariant Metric ($\mathcal{M}_C$) because it is built on the covariance matrix of the data. We emphasize that it is weighted by $n_k/n$ and therefore does not numerically equal standard score or the reduced $\chi^2$, and it incorporates the components described below in a heuristic way.

**Cluster Frequency.** Since we have $n$ observations and $K$ clusters, the "ideal" frequency for each cluster is $\hat{n}_k = n/K$. Assuming Poisson statistics, we can argue that the expected error on each frequency is $\sqrt{n/K}$. Since we have $K$ measurements of the frequencies, we gain another factor of $\sqrt{K}$ in its standard error, giving us $\sigma_{n_k} = \frac{\sqrt{n}}{K}$. and combine the individual z-scores in quadrature to come up with a measure of how far away our clustering result is from the ideal, in terms of the membership frequency.

$$M_{n_k} = \sum_{k=1}^{K} \left( \frac{n_k - \hat{n}_k}{\sigma_{n_k}} \right)^2 \tag{4}$$

$M_{n_k}$ is a standardized measure of how different the clusters are in terms of their frequency. Ideally, we would like to have $M_{n_k}$ as close to zero as possible.

**Cluster Variance.** Once the clustering is done, we have the sum of squared errors **SSE**. If the clusters have the same variance, then **SSE** should be shared among them in proportion to the frequency.

$$\mathbf{SSE}_k = \frac{n_k - 1}{n - K}\mathbf{SSE} \tag{5}$$

$\mathbf{SSE}_k$ is the sum of the squared errors of the observations to their respective centroids. The expected "ideal" variance, therefore, is this sum divided by $n_k - 1$.

$$\hat{S}_k^2 = \frac{\mathbf{SSE}_k}{n_k - 1} = \frac{\mathbf{SSE}}{n - K} \tag{6}$$

The actual variances of the clusters are estimated during the clustering process, and is reported in terms of within standard deviations, but aggregated over all variables. Ignoring the cases where $n_k = 1$,

$$S_k^2 = \frac{1}{n_k - 1}\sum_{i=1}^{n;g_i=k}\sum_{j=1}^{p}(x_{ij} - \mu_{k_j})^2 \tag{7}$$

The standard error in the variance is obtained by recognizing that the sample variance (when multiplied by $(n_k - 1)/\sigma_{S_k^2}^2$) is a $\chi^2$ distribution of $n_k - 1$ degrees of freedom, which itself has a variance of $2(n_k - 1)$. Therefore, the standard error of the variance is [2]

$$\sigma_{S_k^2} = S_k^2\sqrt{\frac{2}{n_k - 1}} \tag{8}$$

Again, we have an "ideal" variance and a measured one, and we can compute the significance of the difference between them (using the standard errors)and combine their significances to come up with a measure of how the cluster variances compare to the ideal equal variance.

$$M_{S_k^2} = \sum_{k=1}^{K}\left(\frac{S_k^2 - \hat{S}_{k^2}}{\sigma_{S_k^2}}\right)^2 \tag{9}$$

In an ideal clustering solution, we will expect to have very small $M_{S_k^2}$.

**Extending the $\chi_R^2$ Metric.** Now that we have the two new components encapsulating the uniformity among the clusters in terms of frequency and variance, we can extend our $\chi_R^2$ with them to obtain the Covariant Metric ($\mathcal{M}_C$) as follows.

$$\mathcal{M}_C = \frac{\chi_R^2}{M_{n_k} + M_{S_k^2}} \tag{10}$$

where $M_{n_k}$ and $M_{S_k^2}$ are defined above in Eqs. (4) and (9) above. We divide by the sum of these two measures corresponding to the frequencies and variances of the clusters because the overall quality of the K-Means clustering is inversely proportional to them. In other words, if we have two clustering solutions with identical $\chi_R^2$, but different values for $M_{n_k}$ and $M_{S_k^2}$, we have to choose the one with the lower $M_{n_k}$ and $M_{S_k^2}$. Note, however, that a more general way to combine them would as a linear combination, $w_1 M_{n_k} + w_2 M_{S_k^2}$, where $w_1$ and $w_2$ are relative weights whose values are not known a priori.

### 3.4  Quantifying Index Performance

Since we will be comparing multiple indexes with our metric, we may get the same right $K$ from several of them. It would then be fair to ask how we quantify the performance of various indexes. For the first five out of the seven indexes listed earlier (namely **VRC**, **AIC**, **BIC**, **Sil. Wid.** and **DB**), the selection of $K$ is based on a maximum or a minimum. The **Gap** statistic and the $f(K)$ index do not determine $K$ by looking for a maximum or minimum in their variation.

For the **Gap** statistic, the best $K$ recommended by this approach is the smallest number of clusters that shows a decrease, while all values of $K$ such that $f(K) < 0.85$ are potential candidates as the right $K$.

The significance of $K$ selection may be quantified using the concept of curvature: the higher the curvature, the more prominent the minimum or maximum signifying the right $K$. For a continuous function of a single variable, the curvature is proportional to the second derivative. For a discrete function $h(K)$ (where $K$ is an integer), we define a new quantity $\Gamma$, similar to the three-point computation of the second derivative for a continuous functions.

$$\Gamma = \left| \frac{h(K+1) - 2h(K) + h(K-1)}{h(K+1) + h(K-1)} \right| \tag{11}$$

The index with the largest $\Gamma$ value has the most clearly defined peak, signifying the right $K$.

## 4  Experiments on Synthetic Data

### 4.1  Data Generation

We use the R package `clusterGeneration` [21], which can generate clusters of specified sizes in spaces of prescribed number of variables. In `clusterGeneration`, we can also specify the separation among the clusters, using a separation index [22]. We will use various values for these three and other parameters as described below.

**Number of Clusters ($G$):** We generate synthetic data sets with different numbers of clusters: $G \in \{5, 10, 15, 20\}$

**Number of Variables ($p$):** We use the values $p \in \{2, 4, 8, 16, 32\}$ for this parameter

**Separation Index ($J^*$):** This parameter controls how well separated the clusters are, and we use the value $J^* = 0.34$ (for cleanly separated clusters), since we are defining and studying the metric for a data set well suited for K-Means clustering.

Since we are studying the metric for a data set perfectly suited for K-Means clustering, we focus on these 20 data sets for detailed analysis, we use the following values for the other parameters in the generation of the synthetic data.

- Number of noisy variables = 0
- Number of outliers = 0
- Equal cluster membership (of $10p$) for all clusters
- Cluster uniformity (= Range for variances of the covariance matrix) = $[1, 10]$, which generates a reasonable variability.

## 4.2    Analysis of Synthetic Data

With the synthetic data, we first compute our metrics $\mathcal{M}_C$, $\chi_R^2$, and seven indexes (**VRC, AIC, BIC, Sil. Wid., DB, Gap** and $f(K)$) discussed earlier. For each run of the K-Means clustering algorithm, we use 100 random sets of initial seeds from which the best run (based on the sum of squared errors) is chosen. It is important to have large number of starting seeds because of the sensitivity of K-Means to initial conditions, especially when we have large number of clusters and relatively small number of variables [25]. For smaller number of starting seeds, we do see a large fraction of K-Means attempts failing to converge. We also set a generous limit on the maximum number of iterations of 1000 and repeat the whole analysis multiple times and ensure that the results reported are stable.

## 4.3    Results and Discussion

First, we focus on the fraction of the times we can detect the right number of clusters using the metrics $\mathcal{M}_C$, $\chi_R^2$, **VRC, AIC, BIC, Sil. Wid., DB, Gap** and $f(K)$. We define this fraction as the accuracy of the metric and scan for the right $K$ in $\frac{G}{2} < K < 2G$. We run the analysis on all our 60 synthetic data sets (20 for each $J^*$ value), and report the average accuracy for our metric and a variety of indexes in Table 1. Also reported are the average $\Gamma$ values when the right $K$ is detected. Since we are developing a metric that will work best for data sets that are particularly suited for K-Means clustering, the column to consider in Table 1 is for well-separated clusters ($J^* = 0.34$). We can see that $\mathcal{M}_C$ performs very well with extremely well-defined peaks ($\Gamma \approx 28$). Although the Variance Ratio Criterion (**VRC**) and the Davies-Bouldin index (**DB**) also detect the right $K$, the significance of the peak for **VRC** or the minimum for **DB** is at much smaller levels ($\Gamma \approx 0.1 - 0.2$).

**VRC, DB** and **Gap** perform better than $\mathcal{M}_C$ when the clusters are generated with more overlaps, by reducing the value of $J^*$ to 0.01, when the clusters are expected to be more realistic. However, their performance in the real data is poorer than out metric. Also of note is that both **AIC** and **BIC** perform very poorly in the synthetic data, as well as in the real data. (The comparisons of the performances of the metrics in real data is summarized in Table 2 in a subsequent section.)

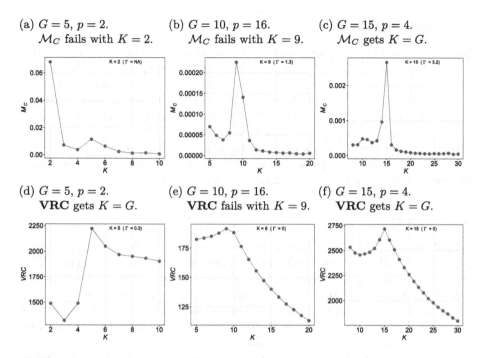

(a) $G = 5$, $p = 2$.
$\mathcal{M}_C$ fails with $K = 2$.

(b) $G = 10$, $p = 16$.
$\mathcal{M}_C$ fails with $K = 9$.

(c) $G = 15$, $p = 4$.
$\mathcal{M}_C$ gets $K = G$.

(d) $G = 5$, $p = 2$.
VRC gets $K = G$.

(e) $G = 10$, $p = 16$.
VRC fails with $K = 9$.

(f) $G = 15$, $p = 4$.
VRC gets $K = G$.

**Fig. 1.** Examples of $\mathcal{M}_C$ (top row) and **VRC** (bottom row) for different $K$, when realistic clusters ($J^* = 0.01$) are used.

A comparison of the shapes of $\mathcal{M}_C$ and **VRC** can be found in Fig. 1, where we see that $\mathcal{M}_C$ typically has a sharper peak. In the real data, $\mathcal{M}_C$ seems to perform even better in detecting the right number of classes.

Table 1 shows that $\mathcal{M}_C$ performs better than $\chi_R^2$, both in terms of the accuracy and the sharpness of the peak. We can therefore conclude that our extension of the $\chi^2$ by incorporating the scores corresponding to ideal cluster frequency and within-standard deviation does add value to the metric.

## 5   Experiments on Real Data

We also perform our experiments in four different real data sets, where we detect the optimal number of clusters automatically using $\mathcal{M}_C$, and compare it to what is known about the data sets independently. In these experiments, we assume that the classes in the data sets form spherical clusters, easily separated by the K-Means algorithm, and, as a consequence, that the ideal number of clusters is the number of classes. If this assumption does not hold true for the data set under consideration, our metric will not work. Indeed, the definition of our metrics would also be invalid in that case.

## 5.1   Variable Selection

Since we are testing our metrics on labeled data sets, we can directly compute the purity of the clusters by counting the number of correctly assigned observations. We assume that the ideal number of clusters (ideal $K$) is the number of distinct values of the label. After selecting the best variables based on purity, we will iterate over various values of $K$ and expect to see a clear peak for the Covariant metric when plotted against various $K$ values. Note that we will use the same "best" variables for all other indexes to which we compare our Covariant metric. Therefore, there is no unfair advantage or biases in using the selected variables in favor of our metrics. The four data sets used are briefly described below.

## 5.2   Data Sets

**Iris Data Set.** The classic Iris data set [11] contains 150 flower measurements along four variables (*Sepal Length, Sepal Width, Petal Length* and *Petal Width*) from three different iris species (*Setosa, Versicolor* and *Virgnica*). Each species has 50 data points in the data set. Since there are three species, we know, beforehand, that the ideal number of clusters should be three. We select the variables *Petal Length* and *Petal Width* as the variables (based on the highest purity) to use when looking for the best $K$.

We can now look at how the Covariant metric ($\mathcal{M}_C$) varies when we cluster with different $K$s. The dependence is shown in Fig. 2a. We can see that the ideal $K = 3$ clearly shows up as a peak in both distributions, much more clearly in $\mathcal{M}_C$.

**Table 1.** Accuracy and $\Gamma$ of various metrics

| Metric | Well-separated ($J^* = 0.34$) | Medium ($J^* = 0.21$) | Realistic ($J^* = 0.01$) |
|---|---|---|---|
| $\mathcal{M}_C$ | 100.0% (28.6) | 90.0% (18.0) | 45.0% (7.0) |
| $\chi^2_R$ | 45.0% (0.1) | 65.0% (0.1) | 35.0% (0.1) |
| **VRC** | 100.0% (0.2) | 100.0% (0.2) | 60.0% (0.1) |
| **AIC** | 0.0% (−) | 0.0% (−) | 0.0% (−) |
| **BIC** | 0.0% (−) | 0.0% (−) | 0.0% (−) |
| **Sil. Wid.** | 95.0% (0.1) | 75.0% (0.1) | 65.0% (0.1) |
| **DB** | 100.0% (0.2) | 85.0% (0.2) | 85.0% (0.1) |
| **Gap** | 55.0% | 70.0% | 70.0% |
| $f(K)$ | 20.0% | 25.0% | 30.0% |

Accuracy of $K$ selection: the fraction of the times when the reconstructed number of clusters is the same as the generated number ($K = G$). The numbers between parentheses are the mean $\Gamma$ (averaged when $K = G$). Note that **Gap** and $f(K)$ do not detect the ideal $K$ using maximum or minimum, and therefore $\Gamma$ is not reported.

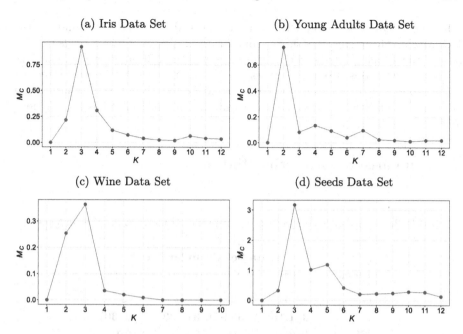

**Fig. 2.** Using our quality metrics to select the optimal number of clusters in various data set. The Covariant Metric $\mathcal{M}_C$ shows clear peaks at the known number of clusters ($K = 3$ for the Iris, Wine and Seeds, and $K = 2$ for the Young Adults).

**Young Adults Data Set.** We collected anonymous data from our students. The data set has 127 observations of four numeric variables (*Height, Weight, Age* and *HairLength*) and a label (M or F for male or female). Note that in Singapore, male university students are expected to be about 2 to 3 years older than their female classmates because of their military service obligation. Therefore, we may expect the *Age* variable to have some differentiating power while clustering the data. Following the same procedure as in the iris data set, we select *Weight* and *HairLength* as the best variables to use, for the best possible purity of 98.4%.

A blind K-Means clustering (with $K = 2$) using the four numeric variables is likely to segment the Young Adults data into male and female students. The ideal number of clusters is indeed two. In Fig. 2b, we have plotted $\mathcal{M}_C$ as a function of $K$, and it shows a clear peak at $K = 2$.

**The Wine Data Set.** The publicly available Wine data set [1], from the UCI Machine Learning Repository [9], has 12 attributes, making the combinatorial problem of selecting the best variables for K-Means clustering challenging with over 8000 possible combinations. From among the multiple variable combinations, we select the combination of *Alcohol, Ash, Flavanoids* and *OD280_OD315* based on the highest purity of 90.5%. The Wine data set also has three classes, and Fig. 2c shows that the Covariant metric ($\mathcal{M}_C$) has a clear peak at $K = 3$.

**The Seeds Data Set.** The publicly available Seeds data set [7] (again from the UCI Machine Learning Repository) contains three classes of wheat seeds with 70 observations each. It has seven attributes, giving us 120 different combinations of variables to choose from. From these combinations, we select *Area*, *Perimeter*, *Compactness* and *Asymmetry* based on the highest purity of 90.0% that we can get. KSelection-Seeds shows that the Covariant metric ($\mathcal{M}_C$) has a clear peak at $K = 3$, as expected.

# 6    Comparison with Other Indexes

**Table 2.** Performance comparison of our proposed metric and other indexes

| | Data Set (Standardized) | | | |
|---|---|---|---|---|
| **Index** | Iris | YA | Wine | Seeds |
| $G$ | **3** | **2** | **3** | **3** |
| $\mathcal{M}_C$ | **3** (1.15) | **2** (65.92) | **3** (2.32) | **3** (13.30) |
| **VRC** | 10 (0.01) | **2** (0.27) | **3** (0.14) | **3** (0.10) |
| **AIC** | 4 (0.04) | 7 (0.02) | 10 (0.00) | 12 (−) |
| **BIC** | **3** (0.24) | 4 (0.07) | 7 (0.01) | 7 (0.01) |
| **Sil. Wid.** | 2 (0.10) | **2** (0.21) | **3** (0.07) | 2 (0.07) |
| **DB** | 2 (0.38) | **2** (0.12) | **3** (0.08) | 2 (0.01) |
| **Gap** | **3** | **2** | 4 | **3** |
| $f(K)$ | 2 | **2** | 2 | 2 |

| | Data Set (Raw) | | | |
|---|---|---|---|---|
| **Index** | Iris | YA | Wine | Seeds |
| $G$ | **3** | **2** | **3** | **3** |
| $\mathcal{M}_C$ | **3** (2.51) | **2** (16.83) | **3** (1.52) | **3** (3.69) |
| **VRC** | 10 (0.03) | **2** (0.08) | **3** (0.25) | **3** (0.11) |
| **AIC** | 5 (0.04) | 12 (−) | 6 (0.01) | 12 (−) |
| **BIC** | 4 (0.05) | 12 (−) | **3** (0.17) | 9 (0.01) |
| **Sil. Wid.** | 2 (0.16) | **2** (0.12) | **3** (0.16) | 2 (0.09) |
| **DB** | 2 (0.41) | **2** (0.20) | **3** (0.22) | 2 (0.10) |
| **Gap** | 5 | 3 | **3** | **3** |
| $f(K)$ | 2 | **2** | 2 | 2 |

The top row is $G$, the number of classes in our data sets. When an index predicts the right $K$, it is highlighted in **bold**. ($\Gamma$ is reported between parentheses. It cannot be calculated at the end of the range $K = 12$.)

Some of the indexes to which we are comparing our metric may perform differently when the data set is standardized (such that all variables zero mean and unit standard deviation). For this reason, we study the performance of the indexes and our metric on both standardized data sets as well as the raw ones. Our proposed metric, however, does not require the data set to be standardized. In fact, since our metric takes into account the full covariance matrix on a per-cluster basis, it can be argued that it should perform as well or better in the *raw* data set.

We can see from Table 2 that our metric $\mathcal{M}_C$ performs very well on standardized data sets, detecting the right $K$ in all four data sets, while the other indexes seem to struggle. Of the seven other indexes considered, the **VRC** index seems to perform best with three right predictions. However, its significance measure ($\Gamma$) is low. When run on the data sets without any normalization, $\mathcal{M}_C$ continues to perform well, as we can see in Table 2. The other indexes seem to perform marginally worse on the raw data sets than on the standardized ones.

Note that the sharpness of the peaks representing the right value of $K$, as measured by $\Gamma$ is significantly higher for the Covariant metric $\mathcal{M}_C$, both in the standardized as well as the raw data sets (Table 2), when compared to any other index. The significance of the peaks ($\Gamma$) for our metric improves with standardization for three data sets, while decreases for the other one, which is consistent with our expectation that standardization should not affect its performance.

## 7   Limitations

The main motivation behind this work, in addition to pure academic interest, is to automate K-Means clustering such that it can be deployed in situations where automatic insight generation is desired. (For example, consider customer segmentation for marketing purposes where new customers are continually added to the database.) Since the impetus behind this work is automated processing, we have not attempted to prepare the data in any fashion.

The mathematical validity of the Covariant Metric ($\mathcal{M}_C$), being a ratio of two entities that may be thought of as $\chi^2$, is not yet fully established. It is similar to the odds ratio calculation commonly used in the data science community, but on shakier theoretical footing. It is hoped that other researchers may be able to find a more theoretically sound way of combining the components (defined in Eq. (4) and (9)) into a better metric than the one in Eq. (10). We can see from our results that there is information in the Covariant Metric when it comes to $K$ selection (Fig. 2).

We may be able to use the significance of the peak, $\Gamma$ as defined in Eq. (11), either directly or in combination with the peak value of $\mathcal{M}_C$ in order to select the right $K$. We have not explored this idea further due to the uncertainty in the mathematical foundation of such an approach. Again, other researchers may be able to come up with theoretically defensible methods of using $\Gamma$.

Lastly, in defining our Covariant Metric, we implicitly assumed the need for a balanced data set (in which distinct classes occur with roughly the same frequency, and with similar within-standard deviation), which may prove to be impractical in unattended deployments. While it is easy to see that the K-Means algorithm works best with balanced data sets, the usability of the Covariant Metric is limited to K-Means because of this assumption.

## 8   Conclusion

In this paper, we proposed a new quality metric for K-Means clustering and benchmarked it against existing indexes. From our comparative studies on synthetic data, we see that the Variance Ratio Criterion [5] (**VRC**) works remarkably well, followed closely by the Davies-Bouldin index [8] (**DB**). Our own index $\mathcal{M}_C$ proposed in this article came in third when tested on synthetic data, but easily outperformed both **VRC** and **DB** in real data. Besides, the significance of the peak indicating the right $K$ was substantially larger for $\mathcal{M}_C$.

All other indexes performed poorly on both synthetic as well as real data. Either $\mathcal{M}_C$ or **VRC** seems to be preferable to the popular "elbow" method (which looks for a kink in the variation of the sum of squared errors, and is very subjective). Furthermore, our results indicate that both the Akaike and the Bayesian Information Criteria (**AIC** [3] and **BIC** [24]) are ineffectual in selecting the right $K$ in K-Means clustering. The Gap Statistic [26] (**Gap**) performs slightly better than the information criteria, but it is prohibitively expensive, computationally.

Although more systematic exploration on more data sets is indicated, our Covariant Metric ($\mathcal{M}_C$) metric does show promise in the real data sets that we studied so far, as well as on an extensive collection of synthetic data. When it comes to discovering the right number of clusters, $\mathcal{M}_C$ performed remarkably well. In fact, in real data, it outperformed the all other commonly used indexes of clustering quality by impressive margins.

Once we have a reliable metric for the quality of clustering, we can automate and build upon the current K-Means clustering algorithm. For instance, we can create scripts that will automatically select the optimal number of clusters (and possibly the best variables to use). Much like the forward selection or backward elimination processes in linear regression, K-Means clustering then becomes amenable to automatic optimizations. Furthermore, with robust metrics enabling automatic discovery of the right number of clusters, it may become possible to deploy K-Means clustering in situations where automated generation of insights without manual supervision is desired.

## References

1. Aeberhard, S., Coomans, D., de Vel, O.: Comparison of classifiers in high dimensional settings. Technical report. 92–02, Department of Computer Science and Department of Mathematics and Statistics, James Cook University of North Queensland (1992). https://doi.org/10.1016/0031-3203(94)90145-7

2. Ahn, S., Fessler, J.A.: Standard errors of mean, variance, and standard deviation estimators (2003)
3. Akaike, H.: A new look at the statistical model identification. IEEE Trans. Autom. Control **19**, 716–723 (1974). https://doi.org/10.1109/TAC.1974.1100705
4. Arbelaitz, O., Gurrutxaga, I., Muguerza, J., Pérez, J.M., Perona, I.: An extensive comparative study of cluster validity indices. Pattern Recogn. **46**(1), 243–256 (2013). https://doi.org/10.1016/j.patcog.2012.07.021
5. Caliński, T., Harabasz, J.: A dendrite method for cluster analysis. Commun. Stat.-Simul. Comput. **3**, 1–27 (1974). https://doi.org/10.1080/03610927408827101
6. Campo, D., Stegmayer, G., Milone, D.: A new index for clustering validation with overlapped clusters. Expert Syst. Appl. **64**, 549–556 (2016). https://doi.org/10.1016/j.eswa.2016.08.021
7. Charytanowicz, M., Niewczas, J., Kulczycki, P., Kowalski, P.A., Łukasik, S., Żak, S.: Complete gradient clustering algorithm for features analysis of X-ray images. In: Piętka, E., Kawa, J. (eds.) Information Technologies in Biomedicine, vol. 69, pp. 15–24. Springer, Cham (2010). https://doi.org/10.1007/978-3-642-13105-9_2
8. Davies, D.L., Bouldin, D.W.: A cluster separation measure. IEEE Trans. Pattern Anal. Mach. Intell. **PAMI-1**, 224–227 (1979). https://doi.org/10.1109/TPAMI.1979.4766909
9. Dheeru, D., Taniskidou, E.K.: UCI machine learning repository (2017)
10. Feng, Y., Hamerly, G.: PG-means: learning the number of clusters in data. In: Schölkopf, B., Platt, J., Hoffman, T. (eds.) Advances in Neural Information Processing Systems, vol. 19. MIT Press (2006)
11. Fisher, R.A.: The use of multiple measurements in taxonomic problems. Ann. Eugen. **7**, 179–188 (1936). https://doi.org/10.1111/j.1469-1809.1936.tb02137.x
12. Fränti, P., Rezaei, M., Zhao, Q.: Centroid index: cluster level similarity measure. Pattern Recogn. **47**(9), 3034–3045 (2014). https://doi.org/10.1016/j.patcog.2014.03.017
13. Hämäläinen, J., Jauhiainen, S., Kärkkäinen, T.: Comparison of internal clustering validation indices for prototype-based clustering. Algorithms **10**, 105 (2017). https://doi.org/10.3390/a10030105
14. Hamerly, G., Elkan, C.: Learning the K in K-means. In: Advances in Neural Information Processing Systems, vol. 17 (2004)
15. Hartigan, J.A.: Clustering Algorithms. Wiley Series in Probability and Mathematical Statistics: Applied Probability and Statistics. Wiley (1975)
16. Ishioka, T.: An expansion of X-means for automatically determining the optimal number of clusters. In: Computational Intelligence (2005)
17. Mahalanobis, P.C.: On the generalized distance in statistics. Proc. Natl. Inst. Sci. India **2**, 49–55 (1936)
18. Pakhira, M.: Finding number of clusters before finding clusters. Procedia Technol. **4**, 27–37 (2012). https://doi.org/10.1016/j.protcy.2012.05.004
19. Pelleg, D., Moore, A.W.: X-means: extending K-means with efficient estimation of the number of clusters. In: ICML (2000)
20. Pham, D., Dimov, S., Nguyen, C.: Selection of K in K-means clustering. Proc. Inst. Mech. Eng. Part C-J. Mech. Eng. Sci. **219**, 103–119 (2005). https://doi.org/10.1243/095440605X8298
21. Qiu, W., Joe, H.: Generation of random clusters with specified degree of separation. J. Classif. **23**(2), 315–334 (2006). https://doi.org/10.1007/s00357-006-0018-y
22. Qiu, W., Joe, H.: Separation index and partial membership for clustering. Comput. Stat. Data Anal. **50**, 585–603 (2006). https://doi.org/10.1016/j.csda.2004.09.009

23. Rousseeuw, P.J.: Silhouettes: a graphical aid to the interpretation and validation of cluster analysis. J. Comput. Appl. Math. **20**, 53–65 (1987). https://doi.org/10.1016/0377-0427(87)90125-7

24. Schwarz, G.: Estimating the dimension of a model. Ann. Stat. **6**, 461–464 (1978). https://doi.org/10.1214/aos/1176344136

25. Sieranoja, S.: How much K-means can be improved by using better initialization and repeats? Pattern Recogn. **93** (2019). https://doi.org/10.1016/j.patcog.2019.04.014

26. Tibshirani, R., Guenther, W., Trevor, H.: Estimating the number of clusters in a data set via the gap statistic. J. R. Stat. Soc.: Ser. B (Stat. Methodol.) **63**, 411–423 (2002). https://doi.org/10.1111/1467-9868.00293

27. Yue, S., Wang, J., Wang, J., Bao, X.: A new validity index for evaluating the clustering results by partitional clustering algorithms. Soft. Comput. **20**(3), 1127–1138 (2015). https://doi.org/10.1007/s00500-014-1577-1

28. Zhao, Q., Fränti, P.: WB-index: a sum-of-squares based index for cluster validity. Data Knowl. Eng. **92**, 77–89 (2014). https://doi.org/10.1016/j.datak.2014.07.008

# Clustering Method for Touristic Photographic Spots Recommendation

Flavien Deseure-Charron, Sonia Djebali$^{(\boxtimes)}$, and Guillaume Guérard

Léonard de Vinci Pôle Universitaire, Research Center,
92 916 Paris La Défense, France
flavien.deseure-charron@edu.devinci.fr,
{sonia.djebali,guillaume.guerard}@devinci.fr

**Abstract.** Tourism and photography have become very complementary, and tourists are constantly seeking the best spots to capture pictures and memorize their vacations. However, the search for the best and unforgettable photographic spots is difficult and time-consuming for tourists, especially when visiting new regions. In this paper, we propose a method for discovering tourist photo spots from geotagged photos using clustering algorithms. The clusters are characterized to determine the type of photos such as selfies or panoramic. We compare our approach to the most used clustering algorithms namely K-Means and DBSCAN. The approach is simulated and experimentally evaluated on a real photographic dataset of the French capital *Paris*. Our approach identifies the best-known, quirky and thematic spots in the reference websites.

**Keywords:** Tourism · Photographic spots · Clustering · HDBSCAN · Knowledge discovery

## 1 Introduction

Nowadays, tourism is considered one of the largest and fastest-growing industries. It is a significant economic sector for many countries in the world. Tourism is deeply related to photography [4], especially because pictures allow travelers to maintain good memories of their destinations [2]. *Deborshee Gogoi* introduces in 2014 this concept as: *"Photographic tourism is that form of special interest tourism in which tourist visits a particular place with the primary aim of photographing subjects that are unique to him. The scope of photography may range from landscapes, portraits, architectures, culture, food and wildlife to even macro subjects"* [10].

With the exponential increase of compact, cheap, or user-friendly cameras, tourists tend to share more and more photographs to immortalize their experience and keep memories. This affluence of photos has led to the development of multiple photo-sharing services such as *Flickr* and *Instagram*. These platforms have redefined the way that people travel [11,14]. According to travel websites like *Expedia*[1], the main priority for young travelers is whether their destination is visually appealing and lends itself to be photographed for posting on photo-sharing networks.

However, finding the *"best"* spots to take photographs remains a tedious task, especially when tourists do not know the region they visit [9]. The studies that have been conducted up to now are focused on the identification and discovery of a specific points that someone may find interesting using geotagged photos. These points are mentioned in literature as *hotspots, points of interest* (POI), and *areas of interest* (AOI) [3].

Our study focuses on the identification of the areas where the photos are captured about a POI and not the identification of POI or AOI. To distinguish those areas, they are named *Touristic Photographic Spots* (TPS) in the rest of the paper. For each TPS, a set of characteristics is determined define the kind of photos. For example, a TPS to have a panorama view, a TPS to acquire a sunset in front of a monument, etc.

This paper proposes an approach to identify the *Touristic Photographic Spots* of a POI and to qualify them. Our contributions can be resumed as the following:

- Clustering methods to determine TPS: this method take into account the various aspect of the photos such as the density of photos, the distance to the POI, the angle to the POI. In this manner, the clustering method is based both on the geographical density of photos and on the variability of the metadata. The choice of clustering algorithms must take into account both the density and the proximity of data in respectively the geographical aspect and the photographic aspect.
- A knowledge extraction to qualify each TPS: from the metadata of the cameras, we compute for each TPS some knowledge such as its popularity, the best time of the day to take photos or its focus.

This paper is organized as follows. Section 2 describes the related work for representing and discovering spots from geotagged photos. Section 3, presents our approach for identifying photographic spots using clustering methods. Section 4 describes and comments the results. Finally, Sect. 5 presents our conclusions and recommendations for future studies.

## 2    Related Work

The studies that have been conducted up to now, in the field of tourists photography are focused on the identification of the *hotspots, points of interest* (POI), and *areas of interest* (AOI). As far as we know, no approach has addressed our problem which is the identification of photographic spots using geotagged photos. The closest works on our problem are the discovery of POI or AOI.

Some studies focus on *density* [6,8]. These aggregated data by hexagons to produce density maps and therefore found the POI from the main peaks. This

---

approach has the advantage to be simple but the obtained density function is too smooth to be used for photo spot discovery.

Some other studies considers a quite different approach, focusing on clustering algorithms. Indeed, discovering POI from geotagged photos can be treated as a clustering problem to identify the most photographed places. More precisely, the method used for geotagged photos are part of the geospatial clustering.

*K-Means clustering algorithm* (centroid-based algorithm) is the most used one to identify clusters from geotagged data [15]. However, K-Means requires the number of clusters as an input parameter, and it detects only spherical clusters. This shape is unsuitable to reality. Indeed, peoples take photos depending on the urban structure and the topology of the region, not as a bird view.

*Density-based clustering methods* are used to identify POI because a high photos activity can be measured by density. These methods don't require the number of clusters as an input parameter. They can handle arbitrary shape clusters, and sparse regions are treated as noise. Some example of algorithms used are *Mean Shift* [5,18], *Ordering Points To Identify the Clustering Structure* (OPTICS) [9] and *Density-Based Spatial Clustering of Applications with Noise* (DBSCAN) [7]. Kisilevich et al. [16] propose *Photo-DBSCAN* (P-DBSCAN) a new density-based clustering algorithm based on DBSCAN, that weight the photo on various metadata. This method has been exploited and enriched during the last decade [17,19].

Some recent studies use the topological structure of photos (spatial and non-spatial) to determine clusters such as GeoSOM based on Self-Organizing map [12]. The structure can also be studied through data mining such as FPGrowth to understand the behaviors of tourists [13]. But those methods are used to determine behaviors and patterns of people considering spatial and temporal as continuous data.

Based on these varied approaches, we can deduce that to determine *Touristic Photographic Spots*, the choice of the algorithm is important and is directly related to the data. If we take the example of the *Eiffel Tower*, we found many streets, bridges, and parks very close to the monument having diverse views and which constitute a different TPS in its own right. However, these are undetected by the most used methods seen previously. Centroid-based don't suit the urban infrastructure and Density-based may regroup excessively vast areas due to the high proximity of those areas.

Our approach will be compare to the most used clustering algorithms for POI identification namely K-Means and DBSCAN.

# 3   Our Approach

Our objective is to propose a method to identify the TPS of each POI and qualify the TPS. Each POI is characterized by its type (hotel, restaurant, attraction) and localization (lat, long). The POI is photographed by several users which are identified by an identifier. Each photo is characterized by the tags of the photo, localization (lat, long), the date and time when the photo is captured. A set of data is available on the camera used for each photograph as ISO, aperture, shutter speed, and focal length.

Since we are looking for TPS as areas of various shape and size, and with a homogeneous data for the characteristics, we propose a double clustering to define TPS. The first algorithm, as a geographical clustering, will determine the shape of the areas as when the second one, a data clustering guarantee the homogeneity of the data. Then, we extract information from the clusters concerning photographic tourism.

Our approach can be resumed in three steps as follows (see Fig. 1):

1. *Global clustering*: to create the continuous density areas, we apply a first clustering algorithm on the Cartesian projection associated with the photos of each POI.
2. *Local clustering*: to define TPS. For each cluster from the global clustering, the local clustering is based on two new parameters, distance to POI and angle with POI. Those greatly transform the view of any photos and are required to refine previous clusters.
3. *TPS qualification*: extraction of knowledge from photos to qualify the found TPS. The qualification is used to recommend the latter to tourists.

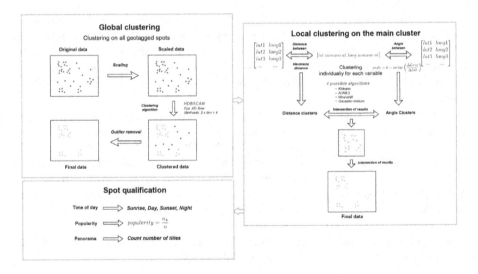

**Fig. 1.** Flowchart of our approach.

## 3.1 Global Clustering

The first step of our approach is to determine the shape of areas where tourists take photos. The first step for any geospatial clustering is to make a Cartesian projection of all photos. The global clustering has two main goals. Firstly, the numerous outliers has to be removed, *i.e.* the photos that cannot constitute a cluster. We define as outliers a small amount of geotagged data with a low density that are not representative of any trend. Secondly, the TPS may be at any size. It is dependent of the topology and the urban infrastructure, which constrains how photos are taken.

To achieve these two goals, we use a density-based clustering algorithm. We choose this kind of algorithms because they separate clusters by contiguous areas of low point density. The data points in the separating areas of low point density are typically considered outliers. Some existing methods like DBSCAN and OPTICS fail to identify clusters with different density levels because they are based on a "flat" (*i.e.* non-hierarchical) representation. One of the methods that solve this problem is HDBSCAN [20]. It is a clustering algorithm that extends DBSCAN by converting it into a hierarchical clustering algorithm. This method works in three steps: first, it estimates the densities around certain data to determine a threshold; then, it selects areas above this threshold density; finally, it combines points in these selected areas.

Most of the density-based methods require the assumption of a density threshold. They compute a threshold and gather the data with densities above the threshold and group theme together to form clusters. To use HDBSCAN algorithm, first, we need to estimate the density around some data to build the density landscape of the dataset. The HDBSCAN algorithm computes the *Core distance* of a random set of data thanks to the *K-th nearest neighbor* (KNN) method. Data in denser regions would have smaller Core distances while data in sparser regions would have larger Core distances. The density landscape is the inverse of the Core distance. Then, HDBSCAN builds a hierarchy to figure the right density for each cluster and how to cut like a hierarchical clustering.

In our approach we compare the HDBSCAN algorithm with the most used clustering algorithms for POI identification namely K-Means and DBSCAN.

### 3.2 Local Clustering

The global clustering provides a set of clusters that contains a continuous density area of geotagged photos. Apart from the location of the photo, two other parameters as angle of view and distance to the POI may vary greatly inside each cluster. These parameters characterize the TPS differently as they alter the purpose of the photos. The angle of view and distance are greatly heterogeneous into the cluster around and near the POI, and with very large clusters. To refine the clusters, we apply a second clustering based on those two parameters.

The second clustering is not applied on all clusters but just on clusters having a threshold of variation of the angle and distance to POI. Those thresholds define the limit of acceptable heterogeneity of a cluster. The two thresholds are defined as follows:

- Clusters with a surface representing a total angle value above a threshold are refined. We fix this threshold to *one hour angle*, which corresponds to 15 degrees. It corresponds to the change of framing in photography.

– Clusters with the maximum distance between two points of the cluster must be superior to a threshold are refined. We fixed this threshold to *four-time* the *epsilon* value of the global clustering (thus two times the diameter to have a sufficient size).

To define the angle formed between all points of the cluster and the coordinates of the POI we use the following formula:

$$angle = \theta = \arctan \left( \frac{\Delta Long}{\Delta Lat} \right) \tag{1}$$

where $\Delta Long$ and $\Delta Lat$ are the difference in longitude and latitude between the data of the cluster and the POI, respectively. This angle is computed in radian.

To compute the distance between all points of the cluster and the coordinates of the POI, we use a geodesic distance *i.e.* harversine distance.

Since we cannot make assumptions about the form and density of the second clustering, we propose using various algorithms with pros and cons and to compare their results to determine which one fits these data. We used internal and external measurements to select the most appropriate method. For comparison purposes, our approach should be deterministic.

In order to choose the best algorithm, we compare the results of four algorithms with their pros and cons from the four main types of clustering:

**Partitional/Centroid-Based: K-Means** [1]**.** The K-Means algorithm is still widely used because of its simplicity and good performance. The most key parameter to set is the number of clusters. To tune this number, we use the *Elbow* method [22] using the *Within-Groups Sum of Squares* (WGSS). Indeed, we want to evaluate the cohesion of the clusters and not the separation, as the clusters are very close to each other.

**Distribution Model-Based: Gaussian Mixture Model (GMM)** [1]**.** This algorithm employs an interesting approach that tries to represent the dataset as a mixture of normal distribution. A GMM tends to group the data points belonging to a single distribution together. While K-Means forms spherical shape cluster, GMM can produce various ellipsoid shapes for the same dataset. For the number of components parameter, in the same way as with K-Means, we implement the Elbow method with WGSS.

**Density-Based: Mean Shift** [1]**.** This deterministic method update potential centroids to be the mean of the points within a given bandwidth. Mean Shift works very well on spherical-shaped data. Furthermore, it automatically selects the number of clusters contrary to other clustering algorithms like K-Means. The bandwidth was estimated by computing the *K-Nearest Neighbors* KNN algorithm as recommended.

**Hierarchical: AGNES** [1]**.** Hierarchical method are deterministic and constructs a hierarchical tree of distances between data, called a dendrogram. This is helpful because the algorithm produces an explicit graphical depiction of the clusters. The AGNES method is one of the most used. It adopts a bottom-up approach. For the number of clusters, we employ the Elbow method with WGSS with a *Ward linkage*.

### 3.3 Indexes and Validation

After applying the global clustering and the local clustering, we intersect the results for both parameters angle and distance to produce new clusters from the main one. The evaluation is essential to our approach, as it will allow us to choose the most appropriate algorithm for the local clustering. The evaluation allows us to find the most efficient algorithm.

We implemented internal measurements for the evaluation based on the Cartesian projection of the data. The difference between algorithms can be very closed and external evaluation would be ineffective and irrelevant since we refine a single cluster. In our context, we need a high cohesion for clusters as there are not well separated. Therefore, we choose the following evaluation:

1. *Ball Hall*: it computes the mean dispersion of a cluster, *i.e.* the mean of the squared distances of the points of the cluster with respect to their center. The lower the value, the better the clustering is.
2. *Banfeld-Raftery index*: it is the weighted sum of the logarithms of the mean of the squared distances between the points in the cluster and their center. The logarithm allows smoothing of the impact of big or small clusters in the number of points. The lower the value, the better the clustering is.

The final clustering is composed of clusters from the global clustering and refined clusters from the local clustering. Since the clusters are computed from the data for a POI, they represent TPS for this POI. The process is done for each POI.

### 3.4 TPS Qualification

Once the TPS of a POI are defined, we qualify these TPS to perform recommendations to tourists. We define three ways to qualify the TPS: first, according to the time of day, then, whether this TPS is a panorama or not, and finally according to the popularity of the TPS.

**Time of Day.** We have broken down the day into four parts: Sunrise, Day, Sunset, Night. We use the date of the photo and the $ISO^2$ of the camera to determine the right part for each photo. We manage the time when the photo was captured with a margin of $\pm10$ min. (1.4% of the day) for sunrise and $\pm20$ min (2.8% of the day) for sunset depending on the timezone and period of the year. We can also deduce the time of day from the $ISO$ used by the photo. The higher the ISO, the more sensitive the camera sensor becomes, and the brighter your photos appear. $ISO100$ is used in a sunny and open area, $ISO400$ is used during a cloudy day, $ISO800$ and higher are used from sunset to night time of

---

[2] https://www.adobe.com/creativecloud/photography/discover/iso.html.

day. From each TPS, we determine the percent of each part of the days from its photos.

**Panorama.** This category indicates TPS that are likely to be panoramas. To perform this, we combined several indicators: the number of POI having this TPS, the aperture, and the focal length of each photo. First, we selected all the photos located in the same cluster and counted the number of different POI taken in each photo (from the tags). We compute the mean and the ratio of photos having several POI to those with one POI. For this second indicator, the aperture value designates the width of the hole within the lens through which the light travels into the camera body. When the aperture is very narrow (below $f/8$), the depth of field is large and therefore there is a chance that the photo is a panorama. We take into account the largest aperture used among the photos and its percentage. Finally, the focal length represents the measure of the optical distance inside the lens from which all light rays converge on the image sensor of the camera. The lower the value, the wider the field of view is, and therefore the more likely it is a panorama. Focal lengths below 55 mm are used to take large-angle photos. We take into account the largest focal length used among the photos and its percent.

**Popularity.** The third way to qualify a TPS is according to its popularity, in number of photos, and to be able to classify this TPS as unusual or unmissable for example. A direct and effective measure is the percentage of importance of the cluster as follows: $popularity = \frac{n_k}{n}$. Where $n$ remains the number of points in the dataset and $n_k$ is the number of points in the $k^{th}$ cluster.

## 4    Experiments

We conducted experiments on the social network *Flickr* over a period from 2007 to 2019. We chose Flickr because it is primarily aimed at professionals and photo enthusiasts. Moreover, it grants us access to the data related to the camera. For our case study, we have chosen the city of *Paris*, because it is one of the most attractive cities in the world; regularly ranking first among the most visited cities in the world. In our dataset gathered from Flickr focused on *Paris* region, we have $2,945,085$ geotagged photos on $1,414,816$ POIs taken by $98,555$ users with $2,948$ different camera sets.

### 4.1    Data Processing

The tags in each photo, which designate the POI taken, have not been pre-filled by the social network Flickr. Thus, it is written in many alternative ways in the dataset (misspellings, translated into other languages). To overcome this problem, we grouped all titles with a name at 85% similar *Cosine similarity* using the vector embedding generated from *Sentence-Bert* [21]. At that time, we removed photos captured in a short time interval by the same user. As we want to extract knowledge from our clusters, especially at the time of day when a POI was taken, we chose a duration of five hours at minimum between two photos. This allows us to take into account the users who took the POI during the day and at sunset for example. Then, for determining the coordinates of the POI, we use the *Open Topo Data API*[3]. Finally, we project the data into a Cartesian plane to perform our method.

To experiment with our approach, we present the result of the *Eiffel Tower* POI. We choose this POI because, with more than 7 million visitors a year, it is the most visited and photographed monument in the world on various social networks like Flickr and Instagram. Since the monument is seen on most websites about *Paris* and lots of media provides good analytics of the photographic tourism of the *Eiffel Tower*, we can easily compare our results to them. In this specific POI, after pre-processing, we get $17,781$ photos taken by $11,372$ users with 674 different cameras.

To validate our approach, we compare the results with the spots referenced on the blogs and touristic websites: lonelyplanet with 3.2 millions monthly visitors[4], lodgisblog with, 90 thousands monthly visitors (french), blog specialized in photographic tourism[5], image banks (Google Images, Instagram, Pinterest).

## 4.2   Global Clustering Comparison

Based on the $17,781$ photos tagged *Eiffel Tower* we produce a Cartesian projection then we implement the clustering algorithm. We initially performed the global clustering using three methods: HDBSCAN, KMeans, and DBSCAN.

**Conventional Methods.** First of all, K-Means has been chosen as a reference as it is one of the most used and simple clustering methods. This method randomly initiates $K$ points in the data as centroids and assigns all points to the nearest centroids. Then the centroid moves to the average of the points assigned to it. And we rehearse this step until convergence. The most critical parameter to determine is the number of clusters. To acquire the most proper value for this number, we compute the algorithm in a range from 2 to 100 and compute two metrics: *Silhouette* and *Davies-Bouldain*. The objective is to determine the value that maximizes the Silhouette score and minimizes the Davies-Bouldain score. Finally, we compute the mean value rounded up to the nearest whole number between the best number of clusters according to each index and employ it as a parameter for our method.

Secondly, DBSCAN has been chosen as the most used density-based method. For the *epsilon* value estimation, we use the *k-dimensional tree* (kd tree) method: we compute the Nearest Neighbors algorithm on our data and get the distances between all neighbors. Then, we select the elbow of the curve of the distance and use the associated distance as the epsilon point. For the *minPoints* parameter,

---

[3] https://www.opentopodata.org/.

[4] https://www.lonelyplanet.fr/article/10-points-de-vue-sur-la-tour-eiffel.

[5] http://blog.lodgis.com/top-10-des-vues-sur-la-tour-eiffel/.

we set the value to 4, which is equivalent to $2 * number\_of\_dimension$ as recommended in the original paper of DBSCAN [7]. Finally, we choose the Euclidean distance for performance reasons.

**Indexes and Validation of the Global Clustering.** To determine the best clustering between K-Means, DBSCAN, and HDBSCAN, we will use the following indexes:

1. Banfeld-Raftery: As reminder, it measures the mean of the squared distances between the points of a cluster and its center. The lower the value, the more points are closed to the center.
2. Davies-Bouldain: The score is the average of the maximum ratio between the distance of a point to the center of its cluster and the distance between two clusters centers. As a result, clusters that are farther apart and less scattered will score higher. The minimum value for this index is 0 and lower values indicate better clustering.
3. Calinski-Harabasz: It is the ratio between the between-cluster variance and the within-cluster variance. The Calinski-Harabasz index varies between 0 (worst clustering) and *infinite* value (best clustering). It increases linearly with the number of points in the sample. Therefore, its order of magnitude can vary considerably from one dataset to another.

From those indexes, we choose the clustering with the highest Banfeld-Raftery value, the lower Davies-Boudlain index, and the highest Calinski-Harabasz index.

**Results.** As a reminder, the cohesion of the clusters is the most important aspect as opposed to the separation. Indeed, two photos spots can be very close or even touch each other and thus decrease the separation value and all associated measurements. The Table 1 shows the results of the three clustering algorithms and the values of the cohesion index *Banfeld-Raftery*, *Davies-Bouldain* and *Calinski-Harabasz* of each algorithm.

**Table 1.** Internal measurements comparison of classical methods (Eiffel Tower).

|                    | K-Means     | DBSCAN       | HDBSCAN     |
|--------------------|-------------|--------------|-------------|
| Banfeld-Raftery    | $-1.86e^5$  | $-1.88e^5$   | $-1.94e^5$  |
| Davies-Bouldain    | $7.3e^{-1}$ | $5.06e^{-1}$ | $5e^{-1}$   |
| Calinski-Harabasz  | $1.3e^4$    | $4.39e^3$    | $4.4e^4$    |

As a reminder, we choose the clustering with the highest Banfeld-Raftery value, the lower Davies-Boudlain index, and the highest Calinski-Harabasz index. HDBSCAN presents the best overall results, close to DBSCAN. Thus, we choose its cluster with $-1.94e^5$ Banfeld-Raftery value, $5e^{-1}$ Davies-Boudlain index and $4.4e^4$ Calinski-Harabasz index. The Fig. 2 presents the results of the three algorithms. K-Means (Fig. 2c) provides clearly some unintelligible clusters. The main difference between HDBSCAN (Fig. 2a) and DBSCAN (Fig. 2b) is the clusters of small size are not discovered by DBSCAN. Moreover, since DBSCAN have no regulation on the density/sparsity of data, it tends to find compact cluster which affects clusters or borders with less density than the fixed threshold of DBSCAN.

The global clustering also allowed identifying the main photographic spots (Fig. 2a). We noticed that most of photographic spots use on the internet were identified with our method. Most of the spots are widely shown is the reference websites about the *Eiffel Tower*: 1) *Arc de Triomphe*, 2) *Place de la Concorde*, 3) *Montmartre*, 4) *Centre Pompidou*, 5) *Printemps Haussman* (rooftop), 6) *Tour Montparnasse*, 7) *Galeries Lafayette* (rooftop). The global clustering also allowed to identify of photographic spots not present on the reference sites like the two bridges a) *Grenelle bridge*, b) *Mirabeau bridge*, etc.

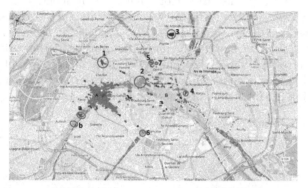

(a) Global clustering with HDBSCAN.

(c) Global clustering with K-Means.

(b) Global clustering with DBSCAN.

**Fig. 2.** Global clustering comparison. (Color figure online)

## 4.3   Local Clustering Comparison

Two clusters are selected for the local clustering: *A* (in dark green) and *B* (in yellow) from Fig. 2a. In the results, we present the local clustering for the first selected cluster.

We test the four algorithms: K-Means, Mean Shift, DBSCAN, Agglomerative clustering. We compare results with Ball Hall and Benfeld-Raftery index

(Table 2). As a reminder, we choose the clustering with the lowest Ball Hall value and Benfeld-Raftery index.

**Table 2.** Internal measurements comparison of algorithms used for local clustering (Eiffel Tower).

|  | KMeans | AGNES | Mean shift | Gaussian mixtures |
|---|---|---|---|---|
| Ball Hall | $8e^{-5}$ | $8e^{-5}$ | $1e^{-4}$ | $8e^{-6}$ |
| Banfeld-Raftery | $-1.65e^5$ | $-1.60e^5$ | $-1.67e^5$ | $-1.66e^5$ |

The results of the algorithms are presented in Fig. 3. The GMM clustering provides the best results which are also close to the urban structure near the *Eiffel Tower*. From top left to bottom right, we can mention clusters located at *Trocadero's Garden*, *Alma's bridge* with *Palais de Tokyo*, *Iéna's bridge* with *Carrousel*, closest northern roads (*Quai Branly*), closest eastern roads (*Bourdonnais avenue and University*), the inner ring represents the various points of view at the feet of the tower, then on the left in light blue we got *Bir Hakeim bridge*, following by the *Emile Antoine stadium*, the *Hotel Pullman* and finally the bottom pink cluster represents the *Champ de Mars*.

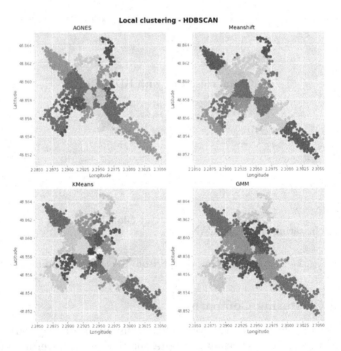

**Fig. 3.** Comparison of the four algorithms for local clustering on the Eiffel Tower in Paris.

As shown, the local clustering provides a more comprehensive view of photographic tourism near the *Eiffel Tower*. The various cited places offer various points of view and ways to handle the tower's environment.

### 4.4   Spot Qualification

*Time of Day.* During our experiments, we discovered good spots for sun dusk as the proportion differs a lot from the reference one (between 7% and 10% of all photos TPS instead of 2.78%), see TPS in Fig. 4. Those spots are on the east, northeast of the *Eiffel Tower*, and they are at a higher altitude. Some spots are close to the POI, tourists capture the sun dusk between the feet of the *Eiffel Tower*. One TPS on the bottom left is the *Grenelle bridge* (number 26), and many others are on various bridges. Tourists can see the sun dusk glare on the water in front of the tower. We conclude those spots are ideal to take the sun dusk close to the *Eiffel Tower*.

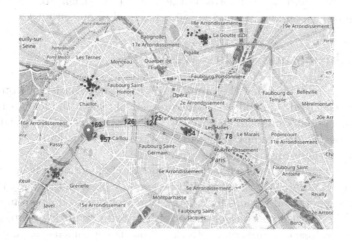

**Fig. 4.** Clusters with a high percent of sun dusk.

*Panorama.* Concerning the panorama, we obtained the following map Fig. 5). We found most of panorama's like *La Défense* (number 1), *Montmartre* (number 2), *Arc de Triomphe* (number 3), some bridges, the *Place de la Concorde* (number 4), *Saint Honoré district* (number 5) and *Les Halles* (number 6). Those places offer an advantageous point of view of some parts of Paris and are at higher altitudes or with vast open views. Those results are biased as most tourists want to place the *Eiffel Tower* in every photo.

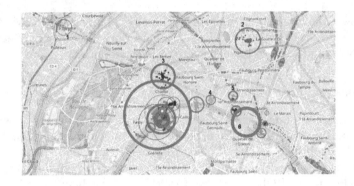

**Fig. 5.** Panorama's detection for Eiffel Tower

*Popularity.* The most popular spots are all around the monument (main global cluster) and in large places (*Place de la Concorde, Montmartre, Arc de Triomphe*). Moreover, most of the photos on the internet are taken from those places.

A deep discussion about our results is presented in our github[6].

## 5    Conclusion and Future Work

Our approach allows solving the problem of identifying Touristic Photographic Spots. It has succeeded in addressing the problem of places "hidden" by the main cluster using a double clustering approach. Our context-free method adapts itself to the dataset. It determines TPS with characteristics thanks to a benchmark of adapted methods and index comparison.

In future works, we will enhance the spot's qualification. Flickr grants us access to the *EXIF* data of the images. It is therefore possible to qualify the spots by using the photo and camera metadata. Moreover, we need to strengthen our methods. We are also considering adopting the Self-Organizing Map, Spectral Clustering and Spacial Clustering with the following five dimensions to find TPS: three spatial dimensions, distances to POI, and angles with POI.

## References

1. Benabdellah, A.C., Benghabrit, A., Bouhaddou, I.: A survey of clustering algorithms for an industrial context. Procedia Comput. Sci. **148**, 291–302 (2019)
2. Berger, H., Denk, M., Dittenbach, M., Pesenhofer, A., Merkl, D.: Photo-based user profiling for tourism recommender systems. In: Psaila, G., Wagner, R. (eds.) EC-Web 2007. LNCS, vol. 4655, pp. 46–55. Springer, Heidelberg (2007). https://doi.org/10.1007/978-3-540-74563-1_5
3. Camilleri, M.A.: The tourism industry: an overview. In: Camilleri, M.A. (ed.) Travel Marketing, Tourism Economics and the Airline Product. THEM, pp. 3–27. Springer, Cham (2018). https://doi.org/10.1007/978-3-319-49849-2_1

---

[6] https://github.com/flaviendeseure/Clustering-Method-for-Touristic-Photographic-Spots-Recommendation.git.

4. Cederholm, E.A.: The use of photo-elicitation in tourism research-framing the backpacker experience. Scand. J. Hosp. Tour. **4**(3), 225–241 (2004)
5. Clements, M., Serdyukov, P., De Vries, A.P., Reinders, M.J.: Using flickr geotags to predict user travel behaviour. In: Proceedings of the 33rd International ACM SIGIR Conference, pp. 851–852 (2010)
6. Da Rugna, J., Chareyron, G., Branchet, B.: Tourist behavior analysis through geotagged photographies: a method to identify the country of origin. In: 2012 IEEE 13th International Symposium CINTI, pp. 347–351. IEEE (2012)
7. Ester, M., Kriegel, H.P., Sander, J., Xu, X., et al.: A density-based algorithm for discovering clusters in large spatial databases with noise. In: KDD, vol. 96, pp. 226–231 (1996)
8. García-Palomares, J.C., Gutiérrez, J., Mínguez, C.: Identification of tourist hot spots based on social networks: a comparative analysis of European metropolises using photo-sharing services and GIS. Appl. Geogr. **63**, 408–417 (2015)
9. Gavric, K.D., Culibrk, D.R., Lugonja, P.I., Mirkovic, M.R., Crnojevic, V.S.: Detecting attractive locations and tourists' dynamics using geo-referenced images. In: 2011 10th, TELSIKS, vol. 1, pp. 208–211. IEEE (2011)
10. Gogoi, D.: A conceptual framework of photographic tourism. IMPACT: IJRANSS **2**, 109–114 (2014)
11. Gretzel, U.: Tourism and Social Media, vol. 2. Sage, Thousand Oaks (2018)
12. Henriques, R., Bacao, F., Lobo, V.: Exploratory geospatial data analysis using the GeoSOM suite. Comput. Environ. Urban Syst. **36**(3), 218–232 (2012)
13. Höpken, W., Müller, M., Fuchs, M., Lexhagen, M.: Flickr data for analysing tourists' spatial behaviour and movement patterns: a comparison of clustering techniques. J. Hospitality Tourism Technol. (2020)
14. Kang, M., Schuett, M.A.: Determinants of sharing travel experiences in social media. J. Travel Tourism Mark. **30**(1–2), 93–107 (2013)
15. Kennedy, L., Naaman, M., Ahern, S., Nair, R., Rattenbury, T.: How flickr helps us make sense of the world: context and content in community-contributed media collections. In: Proceedings of the 15th ACM Multimedia, pp. 631–640 (2007)
16. Kisilevich, S., Mansmann, F., Keim, D.: P-DBSCAN: a density based clustering algorithm for exploration and analysis of attractive areas using collections of geo-tagged photos. In: Proceedings of the 1st COM-GEO International Conference, pp. 1–4 (2010)
17. Kuo, C.L., Chan, T.C., Fan, I., Zipf, A., et al.: Efficient method for POI/ROI discovery using flickr geotagged photos. ISPRS Int. J. Geo Inf. **7**(3), 121 (2018)
18. Lu, X., Wang, C., Yang, J.M., Pang, Y., Zhang, L.: Photo2Trip: generating travel routes from geo-tagged photos for trip planning. In: Proceedings of the 18th ACM International Conference on Multimedia, pp. 143–152 (2010)
19. Majid, A., Chen, L., Chen, G., Mirza, H.T., Hussain, I., Woodward, J.: A context-aware personalized travel recommendation system based on geotagged social media data mining. Int. J. Geogr. Inf. Sci. **27**(4), 662–684 (2013)
20. McInnes, L., Healy, J., Astels, S.: HDBSCAN: hierarchical density based clustering. J. Open Source Softw. **2**(11), 205 (2017)
21. Reimers, N., Gurevych, I.: Sentence-BERT: sentence embeddings using Siamese BERT-networks. arXiv preprint arXiv:1908.10084 (2019)
22. Satopaa, V., Albrecht, J., Irwin, D., Raghavan, B.: Finding a "Kneedle" in a haystack: detecting knee points in system behavior. In: 2011 31st ICDCS, pp. 166–171. IEEE (2011)

# Personalized Federated Learning with Robust Clustering Against Model Poisoning

Jie Ma, Ming Xie, and Guodong Long[✉]

Australian Artificial Intelligence Institute, Faculty of Engineering and IT, University of Technology Sydney, Ultimo, NSW, Australia
{jie.ma-5,ming.xie-1}@student.uts.edu.au, guodong.long@uts.edu.au

**Abstract.** Recently, federated Learning (FL) has been widely used to protect clients' data privacy in distributed applications, and heterogeneous data and model poisoning are two critical challenges to attack. To tackle the first challenge that data of each client is usually not independent or identically distributed, personalized FL (PFL) or clustered FL, which can be seen as a cluster-wise PFL method to learn multiple models across clients or clusters. To detect the anomaly clients or outliers, local outlier factor is a popular method based on the density of data points. Therefore, a nested bi-level optimization objective is constructed, and an algorithm of PFL with robust clustering called FedPRC is proposed to detect outliers and maintain state-of-the-art performance. The breakdown point of FedPRC can be at least 0.5. Our experimental analysis has demonstrated effectiveness and superior performance in comparison with baselines in multiple benchmark datasets.

**Keywords:** Personalized federated learning · Robust clustering · Model poisoning

## 1 Introduction

Recently, federated Learning (FL) [33], which was first proposed in 2017, has been widely used to protect clients' data privacy in distributed applications, such as Google's Gboard on Android [33], Apple's Siri [16], Computer Visions [23,24,30], Smart Cities [48] and Healthcare [29,37,45]. The classical FL method, called FedAvg [33], is to train a global model across all clients using gradients to communicate efficiently and privately. Vanilla FL is apparently vulnerable to model poisoning attacks due to its decentralized nature. Therefore, it is challenging to develop an FL application that has good personalized decision-making ability and is robust against model poisoning attacks.

© The Author(s), under exclusive license to Springer Nature Switzerland AG 2022
W. Chen et al. (Eds.): ADMA 2022, LNAI 13726, pp. 238–252, 2022.
https://doi.org/10.1007/978-3-031-22137-8_18

Furthermore, it has been proposed that the not independent or identically distributed (non-IID) challenge can lower the accuracy and efficiency of the training performance. It indicates that the data distribution of each client can be different due to unique attributes or behaviour. Therefore, a globally shared model may not generalize well and fairly in all clients. Personalized FL (PFL) is the most popular method to address this challenge. Based on granularity, PFL can be categorized into cluster-wise PFL and client-wise PFL. PFL methods, such as Ditto [25] and WeCFL [31], train multiple models client-wise or cluster-wise to better adapt to each client or cluster, while knowledge is still shared to improve the performance.

Model poisoning is another challenge in realistic FL. In a distributed system of FL, some malicious agents may upload fake or dirty gradients to the server in the aggregation step, and then the aggregated model to distribute is poisoned. It is naive to adopt anomaly detection techniques to find these malicious agents or outliers. Local outlier factor (LOF) [5] is an efficient method based on the density of data points.

To tackle the two challenges outlined above at the same time, it is difficult to embed the anomaly detection technique into the PFL. We constructed a nested bi-level optimization problem to combine client-wise PFL, cluster-wise PFL and anomaly detection. An algorithm of PFL with robust clustering (FedPRC) is proposed to detect outliers and maintain state-of-the-art performance. Our contributions are summarized below.

- We formulate the PFL problem with robust clustering into a nested bi-level optimization framework.
- We propose a novel PFL with robust clustering (FedPRC) algorithm to solve the complex optimization problem, and the algorithm can resist Byzantine workers.
- The experimental analysis demonstrates the effectiveness and superior performance in comparison with baselines in multiple benchmark datasets.

The remaining sections of the paper are organized as follows. Section 2 introduces related work. We then formulate the problem of PFL with robust clustering in Sect. 3. The proposed FedPRC algorithm is outlined in Sect. 4. Experimental settings and empirical study are discussed in Sect. 5.1 and 5.2, respectively. Finally, we present the conclusion in Sect. 6.

## 2   Related Work

### 2.1   PFL

PFL is the most popular technique used to address the non-IID challenge in FL, as vanilla FL [33] delivers only one globally shared model that cannot fit all clients' data. Based on granularity, PFL can be categorized into cluster-wise PFL and client-wise PFL. For cluster-wise PFL, also called clustered FL, clients are grouped into several clusters, and then an identical number of models are trained based on these clusters. There are mainly two variants in cluster-wise PFL methods, the representation of a client and the clustering method. The work [44] uses model parameters to represent clients and K-means to do clustering. CFL [39] uses hierarchy clustering to divide clients into two clusters based on the cosine similarity of gradients iteratively. The loss of models is also used to cluster clients by HypCluster [32] and IFCA [20]. The unified formulation and convergence of cluster-wise PFL are studied by [31].

For client-wise PFL, each client has its personalized model, either in model structure or model parameters, even in the loss function. A simple but effective method is to fine-tune the trained global model [10,17]. Ditto [25] proposed a bi-level optimization framework using a penalty term to constrain the distance between the local model and the global model. FedRep [12] divides the network into the backbone and the head, and learns shared parameters for the backbone and unique parameters for the head. FedProto [42] adopts prototypes instead of gradients to communicate and is more privacy-protective and communication-effective. Research by [8,40] aims to train a global hyper-network or meta-learner instead of a global model before sending it to clients for local optimization. Meta learning and multi-task learning are also applied into PFL including [17,41].

## 2.2    Robust Clustering

The objective of clustering is to group similar objects together, and group dissimilar objects into different clusters. And robust clustering is to enhance the robustness of clustering results against outliers [19]. Many works have been conducted in this area including [15,36]. Vanilla robust clustering methods include mixture modeling [46], trimming approach [18]. Recently a number of works in robust clustering have been studied by [1,14,18,22,46,47]. The work [6] researches K-means with the bootstrap of median-of-means (MOM). The MOM estimator can mitigate the influence of outliers, whereas the estimator of mean is not good at addressing outliers. The bootstrap of MOM (bMOM) enhances the robustness against outliers can thus achieve a better breakdown point, which is a measure to quantify the toleration of outliers.

## 2.3    Model Poisoning and Anomaly Detection

The way a malicious agent generates an arbitrary update vector by merely shuffling data labels seems very similar to the standard dirty-label poisoning in the study of [9]. However, in the FL setting, the possibility of an adversary controlling a small number of malicious agents and performing a model poisoning attack to manipulate the learning process so that the jointly trained global model, which turns into misclassification over some data, is much higher. FL is apparently vulnerable to model poisoning attacks due to its decentralized nature. A line of work has been done already [3,4,13]. In contrast to previous work, this work focuses on detecting these malicious agents during the central clustering phase by applying a density method to reduce the impact of those agents' updates on the aggregation of the cluster center.

Anomaly detection can be described as the problem of finding patterns in data that do not conform to expected behavior. Anomalies and outliers are two terms used most commonly in the context of anomaly detection. Clustering can be used as a technique for the training of the normality model, where similar data points are grouped together into clusters using a distance function such as [34]. Additionally, LOF [5] is a widely-used density-based anomaly detection method. However, in the case of our method, we already know that malicious agents are the anomalies that we tried to identify. The outcome after precluding those identified outliers would be benign agents, and then only the benign agents' weight matrix would feed into our clustering algorithm. The identifying outliers stage has no inherited relation to the next clustering phase.

## 3 Methodology

Before outlining the methodology, the notations, which can be separated into three parts, FL, clustering, and LOF, are listed below (Table 1).

**Table 1.** Table of Notations

| Components | Notation | Definition |
|---|---|---|
| FL | $m$ | Number of clients in FL system |
|  | $D_i, \lvert D_i \rvert$ | The dataset and its size on Client $i$ |
|  | $\mathcal{M}_i$ | Model function or structure of Client $i$ |
|  | $\omega_i$ | Model parameters of Client $i$ |
|  | $\mathcal{L}_i$ | Loss function of Client $i$ |
|  | $\lambda_i$ | The importance weight of Client $i$, usually measured by its dataset size |
|  | $E$ | Number of local update steps |
| Clustering | $K$ | Number of clusters |
|  | $r_{i,k} \in \mathbb{R}^{m*K}$ | The assignment matrix, $r_{i,k} = 1$ if $i \in k$ else $r_{i,k} = 0$ |
|  | $i \in k$ | Client $i$ belongs to Cluster $k$ |
|  | $g_i$ | General form to represent Client $i$ depending on $h_i$, $l_i$, $D_i$ or something else, e.g. model parameters or loss |
|  | $G_k$ | General form to represent the centroid of Cluster $k$, and usually a linear combination of $g_i$ with $i \in k$ |
|  | $d(g_i, G_k)$ | The distance function of general representations between Client $i$ and the center of Cluster $k$, e.g. Euclidean distance |
| LOF | $n$ | Number of neighbours |
|  | $c_i$ | Indicator. 1 if Client $i$ belongs to inliers else 0 |

### 3.1   PFL

For the classical FL problem, the objective can be formulated as shown below,

$$\underset{\omega}{\text{minimize}} \sum_{i=1}^{m} \lambda_i \mathcal{L}(\mathcal{M}, D_i, \omega). \tag{1}$$

The framework is shown in Fig. 1. The algorithm FedAvg [33] is also implied in this figure, which can be summarized as four steps, model initialization or distribution from server to clients, local update on clients, gradients upload from clients to the server, and model aggregation on the server.

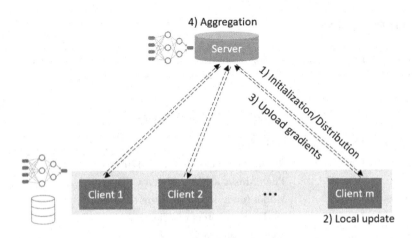

**Fig. 1.** Framework of classical FL

For the client-wise PFL problem, the objective can be formulated as shown below,

$$\underset{\{\omega_i\}}{\text{minimize}} \sum_{i=1}^{m} \lambda_i \mathcal{L}_i(\mathcal{M}_i, D_i, \omega_i), \tag{2}$$

which means an arbitrary client $i$ may have its importance $\lambda_i$, unique dataset $D_i$, model structure $\mathcal{M}_i$, model parameters $\omega_i$, and loss function $\mathcal{L}_i$.

### 3.2   LOF

To understand LOF [5], a density-based anomaly detection method, there are five key definitions step by step. First, $n$-$d$ of an object $o$ is defined as the distance $d(o, p)$ between $o$ and $p \in D$ which satisfies:

- There are at least $n$ objects $o' \in D\setminus\{o\}$, which holds $d(o, o') \leq d(o, p)$, and
- There are at most $n - 1$ objects $o' \in D\setminus\{o\}$, which holds $d(o, o') < d(o, p)$.

Second, the $n$-$d$ neighborhood of an object $o$ can be defined as:

$$N_{n\text{-}d(o)}(o) = \{q \in D | \{o\} \mid d(o, q) \le n\text{-}d(o)\}. \tag{3}$$

Third, the reachability distance of an object $p$ w.r.t. object $o$ is defined as:

$$reach\text{-}d_n(o, p) = max\{n\text{-}d(o), d(o, p)\}. \tag{4}$$

As shown in Fig. 2, the reachability distance of $o, p_1$ and $o, p_2$ equals $n$-$d(o)$ and $d(o, p_2)$, respectively.

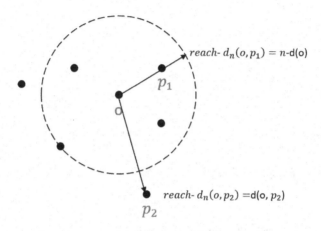

**Fig. 2.** Reachability distance of $o, p_1$ and $o, p_2$, respectively, for $n = 5$

Next, the local reachability density (lrd) of object $o$ is defined as:

$$lrd_n(o) = \frac{|N_n(o)|}{\sum_{p \in N_n(o)} reach\text{-}d_n(o, p)}. \tag{5}$$

Finally, the LOF of object $o$ is defined as:

$$LOF_n(o) = \frac{\sum_{p \in N_n(o)} \frac{lrd_n(p)}{lrd_n(o)}}{|N_n(o)|}. \tag{6}$$

To judge whether an object belongs to outliers, usually yes if its $LOF > 1$, which means it has a lower density than its neighbors, thus an outlier. With the appropriate $n$ chosen, the breakdown point for LOF can be at least 0.5, which means that unless the malicious clients are the majority and behave similarly, LOF will always work.

### 3.3   Proposed Method

The framework of our proposed method of PFL with a robust clustering structure to attack the model poisoning is illustrated in Fig. 3. Its optimization objective can be formulated in the equation below, which is like a nested bi-level optimization problem.

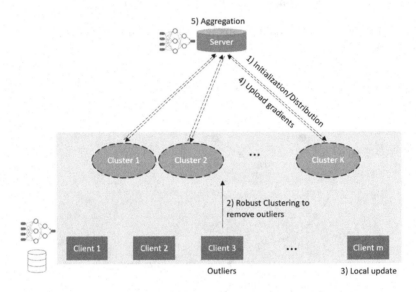

**Fig. 3.** Framework of proposed method

$$\underset{\{\omega_i\}}{\text{minimize}} \ \frac{1}{m} \sum_{k=1}^{K} \sum_{i=1}^{m} \lambda_i r_{i,k} c_i \mathcal{L}(\mathcal{M}, D_i, \omega_i) \tag{7a}$$

$$subject \ to \ \{r_{i,k}\} = \underset{\{r_{i,k}\}}{\text{argmin}} \sum_{k=1}^{K} \sum_{i=1}^{m} \lambda_i r_{i,k} c_i d(g_i, G_k) \tag{7b}$$

$$\{c_i\} = I_{LOF_n(g_i)>1}. \tag{7c}$$

## 4   Algorithm

To solve the complex Objective 7 above, which has three variables, $\Omega$ as the ultimate variable, and $R$ and $C$ as the hidden variables, we need to design an algorithm to solve them step by step carefully. Therefore, the Algorithm 1 named PFL with robust clustering (FedPRC) is proposed as below.

For the initialization, K-means++ [2] is used to set up a more robust initial for the clustering. For the iteration process, it can be merged by two modules, robust clustering and FL. The Robust clustering module is composed of three steps, the Expectation step (E step), the LOF step, and the Maximization step

---

**Algorithm 1:** Personalized FL with robust clustering (FedPRC)

---

*Input:* $\{D_1, D_2, \ldots, D_m\}, K, n$
*Output:* $\{r_{i,k}\}, \{c_i\}, \{\omega_i\}$
*Initialize:*
K-means++ initialization
*Iterate:*
**while** *stop condition is not satisfied* **do**
    **E-Step**:
    Assign each device in $b$ to its closest centroid using updated centroids
    **LOF-Step**:
    Use $LOF_n$ to label outliers
    **M/Aggregation-Step**:
    Recompute the centroids with inliers.
    **Local update-Step**:
    **for** *each cluster $k = 1, \ldots K$* **do**
        Assign centroids to every device in Cluster k.
        **for** $i \in C_k$ **do**
            **for** *E local epochs* **do**
                $\omega_i^{t+1} \leftarrow \omega_i^t - \eta \nabla \mathcal{L}(\mathcal{M}, D_i, \omega_i^t)$
            **end**
        **end**
    **end**
**end**
*End:*
**Fine tuning-Step**
Fine-tuning $\omega_i$ for $E'$ epochs.

---

(M step). And the FL module is composed of three steps either, the Distribution step, the Local update step and the Aggregation step. Due to the M step in robust clustering being the same as the Aggregation step in FL, these two modules can be merged together to form the iteration process. Until convergence or stop condition is satisfied, the output is $K$ models for $K$ clusters. To achieve better performance for each client, a simple but effective personalization technique called fine-tuning is imported as the optimum of one cluster is not the optimum of its clients. Finally, we can obtain $m$ personalized models with robustness against model poisoning for every client.

## 5    Experiments

As a proof-of-concept scenario to demonstrate the effectiveness of the proposed method, we experimentally evaluate and analyse the proposed FedPRC based on the LEAF framework, an FL benchmark [7].

## 5.1   Experimental Settings

*Datasets.* We employed two publicly-available FL benchmark datasets introduced in LEAF [7]. LEAF is a benchmarking framework for learning in federated
settings. The datasets used are Federated Extended MNIST (FEMNIST)[1] [11]
and Federated CelebA (FedCelebA)[2] [28]. We follow the setting of the benchmark data in LEAF. In FEMNIST, the handwritten images are split according
to the writers. For FedCelebA, the face images are extracted for each person and
developed by an on-device classifier to recognize whether the person smiles or
not. A statistical description of the datasets is described in Table 2.

**Table 2.** Statistics of datasets. "#" represents the number of instances.

| DATASET | FEMNIST | CelebA |
|---|---|---|
| # of Data | 805,263 | 200,288 |
| Classes | 62 | 2 |
| # of device | 3,550 | 9,343 |
| Model | CNN | CNN |
| LR | 0.003 | 0.1 |
| Local Epochs | 5 | 10 |

*Local Model.* We use a CNN with the same architecture from [28]. Two data
partition strategies are used: (a) an ideal IID data distribution using randomly
shuffled data, (b) a non-IID partition by use a $\mathbf{p}_k \sim Dir_J(0.5)$. Part of the code
is adopted from [43]. For FEMNIST, the local model's learning rate is 0.003, and
the number of local epochs is 5. For FedCelebA, the learning rate is 0.1, and the
number of local epochs is 10.

*Outliers.* In this work, we evaluate the proposed method using the outliers generated from a poisoning attack tool. The idea of model poisoning adopts from
Krum [4], which is simply boosting each iteration of the learned model in some
worker node. Malicious clients assign wrong labels to each sample in the local
dataset. In other words, explicit boosting works to mimic the benign worker
clients during the learning process; the client tries to perform the same number of epochs on the local dataset via the same training objectives to obtain
an initial gradient update. Since the malicious client wants to ensure the outcome deviates from the true label, it will have to overcome the scaling effect of
gradient updates collected from other nodes. In other words, the final gradient
updates the malicious nodes send back are then scaled by a factor $\Lambda$ by which
the malicious nodes boost the initial update. The $\Lambda$ here is a hyper-parameter

---

[1] http://www.nist.gov/itl/products-and-services/emnist-dataset.
[2] http://mmlab.ie.cuhk.edu.hk/projects/CelebA.html.

which is a multiplier for malicious clients which used to force the trained global model to close its direction. Here we use the number of clients of a subset each iteration, then multiply two as $\Lambda$.

*Baselines.* In the scenario of solving statistical heterogeneity, we choose FL methods outlined below:

1. **NonFed:** We will conduct the supervised learning task at each device without the FL framework.
2. **FedSGD:** uses SGD to optimise the global model.
3. **FedAvg:** is an SGD-based FL with weighted averaging [33].
4. **FedCluster:** is to enclose FedAvg into a hierarchical clustering framework [38].
5. **HypoCluster(K):** is a hypothesis-based clustered-FL algorithm with different $K$ [32].
6. **Robust:** design a framework run in a modular manner, namely, a robust clustering model, and a communication-efficient, distributed, robust optimization over each cluster separately [21].
7. **FedDANE:** is an FL framework with a Newton-type optimization method [27].
8. **FedProx:** adds a proximal term onto an objective function of the learning task on the device [26].
9. **FedDist:** we adapt a distance based-objective function in Reptile meta-learning [35] to a federated setting.
10. **FedDWS:** a variation of FedDist by changing the aggregation to weighted averaging where the weight depends on the data size of each device.
11. **FedPRC(K):** our proposed algorithm FedPRC with different numbers of clusters $K$.

*Training Settings.* We used 80% of each device's data for training and 20% for testing. For the initialization of the cluster centers in FedPRC, we conducted pure clustering ten times with randomized initialization over the gradients matrix which is computed by each client in one epoch local training, and then the "best" initialization, which has the minimal intra-cluster distance, was selected as the initial centers for FedPRC. For the local update procedure of FedPRC, we set $N$ to 1, meaning we only updated $W_i$ once in each local update.

*Evaluation Metrics.* Given numerous devices, we evaluated the overall performance of the FL methods. We used classification accuracy and F1 score as the metrics for the two benchmarks. In addition, due to the multiple devices involved, we explored two ways to calculate the metrics, i.e., micro and macro. The only difference is that when computing an overall metric, "micro" calculates a weighted average of the metrics from devices where the weight is proportional to the data amount, while "macro" directly calculates an average over the metrics from devices.

## 5.2   Experimental Study

*Comparison Study.* As reported in Table 3, we compared our proposed FedPRC with the baselines and found that our proposed FL framework achieves the best performance in most cases. We can see our proposed FedPRC outperforms all baselines in all metrics, which shows the effectiveness and significance of FedPRC. Furthermore, as reported in the last three columns in Table 3, we found that FedPRC with a larger number of clusters empirically achieves a better performance, which verifies the correctness of the non-IID assumption of the data distribution. Due to the experiments on both datasets being very consuming, we use the grid search technique for the number of clusters and only run full experiments with those values, that is, from two to four.

**Table 3.** Comparison of our proposed FedPRC($K$) algorithm with the baselines on FEMNIST and FedCelebA datasets. Note the number in parenthesis following "Fed-PRC" denotes the number of clusters, $K$.

| Datasets | FEMNIST | | | | CelebA | | | |
|---|---|---|---|---|---|---|---|---|
| Metrics (%) | Micro-Acc | Micro-F1 | Macro-Acc | Macro-F1 | Micro-Acc | Micro-F1 | Macro-Acc | Macro-F1 |
| NoFed | 79.4 | 67.6 | 81.3 | 51.0 | 83.8 | 66.0 | 83.9 | 67.2 |
| FedSGD | 70.1 | 61.2 | 71.5 | 46.7 | 75.7 | 60.7 | 75.6 | 55.6 |
| FedAvg | 84.9 | 67.9 | 84.9 | 45.4 | 86.1 | 78.0 | 86.1 | 54.2 |
| FedDist | 79.3 | 67.5 | 79.8 | 50.5 | 71.8 | 61.0 | 71.6 | 61.1 |
| FedDWS | 80.4 | 67.2 | 80.6 | 51.7 | 73.4 | 59.3 | 73.4 | 50.3 |
| Robust(TKM) | 78.4 | 53.1 | 77.6 | 53.6 | 90.1 | 68.0 | 90.1 | 68.3 |
| FedCluster | 84.1 | 64.3 | 84.2 | 64.4 | 86.7 | 67.8 | 87.0 | 67.8 |
| HypoCluster(3) | 82.5 | 61.3 | 82.2 | 61.6 | 76.1 | 53.5 | 72.7 | 53.8 |
| FedDane | 40.0 | 31.8 | 41.7 | 31.7 | 76.6 | 61.8 | 75.9 | 62.1 |
| FedProx | 51.8 | 34.2 | 52.3 | 34.4 | 83.4 | 60.9 | 84.3 | 65.2 |
| FedPRC(2) | 91.3 | 64.9 | 91.7 | 64.1 | 93.8 | 77.2 | 94.1 | 71.5 |
| FedPRC(3) | 91.1 | 63.1 | 91.0 | 62.6 | 93.6 | 77.8 | 93.3 | 70.6 |
| FedPRC(4) | **92.7** | **66.4** | **92.4** | **65.7** | **94.4** | 80.4 | 94.6 | **72.7** |

*Convergence Analysis.* To verify the convergence of the proposed approach, we conducted a convergence analysis by running FedPRC with different cluster numbers $K$ (from two to four) in 100 iterations. As shown in Fig. 4, FedPRC can efficiently converge on both datasets, and it can achieve the best performance with the cluster number $K = 4$. The last step is fine-tuning.

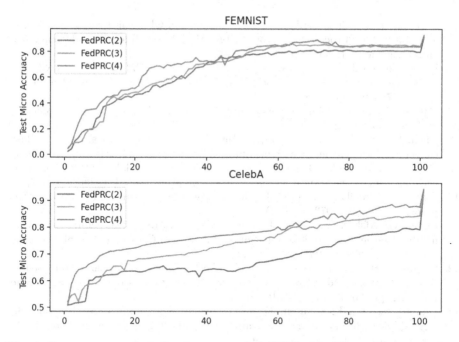

**Fig. 4.** Convergence analysis for the proposed FedPRC with different cluster number (in parenthesis) in terms of micro-accuracy.

## 6    Conclusion

This paper proposed a PFL method with a robust clustered structure to tackle model poisoning attacks in FL while still keeping the state-of-the-art performance. It is novel to combine client-wise, cluster-wise PFL, and robust clustering together to tackle the non-IID and model poisoning challenges in FL.

## References

1. Ana, L.F., Jain, A.K.: Robust data clustering. In: Proceedings of the 2003 IEEE Computer Society Conference on Computer Vision and Pattern Recognition, vol. 2, p. II. IEEE (2003)
2. Arthur, D., Vassilvitskii, S.: K-means++: the advantages of careful seeding. Technical report, Stanford (2006)
3. Bhagoji, A.N., Chakraborty, S., Mittal, P., Calo, S.: Analyzing federated learning through an adversarial lens. In: International Conference on Machine Learning, pp. 634–643. PMLR (2019)
4. Blanchard, P., El Mhamdi, E.M., Guerraoui, R., Stainer, J.: Machine learning with adversaries: Byzantine tolerant gradient descent. In: Advances in Neural Information Processing Systems, vol. 30 (2017)

5. Breunig, M.M., Kriegel, H.P., Ng, R.T., Sander, J.: LOF: identifying density-based local outliers. In: Proceedings of the 2000 ACM SIGMOD International Conference on Management of Data, pp. 93–104 (2000)

6. Brunet-Saumard, C., Genetay, E., Saumard, A.: K-bMOM: a robust Lloyd-type clustering algorithm based on bootstrap median-of-means. Comput. Stat. Data Anal. **167**, 107370 (2022)

7. Caldas, S., et al.: Leaf: a benchmark for federated settings. arXiv preprint arXiv:1812.01097 (2018)

8. Chen, F., Luo, M., Dong, Z., Li, Z., He, X.: Federated meta-learning with fast convergence and efficient communication. arXiv preprint arXiv:1802.07876 (2018)

9. Chen, X., Liu, C., Li, B., Lu, K., Song, D.: Targeted backdoor attacks on deep learning systems using data poisoning. arXiv preprint arXiv:1712.05526 (2017)

10. Cheng, G., Chadha, K., Duchi, J.: Fine-tuning is fine in federated learning. arXiv preprint arXiv:2108.07313 (2021)

11. Cohen, G., Afshar, S., Tapson, J., Van Schaik, A.: EMNIST: extending MNIST to handwritten letters. In: 2017 International Joint Conference on Neural Networks (IJCNN), pp. 2921–2926. IEEE (2017)

12. Collins, L., Hassani, H., Mokhtari, A., Shakkottai, S.: Exploiting shared representations for personalized federated learning. In: International Conference on Machine Learning, pp. 2089–2099. PMLR (2021)

13. Damaskinos, G., El-Mhamdi, E.M., Guerraoui, R., Guirguis, A., Rouault, S.: Aggregathor: Byzantine machine learning via robust gradient aggregation. Proc. Mach. Learn. Syst. **1**, 81–106 (2019)

14. Davé, R.N., Krishnapuram, R.: Robust clustering methods: a unified view. IEEE Trans. Fuzzy Syst. **5**(2), 270–293 (1997)

15. Deshpande, A., Kacham, P., Pratap, R.: Robust $k$-means++. In: Conference on Uncertainty in Artificial Intelligence, pp. 799–808. PMLR (2020)

16. Apple Differential Privacy Team: Learning with privacy at scale (2017). https://machinelearning.apple.com/research/learning-with-privacy-at-scale

17. Fallah, A., Mokhtari, A., Ozdaglar, A.: Personalized federated learning with theoretical guarantees: a model-agnostic meta-learning approach. In: Advances in Neural Information Processing Systems, vol. 33, pp. 3557–3568 (2020)

18. García-Escudero, L.A., Gordaliza, A., Matrán, C., Mayo-Iscar, A.: A general trimming approach to robust cluster analysis. Ann. Stat. **36**(3), 1324–1345 (2008)

19. García-Escudero, L.A., Gordaliza, A., Matrán, C., Mayo-Iscar, A.: A review of robust clustering methods. Adv. Data Anal. Classif. **4**(2), 89–109 (2010)

20. Ghosh, A., Chung, J., Yin, D., Ramchandran, K.: An efficient framework for clustered federated learning. arXiv preprint arXiv:2006.04088 (2020)

21. Ghosh, A., Hong, J., Yin, D., Ramchandran, K.: Robust federated learning in a heterogeneous environment. arXiv preprint arXiv:1906.06629 (2019)

22. Guha, S., Rastogi, R., Shim, K.: Rock: a robust clustering algorithm for categorical attributes. Inf. Syst. **25**(5), 345–366 (2000)

23. He, C., et al.: FedCV: a federated learning framework for diverse computer vision tasks. arXiv preprint arXiv:2111.11066 (2021)

24. Jallepalli, D., Ravikumar, N.C., Badarinath, P.V., Uchil, S., Suresh, M.A.: Federated learning for object detection in autonomous vehicles. In: 2021 IEEE Seventh International Conference on Big Data Computing Service and Applications (BigDataService), pp. 107–114. IEEE (2021)

25. Li, T., Hu, S., Beirami, A., Smith, V.: Ditto: fair and robust federated learning through personalization. In: International Conference on Machine Learning, pp. 6357–6368. PMLR (2021)

26. Li, T., Sahu, A.K., Zaheer, M., Sanjabi, M., Talwalkar, A., Smith, V.: Federated optimization in heterogeneous networks. arXiv preprint arXiv:1812.06127 (2018)
27. Li, T., Sahu, A.K., Zaheer, M., Sanjabi, M., Talwalkar, A., Smithy, V.: FedDANE: a federated newton-type method. In: 2019 53rd Asilomar Conference on Signals, Systems, and Computers, pp. 1227–1231. IEEE (2019)
28. Liu, Z., Luo, P., Wang, X., Tang, X.: Deep learning face attributes in the wild. In: Proceedings of the IEEE ICCV, pp. 3730–3738 (2015)
29. Long, G., Shen, T., Tan, Y., Gerrard, L., Clarke, A., Jiang, J.: Federated learning for privacy-preserving open innovation future on digital health. In: Chen, F., Zhou, J. (eds.) Humanity Driven AI, pp. 113–133. Springer, Cham (2022). https://doi.org/10.1007/978-3-030-72188-6_6
30. Luo, J., et al.: Real-world image datasets for federated learning. arXiv preprint arXiv:1910.11089 (2019)
31. Ma, J., Long, G., Zhou, T., Jiang, J., Zhang, C.: On the convergence of clustered federated learning. arXiv preprint arXiv:2202.06187 (2022)
32. Mansour, Y., Mohri, M., Ro, J., Suresh, A.T.: Three approaches for personalization with applications to federated learning. arXiv preprint arXiv:2002.10619 (2020)
33. McMahan, B., Moore, E., Ramage, D., Hampson, S., Arcas, B.A.: Communication-efficient learning of deep networks from decentralized data. In: Artificial Intelligence and Statistics, pp. 1273–1282. PMLR (2017)
34. Muniyandi, A.P., Rajeswari, R., Rajaram, R.: Network anomaly detection by cascading K-means clustering and C4.5 decision tree algorithm. Procedia Eng. **30**, 174–182 (2012)
35. Nichol, A., Schulman, J.: Reptile: a scalable metalearning algorithm. arXiv preprint arXiv:1803.02999 (2018)
36. Paul, D., Chakraborty, S., Das, S.: Robust principal component analysis: a median of means approach. arXiv preprint arXiv:2102.03403 (2021)
37. Rieke, N., et al.: The future of digital health with federated learning. NPJ Digit. Med. **3**(1), 1–7 (2020)
38. Sattler, F., Müller, K.R., Samek, W.: Clustered federated learning: model-agnostic distributed multi-task optimization under privacy constraints. arXiv preprint arXiv:1910.01991 (2019)
39. Sattler, F., Müller, K.R., Samek, W.: Clustered federated learning: model-agnostic distributed multitask optimization under privacy constraints. IEEE Trans. Neural Netw. Learn. Syst. **32**, 3710–3722 (2020)
40. Shamsian, A., Navon, A., Fetaya, E., Chechik, G.: Personalized federated learning using hypernetworks. In: International Conference on Machine Learning, pp. 9489–9502. PMLR (2021)
41. Smith, V., Chiang, C.K., Sanjabi, M., Talwalkar, A.S.: Federated multi-task learning. In: Advances in Neural Information Processing Systems, vol. 30 (2017)
42. Tan, Y., et al.: FedProto: federated prototype learning over heterogeneous devices. arXiv preprint arXiv:2105.00243 (2021)
43. Wang, H., Yurochkin, M., Sun, Y., Papailiopoulos, D., Khazaeni, Y.: Federated learning with matched averaging. In: International Conference on Learning Representations (2020). https://openreview.net/forum?id=BkluqlSFDS
44. Xie, M., et al.: Multi-center federated learning. arXiv preprint arXiv:2108.08647 (2021)
45. Xu, J., Glicksberg, B.S., Su, C., Walker, P., Bian, J., Wang, F.: Federated learning for healthcare informatics. J. Healthcare Inform. Res. **5**(1), 1–19 (2021)
46. Yang, M.S., Lai, C.Y., Lin, C.Y.: A robust EM clustering algorithm for gaussian mixture models. Pattern Recogn. **45**(11), 3950–3961 (2012)

47. Yang, M.S., Wu, K.L.: A similarity-based robust clustering method. IEEE Trans. Pattern Anal. Mach. Intell. **26**(4), 434–448 (2004)
48. Zheng, Z., Zhou, Y., Sun, Y., Wang, Z., Liu, B., Li, K.: Applications of federated learning in smart cities: recent advances, taxonomy, and open challenges. Connect. Sci. **34**(1), 1–28 (2022)

# A Data-Driven Framework for Driving Style Classification

Sebastiano Milardo[1]([⊠])(ID), Punit Rathore[2](ID), Paolo Santi[1](ID), and Carlo Ratti[1]

[1] Massachusetts Institute of Technology, Cambridge, MA, USA
{milardo,psanti,ratti}@mit.edu
[2] Indian Institute of Science, Bangalore, India
prathore@iisc.ac.in

**Abstract.** Traditional driving behaviour recognition algorithms leverage hand-crafted features extracted from raw driving data and then apply user-defined machine learning models to identify driving behaviours. However, such solutions are limited by the set of selected features and by the chosen model. In this work, we present a data-driven driving behaviour recognition framework that utilizes an unsupervised feature extraction and feature selection algorithm and a deep neural network architecture obtained using an Automated Machine Learning (AutoML) approach. To validate the feasibility of this solution, numerical evaluations were performed on a unique real-world driving datasets collected from 29 professional truck drivers in uncontrolled environments, including supervisor's scoring of driver behavior that is used as ground truth data. Our experimental results show that the proposed deep neural network model achieves up to 95% accuracy for multi-class classification, significantly outperforming five other popular machine learning models.

**Keywords:** Driving behaviour classification · Driving style recognition

## 1 Introduction

Individual driving behaviour plays an important role in traffic safety, as well as energy efficiency and vehicle wear and tear [1]. Therefore, there is an increasing interest in quantitatively characterizing driver behavior leveraging massive amounts of data generated by modern vehicles. This is especially important for professional drivers, who are frequently on the road and in most cases drive very large vehicles.

Many researchers have attempted to analyse, identify, model, and classify driving behaviour [2]. However, as detailed in Sect. 2, the following research gaps can be identified:

- Majority of the existing work on driving behaviour analysis are based on exploratory or rule-based methods, which are difficult to validate without actual ground truth.

W. Chen et al. (Eds.): ADMA 2022, LNAI 13726, pp. 253–265, 2022.
https://doi.org/10.1007/978-3-031-22137-8_19

- Existing techniques, based on supervised learning, rely on traditional classification models which validate their results using individual driving events (e.g., harsh turn, harsh braking, etc.) rather than overall driving behavior of a driver (e.g. safe driver, harsh driver).
- Majority of these studies either conducted their experiments in a controlled and/or simulated environment, or performed their analysis on a naturalistic driving dataset collected by a few drivers.
- Though driving behaviour classification from naturalistic driving data is a well-established practice [3], their characterization is still an open problem. In this sense, given the obvious difficulties in contextualizing driving events, the answer provided by many studies is the identification of absolute metrics of evaluation. This approach can lead to conflicting results. For example, harsh braking can be linked both to an aggressive driving style or to quick reflexes in avoiding an unpredictable obstacle.
- Extensive work has been done on driving behaviour analysis for taxi/car drivers. However, different types of vehicles have different dynamics [4] and specifically trucks have some unique characteristics, such as size, weight, and manoeuvrability, which result in even more different driving dynamics [5]. To our knowledge, very limited work focused on truck drivers' behaviour analysis.

To overcome the limitations of existing approaches and fill the research gaps, we developed an *Automated Machine Learning* (AutoML) framework for driving behaviour classification of drivers using real-world, naturalistic driving data collected in an uncontrolled environment. Our major contributions are:

- The proposed solution employs (i) automatic feature extraction and feature selection from the driving data, and (ii) an AutoML framework for deep neural network, based on AutoKeras, to automatically find the best deep neural network architecture and corresponding hyper-parameters for our problem and dataset. Our framework also integrates external spatial information such as road type and maximum allowable speed and temporal information such as the hour of the day, day of the week, etc.
- We leveraged a real-world driving dataset collected by 29 professional trash truck drivers over a period of 76 days, and we validated our model with the actual safety scores assigned to each driver. These scores were assigned by a group of domain experts based on routine observation over a period of 3 months. To our knowledge, this is one of the few datasets that provide human generated feedback about the driving styles in addition to the driving events.
- We classified the driving behaviours for both individual sessions (hourly) and aggregated sessions (daily, weekly, and monthly) for each driver as short-term and long-term driving behaviour.
- Finally, we compared our model with five other popular classification models for driving behaviour classification.

To the best of our knowledge, among the works trying to classify the driving behavior of professional truck drivers by analyzing large-scale datasets [6], this is the only one that uses completely automated feature engineering and neural network search steps while validating its findings using safety scores assigned to the drivers by a group of experts. The remainder of this paper is organized as follows: Sect. 2 summarizes the recent literature on driving style recognition and discusses their limitations and research gaps. Section 3 formulates our problem for periodic and aggregated driving behaviour classification. Section 4 describes datasets, pre-processing, feature-extraction/selection, and ML model employed for driving behaviour classification in our work. Section 5 presents our experimental results on real-world driving dataset, followed by discussion and conclusions in Sect. 6.

## 2    State of the Art

An extensive survey on driving behaviour analysis and driving style recognition is provided in [2,7,8]. At a high level, existing driving behaviour analysis can be divided in two categories: unsupervised and supervised. In the first case, the classification of driving style is achieved through statistical analysis of the relevant input signals, without the knowledge of actual classification. Whereas, the later requires the knowledge of actual driving style classification of the data used for training.

**Unsupervised Approaches.** Among unsupervised approaches, Gaussian mixture model (GMM), $k$-means, and Bayesian learning techniques have been used extensively in driving behaviour analysis studies. Fugiglando $et$ $al.$ [9,10] employed $k$-means algorithm on CAN bus data to identify groups of similar drivers based on the driving behaviour. However, no semantic explanation for the different resulting classes was provided. Mudgal $et$ $al.$ [11] applied a hierarchical Bayesian regression technique on collected driving data to model instantaneous driving behavior at roundabouts. Similarly, McCall $et$ $al.$ [12] employed a Bayesian learning technique to analyze the driving behaviour for braking assistance and collision avoidance. These studies were either conducted in a simulated/controlled environment or they considered data from few drivers. Wang $et$ $al.$ [13] proposed a framework for driving style classification by utilizing primitive driving patterns with the Bayesian nonparametric approaches. The features used in [13] were the vehicle longitudinal acceleration, speed, and the distance from the preceding vehicle. Although unsupervised algorithms have shown their applicability for driving behaviour analysis [14], their output (e.g. clusters) require interpretation in the absence of ground truth.

**Supervised Approaches.** Among supervised approaches, $k$-nearest neighbours (kNN), SVM, and neural network are the most used techniques for driving style recognition. Johnson and Trivedi [15] utilized kNN and dynamic time warping (DTW) in their system, MIROAD, to detect and recognize various aggressive driving events. Similarly, Vaitkus $et$ $al.$ [16] exploited kNN on 117 features extracted from long-term accelerometer data to classify driving styles into normal and aggressive driving. However, the results were evaluated based on the labels defined by observing route signals. Authors in [17,18] evaluate driving styles quantitatively by normalizing the driving behavior based on personalized driver models, and detect abnormal driving by analyzing normalized driving behavior using neural networks. Liu *et al.* [19] extracted 44 features from the driving data, used information entropy to discretize them, and subsequently applied PCA to further reduce the dimension. Then, fuzzy *c*-means and SVM were leveraged to classify the driving styles.

In the past few years, deep learning based approaches have been quite successful for driving style classification. Li *et al.* [20] applied CNN and long short-term memory (LSTM) on driving operational 2-D pictures, constructed using a nested time window technique on sequential data from naturalistic driving, to classify driving styles. A similar approach has been presented by Milardo *et al.* [21] to link physiological signals and driver images to vehicle kinematics and drivers' behavior. Bejani *et al.* [22] employed convolution neural network (CNN) on smartphone acceleration data to extract the features, and subsequently used them to classify driving styles.

Unlike unsupervised approaches, driving style classification using supervised algorithms are easy to validate against the real (labelled) driving data. However, they require a large number of labelled data for reliable performance. To deal with the need of having lots of labeled training data, Wang *et al.* [23] presented a semi-supervised approach, called semi-supervised SVM (S3VM), in which first some representative data points are selected using *k*-means clustering and manually labelled using a rule-based approach, and finally a quasi-Newton algorithm is employed to assign the optimal label to all of the training data. The same features and the same dataset as used in [23] were utilized in [24] to detect driving styles using a kernel density estimation.

Although extensive work has been done on driving style recognition for car drivers, very limited studies have analyzed the driving style of truck drivers. Linkov *et al.* [25] presented a study on the correlation between professional drivers' driving behavior and their personality traits using a truck simulator. However, the primary focus of this study [25] was on fuel efficiency [26] rather than driving behaviour, which was considered as an auxiliary variable. In another similar study, Ferreira *et al.* [1] collected data from professional bus drivers in Lisbon, and applied Naive Bayes classifier to optimize fuel consumption and provide suggestions. Some suggestions such as "Minimize the use of acceleration" and "Minimize the use of braking" are generally related to both efficient fuel consumption and good driving behaviors.

## 3    Problem Statement

The objective of this work is to explore and classify the behaviour of drivers using a data-driven approach applied on their driving data. Let $F_{d,t_w}$ denotes the feature vector of a driver $d \in \{1, 2, .., N\}$ derived from the raw time-series data e.g. speed, frontal and lateral acceleration etc., of his/her driving for a window of $t_w$ time-period, where $N$ is the number of drivers. The time-window $t_w$ can range from hours to months depending on the user requirement, size of the data, and classification performance.

In this article, we categorize both periodic (separate) and aggregated (combined) driving behaviour. While the periodic driving classification represents a driver's trip behaviour for a period $t_w$, the aggregated classification stands for an individual driver behaviour based on his/her overall driving data. For a periodic classification, at any time instant $t$ of a driving trip of driver $d$, the corresponding feature vector $F_{d,t_w}$ from $t - t_w$ to time $t_w$ can be classified into any class of driving styles such as very bad, bad, less than average, above average, good, and very good. Let $n$ denote the total trips a driver segmented by time-window $t_w$, then the aggregated driving behaviour of a driver is achieved by accumulating $n$ driving behaviour classifications of his/her driving history.

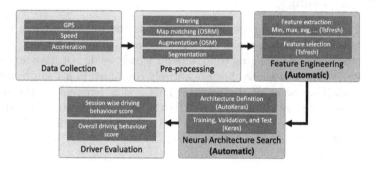

**Fig. 1.** Technical pipeline of the proposed framework.

## 4  Proposed Solution

The proposed solution, as shown in Fig. 1, can be divided into five different steps: data collection, pre-processing, feature engineering, neural network search, and driver evaluation. In the following paragraphs we detail each step.

### 4.1  Dataset Description

The dataset used in this research has been collected by 29 professional truck drivers over a period of 76 days, for a total of 45 million data points. The dataset contains data from 33 different vehicles.

The raw data is collected at 1 Hz and it includes: a timestamp, GPS coordinates, acceleration from a tri-axial accelerometer located in the cabin of the truck, and the speed of the vehicle. In addition to these signals, an identifier for the driver and the truck was also included for each data point.

Additionally, each driver is linked to a score assigned by a group of experts without using the driving data that analyzed in this work. In particular, car crashes and traffic citations are the main factors that were considered to derive the given scores for each driver. The provided scores are categorized into five classes from $A$ to $F$ with $A$ being the best and $F$ being the worst. The distribution of driver scores are shown in Table 1.

**Table 1.** Scores

| Score | A | B | C | D | E | F |
|---|---|---|---|---|---|---|
| # drivers | 10 | 7 | 7 | 3 | 0 | 2 |

### 4.2    Pre-processing

The initial step in the pre-processing analysis is the removal of non relevant data. The original dataset is made of daily sessions, one for each truck. First we remove all the sessions belonging to drivers for which we do not have any rating. Then we remove those sessions that are entirely contained in a 1 km radius area around the starting point of the session. This action removes all the sessions that are recorded during maintenance. After this step, the dataset contains 1276 sessions, with an average length of 207 km and an average duration of 10 h. Fig. 2 shows the final distribution of daily distances.

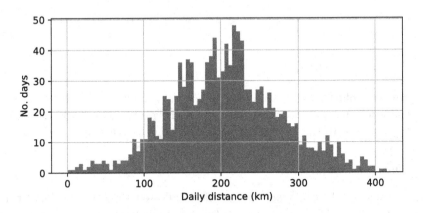

**Fig. 2.** Daily distance distribution

For each of these sessions:

– We correct the errors introduced by the GPS by matching each data point to the road network in the most plausible way according to OSRM [27].

- We augment the raw driving data with road-type classification (motorway, trunk, primary, secondary, service, tertiary, and residential) and maximum allowable speed at each GPS point, using OpenStreetMap (OSM) [28]. For those locations where the maximum speed limit is not reported we use the maximum speed limit retrieved for streets of the same type.
- We remove the data points where the vehicle is not moving and we split the remaining data points on an hourly basis ($t_w = 1$ h).
- Finally, we group the data points based on the road type.

It is important to notice that we have experimented with different time-windows ($t_w = 1$ h, 1 day, 1 week, and 1 month) and corresponding results are reported in Sect. 5.1.

## 4.3 Feature Engineering

Different features can show various types of information about driving style, and initially it is unknown which feature sets most accurately distinguish the different driving styles. Therefore, we decided to extract an extensive set of features from the data and then filter the most informative and predictive ones using a feature selection method.

For each data segments described in Subsect. 4.2, we split the original set of signals (frontal acceleration, lateral acceleration, and speed) into the following derived signals according to the orientation of the accelerometer:

- Accelerating events i.e., instances where the speed of the vehicle is increasing.
- Braking events i.e., instances where the speed of the vehicle is decreasing.
- Right/Left turns i.e., instances with positive/negative lateral acceleration.
- Over-speeding i.e. instances where speed of the vehicle is greater than the speed limit.

Then, we extracted 794 different features from each time-series (signal) utilizing the tsfresh [29] library. Tsfresh is an open-source Python library which can automatically extracts features from time-series data. First, it extracts 72 unique features, as described in the tsfresh documentation [30], then based on these unique features it generates a total 794 features for each time-series by using different parameter settings. Some of the features computed using tsfresh are summary features (min, max, mean, median, mode, variance), quantiles, skewness, kurtosis, average energy, auto-correlation, entropy, binned entropy, FFT coefficients, wavelet coefficients, etc. However, not all the extracted features are relevant for the analysis. Therefore, we reduced the number of features using a feature selection process. This process consists of two phases: first each feature vector is individually and independently evaluated with respect to its significance for predicting the score of the driver. The significance of a feature is addressed by statistical hypothesis testing, and the result of these tests is a vector of p-values, quantifying the significance of each feature for predicting the score. Then the vector of p-values is evaluated on the basis of the Benjamini-Yekutieli procedure [31] in order to decide which features to keep. At the end of this process only 638 features are selected.

### 4.4   Neural Architecture Search

Deep learning based approaches have been quite successful in recent years for classification problems. However, it is difficult to find the optimal configuration of a neural network architecture, which can lead to numerous unsuccessful experiments depending on the dataset and the problem. Typical hyper-parameters that need to be tuned include the type and number of layers, optimizer algorithm (SGD, Adam, etc.), learning rate, and regularization to name a few.

We address the above problem using an Automated Machine Learning (AutoML) approach. In particular, we employed AutoKeras [32] which is an AutoML system which can perform automatic model selection and hyperparameter tuning for a given task. Using AutoKeras, we can search the best neural network architecture for the given learning task and input dataset. AutoKeras [32] utilizes a Bayesian optimization approach to guide through the search space by designing a neural network kernel and selecting the most promising network morphing operations (e.g. inserting a new layer or adding a skip-connection).

## 5   Results

We set AutoKeras to try 100 different models with 80-20 training-validation data ratio and we trained 200 epochs for each model to find the best neural network architecture for our task. The obtained features from the feature extraction step were fed as an input to classify each segment (session) into one of the 5 classes mentioned in Sect. 4.1. All the models were evaluated using mean squared error (MSE) and mean absolute error (MAE) between actual scored and predicted scores. The best neural network architecture that gives us the highest accuracy is shown in Fig. 3.

The layers that can be found in the generated deep neural network are:

- MultiCategoryEncoding: this layer is used to encode categorical features to numerical features.
- Dense: a densely-connected NN layer that implements the operation: $out = activation(dot(in, kernel) + bias)$ where $activation$ is the element-wise activation function passed as the activation argument, $kernel$ is a weights matrix created by the layer, and $bias$ is a bias vector created by the layer.
- ReLU: Rectified Linear Unit activation function. An activation layer that returns element-wise $max(x, 0)$.
- Softmax: an activation layer that converts an input array into a vector of values that follows a probability distribution whose total sums up to 1. The output values are in the range $[0,1]$

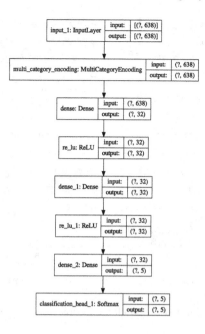

**Fig. 3.** Deep neural network architecture

The different layers are organized as shown in Fig. 3. Each block in the figure shows the name and type of each layer and the input and output size.

## 5.1 Selection of Time-Window for Aggregation

In this experiment, we choose different time-windows to create segments, classify them in one of the 5 classes, and evaluate predicted scores against ground truth scores. Fig. 4 shows the confusion matrices between the predicted scores and ground truth scores for different time-window periods. Specifically: $t_w = 1\,$h, 1 day, 1 week, and 1 month. It can be noted that the results improve as the size of the time-windows increase. We believe that this behavior can be linked to the variability of driving activities. In general, while the behavior of a driver might change significantly from one hour to another, for example if the driver is driving in a crowded urban environment or on an highway, the behavior over longer periods of time seems to be more consistent and the classification results are closer to the labels provided by the ground truth. This can be noted by looking closely at the misclassified data points. While in the 1 h scenario, even drivers that are classified positively can be classified negatively, the weekly and the monthly aggregation windows show a more concentrated distribution of results around the expected outcomes.

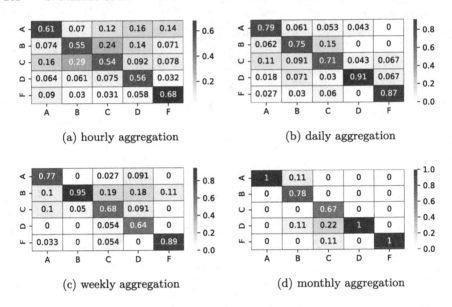

(a) hourly aggregation     (b) daily aggregation

(c) weekly aggregation     (d) monthly aggregation

**Fig. 4.** Effect of aggregation period on classification performance for five class classification.

## 5.2 Comparison of Different Models

To put our model's result (DNN) into the perspective, we compare the deep neural network model's performance with several other classification models such as k-nearest neighbour (kNN), decision tree (J-48), support vector machine (SVM), random forest (RF), and Adaptive Boost (Adaboost) model, which are popular and have been used in literature for driving style recognition. To run a fair comparison, we have tested the proposed solution against the reference models using all four aggregation windows, and reporting for each combination the Accuracy, the Mean Squared Error (MSE), the Mean Absolute Error (MSE), and the F-1 Score. Each model has been trained using the same randomly selected 80% extract of the entire dataset, and tested on the remaining 20%. For the kNN model we tried all values of $k$ in the range $[1, 50]$ and we obtained the best results using $k$ equal to 6, for the random forest model the number of estimators was searched in the range $[1, 1000]$ and we found the best results using 100, similarly for the AdaBoost model after an extensive search the number of estimators was set to 50 and the learning rate to 1.

Results are shown in Table 2, 3, 4 and 5. Although the results change based on the aggregation window, showing the best results using the daily and weekly aggregation windows, it is possible to notice that in all conditions the DNN model outperforms all the other models. A point that we would like to underline is that the worst results are obtained on the hourly and monthly aggregation windows. In the first case a possible justification is the variability of driver's behavior while in the latter the reduced number of data points negatively impacts the training of the models.

**Table 2.** Classification performance (hourly)

|     | DNN   | kNN   | J48   | SVM   | RF    | AdaBoost |
|-----|-------|-------|-------|-------|-------|----------|
| ACC | 68.4% | 24.6% | 44.0% | 46.0% | 63.0% | 42.6%    |
| F1  | 64.6% | 31.4% | 44.3% | 42.5% | 61.6% | 40.6%    |
| MAE | 1.25  | 1.72  | 2.30  | 2.16  | 1.44  | 2.33     |
| MSE | 6.16  | 6.03  | 12.66 | 11.69 | 7.69  | 12.75    |

**Table 3.** Classification performance (daily)

|     | DNN   | kNN   | J48   | SVM   | RF    | AdaBoost |
|-----|-------|-------|-------|-------|-------|----------|
| ACC | 92.0% | 51.7% | 67.3% | 73.9% | 88.0% | 50.3%    |
| F1  | 92.0% | 62.4% | 67.6% | 73.7% | 87.7% | 49.1%    |
| MAE | 0.36  | 0.92  | 1.32  | 1.04  | 0.48  | 1.89     |
| MSE | 2.13  | 2.81  | 7.13  | 5.80  | 2.95  | 9.54     |

**Table 4.** Classification performance (weekly)

|     | DNN   | kNN   | J48   | SVM   | RF    | AdaBoost |
|-----|-------|-------|-------|-------|-------|----------|
| ACC | 94.9% | 46.7% | 74.7% | 82.2% | 88.8% | 56.1%    |
| F1  | 95.0% | 56.8% | 74.8% | 82.0% | 88.3% | 56.1%    |
| MAE | 0.34  | 1.04  | 1.06  | 0.72  | 0.50  | 3.17     |
| MSE | 2.02  | 3.17  | 5.49  | 4.07  | 3.17  | 7.62     |

**Table 5.** Classification performance (monthly)

|     | DNN   | kNN   | J48   | SVM   | RF    | AdaBoost |
|-----|-------|-------|-------|-------|-------|----------|
| ACC | 68.1% | 18.5% | 44.4% | 62.9% | 62.9% | 37.0%    |
| F1  | 67.2% | 23.0% | 42.2% | 61.1% | 54.9% | 29.3%    |
| MAE | 1.06  | 2.03  | 2.29  | 1.55  | 1.62  | 8.29     |
| MSE | 8.01  | 7.89  | 11.11 | 8.74  | 8.29  | 6.66     |

# 6  Conclusion and Future Work

This article presented an automated machine learning (AutoML) based framework for driving behaviour recognition. The developed framework employs an automatic feature extraction technique and AutoKeras based deep neural network architecture to make feature selection and model selection process fully automatic.

We tested our driving style classification framework on real-world driving dataset and experimental results showed that our solution achieves up to 95% classification accuracy for five-class classification, and significantly outperforms five other ML classification models which are popular for driving style classification. As a part of this continued research and to overcome some limitations of the current dataset, we plan to incorporate the effect of weather by leveraging real-time weather data, extend the size of the driving data, and increase the resolution of scoring events.

**Acknowledgment.** We would like to thank all the members of the MIT Senseable City Lab Consortium for supporting this research.

# References

1. Ferreira, J.C., de Almeida, J., da Silva, A.R.: The impact of driving styles on fuel consumption: a data-warehouse-and-data-mining-based discovery process. IEEE Trans. Intell. Transp. Syst. **16**(5), 2653–2662 (2015)
2. Martinez, C.M., Heucke, M., Wang, F., Gao, B., Cao, D.: Driving style recognition for intelligent vehicle control and advanced driver assistance: a survey. IEEE Trans. Intell. Transp. Syst. **19**(3), 666–676 (2018)
3. Lin, X., Zhang, K., Cao W., Zhang, L.: Driver evaluation and identification based on driving behavior data. In: 2018 5th International Conference on Information Science and Control Engineering (ICISCE), pp. 718–722 (2018)
4. Zhu, X., Srinivasan, S.: A comprehensive analysis of factors influencing the injury severity of large-truck crashes. Accid. Anal. Prev. **43**(1), 49–57 (2011)
5. He, Y., Yan, X., Xiao-Yun, L., Chu, D., Chaozhong, W.: Rollover risk assessment and automated control for heavy duty vehicles based on vehicle-to-infrastructure information. IET Intel. Transp. Syst. **13**(6), 1001–1010 (2019)
6. Higgs, B., Abbas, M.: Segmentation and clustering of car-following behavior: recognition of driving patterns. IEEE Trans. Intell. Transp. Syst. **16**(1), 81–90 (2015)
7. Wang, W., Xi, J., Chen, H.: Modeling and recognizing driver behavior based on driving data: a survey. Math. Probl. Eng. **2014** (2014)
8. Chan, T.K., Chin, C.S., Chen, H., Zhong, X.: A comprehensive review of driver behavior analysis utilizing smartphones. IEEE Trans. Intell. Transp. Syst. **21**(10), 4444–4475 (2019)
9. Fugiglando, U., Santi, P., Milardo, S., Abida, K., Ratti, C.: Characterizing the "Driver DNA" through can bus data analysis. In: Proceedings of the 2nd ACM International Workshop on Smart, Autonomous, and Connected Vehicular Systems and Services, CarSys'17, pp. 37–41, New York, NY, USA (2017). Association for Computing Machinery
10. Fugiglando, U., et al.: Driving behavior analysis through can bus data in an uncontrolled environment. IEEE Trans. Intell. Transp. Syst. **20**(2), 737–748 (2019)
11. Mudgal, A., Hallmark, S., Carriquiry, A., Gkritza, K.: Driving behavior at a roundabout: a hierarchical bayesian regression analysis. Transp. Res. Part D: Transp. Environ. **26**, 20–26 (2014)
12. McCall, J.C., Trivedi, M.M.: Driver behavior and situation aware brake assistance for intelligent vehicles. In: Proceedings of the IEEE, vol. 95, no. 2, pp. 374–387 (2007)
13. Wang, W., Xi, J., Zhao, D.: Driving style analysis using primitive driving patterns with bayesian nonparametric approaches. IEEE Trans. Intell. Transp. Syst. **20**(8), 2986–2998 (2019)

14. Milardo, S., Rathore, P., Santi, P., Buteau, R., Ratti, C.: An unsupervised approach for driving behavior analysis of professional truck drivers. In: Martins, A.L., Ferreira, J.C., Kocian, A. (eds.) INTSYS 2021. LNICST, vol. 426, pp. 44–56. Springer, Cham (2022). https://doi.org/10.1007/978-3-030-97603-3_4

15. Johnson, D.A., Trivedi, M.M.: Driving style recognition using a smartphone as a sensor platform. In: 2011 14th International IEEE Conference on Intelligent Transportation Systems (ITSC), pp. 1609–1615. IEEE (2011)

16. Vaitkus, V., Lengvenis, P., Žylius, G.: Driving style classification using long-term accelerometer information. In: 2014 19th International Conference on Methods and Models in Automation and Robotics (MMAR), pp. 641–644. IEEE (2014)

17. Shi, B., et al.: Evaluating driving styles by normalizing driving behavior based on personalized driver modeling. IEEE Trans. Syst. Man Cybern. Syst. **45**(12), 1502–1508 (2015)

18. Hu, J., Xu, L., He, X., Meng, W.: Abnormal driving detection based on normalized driving behavior. IEEE Trans. Veh. Technol. **66**(8), 6645–6652 (2017)

19. Liu, Y., Wang, J., Zhao, P., Qin, D., Chen, Z.: Research on classification and recognition of driving styles based on feature engineering. IEEE Access **7**, 89245–89255 (2019)

20. Li, G., Zhu, F., Qu, X., Cheng, B., Li, S., Green, P.: Driving style classification based on driving operational pictures. IEEE Access **7**, 90180–90189 (2019)

21. Milardo, S., Rathore, P., Amorim, M., Fugiglando, U., Santi, P., Ratti, C.: Understanding drivers' stress and interactions with vehicle systems through naturalistic data analysis. IEEE Trans. Intell. Transp. Syst. **23**, 1–12 (2021)

22. Bejani, M.M., Ghatee, M.: Convolutional neural network with adaptive regularization to classify driving styles on smartphones. IEEE Trans. Intell. Transp. Syst. **21**(2), 543–552 (2020)

23. Wang, W., Xi, J., Chong, A., Li, L.: Driving style classification using a semisupervised support vector machine. IEEE Trans. Human-Mach. Syst. **47**(5), 650–660 (2017)

24. Han, W., Wang, W., Li, X., Xi, J.: Statistical-based approach for driving style recognition using bayesian probability with kernel density estimation. IET Intel. Transp. Syst. **13**(1), 22–30 (2019)

25. Linkov, V., Zaoral, A., Řezáč, P., Pai, C.-W.: Personality and professional drivers' driving behavior. Transp. Res. F: Traffic Psychol. Behav. **60**, 105–110 (2019)

26. Hlasny, T., Fanti, M.P., Mangini, A.M., Rotunno, G., Turchiano, B.: Optimal fuel consumption for heavy trucks: a review. In: 2017 IEEE International Conference on Service Operations and Logistics, and Informatics (SOLI), pp. 80–85 (2017)

27. Osrm: Open source routing machine (2020)

28. Haklay, M., Weber, P.: Openstreetmap: user-generated street maps. IEEE Pervasive Comput. **7**(4), 12–18 (2008)

29. Christ, M., Braun, N., Neuffer, J., Kempa-Liehr, A.W.: Time series feature extraction on basis of scalable hypothesis tests (tsfresh-a python package). Neurocomputing **307**, 72–77 (2018)

30. Christ, M.: Tsfresh: a time-series feature extraction toolbox (2016)

31. Benjamini, Y., Yekutieli, D.: The control of the false discovery rate in multiple testing under dependency. Annals of statistics, pp. 1165–1188 (2001)

32. Jin, H., Song, Q., Hu, X.: Auto-keras: an efficient neural architecture search system. In: Proceedings of the 25th ACM SIGKDD International Conference on Knowledge Discovery & Data Mining, pp. 1946–1956 (2019)

# Density Estimation in High-Dimensional Spaces: A Multivariate Histogram Approach

Pedro Strecht[1]([✉])(iD), João Mendes-Moreira[1](iD), and Carlos Soares[1,2,3](iD)

[1] LIAAD-INESC TEC, Faculdade de Engenharia, Universidade do Porto,
R. Dr. Roberto Frias, 4200-465 Porto, Portugal
{pstrecht,jmoreira,csoares}@fe.up.pt

[2] LIACC, Faculdade de Engenharia, Universidade do Porto, R. Dr. Roberto Frias,
4200-465 Porto, Portugal

[3] Fraunhofer Portugal AICOS, R. Alfredo Allen 455, 4200-135 Porto, Portugal

**Abstract.** Density estimation is an important tool for data analysis. Non-parametric approaches have a reputation for offering state-of-the-art density estimates limited to few dimensions. Despite providing less accurate density estimates, histogram-based approaches remain the only alternative for datasets in high-dimensional spaces. In this paper, we present a multivariate histogram approach to estimate the density of a dataset without restrictions on the number of dimensions, containing both numerical and categorical variables (without numerical encoding) and allowing missing data (without the need to preprocess them). Results from the empirical evaluation show that it is possible to estimate the density of datasets without restrictions on dimensionality, and the method is robust to missing values and categorical variables.

**Keywords:** Density estimate · Multivariate histogram · Missing data

## 1 Introduction

Density is a scalar quantity that measures the concentration of a phenomenon in a unit of space. Common examples are the population density of a country, i.e., how many people on average are in a square meter. If the phenomenon is the concentration of a gas, density refers to the average number of particles contained in a cubic meter. When the concept is applied to the data generated by a phenomenon, density is the average number of observations in a unit of hypervolume (a generalization of length, area, and volume to any number of dimensions).

In statistics, one of the most common uses for density is to calculate probabilities. In machine learning, it is an important tool for exploratory data analysis, an initial investigation in a data collection to understand its shape and features, such as skewness, multimodality, and anomalies [12]. Density analysis can also reveal the need to collect more data [13]. There are many applications using density in various areas of research, described in Sect. 2.3.

© The Author(s), under exclusive license to Springer Nature Switzerland AG 2022
W. Chen et al. (Eds.): ADMA 2022, LNAI 13726, pp. 266–278, 2022.
https://doi.org/10.1007/978-3-031-22137-8_20

Often a dataset is a sample of a population, making it possible to estimate the population's true density by calculating the dataset's density. Density estimation can use parametric, non-parametric and histogram-based approaches, detailed in Sect. 2.2. Parametric approaches are limited to cases where the data distribution fits a known distribution (via parameter tuning). Non-parametric approaches have a broader scope and are widely regarded as offering the best density estimates. However, estimating density becomes a gradually complex and computationally demanding process as the dimensionality (number of variables) of the dataset increases. For instance, Kernel Density Estimators (KDE) are unfeasible, even in datasets with as few as seven dimensions. Histograms, although providing less accurate density estimates when compared to the other approaches, remain the only viable alternative. Current implementations of histogram-based approaches, despite being able to deal with more dimensions than KDE and other non-parametric approaches, still cannot deal with reasonably sized datasets. Another constraint is related to the type of data. Histograms represent the frequency of continuous numerical data and cannot be used with categorical data. Moreover, histograms do not naturally handle missing data. The dataset has to be preprocessed before density estimation by either filling in the missing values with a specific method or removing the observations with missing data.

The main contributions of this paper are: 1) a histogram-based approach to estimate the density of datasets designed to address the aforementioned limitations, i.e., a method that is not limited by the dimensionality of the dataset, is equally able of handling both numerical and categorical variables, and deals with missing data (the only restriction is the size of the dataset constrained by the available memory); 2) an empirical evaluation on a number of datasets with different characteristics, demonstrating the ability to perform density estimation within reasonable computational time; 3) an empirical evaluation of the effects of missing values as well as categorical variables in the density estimates.

The remainder of this paper is structured as follows: Sect. 2 presents the background and related work on density estimation and some of its applications; Sect. 3 presents a detailed description of the proposed method with a few illustrative examples; Sect. 4 presents the empirical evaluations, results and their discussion; Sect. 5 presents the conclusions.

## 2   Background and Related Work

### 2.1   Basic Concepts

A *random variable* is a formalization of a quantity or object which depends on random events or experiments with an unpredictable outcome. If the possible outcomes are finite and can be counted, the variable is *discrete*, otherwise, if endless and impossible to count, it is *continuous* [8]. In either case, it is common for some values to occur more often than others. A *probability distribution* is a non-negative function specifying how likely a random variable is to take each of its possible values, ranging from zero (impossibility) to one (absolute certainty). For discrete variables, a *probability mass function* provides the probability of each specific value. For continuous variables, as there is an infinity of possible values, the probability of taking a specific value is zero, a *probability density function* or *density*, $p(x)$, is used. An example is a Normal distribution (Eq. 1), $N(\mu, \sigma^2)$ with parameters mean ($\mu$) and variance ($\sigma^2$). The probability $P$ that a value of the variable $x$ is within an interval $[a, b]$ can be calculated from $p(x)$ as the area under a density curve (Eq. 2).

$$p(x) = \frac{1}{\sigma\sqrt{2\pi}}e^{-\frac{1}{2\sigma^2}(x-\mu)^2} \quad (1) \qquad P(a \leq x \leq b) = \int_a^b p(x)dx \quad (2)$$

## 2.2 Approaches to Density Estimation

Density usually is estimated, as observations in a dataset are random samples of a larger set or population. A *dataset* ($D$) is a collection of $n$ observations $(X_1, \ldots, X_n \in \mathbb{R}^d)$ described by $d$ variables (or dimensions). *Density estimation* is the process of using a sample dataset to find a *density estimator*, $\hat{p}(x)$, a function that provides an approximation of the unknown density, $p(x)$, of a population [5]. Density estimation can use parametric, non-parametric and histogram-based approaches, described next.

*Parametric approaches* are used when a preliminary analysis of the data can clearly reveal that it follows a well-known distribution or this is assumed, even without preliminary analysis. The distribution parameters are tuned in order to fit it as closely as possible to the data distribution. The advantages are the simplicity of the process and the possibility of understanding the parameters [12]. However, it is limited to scenarios of data fitting a known distribution. For instance, assuming that a dataset with $n$ observations can be fitted to a Normal distribution (Eq. 3), the density estimator uses estimates for the parameters. A simple method is estimating the sample mean, $\hat{\mu}_n$ and the sample variance, $\hat{\sigma}_n^2$ from the observations.

*Non-parametric Approaches* are used when the data does not appear to follow a well-known distribution. Thus, a continuous function from the data distribution is devised, a process analogous to training a prediction model. The advantage is the flexibility to estimate density for any distribution, regardless of its shape. However, the process becomes harder with escalating computational costs, both in available memory and elapsed time as dimensionality increases. Most implementations deal with two dimensions (e.g., *GenKern* [9] and *KernSmooth* [15]), with very few handling more (e.g., three in *sm* [7] and six in *ks* [3]). One of the most used algorithms is Kernel Density Estimators (KDE) [16] (Eq. 4) with parameters *kernel function* ($K$), a smooth and symmetric function (e.g., Gaussian, Epanechnikov), and *smoothing bandwidth* ($h$), a positive value for the amount of smoothing. Each observation ($X_i$) is replaced by a small density cluster (shaped like the kernel function) and all of them are added up.

$$\hat{p}(x) = \frac{1}{\hat{\sigma}_n\sqrt{2\pi}}e^{-\frac{1}{2\hat{\sigma}_n^2}(x-\hat{\mu}_n)^2} \quad (3) \qquad \hat{p}(x) = \frac{1}{nh}\sum_{i=1}^n K\left(\frac{x - X_i}{h}\right) \quad (4)$$

*Histogram-based approaches* are used to summarize data in the early stages of exploratory data analysis. Density estimation is obtained by the frequency of observations when grouped into bins. Due to their simplicity, histograms are an important and popular data analysis tool, with very quick results and low computational costs, even for large datasets. However, an acceptable performance is a trade-off for the quality of the estimate. A count-based method provides a less accurate estimate, when compared to non-parametric approaches. A histogram (Eq. 5) is a function that partitions the domain of a variable into $M$ bins ($x \in B_k$). Each bin ($B_k$) has $N_k$ observations with $h_k$ width. For instance, if $x$ is a numerical variable with domain $[a, b]$, the partition into bins is given by Eq. 6. For categorical variables, each bin is a set of one or more distinct values.

$$\hat{p}(x) = \frac{N_k}{nh_k} \quad (5) \qquad B_k = \begin{cases} [a; a + h_1] & k = 1 \\ (a + \sum_{i=1}^{k-1} h_i; a + \sum_{i=1}^{k} h_i] & 1 < k < M \\ (a + \sum_{i=1}^{k-1} h_i; b] & k = M \end{cases} \quad (6)$$

## 2.3   Applications in Research

There are a number of applications of density estimation. In a study on the causes of sudden infant death [4], a particular type of cell (degranulated mast) in infants who died both of known and unknown causes was counted. The density estimate revealed that between a quarter and a third of the cases, the count was exceptionally high, hinting for further clinical investigation.

Another example is an experiment with measurements of the height on 15000 points of a steel surface [2]. The density estimate presented a unimodal distribution around $30\,\mu m$ with two tails around it. The right one detected parts of the surface in contact with other surfaces, whereas the left one proved the existence of leaks, causing fatigue cracks or grease accumulation points. Both findings led to improvements in the steel surface fabrication process.

In another study, the goal was to understand the direction in which a group of 76 turtles were headed when released to swim freely in the ocean [11]. The density estimate was bimodal with most individuals swimming towards the $60°$ direction while a smaller proportion swam towards the exact opposite.

## 2.4   Example: Density of the Old Faithful Dataset

Density estimation can be illustrated using the Old Faithful Geyser dataset [1], with 272 observations of two variables, namely, the 'eruption time' and 'waiting time to next eruption' (measured in minutes). In Fig. 1 density is represented as a 2-dimensional histogram in a 10×10 grid. The lightest rectangles correspond to areas of greater concentration of data. Figure 2 presents a similar histogram with increased granularity, in a 30×30 grid. The rectangles become smaller and there are many more areas with zero density. Figure 3 depicts a density surface obtained from a non-parametric approach.

**Fig. 1.** 10×10 histogram    **Fig. 2.** 30×30 histogram    **Fig. 3.** Density surface

## 3    A Multivariate Histogram-Based Approach

The proposed approach for estimating the density of a dataset consists of a *multivariate histogram* ($\Gamma$), a generalization of a histogram for more than one variable. Formally, it is defined as a set of $M$ pairs, each with a hyperrectangle ($H_k$) and either a relative frequency ($f_k$) or a density estimate ($\hat{p}_k$) (Eqs. 7 and 8).

$$\Gamma_f = \{(H_k, f_k), k \in [1, M]\} \quad (7) \qquad \Gamma_{\hat{p}} = \{(H_k, \hat{p}_k), k \in [1, M]\} \quad (8)$$

Similarly as a 1-dimensional histogram partition the domain of a variable into bins, a multivariate histogram partitions a hyperspace (defined by the cross product of the domain of all variables) into hyperrectangles, for which the density is estimated separately. The method is presented in Fig. 4 as three separate tasks due to implementation issues. The hypergrid can be specified by the user or it can be automated via configuration parameters (also specified by the user). While counting observations according to a hypergrid is a simple computation ($\Gamma_f$), calculating hypervolumes is a considerably heavier one, carried out only for the hyperrectangles with non-zero relative frequency ($\Gamma_{\hat{p}}$).

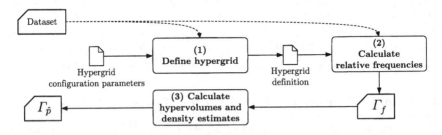

**Fig. 4.** Overview of the multivariate histogram-based approach

### 3.1    Define Hypergrid

A bin of a categorical variable $x_i$ is a set of $m$ values (Eq. 9), while in numerical variables, is a half-open interval (Eq. 10). Formally, a *hyperrectangle* ($H_k$) is a $d$-dimensional subspace defined as a tuple of bins (Eq. 11). The variable's bins are contained in its empirical domain (Eq. 12) which is the set of distinct values observed in a dataset. Partitioning the hyperspace into hyperrectangles defines a *hypergrid*.

$$B_{i,k} = \{v_1, v_2, \ldots, v_m\} \quad (9)$$

$$B_{i,k} = (\alpha, \beta] \quad (10)$$

$$H_k = (B_{1,k}, B_{2,k}, \ldots, B_{d,k}) \quad (11)$$

$$B_{i,k} \subseteq \mathrm{dom}(x_i) \quad (12)$$

Table 1 illustrates these concepts for a dataset with three variables: age ($x_1$), gender ($x_2$) and income ($x_3$). For the numerical variables ($x_1$ and $x_3$), the creation of bins was automated in order to create three bins (with $x_3$ discretized). For the categorical variable ($x_2$), a bin was created for each value. The partitioning resulted in the hypergrid definition of Table 2. Figure 5 depicts the hypergrid and an extract of the list of hyperrectangles ($H_1$ to $H_{18}$).

**Table 1.** Hypergrid configuration

| Var. | Type | Domain | Precision |
|---|---|---|---|
| $x_1$ | discrete | [18; 80] | |
| $x_2$ | categorical | {female, male} | |
| $x_3$ | continuous | [700.5; ...; 2340.7] | 0 |

**Table 2.** Hypergrid definition

| Var. | $B_1$ | $B_2$ | $B_3$ |
|---|---|---|---|
| $x_1$ | (17; 35] | (35; 52] | (52; 80] |
| $x_2$ | {female} | {male} | |
| $x_3$ | (699; 1150] | (1150; 1495] | (1495; 2341] |

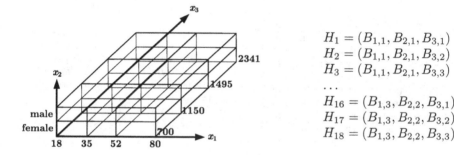

$$H_1 = (B_{1,1}, B_{2,1}, B_{3,1})$$
$$H_2 = (B_{1,1}, B_{2,1}, B_{3,2})$$
$$H_3 = (B_{1,1}, B_{2,1}, B_{3,3})$$
$$\ldots$$
$$H_{16} = (B_{1,3}, B_{2,2}, B_{3,1})$$
$$H_{17} = (B_{1,3}, B_{2,2}, B_{3,2})$$
$$H_{18} = (B_{1,3}, B_{2,2}, B_{3,3})$$

**Fig. 5.** Hypergrid and hyperrectangles

## 3.2 Calculate Relative Frequencies

The *frequency* ($N_k$) is the number of observations ($X_i$) from a dataset $D$ contained in a hyperrectangle $H_k$ (Eq. 13). The *relative frequency* ($f_k$) is the proportion of those in relation to the total number of observations ($n$) in the dataset (Eq. 14).

$$N_k = |\{X_i : X_i \in D \wedge X_i \in H_k\}| \quad (13)$$

$$f_k = \frac{N_k}{n} \quad (14)$$

## 3.3 Calculate Hypervolumes and Density Estimates

The *hypervolume* ($V_k$) is a measure of the volume of a hyperrectangle and a generalization of the *bin width* of 1-dimensional histograms [10]. Formally, a hypervolume is the product of the width ($w_{j,k}$) of all the edges (the bins in this context) of a hyperrectangle $H_k$ (Eq. 17). For numerical variables (either discrete or continuous) the width is the difference between the upper and lower bounds of the edge (Eq. 15). For categorical variables, the width is the number of distinct values in the edge (Eq. 16).

The *density estimate* ($\hat{p}_k$) is the average number of observations in a unit of hypervolume. Formally, it is the relative frequency normalized by the hypervolume (Eq. 18). Consequently, unlike the cumulative sum of relative frequencies, the density estimates of all the hyperrectangles do not add up to one.

$$w_{j,k} = \max B_{j,k} - \min B_{j,k} \quad (15) \qquad w_{j,k} = |B_{j,k}| \quad (16)$$

$$V_k = \prod_{j=1}^{d} w_{j,k} \quad (17) \qquad \hat{p}_k = \frac{f_k}{V_k} \quad (17)$$

An example for the dataset of Table 1 is presented in Table 3, listing only the hyperrectangles with density estimates ($\hat{p}$) greater than zero ($\Gamma_{\hat{p}}$). Table 4 lists all the intermediate calculations, namely, frequencies ($N$), relative frequencies ($f$), width of hyperrectangles edges ($|B|$), and hypervolumes ($V$).

**Table 3.** A multivariate histogram ($\Gamma_{\hat{p}}$)

| $k$ | $B_1$ | $B_2$ | $B_3$ | $\hat{p}$ |
|---|---|---|---|---|
| 1 | (17; 35] | {female} | (699; 1150] | $1.6 \times 10^{-5}$ |
| 4 | (17; 35] | {male} | (699; 1150] | $1.2 \times 10^{-5}$ |
| 8 | (35; 52] | {female} | (1150; 1495] | $5.0 \times 10^{-5}$ |
| 9 | (35; 52] | {female} | (1495; 2341] | $0.3 \times 10^{-5}$ |
| 11 | (35; 52] | {male} | (1150; 1495] | $3.6 \times 10^{-5}$ |
| 14 | (52; 80] | {female} | (1150; 1495] | $0.6 \times 10^{-5}$ |
| 15 | (52; 80] | {female} | (1495; 2341] | $0.2 \times 10^{-5}$ |
| 17 | (52; 80] | {male} | (1150; 1495] | $0.8 \times 10^{-5}$ |
| 18 | (52; 80] | {male} | (1495; 2341] | $0.2 \times 10^{-5}$ |

**Table 4.** Intermediate calculations

| $k$ | $N$ | $f$ | $|B_1|$ | $|B_2|$ | $|B_3|$ | $V$ |
|---|---|---|---|---|---|---|
| 1 | 235 | 0.126 | 18 | 1 | 451 | 8118 |
| 4 | 183 | 0.098 | 18 | 1 | 451 | 8118 |
| 8 | 542 | 0.291 | 17 | 1 | 345 | 5865 |
| 9 | 87 | 0.047 | 17 | 1 | 846 | 14382 |
| 11 | 390 | 0.210 | 17 | 1 | 345 | 5865 |
| 14 | 116 | 0.062 | 28 | 1 | 345 | 9660 |
| 15 | 93 | 0.050 | 28 | 1 | 846 | 23688 |
| 17 | 146 | 0.079 | 28 | 1 | 345 | 9660 |
| 18 | 68 | 0.037 | 28 | 1 | 846 | 23688 |

### 3.4   Estimate Density for Datasets with Missing Values

When a dataset has missing data, the approach includes additional tasks. Missing values are considered unknown values and contained in hyperrectangles with edges unknown. A hyperrectangle having one or more unknown edges is an *undefined hyperrectangle*, giving rise to an *incomplete histogram*. The approach for handling missing values is to replace all undefined hyperrectangles with complete ones (without unknown edges), provided that they cover the same region of the subspace. Thus, for each undefined hyperrectangle, it is defined as a *missing values hypergrid*, including only the variables in the unknown edges.

Figure 6 shows an example of an undefined hyperrectangle, enclosing all observations where $x_1 \in (52; 80]$ and with missing values in both $x_2$ and $x_3$. A missing values hypergrid is defined from the combination of the bins of $x_2$ and $x_3$ (Fig. 7), a basis for a *missing values histogram* using an extract of the dataset with observations without missing values on both $x_2$ and $x_3$. The undefined hyperrectangle is replaced by a set of complete hyperrectangles (Fig. 8). When replaced, the relative frequencies of the complete hyperrectangles are scaled by the relative frequency of the undefined hyperrectangle, yielding a second version of the multivariate histogram. However, replacing all undefined hyperrectangles with complete ones can result in multiple instances of the same hyperrectangles. Therefore, it is necessary to coalesce identical hyperrectangles into one by adding their relative frequencies in a third version of the multivariate histogram. The process then resumes in task #3 as described in Sect. 3.3.

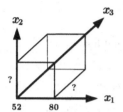

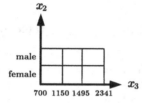

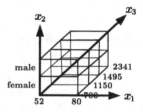

**Fig. 6.** An undefined hyperrectangle

**Fig. 7.** Missing values hypergrid

**Fig. 8.** Complete hyperrectangles

Table 5 is an incomplete histogram, with the undefined hyperrectangle of Fig. 6 (denoted as $k_1 = *1$). Table 6 is the missing values histogram obtained from the observations without missing values in both $x_2$ and $x_3$ (based on the hypergrid in Fig. 7). Figure 8 depicts the combination with the undefined hyperrectangle and subsequent replacement. Table 7 is the completed histogram with the relative frequencies scaled by the relative frequency of the undefined hyperrectangle. The replacement led to the addition of two additional hyperrectangles ($k_2 = 13$ and $k_2 = 16$), while the other four subspaces were initially already covered in Table 5. Hence, all pairs of duplicate hyperrectangles have coalesced into one with the relative frequencies added, resulting in Table 8.

## 4    Evaluation and Results

### 4.1    Computational Performance

The method's computational performance was evaluated by estimating density on a number of datasets from the UCI [6], Kaggle [1], and OpenML [14] repositories, selected according to a diversity of characteristics. We define the *dataset size* ($s = nd$) as the product of $n$ (#observations) with $d$ (#dimensions). The datasets in group #1 have few dimensions ($d \leq 6$), only numerical variables, and no missing values. Group #2 has the same characteristics, with a higher number of dimensions. Group #3 also includes categorical variables and no missing values ($d_c$ specifying the number of categorical variables). Group #4 has datasets with missing values ($n_{mv}$ specifying the number of observations with missing values in at least one variable and $r_{mv}$ the *ratio of missing values*).

**Table 5.** Incomplete histogram

| $k_1$ | $B_1$ | $B_2$ | $B_3$ | $N$ | $f$ |
|---|---|---|---|---|---|
| 1 | (17; 35] | {female} | (699; 1150] | 235 | 0.113 |
| 4 | (17; 35] | {male} | (699; 1150] | 183 | 0.088 |
| 8 | (35; 52] | {female} | (1150; 1495] | 542 | 0.260 |
| 9 | (35; 52] | {female} | (1495; 2341] | 87 | 0.042 |
| 11 | (35; 52] | {male} | (1150; 1495] | 390 | 0.187 |
| 14 | (52; 80] | {female} | (1150; 1495] | 116 | 0.056 |
| 15 | (52; 80] | {female} | (1495; 2341] | 93 | 0.045 |
| 17 | (52; 80] | {male} | (1150; 1495] | 146 | 0.070 |
| 18 | (52; 80] | {male} | (1495; 2341] | 68 | 0.032 |
| *1 | (52; 80] | | | 223 | 0.107 |

**Table 6.** Missing values histogram

| $k_\phi$ | $B_2$ | $B_3$ | $N$ | $f$ |
|---|---|---|---|---|
| 1 | {female} | (699; 1150] | 235 | 0.126 |
| 2 | {female} | (1150; 1495] | 658 | 0.354 |
| 3 | {female} | (1495; 2341] | 180 | 0.097 |
| 4 | {male} | (699; 1150] | 183 | 0.098 |
| 5 | {male} | (1150; 1495] | 536 | 0.288 |
| 6 | {male} | (1495; 2341] | 68 | 0.037 |

**Table 7.** Completed histogram

| $k_2$ | $k_1$ | $k_\phi$ | $B_1$ | $B_2$ | $B_3$ | $f$ |
|---|---|---|---|---|---|---|
| 1 | 1 | | (17; 35] | {female} | (699; 1150] | 0.113 |
| 4 | 4 | | (17; 35] | {male} | (699; 1150] | 0.088 |
| 8 | 8 | | (35; 52] | {female} | (1150; 1495] | 0.260 |
| 9 | 9 | | (35; 52] | {female} | (1495; 2341] | 0.042 |
| 11 | 11 | | (35; 52] | {male} | (1150; 1495] | 0.187 |
| 13 | *1 | 1 | (52; 80] | {female} | (699; 1150] | 0.014 |
| 14 | 14 | | (52; 80] | {female} | (1150; 1495] | 0.056 |
| 14 | *1 | 2 | (52; 80] | {female} | (1150; 1495] | 0.038 |
| 15 | 15 | | (52; 80] | {female} | (1495; 2341] | 0.045 |
| 15 | *1 | 3 | (52; 80] | {female} | (1495; 2341] | 0.010 |
| 16 | *1 | 4 | (52; 80] | {male} | (699; 1150] | 0.011 |
| 17 | 17 | | (52; 80] | {male} | (1150; 1495] | 0.070 |
| 17 | *1 | 5 | (52; 80] | {male} | (1150; 1495] | 0.031 |
| 18 | 18 | | (52; 80] | {male} | (1495; 2341] | 0.032 |
| 18 | *1 | 6 | (52; 80] | {male} | (1495; 2341] | 0.004 |

**Table 8.** Final histogram

| $k_3$ | $B_1$ | $B_2$ | $B_3$ | $f$ |
|---|---|---|---|---|
| 1 | (17; 35] | {female} | (699; 1150] | 0.113 |
| 4 | (17; 35] | {male} | (699; 1150] | 0.088 |
| 8 | (35; 52] | {female} | (1150; 1495] | 0.260 |
| 9 | (35; 52] | {female} | (1495; 2341] | 0.042 |
| 11 | (35; 52] | {male} | (1150; 1495] | 0.187 |
| 13 | (52; 80] | {female} | (699; 1150] | 0.014 |
| 14 | (52; 80] | {female} | (1150; 1495] | 0.094 |
| 15 | (52; 80] | {female} | (1495; 2341] | 0.055 |
| 16 | (52; 80] | {male} | (699; 1150] | 0.011 |
| 17 | (52; 80] | {male} | (1150; 1495] | 0.101 |
| 18 | (52; 80] | {male} | (1495; 2341] | 0.036 |

The method's performance was evaluated by measuring the elapsed time ($t$, in seconds) for density estimation. In all datasets, the hypergrid was created to contain three bins in numerical variables and one separate bin for each value in categorical variables. The results are hardly surprising. The most penalizing factors in density estimation time are the size of the dataset and the ratio of missing values. In smaller datasets ($s < 30000$) (Fig. 9), density is estimated in less that one second (a regression line with nearly zero slope). The exception is a dataset with missing values where $t > 10$, due to the need to perform hyperrectangle replacements. For instance, in the 'Astronauts' dataset, all the observations have at least one variable with a missing value, which greatly increases the calculation time ($\approx 13$ s). In larger datasets (Fig. 10) there is a clear linear relationship between the size of a dataset and density estimation time. However, the slope of the regression line is much higher in datasets with missing values.

## 4.2 Measuring Density with Categorical Variables

In the following experiments, we observed the effect that adding categorical variables has on the density estimates of hyperrectangles. We expect that a new

**Table 9.** Datasets selected for density estimation performance evaluation

| Group | Dataset | $n$ | $d$ | $s$ | $d_c$ | $n_{mv}$ | $r_{mv}$ | $t$ |
|---|---|---|---|---|---|---|---|---|
| #1 | China's population [1] | 73 | 4 | 292 | 0 | 0 | 0.00 | 0.04 |
| | Iris [6] | 150 | 4 | 600 | 0 | 0 | 0.00 | 0.05 |
| | Old faithful [1] | 242 | 2 | 484 | 0 | 0 | 0.00 | 0.06 |
| | Phoneme [14] | 5 404 | 5 | 27 020 | 0 | 0 | 0.00 | 0.16 |
| | 2D elastodynamic metamaterials [6] | 20 520 | 3 | 61 560 | 0 | 0 | 0.00 | 0.37 |
| | Stock market@Kraken [1] | 32 946 819 | 2 | 65 893 638 | 0 | 0 | 0.00 | 264.89 |
| #2 | Wine quality [1] | 1 143 | 12 | 13 716 | 0 | 0 | 0.00 | 0.28 |
| | Dry bean [6] | 13 611 | 16 | 217 776 | 0 | 0 | 0.00 | 1.16 |
| | SGEMM GPU kernel performance [6] | 241 600 | 18 | 4 434 800 | 0 | 0 | 0.00 | 45.85 |
| | Students' dropout/success [6] | 4 424 | 37 | 163 688 | 0 | 0 | 0.00 | 1.62 |
| | Rice MSC [1] | 75 000 | 106 | 7 950 000 | 0 | 0 | 0.00 | 214.05 |
| | Biological response [6] | 3 751 | 1776 | 6 661 776 | 0 | 0 | 0.00 | 78.81 |
| #3 | Height of male and female [1] | 199 | 4 | 796 | 1 | 0 | 0.00 | 0.14 |
| | Students performance in exams [1] | 1 000 | 8 | 8 000 | 5 | 0 | 0.00 | 0.20 |
| | Car evaluation [6] | 1 728 | 7 | 12 096 | 7 | 0 | 0.00 | 0.42 |
| | Auction verification [6] | 2 043 | 9 | 18 387 | 1 | 0 | 0.00 | 0.25 |
| | Diabetes [6] | 29 278 | 2 | 58 556 | 1 | 0 | 0.00 | 0.19 |
| | Airlines train [14] | 10 000 000 | 10 | 100 000 000 | 3 | 0 | 0.00 | 1162.48 |
| #4 | Breast cancer Wisconsin [6] | 699 | 8 | 5 592 | 0 | 16 | 0.02 | 0.14 |
| | HCV [14] | 615 | 13 | 7 995 | 2 | 26 | 0.04 | 0.79 |
| | Astronauts [1] | 952 | 11 | 10 472 | 10 | 952 | 1.00 | 13.77 |
| | Mushroom [6] | 8 124 | 23 | 186 852 | 23 | 2 480 | 0.31 | 4.71 |
| | Job change of data scientists [1] | 19 158 | 12 | 229 896 | 10 | 10 203 | 0.53 | 262.36 |
| | Adult [6] | 32 561 | 13 | 423 293 | 8 | 2 399 | 0.07 | 31.23 |

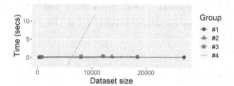

**Fig. 9.** Smaller datasets ($s < 30000$)

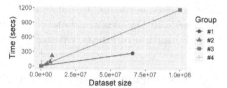

**Fig. 10.** Larger datasets ($s \geq 30000$)

variable causes the observations to spread out over to more hyperrectangles, therefore, decreasing the density of each. Still, as the number of observations is unchanged, adding the densities of the new hyperrectangles must be equal to the total density of the aggregating hyperrectangles.

We selected the Old Faithful dataset [1] (Sect. 2.4) with histogram in Table 10 (for $\hat{p} > 0$). A categorical test variable ($x_3$) was added and randomly filled with two values (c1 and c2). As a result, the observations of each hyperrectangle spread out into pairs of hyperrectangles (Table 11). As expected, the sum of the density estimates of each pair (equal bins in $x_1$ and $x_2$, different bin in $x_3$) in Table 11 is the density estimate of the aggregating hyperrectangle in Table 10. Next, two more categorical test variables were added ($x_4$ and $x_5$), also randomly filled with the same values as $x_3$ (Tables 12 and 13 respectively). Again, the aggregating hyperrectangles of the previous dimension hold the sum of each pair

**Table 10.** Old Faithful histogram

| $k$ | $B_1$ | $B_2$ | $\hat{p}$ |
|---|---|---|---|
| 1 | (1.5; 2.4] | (42; 64] | 0.01904 |
| 2 | (1.5; 2.4] | (64; 80] | 0.00115 |
| 4 | (2.4; 4.3] | (42; 64] | 0.00064 |
| 5 | (2.4; 4.3] | (64; 80] | 0.00617 |
| 6 | (2.4; 4.3] | (80; 96] | 0.00435 |
| 8 | (4.3; 5.1] | (64; 80] | 0.01120 |
| 9 | (4.3; 5.1] | (80; 96] | 0.01379 |

**Table 11.** Adding 1 cat. variable

| $k$ | $B_1$ | $B_2$ | $B_3$ | $\hat{p}$ |
|---|---|---|---|---|
| 1 | (1.5; 2.4] | (42; 64] | {c1} | 0.00810 |
| 2 | (1.5; 2.4] | (42; 64] | {c2} | 0.01094 |
| 3 | (1.5; 2.4] | (64; 80] | {c1} | 0.00029 |
| 4 | (1.5; 2.4] | (64; 80] | {c2} | 0.00086 |
| $\cdots$ | | | | |
| 17 | (4.3; 5.1] | (80; 96] | {c1} | 0.00603 |
| 18 | (4.3; 5.1] | (80; 96] | {c2} | 0.00776 |

**Table 12.** Adding 2 cat. variables

| $k$ | $B_1$ | $B_2$ | $B_3$ | $B_4$ | $\hat{p}$ |
|---|---|---|---|---|---|
| 1 | (1.5; 2.4] | (42; 64] | {c1} | {c1} | 0.00219 |
| 2 | (1.5; 2.4] | (42; 64] | {c1} | {c2} | 0.00591 |
| 3 | (1.5; 2.4] | (42; 64] | {c2} | {c1} | 0.00416 |
| 4 | (1.5; 2.4] | (42; 64] | {c2} | {c2} | 0.00678 |
| $\cdots$ | | | | | |
| 35 | (4.3; 5.1] | (80; 96] | {c2} | {c1} | 0.00287 |
| 36 | (4.3; 5.1] | (80; 96] | {c2} | {c2} | 0.00488 |

**Table 13.** Adding 3 cat. variables

| $k$ | $B_1$ | $B_2$ | $B_3$ | $B_4$ | $B_5$ | $\hat{p}$ |
|---|---|---|---|---|---|---|
| 1 | (1.5; 2.4] | (42; 64] | {c1} | {c1} | {c1} | 0.00131 |
| 2 | (1.5; 2.4] | (42; 64] | {c1} | {c1} | {c2} | 0.00088 |
| 3 | (1.5; 2.4] | (42; 64] | {c1} | {c2} | {c1} | 0.00219 |
| 4 | (1.5; 2.4] | (42; 64] | {c1} | {c2} | {c2} | 0.00372 |
| $\cdots$ | | | | | | |
| 71 | (4.3; 5.1] | (80; 96] | {c2} | {c2} | {c1} | 0.00201 |
| 72 | (4.3; 5.1] | (80; 96] | {c2} | {c2} | {c2} | 0.00287 |

of new hyperrectangles. The results confirm that the more dimensions added to the dataset, the denser the hypergrid and more hyperrectangles are defined. Hence, there is an inevitable tendency for the density associated with each to decrease and the average density also to decrease, as demonstrated by the controlled addition of binary categorical variables.

## 4.3  Measuring Density with Missing Values

In the following experiments, we observed the effect that adding missing values has on the density estimates of hyperrectangles. The method used to handle missing data leads to the inclusion of new hyperrectangles, even in subspaces where there are no observations in the dataset. As the missing data increases, it is expected a variation of density in all hyperrectangles (both existing and new ones).

Missing values were deliberately created by randomly clearing existing ones in the Old Faithful dataset [1]. Table 14 shows the differences ($\Delta\hat{p}$) in density estimates with increases in the rate of missing values for $r_{mv} = 0.01$ ($\hat{p}_1$), $r_{mv} = 0.05$ ($\hat{p}_2$), and $r_{mv} = 0.10$ ($\hat{p}_3$). With missing values representing 1% of the dataset, the differences in the density estimates are about [1]$0-4$ times or less in order of magnitude (including the new hyperrectangles #3 and #7 previously with zero density and, thus, omitted in Table 10). In the subsequent cases (5% and 10% respectively), it is apparent that the absolute differences, while still in the same order of magnitude, increased. Consequently, as the rate of missing values increases, the variation in the density estimates becomes more sizeable.

**Table 14.** Differences in density estimates of adding missing values

| k | $B_1$ | $B_2$ | $\hat{p_1}$ | $\Delta\hat{p_1}$ | $\hat{p_2}$ | $\Delta\hat{p_2}$ | $\hat{p_3}$ | $\Delta\hat{p_3}$ |
|---|-------|-------|-------------|-------------------|-------------|-------------------|-------------|-------------------|
| 1 | (1.5; 2.4] | (42; 64] | 0.01890 | $-1.4 \times 10^{-4}$ | 0.01817 | $-8.7 \times 10^{-4}$ | 0.01796 | $-10.8 \times 10^{-4}$ |
| 2 | (1.5; 2.4] | (64; 80] | 0.00135 | $2.0 \times 10^{-4}$ | 0.00154 | $3.9 \times^{-4}$ | 0.00174 | $5.9 \times 10^{-4}$ |
| 3 | (1.5; 2.4] | (80; 96] | 0.00009 | $0.9 \times 10^{-4}$ | 0.00055 | $5.5 \times 10^{-4}$ | 0.00103 | $1.0 \times 10^{-4}$ |
| 4 | (2.4; 4.3] | (42; 64] | 0.00064 | $0.0 \times 10^{-4}$ | 0.00077 | $1.3 \times 10^{-4}$ | 0.00084 | $2.0 \times 10^{-4}$ |
| 5 | (2.4; 4.3] | (64; 80] | 0.00609 | $-0.8 \times 10^{-4}$ | 0.00613 | $-4.0 \times 10^{-4}$ | 0.00605 | $-1.2 \times 10^{-4}$ |
| 6 | (2.4; 4.3] | (80; 96] | 0.00435 | $0.0 \times 10^{-4}$ | 0.00403 | $-3.2 \times 10^{-4}$ | 0.00394 | $-4.1 \times 10^{-4}$ |
| 7 | (4.3; 5.1] | (42; 64] | 0.00008 | $0.8 \times 10^{-4}$ | 0.00044 | $4.4 \times 10^{-4}$ | 0.00081 | $8.1 \times 10^{-4}$ |
| 8 | (4.3; 5.1] | (64; 80] | 0.01139 | $1.9 \times 10^{-4}$ | 0.01102 | $-1.8 \times 10^{-4}$ | 0.01083 | $-3.7 \times 10^{-4}$ |
| 9 | (4.3; 5.1] | (80; 96] | 0.01359 | $-2.0 \times 10^{-4}$ | 0.01405 | $2.6 \times 10^{-4}$ | 0.01353 | $-2.6 \times 10^{-4}$ |

## 5   Conclusions

Density estimation methods have several applications in the scope of exploratory data analysis. Non-parametric approaches are limited to a small number of dimensions. Histogram-based approaches, despite offering less accurate density estimates, remain the only alternative concerning datasets in high-dimensional spaces. Still, current implementations have dimensionality constraints.

We propose a method without limitation in the number of variables, able to handle categorical variables and missing data. The only limitation of the method is related to the size of the dataset (observations × dimensions). While empirical evaluation confirms that the method is able to deal with high-dimensional datasets it also provides evidence that the method behaves as expected when dealing with categorical variables and missing values. Additionally, the computational performance is closely correlated to both dataset size and missing data.

**Acknowledgments.** This work is financed by National Funds through the Portuguese funding agency, FCT - Fundação para a Ciência e a Tecnologia, within project UIDB/50014/2020.

## References

1. Kaggle (2022). https://www.kaggle.com/datasets
2. Adler, R.J., Firmin, D., Kendall, D.G.: A non-gaussian model for random surfaces. Philos. Trans. R. Soc. Lond. Series A Math. Phys. Sci. **303**, 433–462 (1981)
3. Bowman, A., Azzalini, A.: R package SM: nonparametric smoothing methods (2010). http://www.stats.gla.ac.uk/adrian/sm
4. Carpenter, R., Emery, J.L.: Identification and follow-up of infants at risk of sudden death in infancy. Nature **250**, 729–729 (1974). https://doi.org/10.1038/250729a0
5. Chen, Y.C.: A tutorial on kernel density estimation and recent advances. Biostatistics Epidemiol. **1**, 161–187 (2017)
6. Dua, D., Graff, C.: UCI machine learning repository (2017). http://archive.ics.uci.edu/ml

7. Duong, T.: KS: Kernel smoothing (2011). http://CRAN.R-project.org/package=ks
8. Goodfellow, I., Bengio, Y., Courville, A.: Deep Learning. MIT Press, Cambridge (2016)
9. Lucy, D., Aykroyd, R.: Genkern: functions for generating and manipulating binned kernel density estimates (2010). http://CRAN.R-project.org/package=GenKern
10. Luo, F., Mehrotra, S.: A concentration result of estimating phi-divergence using data dependent partition (2018). http://arxiv.org/abs/1801.00852
11. Mardia, K.V.: Statistics of directional data (1975)
12. Scott, D.W.: Multivariate density estimation and visualization. In: Gentle, J., Härdle, W., Mori, Y. (eds.) Handbook of Computational Statistics. Springer Handbooks of Computational Statistics, Springer, Berlin, Heidelberg (2012). https://doi.org/10.1007/978-3-642-21551-3_19
13. Silverman, B.W.: Density Estimation for Statistics and Data Analisys (1986)
14. Vanschoren, J., van Rijn, J.N., Bischl, B., Torgo, L.: Openml: networked science in machine learning (2014)
15. Wand, M., Ripley, B.: Kernsmooth: functions for kernel smoothing for wand & jones (1995) (2010). http://CRAN.R-project.org/package=KernSmooth
16. Węglarczyk, S.: Kernel density estimation and its application. In: ITM Web of Conferences, vol. 23, p. 00037 (2018)

# Multi-objective, Optimization, Augmentation, and Database

# Correcting Temporal Overlaps in Process Models Discovered from OLTP Databases

Anbumunee Ponniah[1,2(✉)] and Swati Agarwal[1] [iD]

[1] BITS Pilani, K K Birla Goa Campus, Sancoale, India
{p20170418,swatia}@goa.bits-pilani.ac.in
[2] IBM India Private Limited, Bangalore, India

**Abstract.** Event logs extracted from database systems often have overlapping timestamps that interfere with process discovery due to the random nature of such overlaps. Heuristics and object-based analysis of event logs attempt to discover processes using information beyond timestamps. Systems often include an audit function to track compliance of a subset of tasks in a process. However, the logs are not a suitable primary source of event logs since they represent specific tasks instead of entire processes. This paper proposes a mapping and sequence analysis to correlate high-granularity audit records with discovered events. We further present our method's effectiveness in identifying sequences among tasks having the same timestamp.

## 1 Introduction

Process mining (PM) is a discipline of Business Intelligence and Data mining that allows for the discovery, analysis, and enhancement of business processes in the context of Business Process Mining. As described in the Process Mining Manifesto [21], a primary data source (PDS) for PM is event logs, which are a collection of instances of events. These events indicate an activity performed on an entity at a specific time using a set of input data and resulting in a set of outcomes. Process mining primarily relies on attributes and time stamps to detect cases, tasks, and their sequences from event logs [20]. Availability of high-quality event logs is a well-recognised challenge in the field of PM [2]. Events recorded with the same timestamp is one such quality issue that causes task sequencing errors in the discovered process model. Large-scale distributed systems use automated programs to execute business processes that transcend multiple applications. As an example, an Order-To-Cash supply chain business process [11] presents common issues such as missing or inconsistent values, differences in formats, standards, notations, granularity, and many more. The timestamps could hold the same value for multiple steps due to granularity of the logging, delays in the logging system and concurrent batch execution of the process in high-performance systems [5]. Examples of batch processes in the retail domain include end-of-day upload of orders captured by field sales, optimum inventory allocation, warehouse put-away, batch picking, and many more. The batch order creation process involves executing rules such as duplicate orders, availability of inventory, a price calculation, fraud check, and many more. While

W. Chen et al. (Eds.): ADMA 2022, LNAI 13726, pp. 281–296, 2022.
https://doi.org/10.1007/978-3-031-22137-8_21

these steps have a specific purpose and execute in a specific sequence to meet the business requirements, each of these events for a trace could hold identical timestamps when recorded in event logs. Also, such events may be recorded differently for each process instance. Such random order impedes process discovery techniques resulting in a wrong sequence of tasks in the discovered process model. The scope of our work is to address the time overlap issue by analysing a secondary data source (SDS) (e.g., audit logs) for a sequence of related activities and extrapolating the sequence to event logs. We present our work in the context of the retail supply chain order fulfilment sub-process.

We perform our literature review in the context of process mining across log extraction and the use of audit logs. Multiple techniques and tools are available to extract usable event logs from enterprise applications [4], improve the quality of extracted logs through preprocessing [20], and improve process discovery [16]. Existing literature reveals that repairing quality issues in logs is either a manual process [17] or requires a strict reference process model [18]. Other options include using fields from the event logs to deduce sequence such as *case-id, contextual information, transaction information, person information* and many more. [10] presented an approach to extract, transform and store object-centric data, resulting in eXtensible Object-Centric event logs that further allow analysis without case notations. [9] addresses process discovery and conformance checking when the logs contain billions of events (large) and thousands of activities (complex). They propose a Directly Follows Framework using inductive miner. It uses a divide-and-conquer method for process discovery by recursively splitting the process tree and merging their output. For complex processes, [17] presents an approach that uses human inputs to repair non-conforming process models discovered using existing algorithms. The authors use the human-defined process model's hierarchical structure to correct the discovered model's localised flow. Researchers use additional data sources [3], and metadata such as semantic annotations [14] to improve the quality of discovered models. [1] explore the use of additional data sources for process mining in the context of additional fields available for analysis. Audit logs are an essential source of information used in the business intelligence domain. Enterprise Applications (EAs) include an audit function that proves it meets various business, operational, and legal requirements to determine the system's compliance. For example, a retail system responsible for payment will need to record explicit action of the payer and payee as provenance against non-repudiation and for compliance with the payment card industry's data security standard (PCI-DSS). Due to its strict compliance requirements, the audit information could provide insights into the correct sequence of tasks in a process [13]. There are significant differences between logs captured for audit purposes and other logs. Unlike event logs that focus on capturing the occurrence of every event, and audit logs may target a subset of events such as *fraud check, inventory allocation*, and *shipment scheduling* within the *Order-To-Cash* process. Hence audit logs are unsuitable as the primary source of event logs. The captured information establishes compliance and does not directly map to the process. Using audit logs as a supplement source for process mining requires additional steps to map the records to process tasks. Further, compared to regular logs, audit logs may be saved in a separate storage system in a different format and may be subject to additional access and security controls. In our literature review, we found instances of using audit logs for anomaly detection in security [19], health [12] and finance [8] domains. Further, we found instances of process mining used as a tool for determining conformance and compliance in the domains of healthcare [12] and privacy [19]. Using audit logs as a source of event logs requires extensive preprocessing, and the discovered process model likely has gaps in its tasks. However, the audit logs contain important context information recorded in a predictable and consistent order, which may be useful to address conflicts or uncertainties encountered in the model.

**Gaps Identified and Our Objectives:** Current studies treat temporal overlap of tasks as errors similar to missing values and employs mean or binning techniques to correct them. Multiple techniques apply heuristics available in the logs or other contextual information such as design documents or entity relationships to derive a hierarchy of tasks. Thus, there is a need to treat overlapping temporal information as valid data in the context of automated processes. While existing studies use audit logs as a source of anomaly detection, we do not find their instances used as supplementary sources of task descriptions. Further, validating and repairing discovered process models requires the involvement of human experts. A method that augments human involvement could improve the performance and accuracy of the process discovery step. We focus on using audit logs to overcome sequence inconsistencies, validate, and suggest corrections for the discovered model. The proposed work has the following objectives: a) For event logs extracted from an OLTP database, we demonstrate the problem caused in process models due to time overlaps. b) Establish a fundamental basis for utilising an SDS in repairing such process models. c) Design and implement a methodology to correlate sequences between the primary and SDS to repair the analysed process. d) Evaluate the utility of our method by applying it to an OLTP database as the PDS and corresponding audit logs as the SDS. Our contributions include the novel methods of mapping audit activities to business-level process tasks, correlating sequences through analysis of annotations, and using the discovered insights to repair the process model. Our method is transferable to scenarios where an SDS provides contextual information to repair sequence errors in event logs generated from an associated primary source.

## 2    Proposed Method

This section details our proposed method of utilising audit logs to address timestamp-based conflicts in the process model discovered from event logs. The method relies on associating various audit records and the tasks they represent. We first describe the mathematical basis of our proposed Stochastic Sequence Analysis of Secondary Data Source (SA-SDS) method in Sect. 2.1 followed by its detailed solution framework in Sect. 2.2.

## 2.1  Mathematical Basis for the Proposed SA-SDS Method

Let $L$ represent an event log which consist of $N$ events, such that $L = [E_1, E_2, ..., E_N]$ where $E_x = [EID, AY, TS]$ such that $x \in \{1, 2, ..., N\}$ where EID is the unique record identifier, AY represents ACTIVITY and denotes one of $n$ actions being recorded $\{a_1, a_2, ..., a_n\}$ and TS represents the TIMESTAMP which is a standard format system time of the log entry. We define a trace $T_E$ as a sequence of events representing an instance of a process for specific inputs and is a subset of events from $L$. Let the likelihood of activity $a_i$ be followed by $a_j$ denoted as $L(a_i, a_j)$ and is calculated from a training log set such that $L(a_i, a_j) = \frac{N_{(a_i,a_j)}}{(N_{a_i} \cup N_{a_j})}$ where $N_{(a_i,a_j)}$ is the number of times events with activity $a_i$ is followed by events with activity $a_j$ (not necessarily to be consecutive), $N_{a_i}$ and $N_{a_j}$ are the number of events with activity $a_i$ and $a_j$, respectively. For example, activities $a_2$ and $a_7$ in a log containing the following traces with the sequence of activities:

```
[{ {a_5, a_1, a_2, a_8, a_{10}, a_7},
   {a_3, a_2, a_8, a_9, a_7},
   {a_5, a_2, a_{10}, a_7},
   {a_4, a_1, a_7, a_{10}, a_2, a_9} }]
```

Then, we calculate $L(a_2, a_7) = \frac{N_{(a_2,a_7)}}{(N_{a_2} \cup N_{a_7})} = \frac{3}{4} = 0.75$ and $L(a_7, a_2) = \frac{N_{(a_7,a_2)}}{(N_{a_7} \cup N_{a_2})} = \frac{1}{4} = 0.25$ Similarly, let $C$ denote an audit log which consists of $M$ audit records such that $C = [C_1, C_2, ..., C_M]$ each $C_x = [AID, p_i, TS, CX]$ for $x \in \{1, 2, ..., M\}$ where AID is the unique audit record identifier, $p_i$ is PURPOSE which denotes one of the $q$ audit categories $P = \{p_1, p_2, ..., p_q\}$, $TS$ is TIMESTAMP which denotes standard format system time of the audit entry, and $CX$ CONTEXT is a variable length contextual data captured for the purpose of the audit. A trace $T_C$, defined as a sequence of audit records representing an instance of a process for specific inputs, is a subset of audit logs from $C$. Let $L(p_k, p_l)$ be the likelihood of an audit record entry with purpose $p_k$ being followed by a record with purpose $p_l$. $L(p_k, p_l)$ (calculated from a training log set) such that: $L(p_k, p_l) = \frac{N_{(p_k,p_l)}}{(N_{p_k} \cup N_{p_l})}$ where $N_{(p_k,p_l)}$ is the number of times audit records with purpose $p_k$ is followed by records with purpose $p_l$, $N_{p_k}$ and $N_{p_l}$ are the number of audit records with purpose $p_k$ and $p_l$, respectively. Following are the ways to establish an association between purpose fields of audit log records and activity fields of event log records: 1) manual analysis of a test log, 2) a learning mechanism based on an annotated reference training log, or 3) by applying an object-based analysis technique described by [10]. The sequence of records observed in the audit logs will likely preserve the sequence of the actual business process due to its strict compliance needs. In this context, observe the following:

**Observation O1:** Given a set of activities present in the event logs and a set of entries present in the audit logs, there exists an association such that one or more values of *PURPOSE* can be associated with one or more values of *ACTIVITY*, as determined from a pair of the test event and audit logs. We denote the association as, $\{p_1, p_2, ..., p_m\} \rightarrow \{a_1, a_2, ..., a_n\}$   $\forall m, n$

**Observation O2:** The sequence of audit purposes that meet a compliance requirement is available. The likelihood that an audit purpose follows a group of purposes (calculated beforehand) such that

$$L(p_j, (p_1, p_2, ..., p_{(j-1)}))\quad \forall m \neq j, \rightarrow L(a_i, (a_1, a_2, ..., a_{(i-1)}))\quad \forall m \neq i$$

**Inference 1:** Given the above two observations, a similar relationship exists between a sequence of audit purposes and corresponding event activities. Hence the likelihood that an audit purpose follows a group of purposes correlates to the likelihood that an event activity follows a group of activities. We denote the correlation as $L(p_j, (p_1, p_2, ..., p_{(j-1)})), \rightarrow L(a_i, (a_1, a_2, ..., a_{(i-1)}))\quad \forall i, j$.

**Inference 2:** Further, if $L(a_i, a_j) > L(a_j, a_i)\quad \forall i, j$ where activity is associated with purpose, i.e., $(a_i) \rightarrow (p_m)$ and $(a_j) \rightarrow (p_n)\ \exists m, n$ then it can be surmised as $L(p_m, p_n) > L(p_n, p_m)\quad \forall m, n$.

### 2.2 SA-SDS Framework

Building upon the mathematical description, we present the solution framework for the proposed SA-SDS method using the audit logs as the secondary source. Figure 1 depicts the framework as containing the following phases:

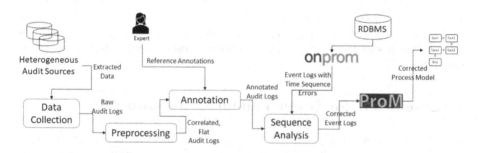

**Fig. 1.** Workflow of modified *onprom* methodology with SA-SDS.

**Data Collection**: Systems capture audit records in heterogeneous sources such as database tables and files. Each source may have different security restrictions, encryption mechanisms, network protocols, and application interfaces. In this phase, we implement the extraction routines to connect to those sources and collect the audit data into a single data repository. At the end of this phase, audit records are available in their raw format for further processing.

**Preprocessing:** The raw audit records from each source may be in a different format (e.g. XML) and include records for many operations. Further, information for operations related to an entity is not available in a single log. In this phase, we parse the collection of raw format logs to filter out just the records related to the analysed business process. We then correlate records across multiple logs to identify records related to the same entity, such as *order* and *product*. We then convert each correlated audit record into a flattened row in a CSV format file, ready for further processing.

**Annotation:** The audit records refer to specific activities which may not directly correlate to the process executed. We employ a domain expert to provide a reference set of audit records annotated with the corresponding process or task. We then expand this annotation to the entire flattened audit logs based on observed patterns of the reference set. A similar approach may be required for the event logs if the task is not clear from the logs.

**Stochastic Sequence Analysis:** For the annotated audit records, we calculate the probabilities of observing a sequence of records to establish the order in which audit activities occur. We identify the events that overlap with the tasks referred to by the annotations. Based on the mathematical basis discussed in Sect. 2.1, the sequence applies to its annotated tasks and the process. At the end of this phase, we list the discrepancies between the sequence assessed from audit logs and the sequence observed in the event logs.

**Correction:** An analyst reviews the discrepancies listed in the previous phase and approves the sequence changes. The log is modified to reflect the new sequence and fed into the PM algorithm to discover the process model. We analyse the generated model for *Precision, Fitness*, and *simplicity*. To validate our methodology, we conduct experiments on *order fulfilment* business process. We discuss the details of our case study and proposed analysis in Sect. 3. Section 4 presents the findings and the evaluation metrics used to validate our proposed approach.

## 3    Case Study for Order Fulfilment Sub-process

In this section, we apply the proposed SA-SDS method to an *order fulfilment* business process. The case study includes data collection, preprocessing, exploratory analysis, audit data annotation, the proposed stochastic sequence analysis, and suggesting corrections to the discovered model. In a retail supply chain, the fulfilment process starts with the creation of an order and culminates with the delivery of the order to the customer. An instance of this process may include tasks such as *check inventory, check serviceability, reserve inventory, get payment details, create order, fraud check, approve order, release to node, schedule to pick, pick items by batch, assess packing needs, pack, generate invoice, notify shipping provider, shipment pickup*, and *delivery*.

We take a subset of the order fulfilment process implemented using a scalable, enterprise-grade intelligent omnichannel OLTP order fulfilment platform that supports a fulfilment pipeline with a series of tasks and rules. The platform records transactions in a relational database and has a configurable audit function recorded in the same relational database.

## 3.1 Data Collection and Preprocessing

We simulate a retail order fulfilment process and collect the resulting logs from OMS. We adopt the onprom toolchain process. Onprom is an Ontology-based Data Association suite of tools to extract event logs from RDBMS. First, we create an *Ontology model* using UML. Later, we use Protege editor's Ontop plugin, which allows us to map ontology classes and objects to data through SQL queries. We introduce annotations to identify cases and events in the data, and thus we extract an eXtensible Event Stream (XES) [7] format log. We provide a masked excerpt from the extracted XES format log in Listing 1.1. As we can infer from the excerpt, for a given order number 15, the timestamps for all tasks are the same. Further, the tasks are recorded in a wrong sequence, with `resolve` coming before `create` and `hold` actions.

**Listing 1.1.** A Concrete Example of a Masked Excerpt from XES Format Log. The snippet shows overlapping timestamp for multiple tasks present in incorrect ordering.

```
...
<trace>
 <string key="concept:name" value="Order15"/>
 <event> <string key="concept:name"  value="RESOLVE"/>
 <date key="time:timestamp"  value="2021−01−17T06:01:01.000"/> </event>
 <event><string key="concept:name"  value="CREATE"/>
 <date key="time:timestamp"  value="2021−01−17T06:01:01.000"/></event>
 <event><string key="concept:name"  value="HOLD"/>
 <date key="time:timestamp"  value="2021−01−17T06:01:01.000"/></event>
 ...
 <event><string key="concept:name,"  value="INVOICE"/>
 <date key="time:timestamp"  value="2021−01−17T06:01:01.000"/></event>
</trace>
...
```

OMS allows for audit information to be captured via a configuration available through tables [22]. OMS stores the audit information across multiple special-purpose tables as a combination of text and XML data. We generate audit records by executing controlled tests for performing fulfilment steps of *create, update, schedule, release, ship,* and *invoice.* We extract the raw audit data from these multiple tables. We then parse the XML audit records and filter them to select records relevant to the analysed process. We then convert the extracted audit records into flat CSV files suitable for further processing using Python scripts. Listing 1.2 (present in Sect. 3.3) gives a snapshot of the raw audit log and the embedded XML format data along with the corresponding flattened log.

## 3.2 Analysis of Raw Data

For extracting event logs, we analyse the OMS installation's Entity-Relationship model to choose six RDBMS tables covering the `Order`, `Shipment`, `Invoice`, and `Enterprise` entities. When modelled as an ontology, the analysis results in seven class properties, seven object properties, 49 data properties, and 56 associations. Table 1 summarises the statistics for the raw XES format log. Table 1 shows that there are 341 events categorised in 11 tasks while total unique cases in the event logs are 64. Further, within audit logs, *create* and *schedule* are the most frequent

**Table 1.** Summary of XES format event log and audit log. OL= Order Level, LL= Line Level, OLL= Order & Line Level

| XES metric | Value | Audit | CREATE | HOLD | RELEASE | SCHEDULE | INVOICE | SHIP |
|---|---|---|---|---|---|---|---|---|
| Events | 341 | Frequency | 472 | 123 | 219 | 472 | 41 | 26 |
| Tasks | 11 | Remarks | OLL | OL | OLL | OLL | OLL | LL |
| Cases | 64 | | | | | | | |

activities, even more than twice of the next most frequent activity, i.e., *release*. Interestingly, the extracted audit logs show a similar distribution of activities as Zipf's law. We visualise the distribution of tasks across the sequence in which they occur. Figure 2 shows the number of occurrences of a task (y-axis) for each sequence step (x-axis). The bar graph shows that sequence steps $T3$ through $T6$ can be one of five or six tasks confirming our problem statement of time overlaps resulting in a random sequence of tasks. The graph reveals that any task can occur at a step. Further, the tasks overlap across different steps with different frequencies. For example, T3 and T4 have overlapping tasks, but `schedule` is present more often in T4 than in T3. Similarly, `create` appears in T1 and T2 but has the highest frequency in T2.

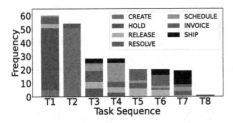
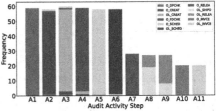

**Fig. 2.** Activities against task sequence.    **Fig. 3.** Audit activities vs audit steps.

Figure 3 shows the frequency of audit activities observed for each audit step. The graph confirms that audit logs provide a clear sequence of the activities in most cases. The actions *OL_SHPD* and *OL_RELEA* are distributed across steps $A8$ and $A9$ indicating parallel execution. Audit records shows presence of `order` level and `line` level records for `create`, `release`, `schedule`, and `invoice`. However, `hold` was observed only at `order` level, indicating that the entire order is placed on hold even if a single line triggers a business rule. These observations are consistent with the expectations of the test run performed.

### 3.3    Annotating Preprocessed Audit Data

In this phase, we obtain a list of keywords and sequences that map audit fields to process tasks from a domain expert. For instance, audit log entries generated during the *changeOrder* API call created audit records with *HoldType* field that indicates the action as *hold* or *release* and the context as *duplicate order check* or *fraud check*. We use the *TransactionId* field to determine the sequence as before or after *create* activity. We captured this knowledge in a mapping file and used it to annotate the flattened CSV. At the end of this step, we have flattened audit logs with annotations that map each record to a process task. 1.3 lists the algorithm, and the code snippet is made available in SA-SDS Github repository. For example, we provide a snapshot of a raw audit log, flattened CSV, and annotated CSV files in the code Listing 1.2. In this sample, the value of the *AuditType* field confirms the corresponding event activity as *schedule* and presence of a *LineID* confirms the activity is performed at the *order line* level. Our mapping for *scheduling of an order line* is the tag *OL_SCHEDULE*.

**Listing 1.2.** Example of Raw Audit Log, Flattened CSV, and Annotated CSV Files

```
Raw Audit Log:
"20210117062401012762 9","20210117060101012762 4","20210117068801012762 8",
"<XDS FIL='audit_detail.del.001.xml' OFF='1002955' LEN='10279'/>",
"2021-01-17-06.01.01.000000", ...
Audit details XML:
<OrderAuditDetail AuditType="Schedule">
<IDs> <ID DataType="class String" Name="OrderID" Value="Order15"/>
<ID DataType="class String" Name="LineID" Value="Order15Line1" /> ...
Flattened Log:
"20210117060101012762 4","Order15",...,"","ORDER_LINE","","Order15Line1"
...
"20210118071042287 83 ","Order30 "...,"","OTHERS || HOLD_TYPE","ORDER",...
Annotated Log:
12,"OL_SCHED","20210117060101012763 0 ","Order15"..,  "ORDER_LINE",...,
2,"O\_YCD_DUPLICATE_ORDER","20210117060101012762 4","Order15",..
...
8,"O_CREATE","20210118071042287 83 ","Order30 ",..." ,"HOLD_TYPE","ORDER "
...
```

**Listing 1.3.** Algorithm for Tagging Audit Log Entries

```
Input: Raw audit log, purpose fields, process mappings rules
Output: Audit log with each entry tagged for a process step
for line in rawAuditLog:
    if line contains contextWords:
        if line contains beginSequence: start new sequence to sequenceLog
        elif: line contains endSequence: mark sequence end to sequenceLog
        else: add line to sequenceLog; record sequenceOffset
for line in sequenceLog:
    for taskKeyword in taskKeywords:
        if line contains taskKeyword:
            use{taskKeyword, sequenceOffset} to get processTag from processMap
            write processTag + line to taggedAuditLog
```

**Listing 1.4.** Algorithm for Sequence Analysis

```
Input: Tagged Audit Log
Output: Matrix of relative occurrence frequencies
initialise frequency to zeros
for processTag in processTags:
    for batch in taggedAuditLog: #batch = audit begin and end sequence
        for tag in batch:#Compare relative tag positions in a batch
            if tagSequence < processTagSequence:
                frequency[tag][processTag] -= 1
            else
                frequency[tag][processTag] += 1
    write frequency to sequenceMatrixLog
```

## 3.4   Stochastic Sequence Analysis of Audit Data

Once the audit data was available as CSV format data, annotated with activity tags, we generated the frequency distribution for each activity using the algorithm listed in 1.4. We generated a matrix capturing the likelihood of activity occurring in the audit logs before or after another activity. The positive cell value indicated the X-axis activity occurring after the Y-axis activity, while a negative value indicated the reverse. A value of zero indicates a parallel occurrence of the activities. The generated matrix provides sufficient insights to create a reference sequence of activities for a domain expert to validate and adjust the discovered process model. Table 2 shows a snapshot of the matrix (both rows and columns) without normalisation. The values in the matrix confirm that our data has more *order level* audit entries than *order line level*, indicating that we can draw higher confidence when resolving *order level* sequencing errors.

**Table 2.** A snapshot of observed likelihood of activities with respect to each other

| ACTIVITY | O_DUPCHK | OL_SHIP | O_INVOICE | OL_INVOICE | O_ADDRCHK | O_FRAUDCHK |
|---|---|---|---|---|---|---|
| O_DUPCHK | −59 | 59 | 58 | 58 | 58 | 58 |
| OL_SHIP | −59 | −59 | −52 | −52 | 58 | 58 |
| O_INVOICE | −58 | 52 | −58 | 58 | 58 | 58 |
| OL_INVOICE | −58 | 52 | −58 | −58 | 58 | 58 |
| O_ADDRCHK | −58 | −58 | −58 | −58 | −58 | 58 |
| O_FRAUDCHK | −58 | −58 | −58 | −58 | −58 | −58 |
| O_CREATE | −58 | −58 | −57 | −57 | −57 | −57 |
| OL_CREATE | −27 | −27 | −27 | −27 | −27 | −27 |
| O_SCHED | −27 | −27 | −27 | −27 | −27 | −27 |
| OL_SCHED | −26 | −26 | −26 | −26 | −26 | −26 |
| O_RELEASE | −20 | −20 | −20 | −20 | −20 | −20 |
| OL_RELEASE | −20 | −20 | −20 | −20 | −20 | −20 |
| NONE | 1 | 1 | 0 | 0 | 0 | 0 |

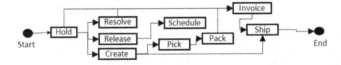

**Fig. 4.** Baseline process model discovered from heuristic miner

**Fig. 5.** Process model corrected after comparing results of SA-SDS method

### 3.5   Correction of Discovered Process Model

We use ProM (an extensible framework for PM tasks) to generate the process model given in Fig. 4. A comparison of the sequence of events between the discovered process model and likelihood matrix (Table 2) reveals that the *duplicate order check* always happens before *create* or any of the other audit records. Also, the likelihood of *invoice generation* happening before *shipment*. It is further apparent that *hold* criteria for *fraud-check* is occurring at the wrong sequence and the *invoice generation* should happen after the *pack* step of the process, given *pack* is part of the *shipment* task. These insights allow us to accurately adjust the model to portray our test system's fulfilment business process. We present the corrected process model in Fig. 5. In the modified process model, *hold* is marked as the root node followed by *create*. The step is important because, on creation, the system places the order on *hold* for "duplicate check". OMS moves the status to *resolve* if the order is not a duplicate. A second *hold-resolve* cycle occurs for "fraud-check" before sending the order to *schedule*. Figure 5 illustrates that the edges between *hold*, *create*, and *resolve* form a loop via *create*. Rest of the states are kept in sequence with *schedule* positioned before *release*. We dropped the nodes for *pick* and *pack* since they were observed for only one event and had no support in the audit log.

## 4   Evaluation and Discussion

We evaluate our method against a popular process mining algorithm; Heuristic miner [15]. Inspired by the current state-of-the-art, we use *Fitness*, *Precision*, *Generalisation*, and *Complexity* to assess the quality of discovered model [6]. **Fitness** is the model's ability to replay all events observed in the log; **Precision** describes the discovered model's ability to accurately represent each event without errors or; **Generalisation** the model's potential to represent unobserved but possible events; and **Complexity** indicates the difficulty for human analysis of the discovered model. [6] quantifies this measure by the structural elements of the model, namely tasks (node), connections (edges), and splits/joins. In the following subsections, we discuss the performance of our model against Heuristic Miner (HM) using each of the metrics above. Table 3 summarizes our comparison of SA-SDS with HM.

**Table 3.** Comparison of process model quality metrics of SA-SDS with HM.

| Metric | SA-SDS | HM | Remarks |
|---|---|---|---|
| Fitness | High | Low | Higher fitness |
| Precision | High | Low | Fewer error flows |
| Generalisation | Medium | Low | More robust |
| Complexity | Simple | Complex | Less edge to node discrepancies |

**Fitness:** We calculate the *Fitness* as the percentage of observed event sequences correctly captured by the model. We interpret the percentage against a scale of low (less than 33%), medium (between 33% and 66%), and high (greater than 66%). We observe that the model discovered using Heuristic Miner has a low baseline *Fitness* factor of 17%, resulting in a rating of *low*. The results indicate an impediment to the algorithm's ability to identify and accommodate tasks and process flow. We note that those events have the same timestamp when analysing the event log entries for which the model got the sequence wrong. In the absence of temporal or activity-based sequences, the algorithm relies on a simple majority of observed sequences to form its model. For example, *create*, *resolve*, and *release* are interpreted as parallel activities when they are not. The model hence cannot replay observed sequences where *create* precedes *schedule*. By choosing *hold* as the root, there is a mismatch in the number of events with an observed state of *create* in comparison to the edge leading from *hold* to *create*.

In comparison, SA-SDS method provides additional insight which qualifies a *hold* related to *check duplicate order* as happening before *order create* while *hold* for *fraud check* happening after. Our method also identifies tasks such as *release* when performed at the order, order line or both levels. Heuristics indicate that *hold* must be *resolved* before being *scheduled*. Overall, the *Fitness* is improved to 86%, resulting in a *high* rating. The experiment could not achieve a higher rating as the audit records did not include *ship* task and analysis of the event logs reveals records that went to *ship* state without an *invoice* generation.

**Precision:** We calculate *Precision* as a percentage of event data for which execution will complete without errors. We use a scale based on low (less than 33%), medium (between 33% and 66%), and high (greater than 66%). HM achieves *high* *Precision* and no errors in the overall flow of the fulfilment process as per the event log content. However, analysis of the model by a domain expert reveals logical errors such as parallelism when the activities are sequential and a loop when there is none. The HM model indicates order creation on an exception due to the negative outcome of one or more rules, such as the duplicate check resulting in duplicate orders in the system.

Overall, 31.6% of the event logs data would complete without errors, lowering its rating to *low*. SA-SDS addresses this issue by noting the need for *duplicate check* before *create* along with better sequencing of activities. We also observe that the model deduced that a *release* is performed directly after *create* in some conditions. Heuristics indicate that resolving holds is an important precondition before releasing an order for fulfilment. Our method correctly identifies this precondition in its analysis. Overall, 89.6% of the event data can replay without errors in our model. The lack of insights into orders that reached *ship* status without an *invoice* being the cause of not getting a better *Precision*.

**Generalisation:** We perform a heuristic assessment of the model's ability to represent different sequences of tasks and sub-tasks. We gave a rating of *high* if the model can accommodate new sequences and additional tasks and sub-tasks. Whereas the rating is *medium* if the model can accommodate either new sequences or additional tasks, and *low* if it can accommodate neither. Due to errors in the HM model, it could not include additional tasks or sequences, resulting in a *Generalisation* rating of *low*. SA-SDS discovers more evidence for discovered process tasks and uncovers further insights that make it better suited for unseen data related to sub-tasks. The model captures *hold* and *release* transitions better than HM. Hence it is better in accommodating valid new sequences of existing tasks. The proposed method is better equipped to update the model accurately when tasks are added or dropped. Overall, the domain expert assessed *Generalisation* as *medium* for the proposed method and model. Our tests were scoped to the *create* to *ship* flow of *fulfilment* sub-process of the *order-to-cash* supply chain process. Thus, we must reevaluate *Generalisation* with a broader dataset that includes more tasks, processes, and flows.

**Complexity:** We propose evaluating the model structure by comparing aspects that lead to poor readability against the total number of nodes and edges in the model. We identified that loops and parallel paths typically make the traversal of the tree more complex than sequential paths. Also, errors such as differences in the number of transitions indicated in the edges compared to the number of events having a state corresponding to a node contribute to *Complexity*. We define *Complexity* C as, $C = \frac{(L+P+D)}{(E+N)}$ where $L$ is the number of loops, $P$ is the number of parallel paths, $E$ is the number of edges, and N is the number of nodes in the model. Further, $D$ is the number of instances where the node values do not equate to the sum of states of incoming edges.

We map the number to a scale of *simple* if $C < 0.5$ and *complex* otherwise, threshold determined by the domain expert. While the HM model created a simple and human-readable model, the *Complexity* score was 0.59 due to a high number of edge-to-node discrepancies. We calculate the corresponding score for the model resulting from SA-SDS as 0.2. Also, since we derived our model from additional insights, it was closely aligned with the domain expert's view of the process. We expect the model's *Complexity* to vary according to the number of tasks and sequences included. Our test data represented one sub-process of the *order-to-cash* process. When dealing with a broader dataset involving large and interdependent processes, we must reevaluate *Complexity*.

## 4.1  Discussion

In this section, we discuss the limitations of SA-SDS. As observed during our evaluation, the audit log entries do not cover all tasks of a process and hence would not help resolve time overlaps of such tasks. Also, the audit logs themselves can face the problem of time overlap. We overcome this problem by relying on a domain expert to provide the correct sequence of audit steps required to meet the compliance needs of the process. The sequence information may not be available in all cases. Further, the domain expert provided reference patterns that allowed our implementation to map the audit log entries to process steps. This mapping needs to be re-implemented for every instance of business-specific audit scenarios, such as auditing a sourcing decision based on financial constraints, product demand, and market dynamics. The algorithm requires re-implementation for standard audits such as PCI-DSS for payment processing, where different implementation strategies generate different audit sequences.

Our choice of applying Heuristic miner on event log generated with *onprom* established a baseline to measure improvement in determining the sequence of tasks. SA-SDS framework augments other process discovery algorithms and log extraction techniques that may reveal additional methods for dealing with time overlaps. Recording high granularity timestamps in the primary event logs can reliably resolve time overlaps. Time overlaps may also indicate the concurrent nature of the process where the sequence is not essential. Further work is required to ascertain the usefulness of audit logs in such situations.

## 5    Conclusion and Future Work

This paper described the challenge of event log entries with an overlapping times-tamp and their impact on the discovered process model. We presented a novel approach to applying insights from audit logs as a secondary data source (SDS) to validate and repair the process model discovered from the event log. Once mapped to process steps, we demonstrated the utility of audit activities for correcting sequence errors in event logs for overlapping steps. Our method achieved $5X$ improvement in *Fitness* and a $2.8X$ increase in *Precision* with the model replaying 197 additional events without errors. Our contribution to the PM community includes the novel method of mapping audit activities to process tasks and further applying analysis to repair sequence errors in event logs.

Future work includes evaluating our approach on a broader set of business processes, entities, and data to effectively evaluate the *Generalisation* and *Complexity* metrics of the discovered process model. It also includes exploring the utility of audit records to provide context.

**Acknowledgments.** We acknowledge Ms Subha Hari, Performance Architect with IBM India Private Ltd, who helped us identify OMS APIs suitable for generating test data, audit data annotation, and cutoff thresholds for evaluation metrics.

## References

1. Alvarez, C., et al.: Discovering role interaction models in the emergency room using process mining. J. Biomed. Inform. **78**, 60–77 (2018)
2. Andrews, R., van Dun, C.G., Wynn, M.T., Kratsch, W., Röglinger, M., ter Hofstede, A.H.: Quality-informed semi-automated event log generation for process mining. Decis. Support Syst. **132**, 113265 (2020)
3. Andrews, R., Wynn, M.T., Vallmuur, K., Ter Hofstede, A.H., Bosley, E., Elcock, M., Rashford, S.: Leveraging data quality to better prepare for process mining: an approach illustrated through analysing road trauma pre-hospital retrieval and transport processes in queensland. Int. J. Environ. Res. Pub. Health **16**(7), 1138 (2019)

4. Calvanese, D., Kalayci, T.E., Montali, M., Tinella, S.: Ontology-based data access for extracting event logs from legacy data: the onprom tool and methodology. In: Abramowicz, W. (ed.) BIS 2017. LNBIP, vol. 288, pp. 220–236. Springer, Cham (2017). https://doi.org/10.1007/978-3-319-59336-4_16

5. Conforti, R., La Rosa, M., ter Hofstede, A.H.M., Augusto, A.: Automatic repair of same-timestamp errors in business process event logs. In: Fahland, D., Ghidini, C., Becker, J., Dumas, M. (eds.) BPM 2020. LNCS, vol. 12168, pp. 327–345. Springer, Cham (2020). https://doi.org/10.1007/978-3-030-58666-9_19

6. Fernández-Cerero, D., Varela-Vaca, Á.J., Fernández-Montes, A., Gómez-López, M.T., Alvárez-Bermejo, J.A.: Measuring data-centre workflows complexity through process mining: the Google cluster case. J. Supercomput. **76**(4), 2449–2478 (2019). https://doi.org/10.1007/s11227-019-02996-2

7. Group, X.W., et al.: IEEE standard for extensible event stream (XES) for achieving interoperability in event logs and event streams. In: IEEE Std, vol. 1849, pp. 1–50 (2016)

8. Hsu, P.Y., Chuang, Y.C., Lo, Y.C., He, S.C.: Using contextualized activity-level duration to discover irregular process instances in business operations. Inf. Sci. **391**, 80–98 (2017)

9. Leemans, S.J.J., Fahland, D., van der Aalst, W.M.P.: Scalable process discovery and conformance checking. Softw. Syst. Model. **17**(2), 599–631 (2016). https://doi.org/10.1007/s10270-016-0545-x

10. Li, G., de Murillas, E.G.L., de Carvalho, R.M., van der Aalst, W.M.P.: Extracting object-centric event logs to support process mining on databases. In: Mendling, J., Mouratidis, H. (eds.) CAiSE 2018. LNBIP, vol. 317, pp. 182–199. Springer, Cham (2018). https://doi.org/10.1007/978-3-319-92901-9_16

11. Li, T., et al.: Flap: an end-to-end event log analysis platform for system management. In: Proceedings of the 23rd ACM SIGKDD, pp. 1547–1556 (2017)

12. Martin, N., et al.: Recommendations for enhancing the usability and understandability of process mining in healthcare. Artif. Intell. Med. **109**, 101962 (2020)

13. Michael, N., Mink, J., Liu, J., Gaur, S., Hassan, W.U., Bates, A.: On the forensic validity of approximated audit logs. In: Annual Computer Security Applications Conference, pp. 189–202 (2020)

14. Okoye, K., Islam, S., Naeem, U., Sharif, S.: Semantic-based process mining technique for annotation and modelling of domain processes. Int. J. Innovative Comput. Inf. Control **16**(3), 899–921 (2020)

15. Porouhan, P., Jongsawat, N., Premchaiswadi, W.: Process and deviation exploration through alpha-algorithm and heuristic miner techniques. In: 12th International Conference on ICT and Knowledge Engineering, pp. 83–89. IEEE (2014)

16. dos Santos Garcia, C.: Process mining techniques and applicationsa systematic mapping study. Expert Syst. Appl. **133**, 260–295 (2019)

17. Schuster, D., van Zelst, S.J., van der Aalst, W.M.P.: Incremental discovery of hierarchical process models. In: Dalpiaz, F., Zdravkovic, J., Loucopoulos, P. (eds.) RCIS 2020. LNBIP, vol. 385, pp. 417–433. Springer, Cham (2020). https://doi.org/10.1007/978-3-030-50316-1_25

18. Song, S., Huang, R., Cao, Y., Wang, J.: Cleaning timestamps with temporal constraints. VLDB J. **30**(3), 425–446 (2021). https://doi.org/10.1007/s00778-020-00641-6

19. Studiawan, H., Sohel, F., Payne, C.: A survey on forensic investigation of operating system logs. Digit. Investig. **29**, 1–20 (2019)

20. Suriadi, S., Andrews, R., ter Hofstede, A.H., Wynn, M.T.: Event log imperfection patterns for process mining: towards a systematic approach to cleaning event logs. Inf. Syst. **64**, 132–150 (2017)

21. van der Aalst, W.: Process mining manifesto. In: Daniel, F., Barkaoui, K., Dustdar, S. (eds.) BPM 2011. LNBIP, vol. 99, pp. 169–194. Springer, Heidelberg (2012). https://doi.org/10.1007/978-3-642-28108-2_19

22. Yesudas, M., Nair, S.K.: High-volume performance test framework using big data. In: 4th International Workshop on Large-Scale Testing, pp. 13–16 (2015)

# WDA: A Domain-Aware Database Schema Analysis for Improving OBDA-Based Event Log Extractions

Anbumunee Ponniah[1,2](✉) and Swati Agarwal[1]

[1] BITS Pilani, K K Birla Goa Campus, Sancoale, India
{p20170418,swatia}@goa.bits-pilani.ac.in
[2] IBM India Private Limited, Bangalore, India

**Abstract.** Ontology-based data access (OBDA) provides a mechanism to extract information from databases through a conceptual ontology and a mapping specification. This paper designs and develops a framework for semi-automated ontology design that improves scalability and reuses compared to manual design. We employ linguistic analysis on a database schema to extract logical relationships and present a domain-aware pre-processing technique to select components for creating the ontology. We present a use case in the context of a business process in the *retail supply chain* domain. The approach is benchmarked against conventional OBDA by evaluating standard metrics for the ontology, event logs, and discovered process model. The results show that the ontology, event log, and process model obtained through the proposed method performs better in most evaluation metrics.

## 1 Introduction

Large companies define, document, implement, monitor, and manage thousands of business processes and assets. The field of Business Process Management (BPM) is the discipline concerned with identifying, organising, and improving the processes related to business operations. The *process model* depicts a sequence and flow of tasks. The information systems implement the processes, recording execution results in logs and data stores [18]. An event log has one or more `traces` with each `trace` consisting of one or more `events`. Each `event` denotes the execution of an atomic `action/task` (denoted by an identifier) at a specific time. Within BPM, process mining (PM) is an active research area on techniques that help enterprises discover, understand, and improve business processes [23]. PM algorithms such as **inductive miner** and **heuristic miner** use the content of the event logs to discover process models [17]. Analysts represent models using notations such as BPMN (Business Process Model and Notation: a graphical notation that depicts the steps in a business process), Workflow (a sequential series of tasks and decisions), Petri-nets (a directed graph model representing state-transition of the process), and many more. Most information systems do not generate quality event logs and require expensive processing of the available data for process mining. Information systems use a database to store configuration, reference, and transaction

© The Author(s), under exclusive license to Springer Nature Switzerland AG 2022
W. Chen et al. (Eds.): ADMA 2022, LNAI 13726, pp. 297–309, 2022.
https://doi.org/10.1007/978-3-031-22137-8_22

details. The database schema design typically uses the Entity-Relationship (ER) model to define entities, key-based relationships, and constraints. Data stored in the database represent the subject, object, conditions, and context required to execute processes. The availability of such comprehensive data makes the database a potential source for event logs. Extracting event logs from databases has challenges due to the data, rules, relationships, and dependencies stored in disparate database objects. OBDA provides a mechanism to extract information from databases through a conceptual ontology and a specification to map ontology components to database columns. However, the ontology creation and the column mapping actions are implementation-specific and require collaborative manual effort between a domain expert and database designer. Our motivation is to address the challenge of extracting event logs from relational database data using database schema descriptions. Our further motivation includes creating an industry-agnostic reusable framework that can be adapted to identify candidate ontology components from a database schema. The framework further reduces the manual work required to create an ontology for an OBDA process to map domain-level process concepts to database tables.

## 1.1   Related Work

This section presents our findings from a comprehensive review of techniques prevalent for log extraction in Process Mining (PM) and its parent field of Business Intelligence. Our review focuses on using a database as a source of logs, creating ontology from data sources, and applying ontology-based techniques for extracting logs and mining processes. In event log extraction, metadata about the table schema and business process provide essential context for log extraction. [7] proposed a simple matching logic using the frequency of entity terms extracted from schema descriptions. [19] followed a semi-automatic bootstrapping technique starting with a pool of keywords and using pre-defined queries on public data sources to extract ontology components. [8,20] automated the ontology bootstrapping by pursuing a rules-based approach which can analyse and extract ontology elements from stored procedures and database constraints. [16] use Association Rules Mining (ARM) to identify the positive and negative association between terms to extract concepts from *big data*. [14] proposed a metamodel to abstract the information system generating the event logs from the data. [21] use a process data model to map the data source and schema components for each process step to extract the log. [1] proposed an object-centric model to capture control flow constraints, data dependencies, activity-class relations, and shared relations. Ontology-based data access (OBDA) provides a framework for mapping an ontology to a base database schema using a mapping mechanism [24]. [13] created a condensed knowledge graph from the raw data to map source data to data analysis outcomes. The knowledge graph was useful to support a query interface to data. [2] provided a meta-mapping strategy to capture enterprise knowledge and use it for mapping to schema elements. They utilised a meta-schema (manually created) to capture concepts and relations.

[15] explored the potential for applying semantic-based annotations to data sources and process models. They employed the conceptual analysis step to annotate data sources with additional information. Their results showed improved classification of traces compared to Fuzzy-BPMN miner and Inductive Miner. [5] presented the OBDA technique by annotating a conceptual model of the available data. The authors developed tools that allow ontology-associated queries to extract XES format data. It relied on a manually created ontology that required high domain expertise. [6] relied on an interactive tool for users to express mappings and map ontology relations for use in OBDA. [9] extract, transform and store object-centric data from a database that considers (*many to many*) relationships. It resulted in eXtensible Object-Centric (XOC) format event logs that allowed process analysis without case notations in XES format.

We found that existing literature relies on manual expertise and thus lacks scalability. Due to their reliance on integrity constraints, current approaches do not discover or utilise logical dependencies in the context of event log extraction. Similarly, ER model does not capture implicit relationships such as parent-child, cause-effect, and many more. Furthermore, the reference information about the domain stored in the database is not being utilised or explored to its fullest. In contrast to the current work, this paper proposes a Word-Association-based Database Analysis (WDA) technique to address the gaps of scalability, logical dependencies, and implicit relationships discovery.

## 2   Novel Research Contributions

We define and implement a novel framework to determine word associations from schema descriptions, identify keywords in the associated word sets, and shortlist ontology components with the help of domain-aware filtering. Our primary innovation is combining the techniques of word similarity and ARM with a Part-of-Speech (POS) based domain-specific filtering strategy. We apply the selected ontology components in creating a process-specific ontology used for OBDA mapping and event log extraction. We use the reference database schema of a popular OLTP-based Order Management Software (OMS) as the context to analyse and present the usefulness of our approach.

## 3   Research Methodology

This paper focuses on creating, implementing, and evaluating a framework that uses an innovative combination of word similarity, ARM, and linguistic techniques. We adapt the well-known design science research (DSR) workflow by adding additional steps for a domain expert to 1) define the relevance of *values* within the problem domain (in the first two phases) and 2) create a reference set of domain-specific entities and intents. This section presents the design, development, and experimental setup. The proposed approach is a multi-step process broadly consisting of four phases, namely, a) *Forming word sets from database schema*, b) *Mining rulesets*, c) *Filtering rulesets*, and d) *Log extraction*. We mine ontology relations from database schema as a preprocessing step for OBDA and event log extraction. To validate the utility of our approach, we consider the order-to-cash process of fulfilling an order against available inventory as executed by OMS. The process contains multiple tasks: a) order creation with one or more product items, b) allocating inventory for the items, c) releasing an order to be fulfilled from a warehouse, d) printing an invoice, and e) shipping using a logistics provider.

## 3.1   Forming Word Sets from Database Schema

ER models define database tables with integrity constraints such as *primary key ::: foreign key* dependencies, uniqueness, and null values. When available, names and descriptions of the tables and columns of the database schema contain metadata meaningful to its purpose within the business context. We hypothesise that columns about the same concepts tend to have similar keywords in their description. The similarity is based on the entities (e.g, *order*) and actions performed (e.g, *pick*). After removing language and domain-specific stopwords, we treat each column description as a sequence of tokens. We create a dataset where each instance represents a set of words from preprocessed column description. We calculate the pairwise similarity between the sets using the Jaccard similarity coefficient. We merge two sets if the coefficient is larger than a heuristically determined cut-off value. We calculate the cut-off value by ensuring the following for the resultant dataset: 1) Average words per set should increase after merging to discard duplicate sets, 2) The number of sets is around half the number of initial input sets to assure an average of one merge per set, and 3) Cut-off measure is higher than the mean similarity to ensure higher quality (high similarity) merges.

At the end of this phase, each dataset instance represents words from similar columns across multiple tables in the database schema. The OMS database schema contained 15262 non-empty column descriptions, of which 8323 were unique. We created an initial dataset where each instance consisted of words extracted from one unique column description. In addition to English language stopwords, we filtered out retail domain and OMS specific stopwords such as *sequence, key, id, code, prefix,* and many more. After preprocessing, the dataset has 4934 sets of words with an average of 5.2 per set. We calculated pairwise Jaccard similarity across sets. The scores fall between 0 and 0.5 with a mean similarity of 0.171. 87.8% of the sets having a similarity score less than 0.05. We merged two sets with similarity scores larger than 0.25, starting with those with higher similarity. We determined the cut-off value of 0.25 by applying the heuristics discussed earlier in the section. After performing the merge operation, the dataset has 2691 sets with 6.72 words per set.

## 3.2   Mining Rulesets

We observe that the word sets are similar to transactions used for Market-Basket Analysis in ARM. We apply Apriori algorithm on the dataset to uncover the rulesets of the form $(X \rightarrow Y)$ where $X$ and $Y$ represent the *antecedent* and *consequent* set of words (referred to as itemsets in ARM), respectively. To evaluate the quality of the rule sets, we further calculate **support** $(s(X \rightarrow Y) = \frac{\sigma(X \cup Y)}{T})$, **confidence** $(C(X \rightarrow Y) = \frac{s(X \rightarrow Y)}{\sigma(X)})$, **lift** $(L\{X \rightarrow Y\} = \frac{C(X \rightarrow Y)}{\sigma(Y)})$, and **conviction** $(\frac{(1-\sigma(Y))}{(1-C(X \rightarrow Y))})$ scores. At the end of this step, we acquire a list of rules for each domain concept along with their lift, support, confidence, and conviction scores. We applied ARM on the OMS dataset of 2691 word sets to generate rules. We discarded the rules (**s**< 0.01 and **C** < 0.01) and thus obtained 72 million rules. In this case study, we extract a glossary of terms related to the supply chain from different sources: Council for Supply Chain Management Professionals, Inbound Logistics Magazine, and IBM Sterling Order Management. We obtain top seven verbs relevant for `fulfilment` sub-process of the `order-to-cash` business process: *create, hold, inventory, invoice, release, schedule*, and *ship*. We filtered the rules by each word and obtained a total of 345156 rules spread across *create* (1274), *hold* (3994), *inventory* (39535), *invoice* (4135), *release* (130405), *schedule* (463), and *ship* (165350).

## 3.3   Tag and Filter Rules

We extract the words from the rules and apply POS tagging to identify the content and function words. Using the identified words, we select rules having a common noun with another noun or verb as they indicate relationships between the columns. With the help of a domain expert, we obtain a reference set of nouns, verbs, and relationships relevant to the *retail supply chain* domain. We use the reference set to select the final list of domain-specific rules, entities, and intents. On the OMS dataset, we identify 228 unique words in the rules across *create* (27), *hold* (16), *inventory* (57), *invoice* (14), *release* (33), *schedule* (11), and *ship* (70). After applying POS tagging on the set of words, we select rules by applying the filtering strategy described above. As an example of a $\{noun_i, verb_i\} \rightarrow \{noun_j, verb_j\}$ rule, we choose *order* (noun) and *hold* (verb) in the antecedent and the words *line* (noun) and *release* (verb) in the consequent part of the rule. Similarly, as an example of a $\{noun_i, noun_j\} \rightarrow \{noun_k\}$ rule we select rules that had words *ship* (noun) and *organization* (noun) in the antecedent and the word *node* in the consequent portion of the rule. We shortlisted a total of 13394 rules across *create* (34), *hold* (300), *inventory* (2323), *invoice* (292), *schedule* (65), and *(ship)* (6402). These rules and corresponding nouns and verbs represented the candidate list of entities and intents from our dataset. We used the reference set to manually select `order`, `line`, `organization`, `node`, `item`, `inventory`, `hold`, `rule`, `ship`, and `invoice` as entities. We also identified `allocate`, `assign`, `configure`, `setup`, `generate`, `place`, `release`, `schedule`, and `update` as intents.

## 3.4   Use in Log Extraction

We use the identified `entities` as ontology classes and `intents` as relationships We use this knowledge to construct an ontology that is used for OBDA by mapping data to ontology classes and objects. We use the onprom toolchain to identify the *case* and *trace* within the ontology. We extract an XES [11] format event log and apply Heuristics Miner [3] for process discovery. We assess the usefulness of our approach using metrics for ontology construction, event logs extraction, and process model discovery (discussed in the Sect. 4). Figure 1 provides the UML view of the resultant ontology for the OMS dataset.

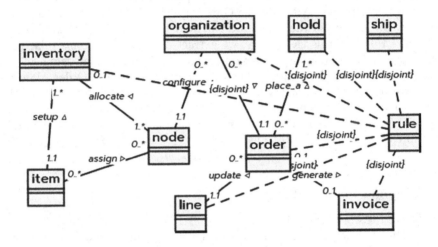

**Fig. 1.** Ontology for order fulfilment with 10 entities and their associations.

We chose the entities `hold`, `inventory`, `invoice`, `item`, `line`, `node`, `order`, `organization`, `rule`, and `ship`. We used relationships such as `allocate` to indicate the allocation of inventory to an item in an order, `create` to indicate the creation of an order, `configure` to represent a node in a supply chain configured as an enterprise organisation capable of shipping orders, `generate` to represent the invoice generation triggered for the order, `release` to initiate the warehouse tasks, and `schedule` to represent the order being made available for shipping. We further chose *Rule* class to represent the business rules such as `duplicate check`, `release` and `schedule`. The resultant ontology correctly depicts `node` as a type of `organization` and `hold` being placed on order for various business conditions set up in `rule`. We used Protege editor's Ontop mapping to create OBDA mapping between the entities and the corresponding database table data. We use the domain concepts and annotations from the ontology to map the data to *traces* and *events*. We save the mapped data as an XES format event log. We apply the Heuristic Miner (HM) algorithm to discover the process model from the extracted event log. We discuss our findings in the next section.

## 4    Evaluation

In this section, we present the results of the proposed WDA approach and benchmark them against the conventional ER-based approach. As detailed in the above section, our approach has three major outcomes: ontology, XES format log, and the discovered process model. OBDA's effectiveness depends on the quality of ontology used for the data access. We observe that in the current state of the art, evaluation of ontologies uses qualitative criteria of accuracy, completeness, cohesiveness, understandability, and computational efficiency. The criteria are measured through the five categories of quantitative metrics, namely *base*, *schema*, *graph*, *knowledgebase*, and *class* [10]. In this paper, we focus on the *completeness* criteria to evaluate the ontology as the process mining use case requires an ontology that models the processes and relationships in their entirety. The parameters of *base* and *schema* metrics indicate the **completeness** of an ontology. The parameters represent the ontology design in terms of attribute richness, inheritance richness, and axiom/class ratio, along with the number of classes and axioms [12]. We compared the ontology created by our approach and ER-based approach on the following: 1) **Axiom count:** The number of asserted statements representing relationships based on class, object, and data properties. 2) **Axiom/class ratio:** The proportion of classes containing axioms among all classes. A higher value indicates the ontology's ability to represent more process dependencies, such as *inventory* allocated to a *node*. 3) **Element count:** The number of classes, relationships, attributes, and individuals. A higher number of elements mean the ontology captures more granular components of the processes. 4) **Attribute richness:** The average number of attributes per class. A minimum of one attribute per class is required to ensure no disjoint classes in ontology. 5) **Inheritance richness:** The average number of sub-classes per class represents the horizontal versus vertical distribution of sub-classes. Values closer to zero indicate flat (more general) ontologies, while higher values indicate deep (more specialised) ontologies. Process mining benefits specialised ontologies that include more layers (granular details) and hierarchical relationships (evidence of task flow within a process).

The ontology section of Table 1 reveals a significant increase in the number of axioms (by 137.6%), elements (by 104%), and subclass associations (by 203.2%). Higher values are due to additional classes and relationships identified in our method. We also observe a higher axiom/class ratio (by 203%), reflecting more relationship assertions in the ontology. Addition of **hold** and **rule** classes allows the *schedule* task to be captured in the ontology. Our ontology has more number of attributes per class (by 150%) to support additional relationships, e.g. it captures tasks involving *line*, *hold*, *organization*, and *invoice* using different attributes of *order*. It also improves the inheritance ratio (by 133%) with a positive non-zero value (6.3), indicating a domain-specialised ontology with greater details. The **rule** class's ability to capture condition-based relationships is one example of such specialisation.

The quality of event logs is an essential criterion for mining good process models. [22] describe widely used quality metrics applicable for event logs: 1) **Completeness:** The log represents all allowed states of the system. 2) **Unambiguity:** Each recorded state is distinct from the others. 3) **Meaningfulness:** All recorded states map back to a state of the system. 4) **Correctness:** The

**Table 1.** Comparison of ontology metrics of proposed model with ER-based model.

| Outcome | Metric | WDA | ER-based | Remarks |
|---------|--------|-----|----------|---------|
| Ontology | Axiom count | 499 | 210 | 137.6% |
| | Entity count | 104 | 51 | 104% |
| | Subclass associations | 94 | 31 | 203.2% |
| | Axiom/class ratio | 33.3 | 9.0 | 270% |
| | Attribute richness | 5.0 | 2.0 | 150% |
| | Inheritance richness | 6.3 | 2.7 | 133% |
| Event log | Cases | 2166 | 2166 | Equal cases |
| | Events | 15583 | 6910 | More events |
| | Types | 9 | 4 | More tasks |
| | Completeness | High | Low | More states |
| | Correctness | High | High | Same |
| | Meaningfulness | High | High | Same |
| | Unambiguity | High | High | Same |
| Process model | Fitness | 0.79 | 0.99 | Lower fitness |
| | Precision | High | Low | More states |
| | Complexity | Simple | Simple | Right complexity |

right state in the log maps to the correct state in the system. The event log portion Table 1 captures the comparison metrics for the extracted data. For the same number of cases, the WDA-based log contained 125.5% more events and 125% more event types than the ER-based log. Additional event types identified by WDA are *hold, line, linerelease, schedule,* and *ship.* These event types led to a better **completeness** score for the WDA-based log. Both WDA and the conventional method rely on the OBDA mapping to determine the state in the log. Log extracted using WDA scores *high* in **meaningfulness** and **correctness** as the OBDA mapping ties the state to the transaction data. Even though the conventional method extracts less number of states, it scores *high* in **meaningfulness** and **correctness** since it too relies on OBDA. [4] describe the metrics adopted by researchers in the process mining domain to evaluate the discovered model. The metrics assess the model's ability to model the sequence of process states from event logs correctly. We use the following metrics to compare the models discovered by WDA and the traditional method: 1) **Fitness**: The model's ability to replay all events observed in the log with values ranging from 0.0 to 1.0. The metric ensures a PM algorithm such as the heuristic miner can recognise the events in the log. 2) **Precision**: Accurately model each event without errors represented in a three-step scale of high, medium, and low. The metric reflects whether the sequence of events is correct. 3) **Complexity**: Measure the ease of human analysis to validate the model against the physical process, represented as either simple or complex. The presence of a higher number of loops and branches impedes human readability.

Figure 2 paints the model discovered by applying **Heuristics Miner** (HM) algorithm on the log generated with the ER model. HM discovers only four process tasks arranged in a linear sequence without any branches or many states. It also erroneously recognises `charge` as a peer step to `invoice` when it should be a sub-task of `invoice`. Figure 3 represents the process model generated from

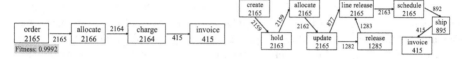

**Fig. 2.** Process model generated with HM on ER definitions.

**Fig. 3.** Process model discovered with HM on WDA generated XES event log.

the event log extracted using WDA. The model presents the expected patterns of fulfilment flow from order creation to shipping. The observed state transitions from **update** to **line-release** and **release** indicates the part where *hold* on orders are *resolved* and the order is *released* to a warehouse for fulfillment. Due to overlapping timestamps, **release** is determined as the prior state to **line-release** for 59% of the events. It is also clear that 41.3% of the orders have reached the **ship** state while just 19.2% of the orders generated an **invoice**. These observations are in line with the expectations of the test execution. As presented in the Process Model part of Table 1, the fitness score for the ER-based model is 0.99. The high fitness score is due to the low number of states identified in the log and the process's sequential structure. WDA-model results in a fitness score of 0.79 due to time overlaps and lack of information about sequencing between multiple rules in the ontology. The proposed WDA-based model scored high on precision by including all the states from the database. ER-based model scores low as it did not include the states *hold, line, linerelease, release, schedule,* and *ship*. Further, the ER model discovers the state *charge* as a separate state when that is a sub-task of the state *invoice* (hence not listed in the WDA-based model). On the **simplicity** metric, both models were simple for human analysts to read and understand due to the absence of loops. In summary, while the WDA-based model scored lower on **fitness**, we found it to be a more comprehensive representation of the **order-to-cash** process. The focus of our paper is to evaluate the utility of Word association-based database analysis on *order fulfilment* part of the **order-to-cash** process. Comprehensively evaluating the **generalisation** metric of the process requires a corpus of more extensive and complex processes. Our proposed model partially addresses this challenge of unavailable event logs by improving the technique to generate logs from the database. We plan to utilise WDA in the future to discover and analyse multi-enterprise supply chain processes.

## 5  Discussion

We observe that the WDA-based approach created an ontology that scored high in the completeness metric measured by the base and schema parameters. Word association analysis on the column descriptions, domain-specific filtering, and reference set of relationship patterns played a crucial role in achieving the outcome. Our technique identified logical dependencies across disjoint tables by choosing word association-based merging of column descriptions. In our experiment, *placing on hold* is a task that blocks the processing of an order based on a business condition. There are multiple types of *hold*: duplicate order, awaiting approval, credit limit, fraud check, and many more. The table that stores the business condition to trigger the *hold* has no schema relationship with the **hold** table. WDA discovered a *hold* rule (R) triggered on *order* and *line* where the R is related to *customer* information validity, *payment* limits, and business-imposed default conditions. The payment relation is uncovered due to co-occurrence of the keywords *hold*, *payment*, and *limit*.

WDA requires database column names and descriptions to be meaningful indicators of their purpose. Schema with cryptic naming conventions or poor-quality descriptions requires other data sources (e.g. table descriptions, design documents) and further processing to extract meaning from the data. The number of word associations and an ARM-based ruleset generation leads to many candidate rules. Filtering the rules by domain-specific keywords reduces the rules from over 72 million to 0.34 million. Further, by choosing the keywords from a widely available glossary of terms for the domain, we can replicate the filtering for any domain where a formal glossary of terms is available.

The role of POS analysis is to focus on the entity (noun) and intent (verb) words. Other NLP techniques such as Named Entity Recognition (NER) and word-embedding based models (GloVe, BERT) can be used in the workflow. We relied on a domain expert to provide a reference set of *antecedent $\rightarrow$ consequent* pairs and POS combinations. The reference set helps in eliminating invalid noun-verb pairs and determines the quality of the final candidate list of entities and intents. A non-expert can convert the final candidate list to an ontology with the expert reviewing the final result. While WDA did not fully automate ontology design, it limited the expert's involvement in creating reference sets and reviewing the ontology design. Thus WDA achieves better scalability than a traditional human-only technique. Further, separating domain-specific dependencies from the core workflow ensures the solution is reusable for other domains.

The ontology creation and OBDA mapping followed the *onprom* toolchain steps. The higher values of *base* and *schema* metrics of the ontology indicate that WDA-based ontology represents the entities and relationships required to completely model a high-level *order fulfilment process*. The ontology is more specialised than a vanilla ER-based approach with more relations (e.g. *node* is a special instance of an *organization*) identified. The ontology is case-specific and will need to change to accommodate use cases involving additional entities or relationships. Creating a comprehensive ontology for the domain covering a larger number of processes and generating business rules by analysing the assertions was not in our current research scope. The extracted event logs had additional events and event types than those extracted by the existing ER-based approach. The OBDA mapping and annotations ensured that the XES format event logs included sufficient content for the HM algorithm to discover the process. The discovered model's ability to depict process variations needs data and ontology for additional workflows, including complex processes from other domains such as Healthcare and Finance. That is part of the future scope of our work.

Overall, WDA provides a scalable framework to generate event logs from a database. It is helpful in information systems that do not generate high-quality event logs but use a data store with a formal schema definition. WDA enables legacy enterprise applications to take advantage of advances in process mining to analyse, optimise and automate their business operations. Our contributions to the community include the innovative approach of combining the techniques of word similarity and ARM with a POS-based domain-specific filtering strategy.

## 6 Conclusion and Next Steps

This article presented a word association-based database schema analysis (WDA) technique to discover entities, intents, and relationships beyond the ER model to improve event log extraction from RDBMS. We presented steps to analyse the database schema through word similarity-based merging, association rule mining, and POS tagging. We further implemented a domain-specific filtering technique to mine valuable relationships that an ontology designer can be used to create an ontology of the business process. Applying ARM on word associations uncovered logical relationships beyond ER design such as calculation of *tax* and *surcharge* being a requisite condition for the *shipping* task. Filtering rules by domain words and POS tagging from 72 million to 13394 improved the time taken for entity and intent identification. Comparative analysis of results across the ontology created and log extracted shows measurable improvement in the effectiveness of our approach in capturing event states and relationships. The model discovered with our approach is better than the ER-based model in nearly all evaluation dimensions and resulted in a comprehensive representation of the business process. Results of WDA can improve *supply chain* process analysis and optimisation. The phases of WDA are transferable to other database schema, domains, and processes.

Future work includes addressing the gaps identified in the Sect. 1.1, evaluating our approach with alternate text analysis techniques on a more extensive set of business processes and applying our technique for intelligent automation outcomes. Incubating a product based on the WDA framework is another work in scope for the future.

## References

1. Artale, A., Montali, M., Tritini, S., van der Aalst, W.M.: Object-centric behavioral constraints: integrating data and declarative process modelling. In: Proceedings of the 30th International Workshop on Description Logics (DL), vol. 1879 (2017)
2. Atzeni, P., Bellomarini, L., Papotti, P., Torlone, R.: Meta-mappings for schema mapping reuse. Proc. VLDB Endowment **12**(5), 557–569 (2019)
3. Ayutaya, N.S.N., Palungsuntikul, P., Premchaiswadi, W.: Heuristic mining: Adaptive process simplification in education. In: 2012 Tenth International Conference on ICT and Knowledge Engineering, pp. 221–227. IEEE, Bangkok (2012)
4. Buijs, J.C., van Dongen, B.F., van der Aalst, W.M.: Quality dimensions in process discovery: the importance of fitness, precision, generalization and simplicity. Int. J. Coop. Inf. Syst. **23**(01), 1440001 (2014)

5. Calvanese, D., Kalayci, T.E., Montali, M., Tinella, S.: Ontology-based data access for extracting event logs from legacy data: the onprom tool and methodology. In: Abramowicz, W. (ed.) BIS 2017. LNBIP, vol. 288, pp. 220–236. Springer, Cham (2017). https://doi.org/10.1007/978-3-319-59336-4_16

6. Gómez, S.A., Fillottrani, P.R.: Materialization of OWL ontologies from relational databases: a practical approach. In: Pesado, P., Arroyo, M. (eds.) CACIC 2019. CCIS, vol. 1184, pp. 285–301. Springer, Cham (2020). https://doi.org/10.1007/978-3-030-48325-8_19

7. Hu, W., Qu, Y.: Discovering simple mappings between relational database schemas and ontologies. In: Aberer, K., et al. (eds.) ASWC/ISWC -2007. LNCS, vol. 4825, pp. 225–238. Springer, Heidelberg (2007). https://doi.org/10.1007/978-3-540-76298-0_17

8. Lakzaei, B., Shamsfard, M.: Ontology learning from relational databases. Inf. Sci. **577**, 280–297 (2021)

9. Li, G., de Murillas, E.G.L., de Carvalho, R.M., van der Aalst, W.M.P.: Extracting object-centric event logs to support process mining on databases. In: Mendling, J., Mouratidis, H. (eds.) CAiSE 2018. LNBIP, vol. 317, pp. 182–199. Springer, Cham (2018). https://doi.org/10.1007/978-3-319-92901-9_16

10. Lourdusamy, R., John, A.: A review on metrics for ontology evaluation. In: IEEE International Conference on Inventive Systems and Control, pp. 1415–1421. IEEE (2018)

11. Mannhardt, F.: Xeslite-managing large xes event logs in prom. BPM Center Report BPM-16-04 8, pp. 224–236 (2016)

12. Masmoudi, M., Lamine, S.B.A.B., Zghal, H.B., Karray, M.H., Archimède, B.: An ontology-based monitoring system for multi-source environmental observations. Procedia Comput. Sci. **126**, 1865–1874 (2018)

13. Mavlyutov, R., Curino, C., Asipov, B., Cudre-Mauroux, P.: Dependency-driven analytics: A compass for uncharted data oceans. In: 7th Conference on Innovative Data Systems Research. CIDR, Chaminade (2017)

14. González López de Murillas, E., Reijers, H.A., van der Aalst, W.M.P.: Connecting databases with process mining: a meta model and toolset. Softw. Syst. Model. **18**(2), 1209–1247 (2018). https://doi.org/10.1007/s10270-018-0664-7

15. Okoye, K., Islam, S., Naeem, U., Sharif, S.: Semantic-based process mining technique for annotation and modelling of domain processes. Int. J. Innovative Comput. Inf. Control **16**(3), 899–921 (2020)

16. Parfait, B., Harrimann, R., André, T.: An efficient approach for extraction positive and negative association rules from big data. In: Holzinger, A., Kieseberg, P., Tjoa, A.M., Weippl, E. (eds.) CD-MAKE 2018. LNCS, vol. 11015, pp. 79–97. Springer, Cham (2018). https://doi.org/10.1007/978-3-319-99740-7_6

17. Peña, M.R., Bayona-Oré, S.: Process mining and automatic process discovery. In: Software Process Improvement (CIMPS), pp. 41–46. IEEE (2018)

18. Pérez-Álvarez, J.M., Gómez-López, M.T., Eshuis, R., Montali, M., Gasca, R.M.: Verifying the manipulation of data objects according to business process and data models. Knowl. Inf. Syst. **62**(7), 2653–2683 (2020). https://doi.org/10.1007/s10115-019-01431-5

19. Qawasmeh, O., Lefranois, M., Zimmermann, A., Maret, P.: Computer-assisted ontology construction system: focus on bootstrapping capabilities. In: Gangemi, A., et al. (eds.) ESWC 2018. LNCS, vol. 11155, pp. 60–65. Springer, Cham (2018). https://doi.org/10.1007/978-3-319-98192-5_12

20. Sbai, S., Chabih, O., Louhdi, M.R.C., Behja, H., Trousse, B.: Using decision trees to learn ontology taxonomies from relational databases. In: 2020 6th IEEE Congress on Information Science and Technology (CiSt), pp. 54–58. IEEE (2021)

21. Schuh, G., Gützlaff, A., Cremer, S., Schmitz, S., Ayati, A.: A data model to apply process mining in end-to-end order processing processes of manufacturing companies. In: 2020 IEEE International Conference on Industrial Engineering and Engineering Management (IEEM), pp. 151–155. IEEE (2020)

22. Suriadi, S., Andrews, R., ter Hofstede, A.H., Wynn, M.T.: Towards a systematic approach to cleaning event logs. Inf. Syst. **64**, 132–150 (2017)

23. van der Aalst, W., et al.: Process mining manifesto. In: Daniel, F., Barkaoui, K., Dustdar, S. (eds.) BPM 2011. LNBIP, vol. 99, pp. 169–194. Springer, Heidelberg (2012). https://doi.org/10.1007/978-3-642-28108-2_19

24. Xiao, G., et al.: Ontology-based data access: a survey. In: Proceedings of the 27th IJCAI International Joint Conferences on Artificial Intelligence (2018)

# A Cricket-Based Selection Hyper-Heuristic for Many-Objective Optimization Problems

Adeem Ali Anwar[1]([⊠])[ID], Irfan Younas[2], Guanfeng Liu[1][ID], Amin Beheshti[1], and Xuyun Zhang[1]

[1] School of Computing, Faculty of Science and Engineering, Macquarie University, Sydney, NSW, Australia
adeem.anwar@students.mq.edu.au, xuyun.zhang@mq.edu.au
[2] Odyssey Analytics, 5757 Woodway Drive, Houston, TX 77057, USA
irfan.younas@odysseyanalytics.net

**Abstract.** While meta-heuristics are usually designed for the optimization problems of the same domain and can achieve superior performance compared with heuristics, their performances suffer severely when dealing with cross-domain problems. Recently, many-objective optimization methods have been proposed to handle the increase of the objectives, and this further renders the cross-domain problems more challenging. A technique known as hyper-heuristic (HH) has been proposed to effectively handle cross-domain optimization problems, without the need to alter the HH extensively. However, existing HHs focus mainly on single- or multi-objective optimization problems, and little work has been done on the many-objective optimization problems (MaOOPs) and lack delta evaluation. Inspired by the sport of cricket, we propose a novel many-objective selection hyper-heuristic technique named cricket-based selection hyper-heuristic (CB-SHH) in this paper, to produce well-diverse and converged optimal solutions for MaOOPs. To the best of our knowledge, we are the first to propose a sports-inspired HH. The proposed technique computes the objective value based on the most recent modification to address one of the problems with HHs, namely the lack of using delta evaluation, which is another contribution of the paper. In CB-SHH, the exploitation and the exploration have been handled using the greedy and randomization mechanism, respectively. Moreover, many-objective meta-heuristics have been used as low-level heuristics to drive the CB-SHH search. CB-SHH has been tested against the benchmark and real-life datasets and has performed significantly better or equal when compared with other meta-heuristics and HHs on 196 out of 200 instances based on Hypervolume (HV) values. Moreover, CB-SHH has the best cross-domain performance measured by $\mu$ norm mean values i.e. producing 234.8% and 76.4% better results than state-of-the-art HH across HV and IGD respectively.

**Keywords:** Hyper-heuristic · Many-objective · Optimization

ⓒ The Author(s), under exclusive license to Springer Nature Switzerland AG 2022
W. Chen et al. (Eds.): ADMA 2022, LNAI 13726, pp. 310–324, 2022.
https://doi.org/10.1007/978-3-031-22137-8_23

# 1   Introduction

Optimization problems produce results based on objective functions. The results are either maximized or minimized depending on the type of problems [9]. There are multiple types of optimization problems based on the number of objective functions i.e., single-objective optimization problems (SOOPs), multi-objective optimization problems (MOOPs) and MaOOPs. In an SOOP, only one objective function is considered. In MOOPs, more than two and less than four objectives functions are considered. Whereas in MaOOPs, the number of objective functions is four or more than four [3].

Many computationally hard optimization problems have been successfully solved by different meta-heuristics which are personalized to each field by their experts. But as the problem changes slightly, meta-heuristics usually fail to produce good results. HH is a cross-domain technique that can be used to solve different optimization problems with a minimum change [6]. The HHs can be divided into two main categories i.e., selection HHs and generation HHs. These categories are made based on the type of search space. For a given optimization problem, the selection HHs deal with the automation of selecting HHs. However, generation HHs deal with the automation of methodologies to generate HHs [6]. In selection HHs, low-level heuristics are selected by high-level approaches [10]. Low-level heuristics are used to drive the search and can be meta-heuristics, recombination operators, etc., while high-level approaches are the selection techniques to select HHs [6, 10, 23, 24, 26].

Heuristics are problem-specific algorithms and meta-heuristics are usually designed for the same domain problems, so their performances suffer while dealing with cross-domain problems. Moreover, the increase in objectives also affects the performances, therefore many-objective algorithms have been created to tackle this problem. However, the issue remains the same and performances continue to suffer in cross-domain problems. Recent literature describes that the current HH techniques are mostly designed to solve SOOPs or MOOPs and other techniques do not produce good results in cross-domain problems. Moreover, only a handful of work has been done in the field of many-objective HHs. Furthermore, the proposed technique computes the objective value based on the most recent modification to address one of the problems with hyper-heuristics, namely the lack of using delta evaluation. Delta evaluation means determining the objective values based on recently changed objective values rather than using the entire set of solutions from the beginning to calculate objective values [6].

To tackle this issue, this paper proposes a novel many-objective selection hyper-heuristic technique named cricket-based selection hyper-heuristic (CB-SHH). The proposed selection HH is inspired by sport of cricket to produce well-diverse and converged optimal solutions for MaOOPs. To the best of our knowledge, the sports-based HH has not been explored before in the literature. In CB-SHH, the exploitation and exploration have been handled using greedy mechanism and randomness respectively. As CB-SHH is developed to solve MaOOPs, hence the three different many-objective meta-heuristics have been used as low-level heuristics to drive the CB-SHH search. The proposed algorithm has been tested against two benchmark datasets i.e., DTLZ and WFG along with one real-life problem i.e., Many-objective pickup and delivery problem (MaOPDP) [3]. The main contributions of the paper are to develop a novel selection HH to solve the MaOOPs effectively, to propose the first sports-inspired HH, to solve the issue of not using delta evaluation in HHs, and to compare the results of proposed technique with other state-of-the-art meta-heuristics and HHs.

The idea is inspired from the sports of cricket. In cricket, two batters play at one time (striker and non-striker) and they try to score as many runs as possible and team with the best scores win the game [17]. Striker is the one who plays the ball and non-striker waits until the striker score certain number of runs to get on strike [17]. Normally the best batters (based on their record/score) are given chances at the beginning and then according to how they score runs in the past their batting position (when they will bat) changes in the future. This idea is incorporated into HH and CB-SHH has been proposed. Meta-heuristics have been used as batters, and applied to the problems and based on how they score (objective values), striker and non-striker have been decided.

In conclusion, a selection hyper-heuristic technique named cricket-based selection hyper-heuristic (CB-SHH) is proposed. CB-SHH has been tested against the benchmark and real-life datasets and has performed significantly better or equal when compared with other meta-heuristics and HHs on 196 out of 200 instances based on Hypervolume (HV) values. Moreover, CB-SHH has the best cross-domain performance measured by $\mu$ norm mean [2] values i.e. producing 234.8% and 76.4% better results than state-of-the-art HH across HV and IGD respectively.

The remaining paper structure is as follows. The related work is being discussed in Sect. 2. The framework of CB-SHH is explained in Sect. 3. Section 4 discusses the empirical studies, whereas the last section presents the conclusion and future work.

## 2 Related Work

This section describes the related work done for many-objective HHs recently. The work done in the field of many-objective HH is very slight. Most of the techniques are based on online feedback along with the perturbation approach [11–15, 18, 20, 29].

An HH known as a HH collaborative multi-objective evolutionary algorithm (HHcMOEA) was presented in [12] for MOOP's and MaOOP's. It was shown that HHcMOEA has produced better results when compared with other MOEAs. A novel many-objective HH was proposed by [15]. MOEA/D and a novel selection technique were mixed in this algorithm. The proposed algorithm experimented against its other modifications. Inverted generational distance (IGD) was used as an evaluation method and WFG was used as a benchmark dataset during the experimental process. For MaOOPs, it exhibited good results.

Many-objective job shop scheduling problem was solved in [18] after the previous work on the same problem [20]. GP along with NSGA-III was used to create the rules called GP-HH. A novel reference point modification technique was introduced. It was proved the novel reference point modification technique showed good results using the proposed technique. A MaOHH named as multi-indicator hyper-heuristic was proposed by [11]. Indicators strengths were used in this HH and their weakness was compensated. Moreover, online learning was used as a feedback mechanism along with the Markov chain. Furthermore, a new framework was presented to overcome the cost of the algorithm. The proposed approach had better results than NSGA-III and MOEA/D.

A population-based HH is based on the combination of different MOEA's and is named as cooperative hyper-heuristic (HH-CO) [13]. The researchers have made two experiments, firstly HH-CO is compared with MOEA's and secondly, HH-CO is compared with the champions of the CEC'18 competition. It was proved that the proposed HH showed better results MaOOPs. In the extension of this work, HH-C was applied on a real-life MaOOP named as wind turbine design problem [14]. With the help of MOEA, Sandra et al. [27] developed a selection HH for many-objective numerical optimisation and obtained promising results when comparing benchmark and real-world datasets. A genetic programming-based HHs was proposed by Atiya et al. [19] for the job-shop scheduling while taking into account the various objectives. Using NSGA-III, Bianca et al. [25] developed an HH for the many-objective quadratic assignment issue. Recently HH named as epsilon-greedy selection HH (HH_EG) was introduced [29]. Different evaluation measures are used by low-level heuristics and it is independent of the parameters. The benchmark datasets have been used and experiments proved that the proposed algorithm showed better results than different MOEAs. However, it is missing the delta evaluation, which determines the objective values using the entire set of solutions from the beginning rather than the most recent changes.

In conclusion, the work done in the field of many-objective HHs is very less as the focus remained on SOOPs or MOOPs and the techniques do not produce good results in cross-domain problems and lacks delta evaluation. To tackle the mentioned issues, CB-SHH is proposed. The proposed algorithm has been tested against two benchmark datasets i.e., DTLZ and WFG along with one real-life problem i.e., MaOPDP. CB-SHH has the best cross-domain performance measured by $\mu$ norm mean values across different evaluation measures.

# 3   Cricket-Based Selection Hyper-Heuristic (CB-SHH)

## 3.1   Framework of CB-SHH

CB-SHH is based on the perturbation method with an online feedback approach known as reinforcement learning outlined in Algorithms 1 and 2. The selection technique (high-level) in CB-SHH is inspired by sport of cricket to produce well-diverse and converged optimal solutions for MaOOPs. Performance indicators i.e., Inverted generational distance (IGD) [22,28] and Hypervolume (HV) [5,22] have been used. Cricket has been used to create CB-SHH. The idea of how two batters remain in the crease, while one being on strike and the other being on

**Algorithm 1.** Framework of CB-SHH

---

$g$: generations $s_c$: the input solutions, $s_n$: new solutions (non-dominated), $g_{max}$: total number of generations, $a$: set of algorithms, $a_{max}$: max number of algorithms, $r$: reference set, $i$: total number of iterations, $p$: population size, $i_{max}$: max number of iterations.

**Input:** $g_{max}, a, i, p$
**Output:** $value$ // HV value
// Reference set is being created for fair comparison
**while** $a_j \leq a_{max}$ **do**
    **while** $g \leq g_{max}$ **do**
        $s_n \leftarrow$ ImplementingAlgorithm $(a_j, i, p, s_c)$;
        $r \leftarrow r + s_n$;
        $g = g + 1$;
    **end**
    $a_j \leftarrow a_{j+1}$;
**end**
// Generating a random value and choosing algorithm based on that value to get maximum range of hypervolume
$d \leftarrow$ GeneratingRandomValue $(min = 0, max = 2)$; $\forall d \in \mathbb{N}$
$s_n \leftarrow$ ImplementingAlgorithm $(a_d, i_{max}, p, s_c)$;
$max \leftarrow$ ApplyHypervolume $(s_n, r)$;
// Generating two random values and choosing two algorithms to start the selection process
$d_1 \leftarrow$ GeneratingRandomValue $(min = 0, max = 2)$;
$d_2 \leftarrow$ GeneratingRandomValue $(min = 0, max = 2)$;
$strike \leftarrow$ ChooseRandom$(d_1, d_2)$;
CricketBasedSelection$(a, i, p, s_c, s_n, d, max, d_1, d_2, strike, r)$;

---

non-strike is applied by selecting two meta-heuristics, one as a striker and the other as a non-striker. The striker is the meta-heuristic which is applied always to the current generation. Whereas the non-striker waits for the striker to score poorly to come as a striker. Based on their scores, either their strike is swapped or the striker can be given out and a new meta-heuristic takes its place based on their previous records (scores). The new meta-heuristic is selected based on the highest scores. The exploitation and exploration have been handled using the greedy mechanism and randomness respectively. As CB-SHH is developed to solve MaOOPs, hence the three different many-objective meta-heuristics have been used as low-level heuristics to drive the CB-SHH search. CB-SHH gets the maximum possible HV value by applying any random meta-heuristic and then initially for a certain generation two random meta-heuristics are selected and one out of them is applied to the problem. In the remaining generations, the meta-heuristic is selected based on the best score among meta-heuristics. The scores of meta-heuristic are calculated based on their performance in the last generations. Moreover, the penalty and reward are being considered while calculating the scores of the algorithm which are computed using the reinforcement learning strategy.

### 3.2 Method

Components of CB-SHH have been discussed in this section.

**Initialization.** In the first generation, the population is created based on randomness, whereas in the remaining generations the population from the previous generation is passed into the next generation. The objectives values and hence the

**Algorithm 2.** CricketBasedSelection($a, i, p, s_c, s_n, d, max, d_1, d_2, strike, r$)

---

// Until end of generations
**while** $g \leq g_{max}$ **do**
    $s_n \leftarrow$ ImplementingAlgorithm ($a_{strike}, i, p, s_c$);
    $value \leftarrow$ ApplyHypervolume ($s_n, r$);
    **if** $value >= 0$ && $value < max * (60/100)$ **then**
        // New algorithm replaces old algorithm
        $score[strike] \leftarrow -2$;
        $temp$.add(getIndex(getMaxValues($score$)));
        $int\ i = 0$;
        **while** $temp.get(i) == d_1 \parallel temp.get(i) == d_2$ **do**
            $temp$.add(getIndex(getMaxValues($score$)));
            $i + +$;
        **end**

        **if** $temp.HasNumberOfValues == 1$ **then**
            $strike \leftarrow temp.get(0)$;
        **else**
            $strike \leftarrow$ ChooseRandom($temp$.getAll());
        **end**
    **end**
    **if** $value >= max * (60/100)$ && $value < max * (70/100)$ **then**
        // Strike changes
        $score[strike] \leftarrow -1$;
        $strike \leftarrow Swap(d_1, d_2)$;
    **end**
    **if** $value >= max * (70/100)$ && $value < max * (80/100)$ **then**
        // Algorithm remains the same
        $score[strike] \leftarrow 2$;
    **end**
    **if** $value >= max * (80/100)$ && $value < max * (90/100)$ **then**
        // Algorithm remains the same
        $score[strike] \leftarrow 4$;
    **end**
    **if** $value >= max * (90/100)$ && $value <= max * (100/100)$ **then**
        // Algorithm remains the same
        $score[strike] \leftarrow 6$;
    **end**
    $s_c \leftarrow s_n$;
    $g = g + 1$;
**end**
// Display the best HV value after last generation
$Display \leftarrow value$;

---

HV of the next generations are calculated using the modification done recently in the solutions. Moreover, only the non-dominated solutions ($s_{nd}$) are allowed to pass to the next generation as shown in (1). As shown in Algorithm 1, the reference set ($r$) is being created for a fair comparison between the meta-heuristics and HHs.

$$func(s_n(g + 1)) = func(s_c(g))$$
$$\forall s_c \in s_{nd} \tag{1}$$

**Algorithm Selection Technique.** CB-SHH selects the better meta-heuristics based on the HV ($value$) in every generation. The meta-heuristics are I-DBEA [4], NSGA-III [7], and MOEA/D [30] as expressed by $a = \{a_1, a_2, a_3, \ldots, a_{max}\}$.

CB-SHH gets the maximum possible HV value ($max$) that can occur for a specific problem among the mentioned meta-heuristics by applying any random meta-heuristic ($a_d$) for $i_{max}$ iterations. For the first generation two random meta-heuristics ($d_1, d_2$) are selected and one out of them ($strike$) randomly is applied to the problem ($a_{strike}$). For every other generations, the best meta-heuristic is applied based on the score of every meta-heuristic as expressed by $score = \{score_1, score_2, score_3, \ldots, score_{max}\}$.

HV ($value$) is calculated for that specific meta-heuristic and if the value is less than the 60% of the maximum value ($max$), a new meta-heuristic is selected based on the best scores and if multiple meta-heuristics have the same score then any random meta-heuristic among the bests is selected. If the value is more than 60% less than 70% of the maximum value ($max$) then the meta-heuristic is swapped with the other randomly chosen meta-heuristic. If the value is more than 70% less than or equal to 100% of the maximum value ($max$) then the meta-heuristic remains the same as expressed in (2). The reason for choosing these percentages is based on the fact that if a meta-heuristic is producing good results then it is better not to change it with the new one and if it slightly produces bad results then it is better to change its strike rather than replacing with new meta-heuristic.

$$strike = \begin{cases} getAnyBestAlgorithm(score), & \text{if } value >= 0 \ \&\& \ value < max * 0.6 \\ swap(d_1, d_2), & \text{if } value >= max * 0.6 \ \&\& \ value < max * 0.7 \\ strike, & \text{if } value >= max * 0.7 \ \&\& \ value < max \end{cases} \quad (2)$$

**Reinforcement Learning Strategy.** The online feedback technique is implemented using reinforcement learning strategy. The algorithm whose HV is higher has received a reward, whereas the algorithm with a lower HV has received a penalty. HV value ($value$) is calculated for every meta-heuristic and depending on the value, the meta-heuristics are awarded or penalized. If the value is less than 60% of the maximum value ($max$) or 70% of the maximum value ($max$) then the meta-heuristics score is penalized by $-1$ and $-2$ respectively. And if the value is less than the 80% of the maximum value ($max$) or 90% of the maximum value ($max$) or 100% of the maximum value ($max$), then the meta-heuristics score is rewarded by 2, 4, and 6 respectively as expressed in (3). The reason for choosing the scores values is inspired by the sport of cricket as the batter can score a maximum of 6 runs [17].

$$score(strike) = \begin{cases} -2, & \text{if } value >= 0 \ \&\& \ value < max * 0.6 \\ -1, & \text{if } value >= max * 0.6 \ \&\& \ value < max * 0.7 \\ 2, & \text{if } value >= max * 0.7 \ \&\& \ value < max * 0.8 \\ 4, & \text{if } value >= max * 0.8 \ \&\& \ value < max * 0.9 \\ 6, & \text{if } value >= max * 0.9 \ \&\& \ value <= max \end{cases} \quad (3)$$

### 3.3  Analysis of CB-SHH

To have optimality in meta-heuristics, convergence and diversity of the solutions in the decision space play an important role [21]. CB-SHH uses many-objective evolutionary algorithms (MaOEAs) as meta-heuristics because MaOOPs fail to produce good results with MOEAs due to the increased number of objectives [21]. Hence, affecting the optimality of solutions [21]. The reason of choosing I-DBEA, MOEA/D, NSGA-III as MaOEAs is because they are state-of-the-art algorithms for MaOOPs and the focus of the work is on the same problems. CB-SHH is handling the offspring generation and environmental selection effectively by considering the non-dominated solutions from the last generation and a novel selection mechanism respectively. Moreover, the balance between exploration and exploitation of an algorithm is important to get an optimal solution globally [21] and one of the main reasons for the proposed technique to work efficiently. As they are useful in exploring the search space and getting closer to the best solutions respectively which are computed using randomness and greedy mechanism respectively. Meta-heuristics are selected randomly as well as based on the best scores. Moreover, a reinforcement learning strategy is being implemented by handling the penalty and reward values in the scores. Hence both randomness and greedy technique is applied. Resulting in solutions being well exploited and explored.

### 3.4  Computational Complexity of CB-SHH (One Generation)

Firstly, for a fair comparison reference set is being created. $O(mg(p^2o))$ is the cost of creating reference set. The assumptions are as follows, $O(p^2o)$ [7] is the cost of implementing the algorithm on a certain problem, where $g$ is the number of generations, $m$ is the number of low-level heuristics, $o$ is the dimension of the objectives, $p$ is the population size.

Secondly, the cost to apply an algorithm to get $max$ value is $O(p^2o)$. The cost to generate random numbers is constant. Lastly, the cost of the CricketBased-Selection is $O(g(p^2o) + (m))$. As it applies an algorithm to a certain problem and then checks the range of $value$ which can cost $O(m)$ in the worst case as it gets the maximum score value among algorithms. Therefore the overall computational cost of one generation of CB-SHH is $O(mg(p^2o))$ i.e., $O(N^3)$, where $N$ represents the loops. The cost of the HH_E is also $O(N^3)$ [29].

## 4  Empirical Studies

CB-SHH has been tested against random HH, recent HH named as HH_E [29] and three MaOEAs (I-DBEA, MOEA/D, NSGA-III). A java framework known as MOEA is used to implement the techniques [1].

### 4.1  Experimental Settings

**Datasets.** Benchmark data sets have been used. The datasets are DTLZ2, DTLZ4, DTLZ5, DTLZ7, WFG2, WFG4, WFG5 and WFG7 [8,16]. DTLZ2 helps in testing the meta-heuristics (algorithms) performance for MaOOPs [8]. DTLZ4, DTLZ5, and DTLZ7 help in testing the meta-heuristics (algorithms) performance on getting diversity, convergence and ability to have diverse sub-populations in the Pareto-front respectively [8]. WFG2 is non-separable, convex and disconnected [16]. WFG4, WFG5, and WFG7 are separable along with concave geometry [16]. MaOPDP has been considered as a real-life problem with six objectives [3]. The MaOPDP is a real-life problem which belongs to the class of vehicle routing problems [3]. In literature, it has many variations but the specific variation with six objectives are considered [3]. This specific variation of MaOPDP is considered as it covers most of the important objectives. The minimization objectives are number of routes, distance, time, distance and time with largest routs, and loss [3]. It has many datasets, but for this study we considered the small dataset known as lc101 with 23 as number of vehicles and 200 as vehicle capacity [3].

**Settings of Parameters.** The evaluation measures for all the meta-heuristics and HHs are calculated with 25 generations ($g$) and 10 iterations ($i$) for the reference set and 1 iteration ($i$) for calculating the meta-heuristics. Iterations ($i_{max}$) for getting the maximum ($max$) are 1000. The considered objectives ($o$) are 4, 5, 7, 10, and 12. The number of seeds is 5, to have fair results. The position variable and distance are set to 5 in WFG. T-test (at 0.05 $\alpha$ value) is used to show the significance of one algorithm over other algorithms.

**Algorithms for Comparative Studies.** MaOEAs (I-DBEA, MOEA/D, NSGA-III), a Random HH, and a recent HH named as HH_EG [29] have been used for experiments. [4] proposed I-DBEA and it is an improved decomposition-based algorithm and found to be used for MaOOPs. MOEA/D was proposed by [30] and is based on the decomposition approach as well and can be used for MaOOPs as well as MOOPs. NSGA-III [7] is based on a reference point-based approach and is useful for solving MaOOPs. Random HH uses MaOEAs randomly throughout the generations and HH_EG is a recently proposed selection HH based on epsilon and greedy methods.

**Performance Indicators.** HV [5,22] is one of the most commonly used performance indicators. Its value range from 0 (worst) to 1 (best). IGD [22,28] is another performance indicator and the lesser value is good. Both IGD and HV consider convergence and diversity of solutions. $\mu$ norm [2] is used widely to compare cross-domain performance.

### 4.2   Experimental Results and Sensitivity Analyses

**Experimental Results.** This section explains the experiments. During experiments mean values of HV and IGD have been calculated for different datasets against multiple algorithms (I-DBEA, MOEA/D, NSGA-III, CB-SHH, R_HH, HH_EG) with 5 different numbers of objectives (4, 5, 7, 10, 12).

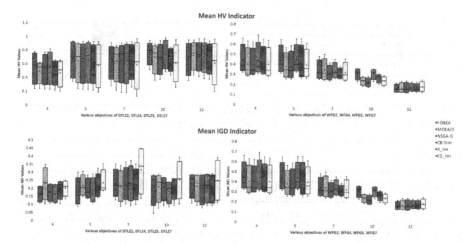

**Fig. 1.** Mean HV, IGD values on DTLZ, WFG while considering 4, 5, 7, 10, and 12 objectives over all algorithms

**Table 1.** $\mu$ norm mean values of different number of objectives and algorithms using HV values

| $\mu$ norm mean on different datasets | $\mu$ norm mean values of different number of objectives and algorithms | | | | | |
|---|---|---|---|---|---|---|
| | I-DBEA | MOEA/D | NSGA-III | CB-SHH | R_HH | HH_EG |
| DTLZ | 0.724518 | 0.571469 | 0.607398 | **0.755652** | 0.495825 | 0.196588 |
| WFG | 0.778608 | 0.277375 | 0.190477 | **0.857067** | 0.408579 | 0.295112 |
| MaOPDP | 0.478381 | 0 | 0.137309 | **1** | 0.326730 | 0.099522 |
| DTLZ, WFG and MaOPDP combined | 0.744900 | 0.414070 | 0.392556 | **0.811082** | 0.449141 | 0.242281 |
| Ranking of algorithms | 2nd | 4th | 5th | **1st** | 3rd | 6th |
| Improvement of CB-SHH over others | 8.9% | 95.9% | 106.6% | – | 80.6% | 234.8% |

**Table 2.** $\mu$ norm mean values of different number of objectives and algorithms using IGD values

| $\mu$ norm mean on different datasets | $\mu$ norm mean values of different number of objectives and algorithms | | | | | |
|---|---|---|---|---|---|---|
| | I-DBEA | MOEA/D | NSGA-III | CB-SHH | R_HH | HH_EG |
| DTLZ | 0.257204 | 0.380509 | 0.251916 | **0.181622** | 0.346625 | 0.852726 |
| WFG | **0.159217** | 0.703679 | 0.785566 | 0.212344 | 0.585429 | 0.763909 |
| MaOPDP | 0.009979 | 0.001216 | 0.013680 | **0** | 0.009930 | 1 |
| DTLZ, WFG and MaOPDP combined | 0.203376 | 0.528902 | 0.506422 | **0.192178** | 0.454903 | 0.812992 |
| Ranking of algorithms | 2nd | 5th | 4th | **1st** | 3rd | 6th |
| Improvement of CB-SHH over others | 5.5% | 63.7% | 62.0% | – | 57.7% | 76.4% |

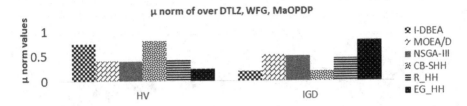

**Fig. 2.** $\mu$ norm mean combined values of DTLZ, WFG, MaOPDP over different number of objectives and algorithms using HV and IGD values

Figure 1 shows the mean HV and IGD values for DTLZ and WFG for all objectives. For DTLZ2, CB-SHH has the best mean HV values for 5 and 10 objectives. For DTLZ4, CB-SHH has the best mean HV values for 4, 7, and 12 objectives. For DTLZ7, CB-SHH has the best mean HV values for 5 and 10 objectives. For WFG2, CB-SHH has the best mean HV values for 4 objectives. For WF4, CB-SHH has the best mean HV values for 5, 7, 10, and 12 objectives. For WFG5, CB-SHH has the best mean HV values for 5 objectives. For WFG7, CB-SHH has the best mean HV values for 10 and 12 objectives. For MaOPDP, CB-SHH has the best mean HV values. For DTLZ2, CB-SHH has the best mean IGD values for 5 and 7 objectives. For DTLZ4, CB-SHH has the best mean IGD values for 4 objectives. For DTLZ5, CB-SHH has the best mean IGD values for 10 objectives. For DTLZ7, CB-SHH has the best mean IGD values for 10 and 12 objectives. For WFG2, CB-SHH has the best mean IGD values for 4 objectives. For WF4, CB-SHH has the best mean IGD values for 5 objectives. For WFG5, CB-SHH has the best mean IGD values for 10 objectives. For WFG7, CB-SHH has the best mean IGD values for 4, 7, 10, and 12 objectives. For MaOPDP, CB-SHH has the best mean IGD values.

To check the cross-domain performance of CB-SHH, the $\mu$ norm mean values have been calculated using HV and IGD mean values. Table 1 shows the values of $\mu$ norm mean values for DTLZ, WFG and MaOPDP. CB-SHH has the best $\mu$ norm mean values for DTLZ, WFG, MaOPDP, and their combined values. The combined $\mu$ norm mean values are represented in Fig. 2 as well. Moreover, it was found to be showing 80.6% and 234.8% improvement against the state-of-the-art HHs.

Table 2 shows the values of $\mu$ norm mean values for DTLZ, WFG and MaOPDP. CB-SHH has the best $\mu$ norm mean values for DTLZ, MaOPDP. And a meta-heuristic i.e. I-DBEA has the best $\mu$ norm mean value for WFG. However overall in the combined $\mu$ norm mean value, CB-SHH outperformed all other algorithms. The combined $\mu$ norm mean values are represented in Fig. 2 as well. Moreover, it was found to be showing 57.7% and 76.4% improvement against the state-of-the-art HHs.

Table 3 shows the t-test values over all datasets, objectives and algorithms using HV values on 0.05 $\alpha$. The significance of CB-SHH has been tested against all other meta-heuristics as well as HHs. CB-SHH has performed significantly better or equal on most of the instances i.e., 196 out of 200.

**Table 3.** HV values comparison over all datasets, objectives and algorithms. Results in terms of significantly worse (-), significantly better (+) and equal are mentioned while alpha value is 0.05

| | Algorithms significance using t-test while considering HV indicator | | | | | |
|---|---|---|---|---|---|---|
| Algorithms | CB-SHH | I-DBEA | MOEA/D | NSGA-III | R_HH | HH_EG |
| CB-SHH | – | +2/37/−1 | +16/24/−0 | +15/25/−0 | +8/30/−2 | +21/18/−1 |
| I-DBEA | +1/37/−2 | – | +14/24/−2 | +15/24/−1 | +12/27/−1 | +19/18/−3 |
| MOEA/D | +0/24/−16 | +2/24/−14 | – | +3/35−/2 | +2/34/−4 | +7/30/−3 |
| NSGA-III | +0/25/−15 | +1/24/−15 | +2/35−/3 | – | +1/35/−4 | +8/28/−4 |
| R_HH | +2/30/−8 | +1/27/−12 | +3/34/−2 | +4/35/−1 | – | +10/28/−2 |
| HH_EG | +1/18/−21 | +3/18/−19 | +3/30/−7 | +4/28/−8 | +2/28/−10 | – |

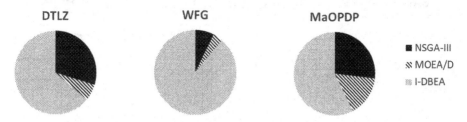

**Fig. 3.** Algorithms utilization over iterations on CB-SHH

The utilization of different algorithms over DTLZ, WFG, and MaOPDP has been shown in Fig. 3. The I-DBEA is the most used meta-heuristic (algorithm) on DTLZ, WFG, and MaOPDP.

**Parametrical Analysis.** Two benchmark datasets have been taken from DTLZ and WFG along MaOPDP to perform parametrical analysis. Two set of parameters have been considered named as $\beta$ and $\gamma$ over all objectives. $\beta$ represents the original parameters done for all the experiments (reference set iterations are set to 10, iterations set to 1, seeds are 5 and generations are 25). Whereas in $\gamma$, the values are set to the following. Reference set iterations are set to 100, iterations set to 10, seeds are 10 and generations are 30. Table 4 represents the best meta-heuristics and HHs for the selected datasets using HV values. It is shown in Table 4 that the parameters setting doesn't affect the results except on one instance where I-DBEA has replaced MOEA/D as the best meta-heuristic for the DLTZ2 on 12 objectives.

**Convergence Analysis.** Convergence analysis has been done by taking one dataset from each of the benchmark datasets i.e. DTLZ2 and WFG4 along with MaOPDP. For DTLZ2, 5 objectives are considered, whereas for WFG4 10 objectives are considered. The convergence over a number of generations driven by HV values is shown in Fig. 4.

**Table 4.** Best meta-heuristics and HHs over DTLZ2, WFG4, MaOPDP on all objectives using HV values. $\beta$ represents original parameters whereas $\gamma$ represents changed parameters.

| Objectives | DTLZ2 | | WFG4 | | MaOPDP | |
|---|---|---|---|---|---|---|
| | $\beta$ | $\gamma$ | $\beta$ | $\gamma$ | $\beta$ | $\gamma$ |
| 4 | R_HH | R_HH | I-DBEA | I-DBEA | — | — |
| 5 | CB-SHH | CB-SHH | CB-SHH | CB-SHH | — | — |
| 6 | — | — | — | — | CB-SHH | CB-SHH |
| 7 | NSGA-III | NSGA-III | CB-SHH | CB-SHH | — | — |
| 10 | CB-SHH | CB-SHH | CB-SHH | CB-SHH | — | — |
| 12 | MOEA/D | I-DBEA | CB-SHH | CB-SHH | — | — |

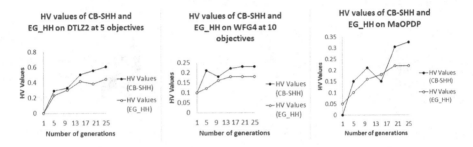

**Fig. 4.** Convergence of HV values of DTLZ2, WFG4 and MaOPDP on 5, 10 and 6 objectives respectively on CB-SHH and EG_HH

## 5    Conclusion and Future Work

A novel many-objective selection hyper-heuristic technique named cricket-based selection hyper-heuristic (CB-SHH) is proposed. CB-SHH is inspired by sport of cricket to produce well-diverse and converged optimal solutions for MaOOPs. The proposed algorithm has been tested against benchmark datasets i.e., DTLZ and WFG along with MaOPDP. CB-SHH has performed significantly better or equal when compared with other meta-heuristics and HH on 196 out of 200 instances based on HV values. It has the best $\mu$ norm mean values for all datasets while calculating using HV values and IGD values except for the WFG while calculated using IGD. In that case, it has the second-best $\mu$ norm mean, only behind I-DBEA and performing better than other HHs. Moreover, it has the best cross-domain performance measured by $\mu$ norm mean values i.e. producing 234.8% and 76.4% better results than state-of-the-art HH across HV and IGD respectively.

In the future, the proposed CB-SHH can be applied to other many-objective benchmark problems and real-life problems.

**Acknowledgement.** Adeem Ali Anwar is the recipient of an iMQRES funded by Macquarie University, NSW (allocation No. 20213183) and Dr. Xuyun Zhang is the recipient of an ARC DECRA (project No. DE210101458) funded by the Australian Government.

# References

1. Moea framework. http://moeaframework.org/. Accessed 25 June 2022
2. Adriaensen, S., Ochoa, G., Nowé, A.: A benchmark set extension and comparative study for the hyflex framework. In: 2015 IEEE Congress on Evolutionary Computation (CEC), pp. 784–791. IEEE (2015)
3. Anwar, A.A., Younas, I.: Optimization of many objective pickup and delivery problem with delay time of vehicle using memetic decomposition based evolutionary algorithm. Int. J. Artif. Intell. Tools $29(01)$, 2050003 (2020)
4. Asafuddoula, M., Ray, T., Sarker, R.: A decomposition-based evolutionary algorithm for many objective optimization. IEEE Trans. Evol. Comput. $19(3)$, 445–460 (2014)
5. Auger, A., Bader, J., Brockhoff, D., Zitzler, E.: Theory of the hypervolume indicator: optimal $\mu$-distributions and the choice of the reference point. In: Proceedings of the Tenth ACM SIGEVO Workshop on Foundations of Genetic Algorithms, pp. 87–102 (2009)
6. Burke, E.K., Hyde, M., Kendall, G., Ochoa, G., Özcan, E., Woodward, J.R.: A classification of hyper-heuristic approaches. In: Handbook of metaheuristics, pp. 449–468. Springer (2010). https://doi.org/10.1007/978-1-4419-1665-5_15
7. Deb, K., Jain, H.: An evolutionary many-objective optimization algorithm using reference-point-based nondominated sorting approach, part i: solving problems with box constraints. IEEE Trans. Evol. Comput. $18(4)$, 577–601 (2013)
8. Deb, K., Thiele, L., Laumanns, M., Zitzler, E.: Scalable test problems for evolutionary multiobjective optimization. In: Evolutionary Multiobjective Optimization, pp. 105–145. Springer (2005). https://doi.org/10.1007/1-84628-137-7_6
9. Derigs, U.: Optimization and Operations Research-Volume IV. EOLSS Publications (2009)
10. Drake, J.H., Kheiri, A., Özcan, E., Burke, E.K.: Recent advances in selection hyper-heuristics. Eur. J. Oper. Res. $285(2)$, 405–428 (2020)
11. Falcón-Cardona, J.G., Coello, C.A.C.: A multi-objective evolutionary hyper-heuristic based on multiple indicator-based density estimators. In: Proceedings of the Genetic and Evolutionary Computation Conference, pp. 633–640 (2018)
12. Fritsche, G., Pozo, A.: A hyper-heuristic collaborative multi-objective evolutionary algorithm. In: 2018 7th Brazilian Conference on Intelligent Systems (BRACIS), pp. 354–359. IEEE (2018)
13. Fritsche, G., Pozo, A.: Cooperative based hyper-heuristic for many-objective optimization. In: Proceedings of the Genetic and Evolutionary Computation Conference, pp. 550–558 (2019)
14. Fritsche, G., Pozo, A.: The analysis of a cooperative hyper-heuristic on a constrained real-world many-objective continuous problem. In: 2020 IEEE Congress on Evolutionary Computation (CEC), pp. 1–8. IEEE (2020)
15. Gonçalves, R., Almeida, C., Lüders, R., Delgado, M.: A new hyper-heuristic based on a contextual multi-armed bandit for many-objective optimization. In: 2018 IEEE Congress on Evolutionary Computation (CEC), pp. 1–8. IEEE (2018)
16. Huband, S., Barone, L., While, L., Hingston, P.: A scalable multi-objective test problem toolkit. In: Coello Coello, C.A., Hernández Aguirre, A., Zitzler, E. (eds.) EMO 2005. LNCS, vol. 3410, pp. 280–295. Springer, Heidelberg (2005). https://doi.org/10.1007/978-3-540-31880-4_20
17. Ian, P., Thomas, J.: Rain rules for limited overs cricket and probabilities of victory. J. R. Stat. Soc. Series D (Stat.) $51(2)$, 189–202 (2002)

18. Masood, A., Chen, G., Mei, Y., Zhang, M.: Reference point adaption method for genetic programming hyper-heuristic in many-objective job shop scheduling. In: Liefooghe, A., López-Ibáñez, M. (eds.) EvoCOP 2018. LNCS, vol. 10782, pp. 116–131. Springer, Cham (2018). https://doi.org/10.1007/978-3-319-77449-7_8

19. Masood, A., Chen, G., Zhang, M.: Feature selection for evolving many-objective job shop scheduling dispatching rules with genetic programming. In: 2021 IEEE Congress on Evolutionary Computation (CEC), pp. 644–651. IEEE (2021)

20. Masood, A., Mei, Y., Chen, G., Zhang, M.: A PSO-based reference point adaption method for genetic programming hyper-heuristic in many-objective job shop scheduling. In: Wagner, M., Li, X., Hendtlass, T. (eds.) ACALCI 2017. LNCS (LNAI), vol. 10142, pp. 326–338. Springer, Cham (2017). https://doi.org/10.1007/978-3-319-51691-2_28

21. Perwaiz, U., Younas, I., Anwar, A.A.: Many-objective bat algorithm. Plos one 15(6), e0234625 (2020)

22. Riquelme, N., Von Lücken, C., Baran, B.: Performance metrics in multi-objective optimization. In: 2015 Latin American Computing Conference (CLEI), pp. 1–11. IEEE (2015)

23. Ross, P.: Hyper-heuristics. In: Search Methodologies, pp. 529–556. Springer (2005). https://doi.org/10.1007/0-387-28356-0_17

24. Sánchez, M., et al.: A systematic review of hyper-heuristics on combinatorial optimization problems. IEEE Access 8, 128068–128095 (2020)

25. Senzaki, B.N.K., Venske, S.M., Almeida, C.P.: Hyper-heuristic based NSGA-III for the many-objective quadratic assignment problem. In: Britto, A., Valdivia Delgado, K. (eds.) BRACIS 2021. LNCS (LNAI), vol. 13073, pp. 170–185. Springer, Cham (2021). https://doi.org/10.1007/978-3-030-91702-9_12

26. SS., V.C., Anand, H.S.: Nature inspired meta heuristic algorithms for optimization problems. Computing, 1–19 (2021). https://doi.org/10.1007/s00607-021-00955-5

27. Venske, S.M., Almeida, C.P., Delgado, M.R.: Comparing selection hyper-heuristics for many-objective numerical optimization. In: 2021 IEEE Congress on Evolutionary Computation (CEC), pp. 1921–1928. IEEE (2021)

28. Wang, L., Ng, A.H., Deb, K.: Multi-objective evolutionary optimisation for product design and manufacturing. Springer (2011). https://doi.org/10.1007/978-0-85729-652-8

29. Yang, T., Zhang, S., Li, C.: A multi-objective hyper-heuristic algorithm based on adaptive epsilon-greedy selection. Complex Intell. Syst. 7(2), 765–780 (2021). https://doi.org/10.1007/s40747-020-00230-8

30. Zhang, Q., Li, H.: Moea/d: a multiobjective evolutionary algorithm based on decomposition. IEEE Trans. Evol. Comput. 11(6), 712–731 (2007)

# Multi-objective Optimization Based Feature Selection Using Correlation

Rajib Das[1], Rahul Nath[1], Amit K. Shukla[2(✉)], and Pranab K. Muhuri[1]

[1] Department of Computer Science, South Asian University, New Delhi, India
rajiblxp@gmail.com, rahul.nath@outlook.com,
pranabmuhuri@cs.sau.ac.in
[2] Faculty of Information Technology, University of Jyväskylä, Box 35 (Agora),
40014 Jyväskylä, Finland
amit.k.shukla@jyu.fi

**Abstract.** The optimal feature selection (FS) problem is widely targeted in the field of machine learning (ML). There are several ways to select the best features when the dataset dimension is small. However, when the dataset and number of features tend to increase, the solution becomes unrealistic as we need to evaluate every subset performance with the model. Various existing heuristics are partially useful as they portray premature convergence and exponential or high computational complexity. To solve this issue, evolutionary approaches-based FS has been extensively used in obtaining the optimal subset of features while maintaining the accuracy of the model. This paper proposes an efficient evolutionary-based multi-objective feature selection approach with a correlation coefficient filter method called Multi-Objective Optimization based Feature Selection (MOOFS). We introduce a two-stage process to select the best optimal features. In the first stage, a subset of features is randomly selected, and then a novel mutual correlation coefficient technique is used to get the important and relevant subset of features. The proposed MOOFS is experimented on several datasets and compared with the classical approach to demonstrate its efficiency.

**Keywords:** Feature selection · Correlation coefficient · Multi-objective optimization · NSGA-II

## 1 Introduction

In the present technology-oriented world, huge number of data are generated intentionally or unintentionally which is stored with some characteristics, mostly referred as features. All the features in the dataset may not be of importance due to dependency implying features dependence on each other and redundancy indicating unwanted and less important features. The process of selecting most optimal subset of feature is commonly termed as feature selection (FS), which is often considered one of the most critical and challenging tasks in machine learning. The reduced feature set improve the computational complexity and further helps in additional analysis. FS reduces feature space ($m \times n$) from very large space to a smaller ($n \times d$) space where d < m. Mathematically, a feature selection problem can be defined as follows: Suppose a dataset $D$ contains $n$ number of features, and the objective is to select the optimal subset of features from $D$ which are relevant. $D$ is represented as follows:

$$D = \{f1, f2, f3, ..., fn\} \tag{1}$$

where $f1, f2, f3, ..., fn$ represents the features of any dataset. In FS, we extract a subset $S = \{f1, f2, f3, ......, fd\}$, where, $d < n$.

In general, FS methods are classified into three categories namely filter, wrapper and embedded (hybrid) approaches. Filter approaches use general characteristics of the data to select features and are independent of learning algorithms. Wrapper methods always include a learning algorithm and according to its performance (increase or decrease), features are selected. The wrapper methods have high computational cost but provide more accurate results. On the other hand, filter approaches have low computational cost with less reliability. Lastly, the embedded methods include of filter and wrapper approaches both where FS is a part of the training process that is held with a learning algorithm.

According to set theory, if any dataset has $n$ number of features, then $2^n$ number of subsets is possible, and our task is to pick best subset by which machine learning model can give best accuracy. To select best subset, we need to evaluate every subset performance with the model, which is unrealistic, when $n$ increase to a huge number. Various existing heuristics are partially useful as they portray premature convergence, exponential or high computational complexity. To solve this issue, evolutionary or multi-objective optimization (MOO) based FS approaches have been extensively used in obtaining the optimal subset of features while preserving the accuracy of the model [1, 5]. These approaches tend to be efficient, effective, and reliable methods. In practice, the FS based on evolutionary, or MOO approaches falls under the wrapper method.

Mathematically, a multi-objective optimization problem (MOOP) can be formulated as follows:

$$minimize\ F(x) = [f1(x), (f2(x)..., fm(x)]^T$$
$$s.t. x \in D, \tag{2}$$

where $D$ is the decision space and $x \in D$ is a decision variable. $F(x)$ consists of $m$ objective functions $fi : D \to O, i = 1, ..., m$, where $O$ is the objective space. The function of two objectives often trade-offs with each other, as, improvement in one objective may lead to degradation of another. A decision maker (DM) whose having expert domain knowledge, implicitly choose a solution which can optimize all objectives simultaneously. The best trade-off solutions are called the Pareto optimal solutions. In this paper, a feature selection problem is formulated under a multi-objective optimization approach where subset of features is selected by simultaneously optimizing more than one objective. Since there are multi-objectives, we propose to use the evolutionary algorithm (EA) of Non dominated sorted genetic algorithm-II (NSGA-II). Although, there are many EA's available in the literature and the selection of "best" algorithm certainly depends upon the characteristics of each problem (No free lunch), NSGA-II is used in this paper due to its scale-up capability [17]. The two contradictory objectives, considered in this paper are: i) the number of selected features, and ii) Accuracy. The main goal of this paper is to propose an efficient evolutionary based multi-objective feature selection approach. As obtaining the maximum accuracy is the prime objective with minimum number of features, we select optimal feature in two stages. In the first stage, subset of features is randomly selected by the initialization step of NSGA-II. In the second stage, mutual correlation coefficient technique is used to get important, and relevant subset of features.

This approach is the hybrid filter-wrapper evolutionary approach where mutual correlation coefficient technique is incorporated inside the considered EA. Using mutual correlation coefficient technique, we calculate pair wise correlation for all features and then calculate average correlation for all features from pair wise correlation matrix. We use a threshold value and for reducing irrelevant and redundant feature we introduce percentile concept which make a novel correlation coefficient technique. In this concept, we only reduce percentage of features whose average correlation coefficient values are above the threshold value. Collectively, the major contribution of this paper is summarized as follows:

1) A novel correlation coefficient filter method is proposed and incorporated with NSGA-II, to obtain optimal subset of features.
2) This paper introduces novel way to reduce only less correlated features based on some threshold value in the filter approach. A novel randomized parameter, in the form of a percentile value, is introduced which decides the number of features to be reduced based on the size of the dataset.
3) The proposed approach enables to easily manage highly correlated dataset.

The remainder of this paper is organized as follows: In Sect. 2, literature review introduces the related existing works in this field. Section 3 details the explanation of the proposed approach of feature selection technique with the novel correlation coefficient technique. In Sect. 4, we will present experimental results of different dataset and compare the proposed approach with traditional NSGA-II. In Sect. 5, we will draw a summary of this article and outline the future research directions.

## 2   Literature Review

MOO approaches have achieved wide attention to solve the FS problems in many applications such as biomedical problems [2], text mining [3], image analysis [4], etc., where more than thousands of features are present. Various feature selection techniques with evolutionary algorithms (EAs) have been proposed in the literature. A survey of evolutionary feature selection techniques can be found in Xue et al. [5]. Binary genetic algorithms (Gas) are popularly used in EA when applied to feature selection. They use N dimensional binary vector for the number of features in the dataset. Here, "1" and "0", shows whether the resultant feature is selected or not [6]. For more than 100 features, exhaustive search techniques of feature selection such as SFS (Sequential Forward Selection), SBS (Sequential Backward Selection), SFFS (Sequential Floating Forward Selection), SFBS (Sequential Floating Backward Selection) algorithms are become computationally infeasible in feature selection. To solve issue of selecting subset of features in large dataset, evolutionary computing (EC) algorithms have drawn attention of the researchers. It has non-exhaustive search procedure which is computationally expensive but not computationally infeasible. To solve FS problems effectively, metaheuristics algorithms are coupled with wrapper methods that search the lower dimensional dataset space by iteratively calling the learning algorithm [7]. Mostly used metaheuristics algorithms focused on single objective of FS problems are GA [7, 17], Genetic Programming (GP) [8], Particle Swarm Optimization (PSO) [9], Differential Evolution (DE) [4], Ant Colony Optimization (ACO) [10]. Multi-objective EAs like NSGA-II [11, 20], multi-objective evolutionary algorithm with domain decomposition (MOEA/D) [12], Multi-objective particle swarm optimization (MOPSO) [13] are mostly used for the multi-objective FS problems. Hu et al. proposed fuzzy multi-objective FS method with particle swarm optimization, called PSOMOFS, where a fuzzy dominance relationship is developed to compare the goodness of candidate particles and global leader of particles are determined by fuzzy crowding distance measure [14]. Xue et al. proposed PSO based multi-objective FS algorithms where feature selection problem addressed by nondominated sorting and applying crowding distance, mutation, and dominance into PSO [15]. Chen et.al. Proposed an efficient ACO for image feature selection [16]. In ACO, graph is made to solve feature selection problem in which each feature is considered a node of the graph. Any feature i.e., node is selected if an ant has visited the node. Hancer et al. [22] developed first multi-objective artificial bee colony (MOABC) framework for feature selection in classification, where a new fuzzy mutual information-based criterion is proposed to evaluate the relevance of feature subsets. Khushaba et al. [18] proposed a novel feature selection algorithm by combining DE with ACO where DE was used to search for the optimal feature subset based on the solutions obtained by ACO.

## 3   Multi-objective Optimization Based Feature Selection (MOOFS)

Finding the optimal feature subset and maintaining the classification accuracy of low and high dimensional dataset are the goal of our proposed approach. For this purpose, we propose a novel Multi-Objective Optimization Based Feature Selection (MOOFS). The candidate solution of MOOFS is encoded as a vector of n bits where each bit can take value of 1 or 0 and number of selected features is decided according to the value of 1 in the vector. Each vector is an individual in the population. For example, binary vector solutions X of a dataset with n features represent as:

$$X = (x_1, x_2, x_3, x_4 \ldots, x_n), x_j \in \{0, 1\} \tag{3}$$

Then, the selected feature subset of X is:

$$X = (x_1, x_2, x_3, x_4 \ldots, x_d) \tag{4}$$

where d < n and $x_j = 1$ represents that the corresponding jth feature is selected.

In our proposed MOOPS approach, we use binary NSGA-II algorithm with mutual correlation coefficient technique where for all feature pair, we calculate average absolute mutual correlation of a feature over k (= n − 1) features, where n is the total number of features and k has (n − 1) features. Since NSGA-II algorithm randomly select initial population and offspring are generated using crossover and mutation, there is high possibility to select unwanted and noisy feature in each generation. To select relevant features efficiently, after generating the population of feature set by NSGA-II, we use mutual correlation coefficient technique (MCCT) before evaluating the selected features. MCCT reduce the uncorrelated and less important features and gives the best subset of features. These subsets of features evaluate by classification model to get the accuracy. To clearly understand the proposed MOOFS, we introduce a novel algorithm (Algorithm 1) and then explain the step-by-step procedure.

---

**Algorithm 1: Multi-Objective Optimization Based Feature Selection**

**Input:** Population Size, Maximum Iteration, Dataset (D), Classification Algorithm,
**Output:** A set of non-dominated solutions.

---

1. Generate Randomly initialize binary solutions with Population Size
2. For it 1 = to Maximum Iteration
   a. Find out Subset-Features-Population according to value 1 in each individual of Population
   b. Reduced Subset-Features-Population: = Call Percentile_Correlation_Coefficient (Subset of Features, D)
   c. Evaluate the whole Subset-Features-Population with KNN Classifier.
   d. Apply Non-Dominate sorting and Calculate Crowding distance.
   e. Do Selection, Crossover, and Mutation
   f. Generate New Population.
3. Return: The non-dominated pareto optimal front (POF).

---

In Step 1, a population with M individuals is randomly initialized in the binary space {0, 1}. The variable size of a population is determined according to the feature size of a dataset. The core process of (MOOFS) algorithm starts from Step 2, which is iterated until the stopping criteria is met or a maximum number of iterations is reached. In Step 2(a), the subset of features is identified with a position of 1 in the chromosome. Then in Step 2(b), individuals are randomly generated, noisy and unwanted feature may be included which can degrade the accuracy of the classifier. After that in Step 2(c), k-nearest neighbors (KNN) classifier is used to evaluate the reduced subset of features with l-fold cross validation (l = 10) technique. KNN is one of the widely used classifier in evolutionary feature selection algorithms. The number of selected subset of features and accuracy with KNN classifier are two objectives of our MOOFS algorithms. The optimal features are selected by simultaneously optimizing the objectives which are number of selected features and accuracy. The objective functions are described as follows:

i)  The number of selected features: The objective of this function is minimizing number of features:

$$\min F_1(X) = |X| \qquad (5)$$

where $|X|$ denotes the cardinality of selected subset of features.

ii) Accuracy: Maximizing the accuracy of the classification model represent the higher performance of classification. In this paper, accuracy is calculated by the KNN classifier with k fold cross-validation method (k = 10) and objective function is defined as follows:

$$\max F_2(X) = \left( \frac{1}{k} \sum_{i=1}^{k} \frac{N_{Cor}}{N_{All}} \right) \times 100 \qquad (6)$$

where $N_{Cor}$ denotes the correctly classified test samples, and $N_{All}$ is the total number of test samples.

In Step 2(d), Non-dominated sorting used on the population and the crowding distance is calculated, which are key process of NSGA-II algorithm. In Step 2(e, f), selection, crossover and mutation is performed. The output of crossover and mutation is used to generate new population for the next generation. Repeat the step 2 until the maximum number of iterations/generations is reached. At the end in Step 3, the non-dominated Pareto optimal front (POF) is returned where POF includes the number of selected features and accuracy.

To improve the performance of the classifier, important, linearly independent features need to be selected by removing noisy and unwanted feature. This task is done using Algorithm 2 (Percentile_correlation_coefficient), which returns the reduced subset of features after removing constant, and highly correlated features. In this algorithm, basic correlation coefficient method is used with some conditional step. The correlation coefficient technique measures the linear dependency or uncorrelation between two features. Two features are uncorrelated when their correlation coefficient is 0 and linearly dependent when their correlation coefficient is +1(positively correlated) or −1(negatively correlated). Generally, the approach of algorithm 2 may be refereed as a filter approach.

---

**Algorithm 2: Percentile_Correlation_Coefficient**

---

**Input:** Subset of Features, Dataset D
**Output:** Reduced Features Subset

---

1. Initialize percentile value p, number of features n, threshold value
2. Calculate w according percentile value with formula w=round(n*p)
3. For t=1 to w
    a. Set k=n-1
    b. Calculate Correlation Coefficient of Data set D according to Eq. (5)
    c. Compute average absolute mutual correlation of all features according to Eq. (4)
    d. Set s= largest average mutual correlation
    e. If s>threshold remove having largest average mutual correlation feature from Dataset D
    f. Set n=n-1
4. Return Reduced Subset of Feature

---

Algorithm 2 shows the pseudocode of the percentile correlation coefficient algorithm. In step 1 of this algorithm 2, we initialize the percentile value p (p = 0.3), number of features (n) according to the input parameter-subset of features, and threshold value (threshold = 0.9). Then, we calculate variable, w, according to the value of p and n. The value of w decides how many times step 3 will iterate to reduce the feature following the condition of threshold value. In step 3(a), we define a variable k = n-1 and in 3(b), the correlation coefficient matrix of Dataset (D) is computed, according to Eq. (7):

$$r_{x,y} = \frac{\sum_{i=1}^{n}(x_i - \bar{x})(y_i - \bar{y})}{\sqrt{\sum_{i=1}^{n}(x_i - \bar{x})^2 \sum_{i=1}^{n}(y_i - \bar{y})^2}} \tag{7}$$

Then, step 3(c) computes the average absolute mutual correlation of one feature to (n-1) feature according to Eq. (8), which is given as follows:

$$r_{i,k(=n-1)} = \frac{1}{k}\sum_{j=1,j\neq i}^{k}|r_{x_i,x_j}| \tag{8}$$

where $i$ is the $i^{th}$ feature, k denotes the all n-1 features except $i^{th}$ feature and $j$ denotes 1 to $k$ features except $i^{th}$ feature.

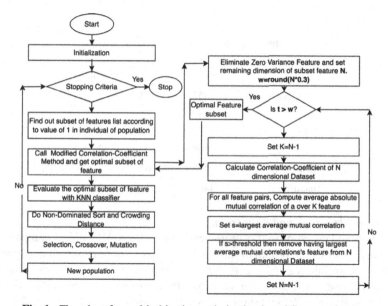

**Fig. 1.** Flowchart for multi-objective optimization based feature selection

Find out largest average mutual correlation value and assign to the variable 's' (step 3(d)). Feature with the largest average mutual correlation will be removed in each iteration, if the value of s is greater than threshold value (step (3(e)). Repeat the step 3 until the t is equated to w. As step 3 will iterate up to the value of w which means we only reduce maximum w feature and minimum zero feature following the condition of step 3(e). Further, the process will be iterated up-to stopping criteria and return the best reduced subset of features (step 4). In this algorithm, we can manage highly correlated dataset, where all feature's average absolute mutual correlation value is greater than the threshold which was impossible in only correlation coefficient technique. Figure 1 pictorially depicts the flowchart of proposed algorithm 1 and 2.

## 4   Experimental Results

The proposed MOOFS uses NSGA-II for the feature selection. To validate the efficiency of the proposed approach, we have used several datasets with various dimensions and compared MOOFS with traditional NSGA-II approach. This section is divided into two subsections. In first we have discussed about the datasets used and parameters settings of the MOO approaches. In the later sub-section, we present the experimentation results.

i.  ***Datasets and parameter settings***
    The details of the datasets used in this paper in mentioned in Table 1 which are taken from UCI dataset [19]. The number of features varies from 33 to 241, while the number of instances goes from 351 to 4464. The maximum iteration in the NSGA-II is 500 while the number of populations is 50. In the percentile_correlation _coefficient method, the parameter threshold is experimented for 0.8 and 0.9 [21, 23].

**Table 1.** Datasets and their dimensions

| Dataset | #Instances | #Features | Classes |
|---|---|---|---|
| Ionosphere | 351 | 33 | 2 |
| Connectionist-Bench (Sonar) | 208 | 60 | 2 |
| Hill-Valley | 606 | 100 | 2 |
| Musk1(Clean1) | 476 | 166 | 2 |
| Tuandromd | 4464 | 241 | 2 |

ii. ***Experimental result and discussion.***
    Our proposed evolutionary Feature selection algorithm (MOOFS) is compared with standard NSGA-II algorithm to validate its efficiency. The first set of our experimental results are Tables 2 and 3, consisting of best accuracy value corresponding to the number of features. These tables also show the comparison among NSGA-II and our proposed MOOFS. Table 2 shows the accuracy and number of feature values when the threshold values is 0.8, whereas Table 3 shows the values for 0.9 threshold.

**Table 2.** Best accuracy of proposed MOOFS (with 0.8 threshold) and NSGA-II

| | MOOFS | | NSGA-II | |
|---|---|---|---|---|
| Dataset | Best accuracy | #Feature | Best accuracy | # Feature |
| Ionosphere | **94.8571** | **7** | 94.2857 | 9 |
| Sonar | **76.5** | **7** | 76 | 7 |
| HillValley | **64.9697** | **10** | 65.8333 | 10 |
| Musk1 | **86.6479** | **26** | 84.5202 | 28 |
| Taundromd | **96.9389** | **39** | 96.8280 | 46 |

From the Table 3, it's evident that best values are obtained when the threshold value is chosen as 0.9. For ionosphere dataset, MOOFS results in a higher accuracy of 94.865 with only six number of features, however, NSGA-II returned nine optimal features with lesser accuracy of 94.286. For Sonar and HillValley datasets, best accuracy is again achieved by MOOFS which is 76.286 and 65.97 with seven and ten features, respectively. The NSGA-II achieved less accuracy but with same number of features.

**Table 3.** Best accuracy of proposed MOOFS (with 0.9 threshold) and NSGA-II

| Dataset | MOOFS | | NSGA-II | |
|---|---|---|---|---|
| | Best accuracy | #Feature | Best accuracy | # Feature |
| Ionosphere | **94.8650** | **6** | 94.2857 | 9 |
| Sonar | **76.2857** | **7** | 76 | 7 |
| HillValley | **65.9696** | **10** | 65.8333 | 10 |
| Musk1 | **85.0622** | **23** | 84.5202 | 28 |
| Taundromd | **97.1384** | **39** | 96.8280 | 46 |

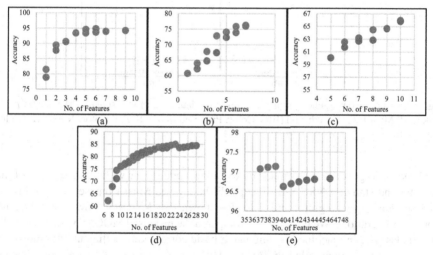

**Fig. 2.** POF of accuracy vs. no. of features of datasets: (a) Ionosphere, (b) Sonar (c) Hill Valley, (d) Musk1, (e) Taundromd (Blue Color: Proposed MOOFS; Red Color: NSGA-II) (Color figure online)

The significant amount of improvement is achieved for the Musk1 and Taundromd datasets, where the accuracy is significantly higher with a smaller number of features as compared to the traditional NSGA-II approach. The improvement in accuracy is achieved with around 18% (Musk1) and 15% (Taundromd) reduction in number of features for the datasets. The marginal improvement in accuracy for HillValley dataset is due to the highly correlation among the features. It is clearly observed that our proposed MOOFS algorithm shows better results compare to standard NSGA-II. Figure 2 shows the POFs of all the datasets. In Fig. 3(a-e), accuracy convergence graph of the five datasets is shown where best accuracy of every iteration is considered.

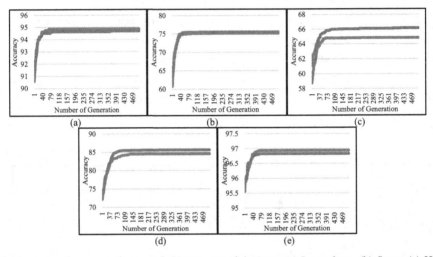

**Fig. 3.** Accuracy convergence graph of Features of datasets: (a) Ionosphere, (b) Sonar (c) Hill Valley, (d) Musk1, (e) Taundromd (Blue Color: Proposed MOOFS; Red Color: NSGA-II) (Color figure online)

This convergence graph is made by five independent runs with 500 iterations of the MOOFS and standard NSGA-II. Except HillValley and Musk1 dataset, we got higher accuracy convergence graph and it is also shown that MOOFS showed stable result around 250 iterations. We have further compared the proposed MOOFS with other well-known classifier in machine learning using basic correlation coefficient with threshold 0.9. The results are compiled in Table 4. Hill Valley data set select only one feature because this dataset is highly correlated. It is observed that except Hill Valley dataset, remaining 5 datasets select more number of features which are 33 for Ionosphere, 57 for Sonar, 106 for Musk1, 154 for Tuandromd. Also, most of the classifier showed less accuracy than our proposed approach. This trade-off of selecting optimal feature with higher accuracy is efficiently achieved by our proposed MOOPS approach.

**Table 4.** Accuracy of different classifier

| Algorithm | Ionosphere | Sonar | Hill Valley | Musk1 | Taundromd |
|---|---|---|---|---|---|
| Gradient boosting | 91.17 | 57.26 | 50.50 | 62.26 | 98.40 |
| Bagging classifier | 91.17 | 57.45 | 51.83 | 66.26 | 98.60 |
| Extra tree classifier | **94.59** | 54.88 | 52.82 | 64.33 | 98.60 |
| Random forest | 93.73 | 55.90 | 51.83 | 65.00 | 98.40 |
| AdaBoost | 92.30 | 66.45 | 50.34 | 67.24 | 98.60 |
| DT | 86.31 | 58.40 | 50.00 | 62.02 | 98.40 |
| Logistics regression | 85.50 | 51.95 | 47.51 | 63.53 | **98.80** |
| KNN | 84.07 | 42.76 | 47.04 | 63.41 | 97.20 |
| Support vector machine | 64.14 | 38.09 | 48.00 | 56.87 | 75.00 |
| NB | 64.41 | 39.47 | 48.00 | 54.73 | 98.40 |
| **NSGA-II with KNN** | **94.28(9)** | **76(7)** | **65.83(10)** | **84.52(28)** | **96.78 (43)** |
| **MOOFS** | **94.86(6)** | **76.28(7)** | **65.96(10)** | **85.06 (23)** | **97.13 (39)** |

## 5  Conclusion and Future Work

This paper introduces a novel feature selection approach formulated under a multi-objective optimization approach where subset of features is selected by simultaneously optimizing more than one objective. Due to the scale-up capability, NSGA-II is utilized where the two considered objectives are optimal number of features, and accuracy. Several datasets with various dimensions are considered for the experimentation. The proposed approach is named Multi-Objective Optimization based Feature Selection (MOOFS), an efficient evolutionary-based multi-objective feature selection approach with a correlation coefficient filter method. The experiments are done on different datasets. From the results, a significant improvement in the accuracy with a considerable reduction in the number of features can be seen. Overall, our approach can reduce irrelevant and redundant features, which helps reduce training time as well as improve the performance of ML algorithms. One of the limitations of the proposed approach is to define the threshold value for reducing the number of features, which can be improved in future work. Further, the capability of the proposed approach for the big datasets shall be experimented. This work can also be extended to self-adapting parameter values where the correlation of the data set, dynamic in nature, can be identified.

## References

1. De La Iglesia, B.: Evolutionary computation for feature selection in classification problems. Wiley Interdisc. Rev. Data Min. Knowl. Disc. **3**(6), 381–407 (2013)
2. Ahmed, S., Zhang, M., Peng, L.: Enhanced feature selection for biomarker discovery in LC-MS data using GP. In: 2013 IEEE Congress on Evolutionary Computation, pp. 584–591. IEEE (2013)

3. Aghdam, M.H., Ghasem-Aghaee, N., Basiri, M.E.: Text feature selection using ant colony optimization. Expert Syst. Appl. **36**(3), 6843–6853 (2009)

4. Ghosh, A., Datta, A., Ghosh, S.: Self-adaptive differential evolution for feature selection in hyperspectral image data. Appl. Soft Comput. **13**(4), 1969–1977 (2013)

5. Xue, B., Zhang, M., Browne, W.N., Yao, X.: A survey on evolutionary computation approaches to feature selection. IEEE Trans. Evol. Comput. **20**(4), 606–626 (2015)

6. Siedlecki, W., Sklansky, J.: A note on genetic algorithms for large-scale feature selection. In: Handbook of Pattern Recognition and Computer Vision, pp. 88–107 (1993)

7. Xue, Y., Tang, Y., Xu, X., Liang, J., Neri, F.: Multi-objective feature selection with missing data in classification. IEEE Trans. Emerg. Topics Comput. Intell. **6**(2), 355–364 (2021)

8. Muni, D.P., Pal, N.R., Das, J.: Genetic programming for simultaneous feature selection and classifier design. IEEE Trans. Syst. Man Cybern. Part B Cybern. **36**(1), 106–117 (2006)

9. Unler, A., Murat, A.: A discrete particle swarm optimization method for feature selection in binary classification problems. Eur. J. Oper. Res. **206**(3), 528–539 (2010)

10. Yan, Z., Yuan, C.: Ant colony optimization for feature selection in face recognition. In: Zhang, David, Jain, Anil K. (eds.) ICBA 2004. LNCS, vol. 3072, pp. 221–226. Springer, Heidelberg (2004). https://doi.org/10.1007/978-3-540-25948-0_31

11. Deb, K., Pratap, A., Agarwal, S., Meyarivan, T.A.M.T.: A fast and elitist multiobjective genetic algorithm: NSGA-II. IEEE Trans. Evol. Comput. **6**(2), 182–197 (2002)

12. Zhang, Q., Li, H.: MOEA/D: a multiobjective evolutionary algorithm based on decomposition. IEEE Trans. Evol. Comput. **11**(6), 712–731 (2007)

13. Zhang, X., Zheng, X., Cheng, R., Qiu, J., Jin, Y.: A competitive mechanism based multi-objective particle swarm optimizer with fast convergence. Inf. Sci. **427**, 63–76 (2018)

14. Hu, Y., Zhang, Y., Gong, D.: Multiobjective particle swarm optimization for feature selection with fuzzy cost. IEEE Trans. Cybern. **51**(2), 874–888 (2020)

15. Xue, B., Zhang, M., Browne, W.N.: Particle swarm optimization for feature selection in classification: a multi-objective approach. IEEE Trans. Cybern. **43**(6), 1656–1671 (2012)

16. Chen, B., Chen, L., Chen, Y.: Efficient ant colony optimization for image feature selection. Sig. Process. **93**(6), 1566–1576 (2013)

17. Muhuri, P.K., Nath, R., Shukla, A.K.: Energy efficient task scheduling for real-time embedded systems in a fuzzy uncertain environment. IEEE Trans. Fuzzy Syst. **29**(5), 1037–1051 (2020)

18. Khushaba, R.N., Al-Ani, A., AlSukker, A., Al-Jumaily, A.: A combined ant colony and differential evolution feature selection algorithm. In: Dorigo, M., Birattari, M., Blum, C., Clerc, M., Stützle, T., Winfield, A.F.T. (eds.) ANTS 2008. LNCS, vol. 5217, pp. 1–12. Springer, Heidelberg (2008). https://doi.org/10.1007/978-3-540-87527-7_1

19. https://archive.ics.uci.edu/ml/datasets.php?format=&task=cla&att=&area=&numAtt=&numIns=&type=&sort=nameUp&view=table

20. Shukla, A.K., Nath, R., Muhuri, P.K., Lohani, Q.D.: Energy efficient multi-objective scheduling of tasks with interval type-2 fuzzy timing constraints in an industry 4.0 ecosystem. Eng. Appl. Artif. Intell. **87**, 103257 (2020)

21. Wang, F., Yang, Y., Lv, X., Xu, J., Li, L.: Feature selection using feature ranking, correlation analysis and chaotic binary particle swarm optimization. In: 2014 IEEE 5th International Conference on Software Engineering and Service Science, pp. 305–309. IEEE (2014)

22. Hancer, E., Xue, B., Zhang, M., Karaboga, D., Akay, B.: A multi-objective artificial bee colony approach to feature selection using fuzzy mutual information. In: 2015 IEEE congress on evolutionary computation (CEC), pp. 2420–2427. IEEE (2015)

23. Xie, Z.-X., Qing-Hua, H., Da-Ren, Y.: Improved feature selection algorithm based on SVM and correlation. In: Wang, J., Yi, Z., Zurada, J.M., Bao-Liang, L., Yin, H. (eds.) ISNN 2006. LNCS, vol. 3971, pp. 1373–1380. Springer, Heidelberg (2006). https://doi.org/10.1007/11759966_204

# SAME: Sampling Attack in Multiplex Network Embedding

Chao Kong[1], Dan Meng[2], Tao Liu[1], Mengfei Li[1], Qijie Liu[1], Liang Zhou[1],
Pingfu Chao[3], and Yi Zhang[1(✉)]

[1] School of Computer and Information, Anhui Polytechnic University, Wuhu, China
{kongchao,liutao,lmf,lqj,lzhou,zhangyi}@ahpu.edu.cn
[2] OPPO Research Institute, Shenzhen, China
[3] School of Computer Science and Technology, Soochow University, Suzhou, China
pfchao@suda.edu.cn

**Abstract.** Network embedding aims to learn the low-dimensional latent representations of vertices in a network. Although there are a few works that combine network embedding with privacy, they do not consider the case of the multiplex network. When a vertex is deleted from the network, it is easy to achieve the deleted relations by remaining embedding vectors. We can also utilize inter-layer and intra-layer information to pay attention to the feature and topological information in a multiplex network. A large amount of auxiliary side information can improve the performance of privacy attack. To address these issues, we propose choosing the vertex degree with selectivity to study the problem of privacy attack in the multiplex network. Our solution SAME, short for *Sampling Attack in Multiplex Network Embedding*, consists of two components. In the embedding component, we leverage adequate information from inter-layer and intra-layer. Then in the privacy attack component, it generates multiple variants of the original network by removing different single vertex. We leverage these variants to train a classifier to recover the deleted relations. We also conduct extensive experiments on several real-world datasets covering the task of link prediction. Both quantitative results and qualitative analysis verify the effectiveness and rationality of our methods.

**Keywords:** Privacy attack · Multiplex network · Representation learning · Membership inference

## 1 Introduction

With the popularization of search engines, recommender systems, and other online applications [1], a huge volume of network data from users has been generated. To perform predictive analytics on network data, it is crucial to obtain the representations (i.e., feature vectors) for vertices. Network embedding aims to learn the low-dimensional latent representations of vertices in a network. Based on the vertex embeddings, standard machine learning techniques can be applied to address various predictive tasks such as link prediction, clustering, and so on. It is also well known in many research fields, including data mining [2], information retrieval [3], natural language processing [4], etc. However, there is an increasing demand for data protection and data recovery, especially in social network scenarios. Therefore, it is imperative and challenging to perform such a model to recover the relations which are removed in the network. To Distinguish from the traditional embedding of a single-layer network, we define the embedding of multiple networks as **multiplex network embedding**. In the multiplex network embedding, there are more networks than single-layer network, which can offer more associated information. As shown in Fig. 1, given a multiplex social network that covers three different social platform information with all embedding vectors, one user $u_7$ in Twitter has been removed. The successful privacy attack attempts to recover the relations by remaining information in the Twitter network and other two platforms' information.

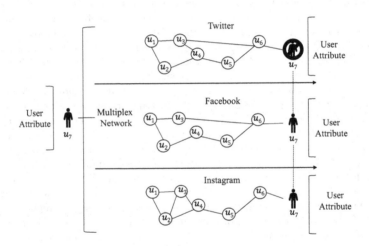

**Fig. 1.** The scenario of privacy attack in multiplex social network. We illustrate the data structure and dependence between vertices and each layers' topological structure from three-layer network. From left to right are vertices attribute, multiplex network structure, and their dependence between each layer.

Early studies mainly focus on the extraction of personal information due to the increased overall awareness of privacy protection [5]. In general, one type of attack is named model inversion attacks [6], and the other type is the membership inference attack [7]. They aim to extract differences in the confidence of the outputs between data used during the training process and data that was not used. Our privacy attack method in multiplex network embedding is the transformation of the inference attack. Despite effectiveness and prevalence, we argue that these methods can be suboptimal for privacy in multiplex network embedding due to three primary reasons: 1) existing privacy attack methods are not tailored for network embedding, and there is no optimization for network and embedding vectors; 2) existing attack techniques do not consider the differences between vertices in a network, which lead to meaningless computing cost; 3) current works in privacy attack of network embedding do not consider the situation in multi-layer network, i.e. multiplex network, which lead to a large amount of valid information being ignored.

Is there a principled way to resolve the challenges mentioned above? To the best of our knowledge, none of the existing works has paid special attention to combining multiplex network embedding with privacy attack simultaneously. In this paper, we propose SAME which utilizes vertex degrees to select samples to train the classifier to predict the removed privacy relations in the multiplex network. Compared with the single-layer network, the multiplex network embedding involves more inter-layer information which can leverage other networks' features and topological information to enhance the performance of the model. We summarize the main contributions of this work as follows.

- In order to efficiently utilize the remaining embedding vectors in the original network after deleted vertex, we proposed SAME to construct a classifier to obtain the relations between the removed vertex and other vertices. Specifically, we remove additional vertices from the network, compute the respective embedding, and compare them with original embedding rather than retraining.
- We simply sample the different vertices with different vertex degrees into multiple bins, which can effectively distinguish the information density in a network to reduce the computation cost.
- To leverage more auxiliary side information, we introduce the multiplex network embedding to capture more features and topological information about deleted data.
- We conduct extensive experiments on five real-world networks covering several prevalent network embedding algorithms. Both quantitative results and qualitative analysis verify the effectiveness and rationality of our SAME method.

The remainder of the paper is organized as follows. We shortly discuss the related work in Sect. 2. We formulate the problem in Sect. 3, before delving into details of the proposed method in Sect. 4. We perform extensive empirical studies in Sect. 5 and conclude the paper in Sect. 6.

## 2  Related Work

This section presents a brief overview of existing literature related to our work, in particular, network embedding, multiplex network embedding, and membership inference attack.

## 2.1 Network Embedding

As an effective and efficient network analytic method, network embedding (a.k.a. graph embedding) aims at converting the network data into a low dimensional space in which the network structural information and network properties are maximally preserved [8]. The study of network embedding problem has become a hot topic in recent years and drawn more attention from both academia and industry. In recent years, there are three main types of network embedding methods including random walk-based methods [9], factorization-based methods [10], and deep neural network-based methods [11].

Following the pioneering work of [9], the random-walk-based methods typically apply a two-step solution: first performing random walks on the network to obtain a "corpus" of vertices, and then employing word embedding methods to obtain the embeddings for vertices. Grover et al. [12] introduce the breadth-first sampling and depth-first sampling to change the method of random walk named node2vec. Tang et al. [13] proposed a method called LINE which considers the 2nd-order proximity in the random walk to catch more implicit relations. However, Cao et al. [10] doubt whether the LINE can not obtain the deeper relations which just utilize the 2nd-order proximity. They proposed the GraRep which used a factorization-based method to obtain high-order relations. Deep neural network-based methods are the state-of-the-art network embedding techniques. Tu et al. [11] proposed a novel deep semi-supervised algorithm for simultaneous graph embedding and node classification, utilizing dynamic graph learning in neural network hidden layer space. DeepEmLA [14] smoothly projected different types of attributed information into the same semantic space, while maintaining the topological structures. Besides capturing high-order proximities, there are several proposals to incorporate auxiliary side information into network embedding, such as user profiles [15], events (hyper-edge) [16], spatio-temporal data [17], etc.

## 2.2 Multiplex Network Embedding

The goal of multiplex network embedding is to achieve the information fusion of multiple features of networks. It can be divided into two categories, joint representation learning and coordinated representation learning [18]. Zhang et al. [19] assume that the same vertex in multiplex networks preserves certain common and unique feature of each layer. Thus, they proposed a scalable multiplex network embedding to learn vertex embeddings in each layer by Deepwalk [9]. Ma et al. [20] simply increased vertex embeddings in each layer of the multiplex network. Recently, Yuan et al. [21] leverage similarity of vertices' ensembles which are selected by information density to propose a multi-view network embedding model. In a departure from joint representation learning, coordinated representation learning ignores the information across layers. Qu et al. [22] utilized a few labeled data of different vertices to learn the weight of views with the attention-based method. Liu et al. [23] extended the method to the multiplex network by network aggregation and layer co-analysis. It is worth mentioning that Ning et al. [18] leverage high-order vertex dependence to explore the problem of multiplex network embedding with coordinated representation learning in recent. It resolves the over smoothing and missing information problems by jointly considering inter-layer and intra-layer. Although our proposed approach varies depending on the ways of representation, its performance is independent of various algorithms used for learning the embeddings.

### 2.3 Membership Inference Attack

As mentioned above, in this paper, the privacy attack in network embedding is regarded as the transfer of general membership inference attack. Li et al. [24] first defined the problem of inference attacks, which meant an adversary could infer the real value of a sensitive entry with high confidence. The basic idea is to extract differences in the confidence of the outputs for data used during training and data that was not used. In social networks, Shokri et al. [25] used the shadow model to construct similar training sets and target sets to determine whether the samples are in the training sets. Zhang et.al [26] introduce a membership inference framework based on representation learning. This framework is independent of the assumptions of the method involved in the synthetic data generation process. Chen et al. [7] devise the first membership inference attack against collaborative inference, to infer whether a particular data sample is used for training the model of industrial Internet of Things. Although these methods work well in their respective fields, they are not tailored for learning on multiplex networks.

## 3  Problem Formulation

As mentioned above in Fig. 1, it is appropriate to represent these data from three platforms that have multiple views and sources as the multiplex network. Hence, three layers can not only represent the intra-layer relations but also can obtain the dependencies and interactions between networks, i.e. inter-layer information. We first give notations used in this paper, and then formalize the privacy attack in multiplex network embedding problem to be addressed.

### 3.1  Notations

The important notations used in this paper are summarized in Table 1. We assume $G$ to be a multiplex network with $|V|$ vertices and $|L|$ layers. In this network, each vertex can interact with others through $|L|$ types of edges ($|L| \geq 2$). Hence, we can simply define the multiplex network architecture as $G = \{G^l(\mathcal{V}, \mathcal{E}^l), l \in L\}$ which is made up of $|L|$ layers with $|V|$ vertices and $|\sum_{l \in L} \mathcal{E}^l|$ edges. Besides, we also define a situation in which each layer of multiplex network has the same vertices set and different edge sets as illustrated in Fig. 1. Let $i, j \in \mathcal{V}$ be two vertices, it is simple to represent the $i^l$ which is the vertex $i$ in layer $l$, and $e_{i,j}^l$ denotes the different vertices in the same layer. If it is

**Table 1.** Important notations used in this paper.

| Notations | Explanations |
|---|---|
| $G$ | A multiplex network |
| $\mathcal{V}, \mathcal{E}^l$ | The sets of vertices and edges in layer $l$ respectively |
| $|V|, |E^l|$ | The size of vertices and edges in layer $l$ respectively |
| $e_{i,j}^{l,l'}$ | The edge between vertex $i$ in layer $l$ and vertex $j$ in layer $l'$ |
| $\mathcal{E}(G)$ | The whole embedding sets of a multiplex network $G$ |
| $\mathcal{E}_{v_i}$ | The remaining representation after removed vertex $v_i$ in one layer |
| $\mathcal{E}'_{v_i}$ | The retraining representation after removed vertex $v_i$ in one layer |

the same vertex in different layers, it can be expressed as $i^l$ and $i^{l'}$ respectively. Meanwhile, there is also an interaction between different vertices in different layers, i.e. $j^{l'}$ can be linked to $i^l$ by the duplicates of $i$ in $l'$ implicitly. For example, in the social network as shown in Fig. 1 which has three layers, $e_{u_6, u_7}^{Twitter, Facebook}$ is a cross-layer edge between $u_6^{Twitter}$ and $u_7^{Facebook}$ through an anchor link.

### 3.2   Problem Definition

The task of privacy attack aims to recover the removed relations in multiplex network by remaining embeddings. In this paper, we employ the coordinated representation learning for multiplex network embedding. Further, when the vertex $v_i$ in one layer is removed, we denote the remaining network as $G_{v_i}$. Therefore, the remaining representation of network is $\mathcal{E}(G_{v_i})$. Because we only consider one vertex within a network, it can be short for $\mathcal{E}_{v_i}$. The $\mathcal{E}_{v_i}$ does not need to retrain, so there is no explicit information in removed vertex $v_i$. However, the implicit information remains in the $\mathcal{E}_{v_i}$ which has been influenced by previously existing edges with $v_i$. When the remaining vertices retraining, we denote the representation of the network as $\mathcal{E}'_{v_i}$.

## 4   Sampling Attack in Multiplex Network Embedding

In this section, we present the proposed SAME model to address the three major challenges mentioned in Sect. 1.

### 4.1   Multiplex Network Embedding

As shown in Algorithm 1, we learn the multiplex network embeddings by coordinated representation learning method before privacy attack. Following the pioneering work [23], the global loss function of the embeddings consists of three different components. They are intra-layer loss function, inter-layer loss function and attribute loss function respectively. As shown in Eq. 1, the loss function of multiplex network embedding method is:

$$L = L_{intra} + L_{inter} + L_{attr}. \tag{1}$$

The inter-layer loss is expressed by $k$ convolution layers and $k$ deconvolution layers as :

$$L_{intra} = \sum_{j=1}^{K/2} \|\hat{\mathbf{Z}}^j - \mathbf{Z}^{K-j}\|_2^2, \tag{2}$$

in which $K$ is the total layer number of convolution-deconvolution layers. The input $\mathbf{Z}^k$ is the vertices attribute $\mathcal{X}$, hence the output $\hat{\mathbf{Z}}^k$ is a reconstruction matrix in respect of $\mathcal{X}$. These two matrices can simply be calculated by the trick of Kipf and Welling. A general definition of graph convolution is given as follows:

$$\mathbf{Z} = \hat{\mathbf{D}}^{-1/2} \hat{\mathbf{A}} \hat{\mathbf{D}}^{-1/2} \mathbf{X} \Theta, \tag{3}$$

where $\mathbf{A}$ is an adjacency matrix, $\hat{\mathbf{A}}$ is the degree matrix, and $\hat{\mathbf{D}}_{ii} = \sum_j \mathbf{A}_{ij}$. $\Delta_{V_i}$ and $\Delta'_{V_i}$ represent the distance matrices of the remaining embeddings of the network and the retraining embeddings of the network respectively. We denote $\Theta$ as the convolutional kernel, and $\Theta_d$ as the deconvolution kernel. Hence, the graph deconvolution layer can be formulated as:

$$\hat{\mathbf{Z}} = \hat{\mathbf{D}}^{-(1/2)} \hat{\mathbf{A}} \hat{\mathbf{D}}^{-(1/2)} \mathbf{Z}^k \Theta_d, \tag{4}$$

where $\hat{\mathbf{Z}}^k$ is an output of the $k$-th layer of deconvolution layer in a neural network.

For inter-layer dependence loss, to save the vertex inter-layer dependence property, the loss function is determined by local dependence measure $P_{pred}(.|l, i)$ and true underlying connecting distribution $P_{true}(.|l, i)$ of vertex $i$ in layer $l$, which can be formulated as:

$$L_{inter} = \frac{1}{N} \sum_{i \in \mathcal{V}} \sum_{l'=1}^{l} -[P_{true}(l'|l, i) log \hat{P}_{pred}(l'|l, i) \\ + (1 - P_{true}(l'|l, i)) log(1 - \hat{P}_{pred}(l'|l, i))], \tag{5}$$

where $P_{pred}(.|l, i)$ is the concatenate of score function $Score(l'|l, i)$:

$$Score(l'|l, i) = \sigma(h_i^l \mathbf{W} \sigma(\frac{1}{N} \sum_{i=1}^{N} h_i^{l'})), \tag{6}$$

in which the $\mathbf{W}$ is the trainable scoring matrix, $\sigma$ is the logistic sigmoid non-linearity and $h_i^l$ is the embedding of the vertex $i$ in layer $l$. The $P_{true}(.|l, i)$ is easy to be obtained by the KL-divergence with structural similarity. To concatenate each element with symbol $\Delta$, the distribution is shown as:

$$P_{true}(.|l, i) = \Delta_{l' \in L} \frac{S_{struc}(l', l|i)}{\sum_{r \in L} S_{struc}(l', l|i)}. \tag{7}$$

---

**Algorithm 1.** The Multiplex Network Embedding Algorithm

---

**Input:** The network $G = < \mathcal{V}, \mathcal{E}, L, \mathcal{X} >$ ; Convolution-deconvolution neural network
with kernal $\Theta$, $\Theta_d$; Iteration times $T$;
**Output:** $\mathcal{E}(G)$: The vertices embeddings of multiplex network $G$ ;
 1: Initialize the parameters of convolution-deconvolution neural network;
 2: **for** Each iteration $t \in T$ **do**
 3:     **for** $l$ in $L$ **do**
 4:         Calculate the Score function in each layer;
 5:         Generate the convolution embedding by Equation 3;
 6:         Generate the deconvolution embedding by Equation 4;
 7:     **end for**
 8:     Calculate $P_{pred}$ by Equation 6;
 9:     Calculate $P_{true}$ by Equation 7;
10:     Update the convolution-deconvolution kernel $\Theta$, $\Theta_d$;
11:     Reconstruct the attribute to obtain $\hat{X}$;
12:     Update the attribute by Equation 8;
13: **end for**
14: Incorporate the vertices embedding;
15: **return** $\mathcal{E}(G)$.

---

It is also important to introduce the reconstruction attribute $\hat{x}_i = \sigma(W \Delta_{l'=1}^{L} \hat{Z}_i^{l'})$ to calculate the attribute loss function as:

$$L_{attr} = \frac{1}{N} \sum_{i \in \mathcal{V}} \sum_{l \in L} -((x_i log \hat{x}_i^l + (1 - x_i) log(1 - \hat{x}_i^l))), \tag{8}$$

where $L$ is the layer number of multiplex network, $x_i$ is the attribute of vertex $i$.

### 4.2   Overview of Privacy Attack

The privacy attack approach consists of three components as shown in Fig. 2.

***Step1. Difference Matrix Calculation.*** First, we calculate the comprehensive network embedding $\mathcal{E}$ of the multiplex network, then remove a vertex $v_i$ in one layer, we obtain the remaining embedding $\mathcal{E}_{v_i}$. After the retraining, a new representation $\mathcal{E}'_{v_i}$ of the whole network is also obtained. Both of the distance matrices $\Delta_{v_i}$ and $\Delta'_{v_i}$ can be calculated between each vertex pair. We finally obtain the difference matrix by $Diff(\mathcal{E}_{V_i}, \mathcal{E}'_{V_i}) = \Delta_{V_i} - \Delta'_{V_i}$.

***Step2. Feature Vectors Construction.*** In order to determine whether there is a relation between attacked vertex and other vertices which are in same layer, we need feature vectors to train a classifier. For each vertex in the remaining network, we calculate the distance changes to other vertices. The changes interval can be divided into different bins which means the dimension of the feature vector. The number of each bin represents the feature vector's value in the fixed dimension.

***Step3. Classifier Training.*** The node degree is also useful in the multiplex network to conduct sampling. In reality, vertex with a high node degree contains more information and is more vulnerable to attack. Hence, we sort the node degree in order of high, medium and low to obtain different numbers of samples by $N_s = (\alpha + \beta + \gamma) * N$. For the retraining representation $\mathcal{E}'_{v_i}$ of whole network which has removed vertex $v_i$, we temporarily remove the sample vertex $v_j$ to retraining a new representation $\mathcal{E}''_{v_j}$. Step1 and step2 are performed for each sample to obtain the final training dataset for constructing the classifier.

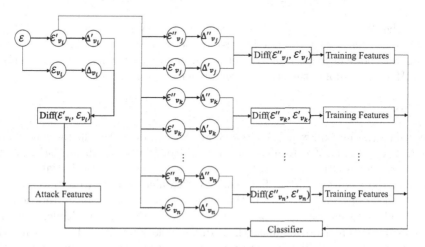

**Fig. 2.** Workflow of privacy attack. When the user has been attacked, we construct the attack feature first. Then, through the shadow model, multiple networks have been constructed to train the classifier. Finally, the deleted relations can recover through the classifier.

### 4.3   Sampling Attack via Vertex Degree

As we described before, inference attack is independent of the algorithm used for training the embedding [1]. However, the auxiliary side information which derives from the multiplex network is vital for privacy attack. Hence, following the workflow of Fig. 2, we first calculate the difference matrix $Diff(\mathcal{E}_{V_i}, \mathcal{E}'_{V_i})$ by the distance matrices between each vertex by Eq. 9:

$$Diff(\mathcal{E}_{V_i}, \mathcal{E}'_{V_i}) = \Delta_{V_i} - \Delta'_{V_i}, \tag{9}$$

where $\Delta_{V_i}$ and $\Delta'_{V_i}$ are the distance matrices of remaining embeddings of network and the retraining embeddings of network respectively.

After we obtain different matrices, we start to construct feature vectors by calculating the distance changes. This step is to decide the dimension of the feature vector. It divides the range of changes into $n$ bins. Through the hyper-parameter $b$, we create bins $b_1, b_2, ..., b_n$ that contain the same number of values. We repeat the above steps to create a training dataset for the classifier. In this step, sampling is necessary to decrease the meaningless computation. Node degree is the key to whether a vertex is vulnerable to attack. The vertices are arranged according to the node degree and divided into three parts: high, medium and low with three hyper-parameters $\alpha, \beta, \gamma$ as shown in Eq. 10:

$$N_s = (\alpha + \beta + \gamma) * N \quad s.t. \quad (\alpha + \beta + \gamma \leq 1), \tag{10}$$

where the $N_s$ is the total number of samples, and $N$ is the number of reaming vertices. By this equation, three hyper-parameters decide the number of samples in different node degrees. Finally, a classifier can be performed by these training data. Here, we employ the support vector machines as our classifier.

### 4.4   Discussion

We utilize the multiplex network embedding method to retrain the whole network. The training algorithm is shown as Algorithm 1. In this algorithm, the input are the multiplex network, neural networks with their initial parameters and the iteration times. The result of this algorithm is to obtain the vertices embeddings. In line 1, we initialize the parameters. For each iteration, we firstly calculate intermediate parameters in each layer of a multiplex network at lines $4 - 6$. Then we calculate three different loss functions by updating the convolution-deconvolution kernel at lines $8 - 12$. At last, we incorporate the vertices embedding into the embeddings of multiplex network $G$.

Hence, we deduce from the algorithm that when the SAME has $S$ samples, the total complexity of SAME is $O(STNE|L|^2)$ where $T$ is the number of iterations, $N$ is the number of vertices in each layer, $E$ is the number of edges of the whole network, and $L$ is the number of layers.

## 5   Experiments

To evaluate the performance of SAME, we employ it to a representative application on five real-world networks - link prediction. Link prediction is usually approached as a classification task that predicts whether a link exists between two vertices. Through empirical evaluation, we aim to answer the following research questions:

**RQ1**: How does SAME perform across different state-of-the-art network embedding methods in link prediction?

**RQ2**: How does SAME perform when we reduce the auxiliary side information through reducing the layer of multiplex network?

In what follows, we first introduce the experimental settings, and then answer the above research questions in turn to demonstrate the rationality of our methods.

**Table 2.** Basic statistics about different multiplex networks used in the link prediction task.

| Datasets | #Nodes | #Edges | Layer descriptions with edges number |
|----------|--------|--------|--------------------------------------|
| Vickers | 29 | 740 | Class: 316; Best friend: 226; Work: 198 |
| CS-Aarhus | 61 | 620 | FaceBook: 193; Leisure: 124; Work: 21; Coauthor: 87; Lunch: 195 |
| London | 369 | 441 | Tube:312; Overground: 82; DLR: 46 |
| CKM | 246 | 1,551 | Advice: 480; Discussion:565; Friend: 506 |
| Celegans | 279 | 5,863 | ElectrJ: 1,031; MonoSyn: 1,639; PolySyn: 3193 |

## 5.1 Experiment Settings

**Datasets.** For the link prediction task, we utilize several types of datasets involving social, biological and transportation. All of these datasets are multiplex networks and public accessible[1]. We list the statistic information about them in Table 2.

- *Vickers* is collected by Vickers from 29 seventh grade students in a school in Victoria, Australia. There are 3 layers in this network constructed by three social questions (get along with, friends, work with).
- *CS-Aarhus* is a multiplex network consisted of five kinds of online and offline relationships (Facebook, Leisure, Work, Co-authorship, Lunch) between the employees of Computer Science department at Aarhus.
- *London* is collected in 2013 from the official website of Transport for London and manually cross-checked. Vertices are train stations in London and edges encode existing routes between stations. Underground, overground and DLR stations are considered.
- *CKM* is collected by Coleman, Katz and Menzel on medical innovation, considering physicians in four towns in Illinois, Peoria, Bloomington, Quincy and Galesburg. They were concerned with the impact of network ties on the physicians' adoption of a new drug, tetracycline. Three sociometric matrices (layers) were generated, based on three questions (advice, discussion, friend).
- *Celegans* contains different types of genetic interactions for organisms in the Biological General Repository (BioGRID). The present folder concerns Caenorhabditis Elegans. The multiplex network used in the paper makes use of the following layers: direct interaction, physical association, additive genetic interaction defined by inequality, suppressive genetic interaction defined by inequality, association, co-localization.

**Baselines.** As we introduced in Sect. 2, there are three main types of embedding algorithms. We choose the *LINE* [13] and *node2vec* [12] as the random-walk-based algorithm, the *GraRep* [10] and *TADW* [8] as the factorization-based algorithm, and *DAGE* [11] as the deep-neural-network-based algorithm. If not specified, we use the open-source tools OpenNE[2] to imply the network embedding algorithms with default parameters.

---

[1]  https://manliodedomenico.com/data.php
[2]  https://github.com/thunlp/openne

**Table 3.** Performance of the SAME on different networks and network embedding algorithms with and without the multiplex network information. The value without parenthesis is the performance of multiplex network embedding, and the value with parenthesis is the performance of embeddings which are only leverage one-layer network information.

| Network | AUC | | | | | |
|---|---|---|---|---|---|---|
| | (LINE) | (node2vec) | (GraRep) | (TADW) | (DAGE) | (SAME) |
| Vickers | 0.82 (0.74) | 0.82 (0.67) | 0.82 (0.76) | 0.82 (0.79) | 0.82 (0.79) | 0.85 (0.81) |
| CS-Aarhus | 0.93 (0.92) | 0.93 (0.90) | 0.93 (0.70) | 0.93 (0.89) | 0.93 (0.89) | 0.93 (0.90) |
| London | 0.78 (0.53) | 0.78 (0.68) | 0.78 (0.52) | 0.78 (0.51) | 0.78 (0.51) | 0.80 (0.51) |
| CKM | 0.87 (0.82) | 0.87 (0.80) | 0.87 (0.76) | 0.87 (0.65) | 0.87 (0.65) | 0.89 (0.65) |
| Celegans | 0.88 (0.86) | 0.88 (0.78) | 0.88 (0.76) | 0.88 (0.73) | 0.88 (0.75) | 0.91 (0.75) |
| Network | Precision@10 | | | | | |
| | (LINE) | (node2vec) | (GraRep) | (TADW) | (DAGE) | (SAME) |
| Vickers | 0.46 (0.41) | 0.46 (0.37) | 0.46 (0.39) | 0.46 (0.40) | 0.46 (0.39) | 0.46 (0.41) |
| CS-Aarhus | 0.63 (0.55) | 0.63 (0.57) | 0.63 (0.51) | 0.63 (0.56) | 0.63 (0.56) | 0.64 (0.56) |
| London | 0.44 (0.29) | 0.44 (0.31) | 0.44 (0.30) | 0.44 (0.31) | 0.44 (0.31) | 0.44 (0.32) |
| CKM | 0.51 (0.44) | 0.51 (0.45) | 0.51 (0.45) | 0.51 (0.46) | 0.51 (0.46) | 0.66 (0.45) |
| Celegans | 0.57 (0.53) | 0.57 (0.50) | 0.57 (0.51) | 0.57 (0.51) | 0.57 (0.52) | 0.59 (0.53) |
| Network | Macro-$F_1$ | | | | | |
| | (LINE) | (node2vec) | (GraRep) | (TADW) | (DAGE) | (SAME) |
| Vickers | 0.41 (0.37) | 0.41 (0.32) | 0.41 (0.35) | 0.41 (0.37) | 0.41 (0.35) | 0.41 (0.35) |
| CS-Aarhus | 0.45 (0.40) | 0.45 (0.41) | 0.45 (0.44) | 0.45 (0.41) | 0.45 (0.39) | 0.51 (0.39) |
| London | 0.37 (0.31) | 0.37 (0.32) | 0.37 (0.31) | 0.37 (0.33) | 0.37 (0.30) | 0.41 (0.31) |
| CKM | 0.54 (0.49) | 0.54 (0.47) | 0.54 (0.44) | 0.54 (0.45) | 0.54 (0.46) | 0.60 (0.44) |
| Celegans | 0.56 (0.51) | 0.56 (0.39) | 0.56 (0.42) | 0.56 (0.43) | 0.56 (0.43) | 0.56 (0.40) |
| Network | Micro-$F_1$ | | | | | |
| | (LINE) | (node2vec) | (GraRep) | (TADW) | (DAGE) | (SAME) |
| Vickers | 0.49 (0.43) | 0.49 (0.41) | 0.49 (0.44) | 0.49 (0.42) | 0.49 (0.41) | 0.53 (0.40) |
| CS-Aarhus | 0.44 (0.41) | 0.44 (0.37) | 0.44 (0.36) | 0.44 (0.34) | 0.44 (0.33) | 0.47 (0.34) |
| London | 0.32 (0.29) | 0.32 (0.30) | 0.32 (0.27) | 0.32 (0.28) | 0.32 (0.30) | 0.33 (0.30) |
| CKM | 0.45 (0.40) | 0.45 (0.39) | 0.45 (0.37) | 0.45 (0.38) | 0.45 (0.40) | 0.44 (0.40) |
| Celegans | 0.50 (0.43) | 0.50 (0.45) | 0.50 (0.46) | 0.50 (0.43) | 0.50 (0.42) | 0.46 (0.46) |

**Evaluation Measures.** For the link prediction task, there are established evaluation measures. We utilize *AUC, Precision@10, Macro-$F_1$* and *Micro-$F_1$* as metrics. We fine tune the $\alpha$, $\beta$, $\gamma$ as 0.005, 0.0025, 0.0025 with the trade-off between the effectiveness and the efficiency in our method.

**Table 4.** Performance Comparison with different layer sets. The first four sets in $CS - Aarhus$, we leverage SAME to validate. The last single layer network, we utilize the LINE method to perform.

| Networks | AUC | Precision@10 | Macro-$F_1$ | Micro-$F_1$ |
|---|---|---|---|---|
| FaceBook-Leisure-Work-Coauthor-Lunch | 0.93 | 0.63 | 0.45 | 0.44 |
| FaceBook-Leisure-Work-Coauthor | 0.93 | 0.63 | 0.44 | 0.44 |
| FaceBook-Leisure-Work | 0.93 | 0.62 | 0.44 | 0.43 |
| FaceBook-Leisure | 0.93 | 0.59 | 0.43 | 0.43 |
| FaceBook | 0.92 | 0.55 | 0.40 | 0.41 |

### 5.2  Performance Comparison Under Privacy Attack (RQ1)

We evaluate the privacy attack on several different networks and embedding algorithms to predict whether a vertex is connected with others, i.e., link prediction. The values outside bracket in Table 3 denote the performance of link prediction task after attacking by SAME which are utilized with multiplex network information. The corresponding values in bracket illustrate the results which only use the most edge number's one-layer information.

Table 3 illustrates the performance comparison with and without multiplex network information after performing privacy attack by SAME on different networks and network embedding algorithms in link prediction task, where we have the following key observations: 1) the attack can recover substantial information of the removed vertex on many networks across several network embedding algorithms. Although in some networks it does not achieve good performance, it also obtains excellent value in several networks such as $AUC = 0.93$ in $CS$-$Aarhus$ with SAME. It means that in practical situations, it is enough to identify an individual; 2) the embedding algorithm is independent of the embedding algorithm, because the key to the performance of attack lies in the structure of the network rather than the effectiveness of the embedding algorithms. There is no embedding algorithm that can perform best over all networks. Hence, we introduce multiplex network embedding to capture more vertex attribute and network topological structure information. It can improve the privacy attack performance impressively.

### 5.3  Performance Comparison with Different Layer Number (RQ2)

In the last study, we observe that it is useful to capture more information in privacy attack. However, how does the layer number affect the performance of privacy attack is also a problem. We choose $CS - Aarhus$ as the dataset in this study because it has the most number of layers. We evaluate the privacy attack on several layers setting by SAME and LINE. The results are shown in Table 4, it validates what we suspected about layer number on privacy attack. It means that the more layers we utilize, the better privacy attack performance we achieve. Besides, it will eventually converge finally.

## 6   Conclusions

In this paper, we have studied the problem of privacy attack in multiplex network embedding model. It is a challenging task due to the difference of vertices' attribute and network structure. We propose a method to demonstrate the importance of defense in network embedding and the effectiveness of auxiliary side information in privacy attack. We have illustrated our proposed method on five real-world networks. Experimental results indicate that multiplex network embedding methods are easier to attack than one-layer network embedding.

In our future work, we plan to extend our work to handle the dynamic multiplex network embedding problems and deploy a distributed algorithm to support more efficient computation.

**Acknowledgment.** This work was supported in part by the National Natural Science Foundation of China Youth Fund (No. 61902001), the Open Project of Shanghai Big Data Management System Engineering Research Center (No. 40500-21203-542500/021), the Industry Collaborative Innovation Fund of Anhui Polytechnic University-Jiujiang District (No. 2021cyxtb4), and the Science Research Project of Anhui Polytechnic University (No. Xjky072019C02, No. Xjky2020120).

## References

1. Kong, C., et al.: Privacy attack and defense in network embedding. In: Chellappan, S., Choo, K.-K.R., Phan, N.H. (eds.) CSoNet 2020. LNCS, vol. 12575, pp. 231–242. Springer, Cham (2020). https://doi.org/10.1007/978-3-030-66046-8_19
2. Gao, M., He, X., Chen, L., Liu, T., Zhang, J., Zhou, A.: Learning vertex representations for bipartite networks. IEEE Trans. Knowl. Data Eng. **34**(1), 379–393 (2022)
3. Wu, J., Xu, Y., Zhang, Y., Ma, C., Coates, M., Cheung, J.C.K.: TIE: a framework for embedding-based incremental temporal knowledge graph completion. In: SIGIR 2021, pp. 428–437 (2021)
4. Cui, L., et al.: Refining sample embeddings with relation prototypes to enhance continual relation extraction. In: ACL 2021, pp. 232–243
5. Yao, J., Dou, Z., Wen, J.: FedPS: a privacy protection enhanced personalized search framework. In: WWW 2021, pp. 3757–3766 (2021)
6. Mo, K., Liu, X., Huang, T., Yan, A.: Querying little is enough: model inversion attack via latent information. Int. J. Intell. Syst. **36**(2), 681–690 (2021)
7. Chen, H.: Practical membership inference attack against collaborative inference in industrial IoT. IEEE Trans. Ind. Inf. **18**(1), 477–487 (2022)
8. Cai, H., Zheng, V.W., Chang, K.C.: A comprehensive survey of graph embedding: problems, techniques, and applications. IEEE Trans. Knowl. Data Eng. **30**(9), 1616–1637 (2018)
9. Perozzi, B., Al-Rfou, R., Skiena, S.: Deepwalk: online learning of social representations. In: SIGKDD 2014, pp. 701–710 (2014)
10. Cao, S., Lu, W., Xu, Q.: GraREP: learning graph representations with global structural information. In: CIKM 2015, pp. 891–900 (2015)
11. Tu, E., Wang, Z., Yang, J., Kasabov, N.K.: Deep semi-supervised learning via dynamic anchor graph embedding in latent space. Neural Netw. **146**, 350–360 (2022)

12. Grover, A., Leskovec, J.: node2vec: Scalable feature learning for networks. In: KDD 2016, pp. 855–864 (2016)
13. Tang, J., Qu, M., Wang, M., Zhang, M., Yan, J., Mei, Q.: Line: large-scale information network embedding. In: WWW 2015, pp. 1067–1077 (2015)
14. Zhao, Z., Zhou, H., Li, C., Tang, J., Zeng, Q.: Deepemlan: deep embedding learning for attributed networks. Inf. Sci. **543**, 382–397 (2021)
15. Zhou, S., et al.: Direction-aware user recommendation based on asymmetric network embedding. ACM Trans. Inf. Syst. **40**(2), 29:1–29:23 (2022)
16. Pham, P., Do, P.: W-mmp2vec: topic-driven network embedding model for link prediction in content-based heterogeneous information network. Intell. Data Anal. **25**(3), 711–738 (2021)
17. Zhang, B., Yuan, C., Wang, T., Liu, H.: STENET: a hybrid spatio-temporal embedding network for human trajectory forecasting. Eng. Appl. Artif. Intell. **106**, 104487 (2021)
18. Ning, N., Long, F., Wang, C., Zhang, Y., Yang, Y., Wu, B.: Nonlinear structural fusion for multiplex network. Complex. 2020, 7041564:1–7041564:17 (2020)
19. Zhang, H., Qiu, L., Yi, L., Song, Y.: Scalable multiplex network embedding. In: IJCAI 2018, pp. 3082–3088 (2018)
20. Ma, Y., Ren, Z., Jiang, Z., Tang, J., Yin, D.: Multi-dimensional network embedding with hierarchical structure. In: WSDM 2018, pp. 387–395 (2018)
21. Yuan, W., et al.: Multi-view network embedding with node similarity ensemble. World Wide Web **23**(5), 2699–2714 (2020). https://doi.org/10.1007/s11280-020-00799-7
22. Qu, M., Tang, J., Shang, J., Ren, X., Zhang, M., Han, J.: An attention-based collaboration framework for multi-view network representation learning. In: CIKM 2017, pp. 1767–1776 (2017)
23. Ning, N., Li, Q., Zhao, K., Wu, B.: Multiplex network embedding model with high-order node dependence. Complex. 2021, 6644111:1–6644111:18 (2021)
24. Li, C., Shirani-Mehr, H., Yang, X.: Protecting individual information against inference attacks in data publishing. In: Kotagiri, R., Krishna, P.R., Mohania, M., Nantajeewarawat, E. (eds.) DASFAA 2007. LNCS, vol. 4443, pp. 422–433. Springer, Heidelberg (2007). https://doi.org/10.1007/978-3-540-71703-4_37
25. Shokri, R., Stronati, M., Song, C., Shmatikov, V.: Membership inference attacks against machine learning models. In: SSP 2017, pp. 3–18 (2017)
26. Zhang, Z., Yan, C., Malin, B.A.: Membership inference attacks against synthetic health data. J. Biomed. Inf. **125**, 103977 (2022)

# Cycles Improve Conditional Generators: Synthesis and Augmentation for Data Mining

Alexander M. Moore[1]([✉]) [iD], Randy Clinton Paffenroth[1] [iD], Ken T. Ngo[2], and Joshua R. Uzarski[2]

[1] Worcester Polytechnic Institute, Worcester, MA 01609, USA
{ammoore,rcpaffenroth}@wpi.edu
[2] Chemical/Biological Innovative Material and Ensemble Development Team, U.S. Army DEVCOM-SC, Natick, MA 01760, USA
{ken.a.ngo.civ,joshua.r.uzarski.civ}@army.mil

**Abstract.** Conditional Generative Adversarial Networks (CGANs) are diversely utilized for data synthesis in applied sciences and natural image tasks. Conditional generative models extend upon data generation to account for labeled data by estimating joint distributions of samples and labels. We present a family of modified CGANs which demonstrate the inclusion of reconstructive cycles between prior and data spaces inspired by BiGAN and CycleGAN improves upon baselines for natural image synthesis with three primary contributions. The first is a study proposing three incremental architectures for conditional data generation which demonstrate improvement on baseline generation quality for a natural image data set across multiple generative metrics. The second is a novel approach to structure latent representations by learning a paired structured condition space and weakly structured variation space with desirable sampling and supervised learning properties. The third is a proposed utilization of conditional image synthesis for supervised learner data set augmentation as an alternative generation metric. Additional experiments demonstrate the successes of inducing cycles in conditional GANs for both image synthesis and image classification over comparable models with no additional tweaks or modifications. We release our source code, models, and experiments here: https://github.com/alexander-moore/Cycles-Improve-Conditional-Generators.

**Keywords:** Deep learning · GANs · Natural image synthesis

## 1 Introduction

Beyond demonstrations producing high-quality synthetic images, training generative models may improve downstream data efficiency, generalization, and robustness of deep learning models across many domains [2]. Generative Adversarial Networks (GANs) have received extensive ongoing study since their inception, and extensions proposed in the seminal introduction of GANs hinted at the subsequent development of conditional GANs which specify the desired class of

© The Author(s), under exclusive license to Springer Nature Switzerland AG 2022
W. Chen et al. (Eds.): ADMA 2022, LNAI 13726, pp. 352–364, 2022.
https://doi.org/10.1007/978-3-031-22137-8_26

generated samples [6,11]. Conditional image synthesis with GANs learns functional approximations from a joint prior-condition space to a joint condition-data natural image space given labeled training samples [11,12].

Accounting for conditional distributions in the training of data synthesis models allows for selection of generated data classes which may improve data efficiency and model robustness during downstream training (Sect. 4.3, [2]). The utilization of downstream supervised learners may additionally be a promising metric in the training of conditional generators, particularly in non-natural image tasks which lack standards of evaluating synthesized samples (Table 3).

This work proposes alterations to the CGAN architecture and training procedures which improve upon baseline conditional image synthesis according to a variety of established and proposed evaluation metrics with a corresponding incremental study isolating contributions to model improvement. We demonstrate that incremental improvements on conditional image synthesis as measured by generative quality metrics correlate to improved downstream learner accuracy for models trained on synthesized samples (Sect. 4.3). This finding begets ongoing study of the role of conditional data synthesizers for data set augmentation to improve model robustness, data efficiency, and performance on low-data paradigms.

We study three models which improve upon baseline conditional generative GANs by inducing cycles inspired by unpaired image-to-image translation from CycleGAN [18] with extensions to conditional data generation (Sect. 3). We define cycles as a composition of functions from an initial space to an intermediate space and back with low reconstructive error such that each function image fulfills some distributional requirement: in the case of CycleGAN ([18], Sect. 2.2), both spaces are image spaces with an adversarial discriminator. By setting one of these spaces to be a multivariate normal latent prior, we induce an autoregressive model with an added adversary, here called the conditional autoencoder-GAN outlined in Sect. 3.1.

Two incremental modifications to this design are introduced in Sects. 3.2 and 3.3 which isolate the contributions of cycles versus autoencoding for encoder-decoder models. A study comparing equivalent baseline conditional GANs to cyclical models is performed in Sect. 4.1. We find that enforcing conditional, reconstructive, and cyclical losses on the proposed models improves image synthesis outcomes.

## 1.1   Contributions

This research contributes to the improvement of conditional natural image synthesis with GANs and the utilization of conditional generation for simultaneous or downstream supervised learning as follows:

1. A novel formulation of three conditional cyclical GANs to incorporate cycling between spaces as well as a bipartite latent space for conditioning (Sect. 3).
2. An incremental study on the inclusion of cycles to conditional generators demonstrating improvement on baseline conditional generators across a variety of experiments and metrics (Sect. 4).

3. A proposed utilization of conditional image synthesis for supervised learning data set augmentation as an alternative generation metric (Sect. 4.3).

## 2   Related Work

### 2.1   Generative Adversarial Networks

GANs have received extensive research for their promising capabilities to learn unique solutions to training data distributions given a min-max game [6]. Optimization uses expectation over samples from the training set $x \sim p_{data}$ of the error of the discriminator given by $\log D(x)$. The discriminator evaluation of real samples is balanced against generated samples given by the expectation over samples $z \sim p_z(z)$ decoded by the generator $G(z)$, where $z$ is the multivariate standard normal prior. The agents update in turn, where the generator tries to minimize the objective against a maximizing discriminator solved by

$$L_{GAN} = \min_G \max_D V(D, G) = \mathbb{E}_{x \sim p_{data}}[\log D(x)]$$
$$+ \mathbb{E}_{z \sim p_z(z)}[\log 1 - D(G(z))]. \quad (1)$$

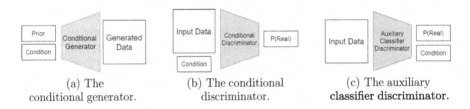

(a) The conditional generator.

(b) The conditional discriminator.

(c) The auxiliary classifier discriminator.

**Fig. 1.** Baseline conditional GAN architectures.

Conditional GANs (Fig. 1) impart structure on the inputs to the generator and discriminator by concatenating some label $y$ to the input vectors of each model [11]. This conditions the min-max optimization as:

$$\min_G \max_D V(D, G) = \mathbb{E}_{x \sim p_{data}}[\log D(x|y)]$$
$$+ \mathbb{E}_{z \sim p_z(z)}[\log 1 - D(G(z|y))]. \quad (2)$$

The Auxiliary Classifier GAN (ACGAN) modifies the concatenated-input discriminator of the CGAN to instead predict the corresponding class of input training and synthetic samples [12]. This modification is discussed in Sect. 3 as a potential benefit of training mixed-loss adversarial models. The ACGAN alters GAN training by incorporating an auxiliary classifier head on the output of the discriminator, with a conditional component given by:

$$L_C = \mathbb{E}_{x \sim p_{data}}[\log P(C = c)] + \mathbb{E}_{x \sim p_z(z)}[\log P(C = c)]. \quad (3)$$

Using this conditional loss, the discriminator is trained to maximize $L_{GAN} + L_C$ (1) while the Generator is trained to maximize $L_C - L_{GAN}$ [12]. In addition to producing labeled data by specifying conditions, approximating conditional distributions of training data with conditional generators is demonstrated to improve generative quality and discriminability of samples over unconditional GANs when labels are available [12].

## 2.2 CycleGAN

The family of models proposed in this work draws from the CycleGAN unpaired image-to-image translation model [18]. CycleGAN (Fig. 2) training uses two unpaired image sets drawn from distinct training distributions $X$ and $Y$, typically natural images. Two models comparable to GANs are trained simultaneously: One generator learns $F(x) = \hat{y}$ for training data $x \in X$ directed by a discriminator trained to distinguish training from synthesized samples in $Y$-space, and another function which learns $G(y) = x$ for training data $y \in Y$ taught by a discriminator who learns to distinguish training from synthesized samples in $Y$-space. These generators are trained under the *cycle* constraint enforced by the reconstructive penalty $F(G(Y)) \approx y$ and $G(F(x)) \approx x$ (Eq. 6).

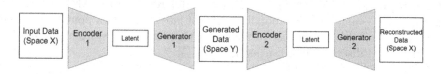

**Fig. 2.** CycleGAN Architecture. Each space $X, Y$ is accompanied by a discriminator returning the adversarial generation loss Eq. 1.

$$\mathcal{L}_{GAN}(G, D_Y, X, Y) = \mathbb{E}_{y \sim p_{data}(y)}[\log D_Y(y)] \\ + \mathbb{E}_{x \sim p_{data}(x)}[\log D_Y(x)] \tag{4}$$

Given generative models $G, F$ with images in spaces $X, Y$ respectively and discriminators $D_X, D_Y$ on spaces $X, Y$ respectively, the full optimization objective $\mathcal{L}(G, F, D_X, D_Y)$ is given by:

$$\mathcal{L}_{GAN}(G, D_Y, X, Y) + \mathcal{L}_{GAN}(F, D_X, Y, X) + \lambda \mathcal{L}_{cyc}(G, F) \tag{5}$$

where

$$\mathcal{L}_{cyc}(G, F) = \mathcal{L}_{recon}(F, G, X) + \mathcal{L}_{recon}(G, F, Y) \\ = \mathbb{E}_{x \sim p_{data}(x)}[||F(G(x)) - x||_1] \\ + \mathbb{E}_{y \sim p_{data}(y)}[||G(F(y)) - y||_1]. \tag{6}$$

The hyperparameter $\lambda$ tunes the relative importance of reconstruction and adversarial components. This system optimizes the following:

$$G^*, F^* = \arg\min_{G,F} \max_{D_x,D_y} \mathcal{L}(G, F, D_X, D_Y). \tag{7}$$

Though CycleGAN emphasizes natural image style transfer, this research investigates the usage of cycles between an unpaired multivariate normal latent space and a natural image data space for conditional synthesis and demonstrates that the regularization imparted by cycles may improve training outcomes over non-cyclical baseline models via regularization imparted by additional optimization objectives.

## 3    Cycles for Conditional Generation

This research proposes the Conditional Autoencoder-GAN (CAEGAN, Sect. 3.1), Inverse Conditional Autoencoder GAN (ICAEGAN, Sect. 3.2), and Cycle-Conditional Autoencoder GAN (CCAEGAN, Sect. 3.3) as incremental cyclical alterations to the conditional GAN (CGAN, [11]) for conditional data synthesis.

Training paradigms comparable to cycles have been proposed in the GAN literature previously [5]. Though implemented differently in BiGAN in which the discriminator evaluated corresponding pairs of latent and data points, Donahue et. al. imparted the significance of reconstructing latent codes from samples for the purpose of disentangling learned features [5].

### 3.1    Conditional Autoencoder-GAN

The CAEGAN (Fig. 3) combines reconstructive cycles with a GAN generator by adding a reconstructive loss term to the CGAN. Loss contribution given by discriminator evaluation of generated samples structures the variation space without an explicit prior divergence penalty as in the VAE [9].

The autoencoder-GAN collapses the GAN generator and autoencoder decoder into a shared parameter model, where gradients are summed and back-propogated to both the encoder and decoder. The generative and reconstructive tasks share parameterizations, leveraging the assumption that the training distribution and the learned approximation share a latent representation space.

Equation (8) gives the triple-criterion optimized by the CAEGAN:

$$\mathcal{L} = \mathcal{L}_{GAN} + \lambda(\mathcal{L}_C + \mathcal{L}_{recon}) \tag{8}$$

Two of the loss elements are directly borrowed from the nominal models: the adversarial $\mathcal{L}_{GAN}$ (1) and the autoencoding pixel-wise reconstruction loss $\mathcal{L}_{recon}$ given by mean squared error between the input sample and model output. $\mathcal{L}_C$ (Eq. 3) is given by a supervised loss from the encoder's prediction of the input sample label, meaning the model must *predict* the corresponding conditions of the input sample in the encoding step.

## 3.2    Inverse Conditional Autoencoder GAN

The Inverse Conditional Autoencoder GAN (ICAEGAN, Fig. 4) serves as the foil to the conditional GAN for an incremental study with the addition of a cycle which recovers latent codes from generated samples. The inverse mapping which returns latent sample estimations of input images not provided by a typical GAN – though it is known to be useful for auxiliary supervised feature learning [12]. Methods such as contrastive learning [3] and BiGAN [5] emphasize the importance of recovering the sampled latent code which led to the generation of an image. The reconstruction loss is taken between the latent sample $v$ and the reconstruction $\hat{v}$.

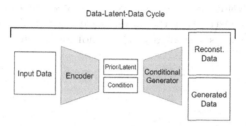

**Fig. 3.** Conditional Autoencoder-GAN (CAEGAN) architecture. Training samples are encoded to a paired latent space given by a variation vector in the prior distribution, and a condition in the label space. From this joint space encoded training samples may be reconstructed, or new samples synthesized from prior sampling.

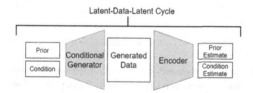

**Fig. 4.** Inverse Conditional Autoencoder GAN (ICAEGAN) Architecture. Pairs sampled from the joint latent-condition space are decoded by a conditional generator, then re-encoded by an encoding function, resulting in a reconstructive loss on the prior and a supervised loss on the condition estimate.

$$\mathcal{L} = \mathcal{L}_{GAN} + \lambda(\mathcal{L}_C + \mathcal{L}_{v\ recon}) \tag{9}$$

The reconstructive loss $\mathcal{L}_{recon}$ is given by recovering the latent sample $v$ in the latent variation space $V$ given decoded samples $G(v)$. The multivariate standard normal prior and condition space are sampled and decoded by the conditional generator. This decoded random latent sample is evaluated by a discriminator, and each model receives an adversarial loss $\mathcal{L}_{GAN}$ given by (1). Departing from the CGAN, by inducing cycles by reconstructions of encoded or decoded points the generated sample is now re-encoded to the condition-variance space which defines the reconstructive and predictive losses on the recovered variation and condition samples $\mathcal{L}_C$ and $\mathcal{L}_{recon}$ (Eq. 3).

### 3.3   Cycle Conditional Autoencoder-GAN

The Cycle Conditional Autoencoder-GAN (CCAEGAN, 5) is the extension of the training algorithms proposed by the CAEGAN and ICAEGAN. Where the CAEGAN and ICAEGAN perform reconstruction of a space under the image of another space (data-latent-data and latent-data-latent, respectively), the cycle autoencoder-GAN performs both tasks using shared coefficients. This leads to a four part loss, given by:

$$\mathcal{L} = \mathcal{L}_{GAN} + \lambda(\mathcal{L}_C + \mathcal{L}_{recon} + \mathcal{L}_{v\ recon}) \tag{10}$$

where the $\mathcal{L}_C$ now comes from two sources: supervised learning of training data labels $\mathcal{L}_{c|x}$, and reconstructing the random samples of $C$-space by generating and re-encoding latent samples from the condition given the decoding of the $(v, c)$ pair:

$$\mathcal{L}_C = \mathcal{L}_{c|x} + \mathcal{L}_{c|G(v,c)} \tag{11}$$

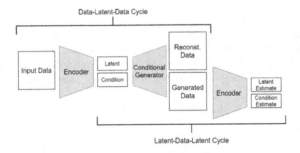

**Fig. 5.** Cycle Conditional Autoencoder-GAN Architecture. Here, two cycles are performed by a shared encoder and generator which incorporate losses from both CAE-GAN and ICAEGAN components.

The CCAEGAN serves as a third incremental step to induce cycles for conditional GANs. The CAEGAN may perform well due to the advantage of reconstructing training samples, which baseline GANs cannot do. This could lead to the model performing well on quantitative metrics by reproducing training samples while failing to produce novel samples [4,12]. The ICAEGAN and CCAE-GAN contrast the CAEGAN as the ICAEGAN only learns from training labels in the same manner as a GAN: indirectly through the lens of the discriminator's feedback.

## 4   Experiments

This section proposes experiments comparing the incremental cyclical models and baselines for generative and supervised learning tasks for a natural image data set. Section 4.1 compares experiments in natural image synthesis quality for the CIFAR-10 data set quantified by Fréchet Inception and Fréchet Joint Distances, two metrics in image synthesis and conditional image synthesis evaluation [4, 10]. We include Sect. 4.3 as a proposed alternative metric for the quality of a trained conditional generator.

Last, we include 4.4 as a proposed direction of research combining generative, supervised, and cyclical models into a united family of distribution-comprehension algorithms which demonstrate adversarial training of natural image synthesis models may improve upon baseline supervised learners.

In order to compare different conditional image synthesis models we emphasize standardized architecture, hyperparameter, training procedures consistent with building blocks from DCGAN [13]. The DCGAN design is marked by the following characteristics:

- Stride convolutions instead of pooling layers.
- Batch normalization between layers in both generator and discriminator.
- ReLU activations in the generator with a tanh output.
- leakyReLu activation in the discriminator.

Each model used the same architectures and hyperparameters: learning rate $\epsilon = 2e - 4$ , Adam optimizer with $\beta_1 = 0.5, \beta_2 = 0.999$, batch size of 16, and 1, 000 training epochs, resulting in 3, 125, 000 training updates per adversary. Each component utilizes five 2-dimensional convolution layers of size 4, stride 2, and padding 1, with ReLU or leakyReLU activations and batchnorm mirroring [13].

There exist numerous large GANs which significantly outperform the DCGAN-based models considered here for the FID metric [1, 11, 14, 15]. The models considered here serve as a study on how the minimal addition of cycles to a DCGAN alter FID and FJD outcomes given three possible cycles without changing the generative architecture. Lastly, models such as BigGAN include substantial amounts of training modifications including but not limited to spectral normalization, self attention modules, hinge losses, skip-z connections, orthogonal regularization, and truncation tricks [1]. Though these changes improve model training stability and FID scores, our goal is to directly compare minimal changes between conditional GANs without these design tweaks.

## 4.1 Quantifying Generative Quality

This section evaluates the generative quality of the proposed and baseline models using the CIFAR-10 [10] data set, a collection of 60, 000 natural images evenly distributed across 10 content classes. A predefined train-test split is used to evaluate supervised learners on the unseen partition as well as compare distributional distance between generated samples and unseen testing examples to evaluate generative quality using the Fréchet Inception and Fréchet Joint Distances.

The Fréchet Inception Distance (FID, [7]) quantifies the quality of a generated distribution with respect to a target distribution by encoding each in a learned representation in the penultimate layer of the Inception-v3 model, a pretrained natural image classifier [16]. The FID is a standard metric for evaluating generative models [4, 8, 17].

The FID does not account for the joint distribution of samples and classes for conditional data to measure class adherence. The Fréchet Joint Distance (FJD, [4]) accounts for joint distributions of images and conditions to express generated sample quality, adherence to the intended class, and distance from other classes. The FJD quantifies the distance between maximum likelihood Gaussian estimation of the conditional distributions in the penultimate layer of Inception-v3.

Table 1 demonstrates the generative quality for baseline and proposed conditional generators. Training and evaluation were performed ten times for each model class. At the conclusion of training, the FID and FJD of the trained model are measured. The variability in the generative quality of these trained models is recorded in the standard error reported for each experiment.

**Table 1.** Fréchet distances across model architectures for CIFAR-10 synthesis. Lower is better.

|                 | FID              | FJD              |
| --------------- | ---------------- | ---------------- |
| CGAN            | $44.37 \pm 0.05$ | $51.97 \pm 0.05$ |
| ACGAN           | $43.91 \pm 0.03$ | $61.59 \pm 0.06$ |
| CAEGAN (OURS)   | $37.67 \pm 0.37$ | $46.42 \pm 0.04$ |
| ICAEGAN (OURS)  | $39.06 \pm 0.02$ | $48.16 \pm 0.02$ |
| CCAEGAN (OURS)  | $\mathbf{35.72} \pm \mathbf{0.02}$ | $\mathbf{43.22} \pm \mathbf{0.03}$ |

It is worth noting that with the exception of the additional cycle the ICAEGAN and CGAN are identical models, as are the CAEGAN and CCAEGAN. There is no increased model capacity or architectural difference beyond the additional loss components introduced by reconstructive cycles. When quantifying the quality of the generated distribution, the ICAEGAN outperforms the CGAN to a substantial degree. By contrasting these models we demonstrate cycles being a substantial benefit to the training of generative models given a fixed decoder capacity.

The discrepancy between the baseline and CAEGAN models cannot be explained only by the autoencoders generating samples based on leaking reconstruction of the training distribution as described in Sect. 3.3 as the ICAEGAN does not perform data-space reconstruction of training samples, and the CCAEGAN has diluted the contribution of the data-space reconstruction compared to the CAEGAN. Rather, generative performance is consistently improved when the learned generative weights are updated in part by the addition of non-adversarial cycle losses.

## 4.2 Data Mining in a Low-Data Regime

Often, data mining application data sets contain few samples. Shortcomings of adversarial training such as modal collapse and divergence are exacerbated when models (typically the discriminator) overfit to the training data due to the reduced problem difficulty.

In this section we investigate how the performances reported in Table 1 change with a substantially smaller data set. One quarter of the training dataset of CIFAR-10 (12, 500 images) are used to evaluate potential collapses or differences more apparent in the model designs for this more challenging image synthesis task (Table 2).

**Table 2.** Small-data Fréchet distances across model architectures. Lower is better.

|         | FID             | FJD             | Difference |
|---------|-----------------|-----------------|------------|
| CGAN    | $79.70 \pm 0.83$ | $90.56 \pm 0.90$ | 10.86      |
| ACGAN   | $\mathbf{71.02 \pm 0.40}$ | $88.58 \pm 0.41$ | 17.56      |
| CAEGAN  | $71.77 \pm 0.67$ | $\mathbf{83.00 \pm 0.68}$ | 11.23      |
| ICAEGAN | $86.04 \pm 1.04$ | $99.01 \pm 1.19$ | 12.93      |
| CCAEGAN | $73.00 \pm 0.80$ | $84.11 \pm 0.85$ | 11.11      |

## 4.3 Augmenting Training Data

The quality of a conditional image synthesizer may be quantified by the downstream performance of a data mining supervised learner trained using synthesized samples. Synthetic data sets corresponding to a higher classification accuracy may indicate higher quality conditional synthesis, particularly for non-natural image tasks without explicit metrics such as the FID and FJD.

Table 3 reports the down-stream test accuracy of a multi-classification neural network according to a swathe of synthetic sample data set proportions. In each cell, ten trials are performed in which a baseline supervised learner is trained using a training set which is the indicated percentage of the CIFAR-10 training set. The remainder of the 50, 000 images are filled in with synthesized labeled samples given by the row name. This means that for the 75% column, 37, 500 CIFAR-10 samples are chosen from a pre-determined shuffle of the training data constant for each experiment, and the image of 12, 500 latent samples are drawn from the corresponding generator and concatenated to the data set.

Results in Table 3 demonstrate how using conditional sample synthesis contributes to model testing performance for a variety of ratios of training data to GAN-augmented data, measuring the trained generator's adherence to semantic content present in CIFAR-10 testing samples as determined by the testing loss of a model trained on the conditional synthesized samples.

**Table 3.** Test accuracy when training data is augmented with different proportions of generated samples.

| Model | 75% Real | 25% Real | 10% Real | 5% Real | 0% Real |
|---|---|---|---|---|---|
| CGAN | 69.89 ± 0.05 | **64.06 ± 0.38** | 34.13 ± 0.16 | 24.91 ± 0.09 | 18.55 ± 0.12 |
| ACGAN | 69.45 ± 0.04 | 44.72 ± 0.20 | 27.93 ± 0.17 | 21.24 ± 0.17 | 16.13 ± 0.15 |
| CAEGAN | 69.45 ± 0.04 | 51.18 ± 0.18 | 34.45 ± 0.19 | 25.89 ± 0.15 | 19.45 ± 0.16 |
| ICAEGAN | 69.44 ± 0.05 | 49.61 ± 0.14 | 32.78 ± 0.17 | 25.42 ± 0.13 | 19.25 ± 0.15 |
| CCAEGAN | **70.00 ± 0.05** | 52.41 ± 0.19 | **36.49 ± 0.21** | **26.46 ± 0.10** | **19.95 ± 0.10** |

## 4.4   Cyclical Models Perform Classification

The final result on the benchmarking contribution of research is the demonstration that cyclical models do not just improve on comparable CGANs for image synthesis, but situationally outperform comparable supervised learners on image classification. One motivation for the study of generative models is improved utilization of training data for robust, efficient models. Efficiency of training data becomes increasingly vital for deep learning success as the number of samples decreases. The encoding component of each of the cycle models CAEGAN, ICAEGAN, and CCAEGAN perform the multiple attention task of encoding samples to the variation and code spaces. To measure supervised learning outcomes each cell of Table 4 indicates ten models trained from scratch on a CIFAR-10 subset.

Instead of training only a predictive model, the three cycle models each use the encoding portions of their architectures to encode the testing samples alongside adversarial generative and reconstructive learning.

**Table 4.** Top-1 test accuracy of classification heads of conditional cycle models. Higher is better.

| Model | 75% of Train | 25% of Train | 10% of Train | 5% of Train |
|---|---|---|---|---|
| Predictor | **69.87 ± 0.23** | **59.12 ± 0.06** | **47.09± 0.07** | 32.60 ± 0.12 |
| CAEGAN | 68.55 ± 0.04 | 56.34 ± 0.08 | 45.78 ± 0.07 | **41.07 ± 0.17** |
| ICAEGAN | 18.60 ± 0.80 | 20.71 ± 0.56 | 15.30 ± 0.89 | 14.05 ± 0.41 |
| CCAEGAN | 65.25 ± 0.25 | 50.86 ± 0.25 | 41.21 ± 0.18 | 34.30 ± 0.13 |

Table 4 demonstrates that the complexity of handling autoencoding, generative, and supervised tasks is a burden for the cyclical models for large volumes of data. Though the CAEGAN and CCAEGAN testing accuracy is in stride with the simple predictor for 75% and 25% of the training set, the results fall off for the CCAEGAN when using 10% (5,000) of training samples. However, a turning point exists between 10% (5,000 samples) and 5% (2,500 samples) of the original set: the regularization imparted upon the encoder by managing the generative, reconstructive, and predictive tripartite loss improves the model's testing accuracy, as both the CAEGAN and CCAEGAN pull ahead of the comparable supervised learner. The ICAEGAN performance remains poor throughout due to the encoder training without direct access to the training data set.

## 5   Conclusions and Future Work

We have demonstrated how the inclusion of cycles improves upon comparable CGANs. We demonstrate that the relationship between conditional image synthesis and supervised learning may benefit both tasks as performance in one area may be used as a metric for the other. These results were consistent across the FID, FJD and proposed augmentation metrics for CIFAR-10. The consistency with which the three highly different cyclical models outperformed the baselines lends itself to the notion that alternative training metrics may regularize against shortcomings of adversarial training by limiting the space of viable parameters.

Generalizations on the theme of cycles for reconstruction may be extended to any model architectures which rely on learning representations of data, and may be beneficial beyond conditional GANs for image synthesis including extensions to autoregressive and supervised learners. Contemporary research including [8] investigates the utilization of augmentations for stabilizing GAN training, particularly for limited-data domains. The study of regularization effects including but not limited to the corresponding quality of synthesized images and model inference during training may be a productive area of research for the stabilization of GAN training. Despite notorious instability in GAN training particularly for low-$n$ tasks, concurrent training of generative and supervised models may be a promising direction for data-efficient multi-task deep learning models.

**Acknowledgements.** This manuscript has been authored with funding provided by the Defense Threat Reduction Agency (DTRA). The publisher acknowledges that the US Government retains a nonexclusive, paid-up, irrevocable, worldwide license to publish or reproduce the published form of this manuscript, or allow others to do so, for US government purposes. Approved for public release. Distribution unlimited.

## References

1. Brock, A., Donahue, J., Simonyan, K.: Large scale gan training for high fidelity natural image synthesis (2018). https://doi.org/10.48550/ARXIV.1809.11096, https://arxiv.org/abs/1809.11096
2. Child, R.: Very deep vaes generalize autoregressive models and can outperform them on images (2021)
3. Dai, B., Lin, D.: Contrastive learning for image captioning. arXiv preprint arXiv:1710.02534 (2017)
4. DeVries, T., Romero, A., Pineda, L., Taylor, G.W., Drozdzal, M.: On the evaluation of conditional gans. arXiv preprint arXiv:1907.08175 (2019)
5. Donahue, J., Krähenbühl, P., Darrell, T.: Adversarial feature learning. arXiv preprint arXiv:1605.09782 (2016)

6. Goodfellow, I.J., et al.: Generative adversarial networks. arXiv preprint arXiv:1406.2661 (2014)
7. Heusel, M., Ramsauer, H., Unterthiner, T., Nessler, B., Hochreiter, S.: Gans trained by a two time-scale update rule converge to a local nash equilibrium. arXiv preprint arXiv:1706.08500 (2017)
8. Karras, T., Aittala, M., Hellsten, J., Laine, S., Lehtinen, J., Aila, T.: Training generative adversarial networks with limited data. arXiv preprint arXiv:2006.06676 (2020)
9. Kingma, D.P., Welling, M.: Auto-encoding variational bayes (2014)
10. Krizhevsky, A., Hinton, G., et al.: Learning multiple layers of features from tiny images (2009)
11. Mirza, M., Osindero, S.: Conditional generative adversarial nets (2014)
12. Odena, A., Olah, C., Shlens, J.: Conditional image synthesis with auxiliary classifier gans (2017)
13. Radford, A., Metz, L., Chintala, S.: Unsupervised representation learning with deep convolutional generative adversarial networks. arXiv preprint arXiv:1511.06434 (2015)
14. Sauer, A., Schwarz, K., Geiger, A.: Stylegan-xl: scaling stylegan to large diverse datasets (2022). https://doi.org/10.48550/ARXIV.2202.00273, https://arxiv.org/abs/2202.00273
15. Suman Ravuri, O.V.: Seeing is not necessarily believing: limitations of biggans for data augmentation. In: International Conference on Learning Representations Workshop 2019 (2019)
16. Szegedy, C., Vanhoucke, V., Ioffe, S., Shlens, J., Wojna, Z.: Rethinking the inception architecture for computer vision (2015)
17. Zhang, H., Zhang, Z., Odena, A., Lee, H.: Consistency regularization for generative adversarial networks. arXiv preprint arXiv:1910.12027 (2019)
18. Zhu, J.Y., Park, T., Isola, P., Efros, A.A.: Unpaired image-to-image translation using cycle-consistent adversarial networks. In: Proceedings of the IEEE International Conference on Computer Vision, pp. 2223–2232 (2017)

# Using the Strongest Adversarial Example to Alleviate Robust Overfitting

Ce Xu$^{(\boxtimes)}$, Xiu Tang, and Peng Lu

Zhejiang University, Hangzhou, China
{xucce,tangxiu}@zju.edu.cn, lupeng@zjuici.com

**Abstract.** Overfitting is considered to be one of the dominant phenomena in machine learning. A recent study suggests that, just like standard training, adversarial training(AT) also suffers from the phenomenon of overfitting, which is named robust overfitting. It also points out that, among all the remedies for overfitting, early stopping seems to be the most effective way to alleviate it. In this paper, we explore the role of data augmentation in reducing robust overfitting. Inspired by MaxUp, we apply data augmentation to AT in a new way. The idea is to generate a set of augmented data and create adversarial examples(AEs) based on them. Then the strongest AE is applied to perform adversarial training. Combined with modern data augmentation techniques, we can simultaneously address the robust overfitting problem and improve the robust accuracy. Compared with previous research, our experiments show promising results on CIFAR-10 and CIFAR-100 datasets with PreactResnet18 model. Under the same condition, for $l_\infty$ attack we boost the best robust accuracy by **1.57%–2.89%** and the final robust accuracy by **7.51%–9.42%**, for $l_2$ attack we improve the best robust accuracy by **1.64%–1.74%** and the final robust accuracy by **3.80%–5.99%**, respectively. Compared to other state-of-the-art models, our model also shows better results under the same experimental conditions. All codes for reproducing the experiments are available at https://github.com/xcfxr/adversarial_training.

**Keywords:** Robust overfitting · Data augmentation · Adversarial training

## 1 Introduction

Despite deep neural models have made an unprecedented progress on a wide range of computer vision tasks, they can be easily fooled by adversarial examples(AEs) [1] , which can be crafted by adding small and invisible perturbation to original images. With such intentional changes called adversarial attack to inputs, many models fail to provide a satisfied performance. To prevent these attacks, a whole lot of defense methods are being proposed. Adversarial training(AT) [2], which creates AEs and then treats them as training sets, is considered as the most efficient approach against adversarial attack.

© The Author(s), under exclusive license to Springer Nature Switzerland AG 2022
W. Chen et al. (Eds.): ADMA 2022, LNAI 13726, pp. 365–378, 2022.
https://doi.org/10.1007/978-3-031-22137-8_27

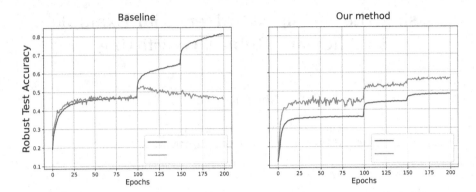

**Fig. 1.** This is the robust test on accuracy between baseline [2] and our method under the $PGD^{10}$ attack. The baseline method suffers serious robust overfitting, but our method doesn't. Further more, our final model surpasses the best checkpoint of the baseline, which can be achieved by early-stopping [3].

However, recently a study [3] finds something intriguing in AT. It observes that just like standard training, AT also suffers from the phenomenon of overfitting (see left picture of Fig. 1). Namely, after several epochs of training, especially after the adjustment of learning rate, the robust test accuracy begins to decrease while robust train accuracy still increases. Various technologies are proposed to address the problem, among which early stopping seems to be the most effective way to alleviate the problem, while other tricks, such as regularization effect of data augmentation, including mixup [4] and cutout [5], seem to be ineffective.

In this paper, we use data augmentation to counter this robust overfitting phenomenon and to achieve better robust accuracy. As shown in the right picture of Fig. 1, throughout the whole training process, the robust test accuracy and the robust train accuracy rises continuously.

Inspired by MaxUp [6], we apply data augmentation to adversarial training process. In our approach, we first generate a set of augmented data and then create adversarial examples(AEs) based on them. The AE which causes the maximal loss is used to perform AT. While MaxUp [6] minimizes the average risk of the worst augmented data, we use the attack method to create AEs and then minimize the average risk of the worst AEs. Our experiments demonstrate that combined with our approach, augmentations including mixup and cutmix can neutralize robust overfitting partially and meanwhile achieve a better prediction result than early stopping scheme. As shown in Fig. 2, compared with the early stopping approach [3], our approach still produces a correct label for perturbed image, while the baseline approach returns a false one.

Our experiments achieve promising results on the CIFAR-10 and CIFAR-100 datasets with the PreactResnet18 model. Under the same condition, for $l_\infty$ attack we boost the best robust accuracy by $1.57\% - 2.89\%$ and the final robust accuracy by $7.51\% - 9.42\%$, for $l_2$ attack we improve the best

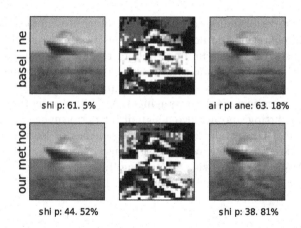

**Fig. 2.** We show the gap between the early stopping method [3] and our method. **Left column**: the original images. **Middle column**: the $l_\infty$ adversarial noises by applying $PGD^{10}$ for 10 iterations. We normalize the noise into $[0, 255]$. **Right column**: the generated adversarial images. We also show the predicted labels and probabilities of these images.

robust accuracy by $1.64\% - 1.74\%$ and the final robust accuracy by $3.80\% - 5.99\%$, respectively. All codes for reproducing the experiments are available at https://github.com/xcfxr/adversarial_training.

## 2    Related Work

Szegedy et al. [1] observe that deep neural models are vulnerable to imperceptible perturbations. With such perturbations, vanilla images become adversarial examples(AEs) which can successfully fool the models. The approaches of generating AE are known as an adversarial attack. Some early approaches adopt the fast gradient sign method(FGSM) [7], which crafts AE with a single gradient step. BIM [8] on the other hand, extends FGSM to iterative small gradient steps. DeepFool [9] declare that changing one pixel is enough to fool the classifier [10]. Among all approaches, projected gradient decent(PGD) [2] is considered as one of the strongest first-order attack. As a result, a lot of PGD-based work was studied, e.g. PGD combined with momentum [11] and logit pairing [12].

To address the problem of adversarial attack, many defense-related work have been proposed. Some defense approaches are not always effective, such as distillation [13] and generator [14,15]. Normally, adversarial training(AT) [2] is considered as the most successful defensive approach. AT has attracted a series of research efforts [10,16,17], among which Trades [17] is a notable work, achieving a trade-off between the efficiency and robust accuracy.

Recently, Rice et al. [3] demonstrate that there is a serious overfitting phenomenon called robust overfitting during AT and there is no effective way as good as early-stopping to tackle it. Another research [18] also points out that despite of their excellent performances on improving robustness in standard training, data-driven augmentations do not improve robustness to $l_p$ norm bounded perturbations. Contrary to their finding, Rebuffi et al. [19] show that data augmentation has the potential to alleviate robust overfitting. In their experiment, although data augmentation alone cannot improve robustness, with smoothing the weights by model weight averaging(WA) [20], augmentation techniques can alleviate overfitting and achieve a significant performance improvement. In another research, Chen et al. [21] show that schemes of learned smoothening is supposed to be a possible way to resist robust overfitting. They smooth the weights by WA and the logits via self-training. In the latest research, Rebuffi et al. [19] incorporate a large number of tricks to achieve the state-of-the-art robust accuracy.

## 3    Preliminaries

### 3.1    Notations

In this section, we use $x \in \mathbb{R}^{W \times H \times C}$ and $y$ to denote a training image and its label, respectively. $(x', y')$ is obtained from $(x, y)$ by data transformation.Let $\mathcal{D}_n$ and $\mathcal{L}$ denote the training dataset with N input-label pairs and the loss function, respectively. The neural network parameterized by $\theta$ is represented as $f_\theta$. So the empirical risk minimization(ERM) can be denoted as $\min_\theta \mathbb{E}_{(x,y) \sim \mathcal{D}_n} [\mathcal{L}(f_\theta(x), y)]$. And we use $\delta$ to represent the perturbation created by adversarial attack and $\delta$ are limited to the range of $S$, where $S$ is chosen to be a $l_p$-norm ball and represents a closed interval $[-\epsilon, \epsilon]$($\epsilon$ defines the maximum perturbation allowed).The letter $m$ is the hyper-parameter of our algorithm, which represents the number of AEs generated for each sample. We denote the accuracy rate on the adversary as "robust accuracy", so the accuracy rate on the training adversary and test adversary are called "robust train accuracy" and "robust test accuracy", respectively.

### 3.2    MaxUp

The key idea of MaxUp is that for each $(x, y)$, Gong et al. [6] generates $m$ samples by applying data transformations, which can be Gaussian Sampling $\mathcal{N}(x, \sigma^2 I)$ or data augmentations. In the next step, among the $m$ data points, they choose the one that maximizes the loss function as a new training sample. The method can be summarized as:

$$MaxUp \quad \arg\min_\theta \mathbb{E}_{(x,y) \sim \mathcal{D}_n} \left[ \max_{i \in [m]} \mathcal{L}(f_\theta(x_i'), y_i') \right] \quad (1)$$

Gong et al. consider MaxUp as a smoothness Regularization. They define

$$
\begin{aligned}
\mathcal{L}_m^{max}(f_\theta(x), y) &= \mathbb{E}\left[ \max_{i \in [m]} \mathcal{L}(f_\theta(x_i)) \right] \\
\mathcal{L}_m^{avg}(f_\theta(x), y) &= \mathbb{E}\left[ \frac{1}{m} \sum_{i=1}^{m} \mathcal{L}(f_\theta(x_i)) \right]
\end{aligned}
\quad (2)
$$

and prove the following equal equation:

$$\mathcal{L}_m^{max}\left(f_\theta\left(x\right),y\right) = L_m^{avg}\left(f_\theta\left(x\right),y\right) + \Phi\left(x,\theta\right) + \mathbf{O}\left(\sigma^2\right)$$
$$c_m^-\|\nabla_x\mathcal{L}\left(f_\theta\left(x\right),y\right)\| \leq \Phi\left(x,\theta\right) \leq c_m^+\|\nabla_x\mathcal{L}\left(f_\theta\left(x\right),y\right)\|, \tag{3}$$

where $c_m^+ \geq c_m^- \geq 0$ and $\sigma^2$ bounds the range of changes in x caused by transformation or data augmentation.

### 3.3 Projected Gradient Descent

Projected Gradient Descent(PGD) [2] is a method for generating reliable first-order adversaries. An $l_p$ PGD adversarial example would start at some random initial perturbation $\delta^{(0)}$, where $\delta^{(0)} \sim \mathcal{U}\left(-\epsilon,\epsilon\right)$ . Then the perturbation will be iteratively adjusted with the following gradient steps while projecting back onto $l_p$ ball with radius $\epsilon$. The whole process can be described as:

$$\tilde{\delta} = \delta^{(t)} + \alpha \cdot \text{sign}\left(\nabla_x\mathcal{L}\left(f_\theta\left(x+\delta\right),y\right)\right)$$
$$\delta^{(t+1)} = \max\left(\min\left(\tilde{\delta},\epsilon\right),-\epsilon\right). \tag{4}$$

Madry et al. [2] treat these newly generated adversaries as datasets and train the robust model to defense adversarial attack, which is known as Adversarial training(AT).

## 4   Methods

### 4.1   Algorithm

Our approach extends MaxUp and AT as follows. For each training example $(x, y)$, we first use augmentation techniques to generate $m$ augmented data $X \in \mathbb{R}^{m \times W \times H \times C1}$. Then we apply PGD-attack to those generated data points. By adding restricted perturbation, we can create $m$ adversarial examples(AEs).

Among the $m$ adversarial examples, we choose the one that generates the maximal loss in our new training dataset. In this way, we can gain new and more complex adversarial examples as our samples. The final step is utilizing the new generated AEs to perform adversarial training. Overall, we propose a new way of applying data augmentations to adversarial training(AT), which can be summed up as:

$$\arg\min_\theta \mathbb{E}_{(x,y)\sim\mathcal{D}_n}\left[\max_{i\in[m]}\mathcal{L}\left(f_\theta\left(x_i+\delta_i\right),y_i\right)\right]. \tag{5}$$

The complete process of our method is described in Algorithm 1. The first six lines are about the input and output. Lines 10 to 15 specifically describe how we create AEs and apply data augmentation into AT at the formula level.

---

[1] In this paper, we adopt mixup [4] and cutmix [22] as our data augment approaches.

**Algorithm 1.** Using Data Augmentation in Adversarial Training

1: **Input:**
2:      training one training sample: $(x, y)$
3:      augmentation techniques: $aug\,(x, y)$
4:      attack method: $attack\,(f_\theta\,(x)\,, y)$
5: **Output:**
6:      new training sample $(X', Y')$
7: $i \leftarrow 0, \quad loss \leftarrow 0$
8: **while** $i \neq m$ **do**
9:      $X, Y = aug\,(x, y)$
10:      $\delta = attack\,(f_\theta\,(X)\,, Y)$
11:      **if** $\mathcal{L}\,(f_\theta\,(X + \delta)\,, Y) > loss$ **then**
12:          $X' \leftarrow X + \delta$
13:          $Y' \leftarrow Y$
14:          $loss \leftarrow \mathcal{L}\,(f_\theta\,(X + \delta)\,, Y)$
15:      **end if**
16:      $i \leftarrow i + 1$
17: **end while**
18: **return** $X', Y'$

## 4.2  Analysis

When $m$ is equal to 1, our algorithm degenerates into ordinary adversarial training. When $m$ become larger, more powerful and more sophisticated adversarial examples(AEs) can be crafted, so that both in quantity and intensity our AEs is more dominant. A trade-off of the approach is that, although parallel computing is possible, from the perspective of computing resources, resources consumed and memory occupied per epoch will increase linearly as $m$ increases. A larger $m$, on the other hand, can result in a faster convergence.

Here is a plausible explanation of why our approach works. With the conclusion of MaxUp(3.2) and our methods, the empirical risk in the AT turns into

$$\mathbb{E}_{(x,y)\sim\mathcal{D}_n}\left[\frac{1}{m}\sum_{i=1}^{m}\mathcal{L}_m^{avg}\,(f_\theta\,(x_i + \delta_i)\,, y_i)\right.$$

$$\left. c_m\|\nabla_{x_i}\mathcal{L}\,(f_\theta\,(x_i + \delta_i)\,, y_i)\,\| + \mathbf{O}\,(\sigma^2)\right], \tag{6}$$

where $c_m^+ \geq c_m \geq c_m^- \geq 0$. So when we perform AT with the worst AE that costs the maximal loss, the loss function has become a combination of a loss term which measures how well the model fits the AE, a regularization term related to the norm of $\nabla_x$ and a high-order infinitesimal term which can be ignored. The expectation of the loss term equals the expectation with normal AT, so our algorithm essentially adds a penalty which restricts the magnitude of $\nabla_x$ in the process of AT. Because of the regularization term, the first step of generating AEs(3.3) has also changed. It turns into

$$\tilde{\delta} = \delta^{(t)} + \alpha \cdot \text{sign} \left( \nabla_{x_i} \left[ \mathcal{L} \left( f_\theta \left( x_i + \delta_i \right), y_i \right) + \Phi(x_i, \theta) \right] \right), \tag{7}$$

where $\Phi(x_i, \theta)$ equals $c_m \| \nabla_{x_i} \mathcal{L} \|$. It is clear that the sign of the gradient may change due to the extra term $\Phi(x, \theta)$, so our methods can affect the process of making the AEs to some extent.

In summary, our approach is a crafty combination of adversarial training and data augmentation, by making sophisticated AEs in parallel to train a more robust model.

# 5   Experiments

## 5.1   Experimental Settings

For a complete experimental comparison, most of our experimental setups follow the original study [3], including the weight decay, the learning schedule and epochs of training, etc.

### 5.1.1   Datasets and Architecture

Our experiments are conducted across two datasets: CIFAR-10, CIFAR-100 [23]. The CIFAR-10 dataset consists of 60000 $32 \times 32$ colour images in 10 classes, with 6000 images per class. There are 50000 training images and 10000 test images. The CIFAR-100 dataset is just like the CIFAR-10, except it has 100 classes containing 600 images each. There are 500 training images and 100 testing images per class. And most of the experiments are implemented on CIFAR-10. In order to observe the whole process of the robust accuracy change and pick the checkpoint of the best performance, after each training epoch, we output the robust loss and robust accuracy on the test set. Because of hardware and time costs during training, all of our experiments are based on ResNet-18 [24].

### 5.1.2   Attack Methods

During the adversarial training, we use $PGD^{10}$ with random initialization and the step size of attack is 2/255. We consider two mainstream types of adversarial perturbation $l_\infty$ and $l_2$, and the norm of them are 8/255 and 128/255 respectively. For evaluation, we keep the same settings as training.

### 5.1.3   Other Setup

We use a fairly common learning schedule: for 200 epochs, the learning rate begins with the rate of 0.1 and decays by a factor of 10 at the 100th and 150th. We also adopt the SGD optimizer in a common way, with a momentum of 0.9 and weight decay of $5 \times 10^{-4}$. For all datasets, we set batch size as 128 for PreActResNet-18. When applying augmentation methods like cutmix [22] and mixup [4] in our method, we default the $\alpha$ to 1.

## 5.2   Experimental Results

### 5.2.1   Across Datasets and Perturbations

Table 1 shows the improvement brought by our methods across different datasets and perturbations. We report robust test accuracy(RA) at two periods to numerically demonstrate the phenomenon. The final-RA indicates the average robust accuracy of last five epochs, the best-RA indicates the best robust accuracy in the whole process of training and the diff-RA equals the best-RA minus final-RA, which can measure the degree of robust overfitting. We consider adversarial training(AT) [2] as baseline. The overfitting shows up across all datasets and perturbations in baseline cases, with the gap between final and best reaching as large as 7.05%(CIFAR-10). Compared with $l_\infty$ perturbation, the overfitting of $l_2$ are much less serious, espeicially in the CIFAR-10, the diff-RA is only 2.72%. We use the code provided by Rice et al. [3] to reproduce the baseline results[2].

**Table 1.** Robust test accuracy under attack with $PGD^{10}$ against $l_\infty$ with radius $\epsilon = 8/255$ across CIFAR-10 and CIFAR-100. Both experiments are based on ResNet-18. The Final equals the average robust test accuracy of last five epochs and the Best is the checkpoint with best robustness during the whole training process. The best results and the smallest difference between best and final are marked in bold.

| Dataset | Norm | Radius | Settings | Robust test accuracy (%) | | |
|---------|------|--------|----------|-------|------|------|
| | | | | Final | Best | Diff. |
| CIFAR-10 | $l_2$ | $\epsilon = \frac{128}{255}$ | baseline | $68.90 \pm 0.68$ | 71.62 | 2.72 |
| | | | Our Methods | $\mathbf{72.70 \pm 0.36}$ | **73.36** | 0.66 |
| CIFAR-10 | $l_\infty$ | $\epsilon = \frac{8}{255}$ | baseline | $46.23 \pm 0.65$ | 53.28 | 7.05 |
| | | | Our Methods | $\mathbf{55.65 \pm 0.34}$ | **56.17** | **0.52** |
| CIFAR-100 | $l_2$ | $\epsilon = \frac{128}{255}$ | baseline | $37.50 \pm 0.12$ | 43.15 | 5.65 |
| | | | Our Methods | $\mathbf{43.49 \pm 0.38}$ | **44.79** | 1.30 |
| CIFAR-100 | $l_\infty$ | $\epsilon = \frac{8}{255}$ | baseline | $21.43 \pm 0.44$ | 28.15 | 6.72 |
| | | | Our Methods | $\mathbf{28.94 \pm 0.43}$ | **29.72** | **0.78** |

With our method, both the final-RA and the best-RA are boosted a lot. For $l_\infty$ attack, we observe the best RA is pushed higher by 1.57%–2.89%. For example, the best robust accuracy on CIFAR-10 rises from 53.28% to 56.17%. Further, the difference between best-RA and final-RA is reduced to only 0.52%(CIFAR-10) and 0.78%(CIFAR-100) respectively, where the overfitting problem is almost solved. Unlike baseline cases whose best robust accuracy is nearly the first decay of learning rate, the checkpoint which has the best-RA in ours is close to the end, which also means robust overfitting phenomenon is mitigated. And for $l_2$ attack, the final-RA was boosted from 68.90% to 72.70% on CIFAR-10 and from 37.50% to 43.49% on CIFAR-100 respectively.

---

[2] https://github.com/locuslab/robust_overfitting.

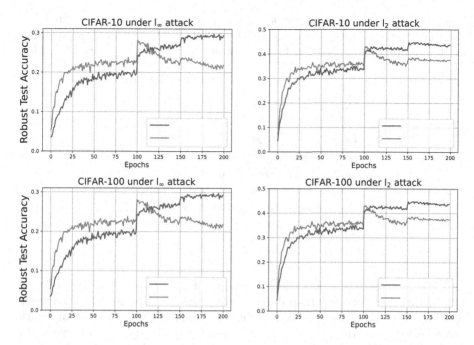

**Fig. 3.** Results of robust test accuracy under attack of $l_2$ and $l\infty$ over epochs for ResNet-18 trained on CIFAR-10, CIFAR-100. Blue/Yellow lines show the our best model and baseline. (Color figure online)

Figure 3 further plots the robust test accuracy curves during training, from which we can clearly observe the diminishing of robust overfitting. The training curve is robustly improved until the end without compromising training accuracy.

### 5.2.2 With Different Augmentation Techniques

Table 2 demonstrates the effectiveness of our methods among various regularization. The baseline is still adversarial training. In Rice et al.'s [3] experiments, data augmentation mitigates overfitting to some degree at the expense of losing accuracy, and early-stopping seems to be the best way to fight against robust overfitting. As Rebuffi et al. [19] point out, cutmix have powerful ability to increase the robustness of model. And in our experiment, mixup [4] mitigates the overfitting but loses a lot of accuracy. Cutmix [22] achieves better result than early stopping under the same conditions and helps to alleviate overfitting well. Combined with our method, both cutmix and mixup acquire significant power. When $m$ is set to 4, cutmix push the final accuracy of normal AT from 46.23% to 55.65% and best accuracy from 53.28% to 56.17%, which also surpass early stopping and normal cutmix a lot. And with $m$ setting to 3, mixup can also be better than normal cutmix, let alone early stopping.

**Table 2.** Robust test accuracy with different regularization methods on CIFAR-10 based on PreActResNet18 under the attack of $PGD^{10}$. The perturbation type is $l_\infty$ and the radius of it is 8/255. The Final equals the average robust test accuracy of last five epoch and the Best is the checkpoint during the whole training period. The best results and the smallest difference between best and final are marked in bold.

| Method | Robust test accuracy (%) | | |
|---|---|---|---|
| | Final | Best | Diff. |
| Baseline | $46.23 \pm 0.65$ | 53.28 | 7.69 |
| Early stopping | 53.10 | 53.30 | **0.20** |
| Mixup($\alpha = 1$) | $49.49 \pm 0.70$ | 51.14 | 2.76 |
| Cutmix($\alpha = 1$) | $53.53 \pm 1.33$ | 55.00 | 1.47 |
| Our method[a] | $\mathbf{55.65 \pm 0.34}$ | **56.17** | 0.52 |
| Our method[b] | $54.10 \pm 0.53$ | 55.73 | 1.63 |

[a] use cutmix and set $m = 4$
[b] use mixup and set $m = 3$

Figure 4 shows the whole training process of robust accuracy with different regularizations. Though mixup doesn't achieve as good results as other methods, all data augmentations seem to help to resist the phenomenon of robust overfitting because the robust accuracy doesn't decrease significantly except for baseline. And we can also observe, without our methods, the fluctuation of robust accuracy curve is large, which indicates that our schemes can make the training process more smooth and stable. Near the end of training, the vibration of our accuracy curves(green and red curves) is much smaller than others.

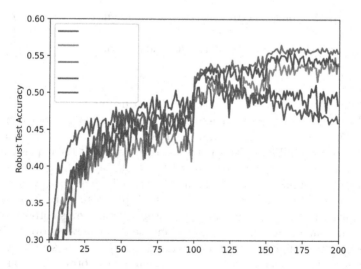

**Fig. 4.** Robust test accuracy against $\epsilon_\infty = 8/255$ on CIFAR-10 with different data augmentation schemes(method[1] and method[2] correspond to Table 3). The model is a ResNet-18 and the panel show the evolution of robust accuracy as training progresses(against $PGD^{10}$). The jump in robust accuracy half and two-thirds through training is due to a drop in learning rate.

### 5.2.3   Ablation Study of $m$

We test our methods with different sizes of $m$ and observe their performance against $l_\infty$ attack. The experiments are based on ResNet-18 [24] and incorporate data augmentation including mixup [4] and cutmix [22]. We vary the size in 1, 2, 3, 4. Note that when $m = 1$, our methods degenerate into normal adversarial training with cutmix or mixup. Against adversarial attack, naive cutmix get stronger performance than mixup, this rule stays the same when $m$ increase. As shown in Table 3, in our experiments, cutmix gets best result when $m$ equals 4, achieving the 56.17% best-RA and 55.65% final-RA respectively. When using mixup, we get similar excellent results when $m \in [2, 4]$. The difference is the smallest when $m$ is equal to 2 for both cutmix and mixup. But when $m$ continues to be larger, the performance begins to degrade, especially the model trained with mixup. Combined with the conclusions in MaxUp [6], we consider the reason is that ResNet-18 is not complex enough.

Figure 5 demonstrates the whole training process when $m$ varies. With our schemes, we can see both cutmix and mixup achieved great improvement. Especially when the learning rate drops for the second time, our accuracy curves continue to rise while normal methods have not changed.

### 5.3   Other Attempts

Drawing on the method of Rebuffi et al. we used the synthetic dataset [18] provided by them and the model weighted average method [20], which improved the robust accuracy by 4.66% to 61.18%. We control the ratio of synthetic data to original data to be 7:3, which is the same as Rebuffi et al. [19]. Here we compare our model with other state-of-the-art methods under the same experimental conditions. We directly used the pretrained model[3] provided by the authors.

**Table 3.** Ablation studies on CIFAR-10 with ResNet-18 when $m$ varies. The attack type is $l_\infty$ with radius 8/255. Experiments are performed with cutmix and mixup respectively and they yield the best robust test accuracy observed during training, the final robust test error averaged over the last five epochs, and the differences between them.

| $m$ | Robust test accuracy with cutmix | | | Robust test accuracy with mixup | | |
|---|---|---|---|---|---|---|
| | Final | Best | Diff. | Final | Best | Diff. |
| 1 | $53.53 \pm 1.33$ | 55.00 | 1.47 | $49.49 \pm 0.70$ | 51.14 | 1.65 |
| 2 | $55.70 \pm 0.24$ | 55.70 | **0.53** | $54.05 \pm 0.66$ | 55.12 | **1.07** |
| 3 | $55.15 \pm 0.50$ | 56.10 | 0.95 | $\mathbf{54.10 \pm 0.53}$ | **55.73** | 1.63 |
| 4 | $\mathbf{55.65 \pm 0.34}$ | **56.52** | 0.87 | $54.01 \pm 0.46$ | 55.55 | 1.54 |

---

[3] https://github.com/deepmind/deepmind-research/tree/master/adversarial_robustness.

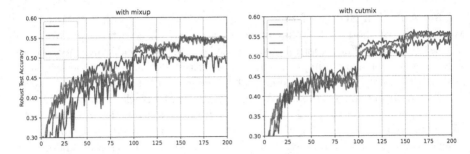

**Fig. 5.** Results of robust test accuracy over epochs with the changing of $m$. All experiments are training on CIFAR-10 with PreActResNet18 against $\epsilon_\infty = 8/255$. The left and right pictures represent the performance of our methods, combined with mixup and cutmix respectively.

As shown in Table 4, under the same conditions of Res-18 and PGD-10, our model outperforms the second model on attacked data and get a similar result on clean data. The third method gets the best results because they use all the 100M synthetic data.

**Table 4.** Robust test accuracy on CIFAR-10 based on PreActResNet18 under the attack of $PGD^{10}$. The perturbation type is $l_\infty$ and the radius of it is 8 /255.We use augmentation to denote the Rebuffi et al.'s model [19] and generation to denote Gowal's model [18].

| Model | Standard test accuracy (%) | Robust test accuracy (%) |
|---|---|---|
| Ours | 83.36 | 61.18 |
| Augmentation | 83.54 | 60.43 |
| Generation | 87.61 | 62.04 |

We have also tried other methods to prevent robust-overfitting in our experiments. Like MaxUp [6] we replace the augmentation step with $\mathcal{N}\left(X, \sigma^2 I\right)$, our attempt in this experiment doesn't get desired result, it does prevent the overfitting but loses a lot of accuracy. And inspired by the procedure of SoftPatchup [25], we change the mix step of cutmix [22]. In original cutmix, $\tilde{x} = M \odot x_A + (1 - M) \odot x_B, \quad \tilde{y} = \lambda y_A + (1 - \lambda)y_B$, we try the transformation of $\tilde{x} = M \odot x_A + (1 - M) \odot x_A * \lambda_2 + (1 - \lambda_2) * (1 - M) \odot X_B, \quad \tilde{y} = \lambda_1 y_A + (1 - \lambda_1)(\lambda_2 y_A + (1 - \lambda_2)y_B)$, where $1 - \lambda_1$ denotes the portion of the cut area and $\lambda_2$ is sampled from the uniform distribution$(0, 1)$. This attempt prevents the overfitting and is better than baseline but the gap with the best result of ours is not small. Because data augmentations help to create complex adversaries, we try to use $PGD^{20}$ to make adversarial examples during training and $PGD^{10}$ to test robustness during testing, but it doesn't work and still suffers overfitting.

# 6 Conclusion

This paper proves that, combining with modern data augmentation techniques, we can improve the robustness of models and alleviate robust overfitting. Compared with Rebuffi et al. we propose a stronger augmentation technique and explore the ability of it. By making more sophisticated and more strong adversarial examples, our methods seem to overcome the classifier's weakness of robust overfitting and get a more promising result. Although it seems to work well, the reason of robust overfitting is still hard to explain. Our future work will delve into the causes of robust overfitting and try to give some reasonable explanation. We will also explore other useful tricks that can improve the robustness of deep neural models to get a more powerful and strong robust model.

# References

1. Szegedy, C., et al.: Intriguing properties of neural networks. In: Bengio, Y., LeCun, Y. (eds.) 2nd International Conference on Learning Representations, ICLR 2014 (2014)
2. Madry, A., Makelov, A., Schmidt, L., Tsipras, D., Vladu, A.: Towards deep learning models resistant to adversarial attacks. In: 6th International Conference on Learning Representations, ICLR 2018 (2018)
3. Rice, L., Wong, E., Kolter, J.Z.: Overfitting in adversarially robust deep learning. In: Proceedings of the 37th International Conference on Machine Learning, ICML 2020. Proceedings of Machine Learning Research, pp. 8093–8104 (2020)
4. Zhang, H., Cissé, M., Dauphin, Y.N., Lopez-Paz, D.: mixup: beyond empirical risk minimization. In: 6th International Conference on Learning Representations, ICLR 2018 (2018)
5. Devries, T., Taylor, G.W.: Improved regularization of convolutional neural networks with cutout. CoRR (2017)
6. Gong, C., Ren, T., Ye, M., Liu, Q.: Maxup: a simple way to improve generalization of neural network training. CoRR (2020)
7. Goodfellow, I.J., Shlens, J., Szegedy, C.: Explaining and harnessing adversarial examples. In: Bengio, Y., LeCun, Y. (eds.) 3rd International Conference on Learning Representations, ICLR 2015 (2015)
8. Kurakin, A., Goodfellow, I.J., Bengio, S.: Adversarial examples in the physical world. In: 5th International Conference on Learning Representations, ICLR 2017 (2017)
9. Moosavi-Dezfooli, S., Fawzi, A., Frossard, P.: Deepfool: a simple and accurate method to fool deep neural networks. In: 2016 IEEE Conference on Computer Vision and Pattern Recognition, CVPR 2016, pp. 2574–2582 (2016)
10. Su, J., Vargas, D.V., Sakurai, K.: One pixel attack for fooling deep neural networks. IEEE Trans. Evol. Comput. **23**(5), 828–841 (2019)
11. Dong, Y., Liao, F., Pang, T., Su, H., Zhu, J., Hu, X., Li, J.: Boosting adversarial attacks with momentum. In: 2018 IEEE Conference on Computer Vision and Pattern Recognition, CVPR 2018, pp. 9185–9193 (2018)
12. Mosbach, M., Andriushchenko, M., Trost, T.A., Hein, M., Klakow, D.: Logit pairing methods can fool gradient-based attacks. CoRR (2018)

13. Papernot, N., McDaniel, P.D., Wu, X., Jha, S., Swami, A.: Distillation as a defense to adversarial perturbations against deep neural networks. In: IEEE Symposium on Security and Privacy, SP 2016, pp. 582–597 (2016)
14. Samangouei, P., Kabkab, M., Chellappa, R.: Defense-gan: protecting classifiers against adversarial attacks using generative models. In: 6th International Conference on Learning Representations, ICLR 2018 (2018)
15. Lin, W., Balaji, Y., Samangouei, P., Chellappa, R.: Invert and defend: model-based approximate inversion of generative adversarial networks for secure inference. CoRR (2019)
16. Wong, E., Rice, L., Kolter, J.Z.: Fast is better than free: revisiting adversarial training. In: 8th International Conference on Learning Representations, ICLR 2020 (2020)
17. Zhang, H., Yu, Y., Jiao, J., Xing, E.P., Ghaoui, L.E., Jordan, M.I.: Theoretically principled trade-off between robustness and accuracy. In: Chaudhuri, K., Salakhutdinov, R. (eds.) Proceedings of the 36th International Conference on Machine Learning, ICML 2019. Proceedings of Machine Learning Research, vol. 97, pp. 7472–7482 (2019)
18. Gowal, S., Rebuffi, S., Wiles, O., Stimberg, F., Calian, D.A., Mann, T.A.: Improving robustness using generated data. In: Ranzato, M., Beygelzimer, A., Dauphin, Y.N., Liang, P., Vaughan, J.W. (eds.) Advances in Neural Information Processing Systems 34: Annual Conference on Neural Information Processing Systems 2021, NeurIPS 2021, pp. 4218–4233 (2021)
19. Rebuffi, S., Gowal, S., Calian, D.A., Stimberg, F., Wiles, O., Mann, T.A.: Data augmentation can improve robustness. In: Ranzato, M., Beygelzimer, A., Dauphin, Y.N., Liang, P., Vaughan, J.W. (eds.) Advances in Neural Information Processing Systems 34: Annual Conference on Neural Information Processing Systems 2021, NeurIPS 2021, pp. 29935–29948 (2021)
20. Izmailov, P., Podoprikhin, D., Garipov, T., Vetrov, D.P., Wilson, A.G.: Averaging weights leads to wider optima and better generalization. In: Globerson, A., Silva, R. (eds.) Proceedings of the Thirty-Fourth Conference on Uncertainty in Artificial Intelligence, UAI 2018, pp. 876–885 (2018)
21. Chen, T., Zhang, Z., Liu, S., Chang, S., Wang, Z.: Robust overfitting may be mitigated by properly learned smoothening. In: 9th International Conference on Learning Representations, ICLR 2021 (2021)
22. Yun, S., Han, D., Chun, S., Oh, S.J., Yoo, Y., Choe, J.: Cutmix: regularization strategy to train strong classifiers with localizable features. In: 2019 IEEE/CVF International Conference on Computer Vision, ICCV 2019, pp. 6022–6031 (2019)
23. Krizhevsky, A., Hinton, G., et al.: Learning multiple layers of features from tiny images (2009)
24. He, K., Zhang, X., Ren, S., Sun, J.: Deep residual learning for image recognition. In: 2016 IEEE Conference on Computer Vision and Pattern Recognition, CVPR 2016, pp. 770–778 (2016)
25. Faramarzi, M., Amini, M., Badrinaaraayanan, A., Verma, V., Chandar, S.: Patchup: a regularization technique for convolutional neural networks. CoRR (2020)

# Deduplication Over Heterogeneous Attribute Types (D-HAT)

Loujain Liekah[1]([✉]) [ID] and George Papadakis[2] [ID]

[1] University of Claude Bernard Lyon 1, Villeurbanne, France
loujain.liekah5@gmail.com
[2] National and Kapodistrian University of Athens, Athens, Greece
gpapadis@di.uoa.gr

**Abstract.** Deduplication is the task of recognizing multiple representations of the same real-world object. The majority of existing solutions focuses on textual data, this means that data sets containing boolean and numerical attribute types are rarely considered in the literature, while the problem of missing values is inadequately covered. Supervised solutions cannot be applied without an adequate number of labelled examples, but training data for deduplication can only be obtained through time-costly processes. In high dimensional data sets, feature engineering is also required to avoid the risk of overfitting. To address these challenges, we go beyond existing works through D-HAT, a clustering-based pipeline that is inherently capable of handling high dimensional, sparse and heterogeneous attribute types. At its core lies: (i) a novel matching function that effectively summarizes multiple matching signals, and (ii) *MutMax*, a greedy clustering algorithm that designates as duplicates the pairs with a mutually maximum matching score. We evaluate D-HAT on five established, real-world benchmark data sets, demonstrating that our approach outperforms the state-of-the-art supervised and unsupervised deduplication algorithms to a significant extent.

**Keywords:** Clustering · Entity matching · Data quality

## 1 Introduction

Integrating overlapping and complementary data sets is a common process that creates new and valuable knowledge [3]. The main task of integration is to identify *duplicate records*, which represent the same real-world entity, such as products, institutes, or patients. This task is called *deduplication* [8], entity matching [13], entity resolution [19] or record linkage [10]. It constitutes a crucial task that improves the data quality by repairing and curating data sources [9], reducing the storage size, and preparing data for downstream applications [8].

Existing solutions for deduplication are based on calculating pairwise similarity scores from one or more attributes [6]. The *unsupervised* methods create

This project has received funding from the European Union's Horizon 2020 research and innovation program under grant agreement No 875171.

a similarity graph, where the nodes correspond to records and the edges are weighted by the matching scores of the adjacent nodes [12]. The graph is then partitioned into clusters such that all nodes within each cluster correspond to duplicate records. These approaches typically calculate matching scores by treating all attributes as textual data [6]. However, real-world data sets involve heterogeneous attribute types, i.e., numerical, categorical and boolean attributes. Casting these types as strings disregards important information and possibly leads to inaccurate matching scores. For example, the prices "14" and "14.00" are identical as numbers, but partially similar when compared as sequences of characters and totally dissimilar when treated as tokens. Hence, unsupervised techniques need to correctly model and support heterogeneous attribute types.

On the other hand, *supervised* methods typically model deduplication as a binary classification task [13]. They convert each pair of records into a feature vector by applying similarity metrics on different attributes. The vectors are then labelled to train a classifier that predicts the matching status for unlabelled pairs. However, these approaches face multiple challenges: (i) The curse of dimensionality, i.e., tasks become exceedingly difficult with a higher number of dimensions. (ii) Labeled data is scarce, but obtaining it through crowd-sourcing is costly and time-consuming [22]. Moreover, its size and quality affects the end result to a significant extent [17], but are hard to ensure, due to the heavy class imbalance. (iii) Supervised methods require long training times [17].

To address these shortcomings, we introduce D-HAT (Deduplication with Heterogeneous Attribute Types), a novel clustering-based pipeline for end-to-end deduplication. D-HAT goes beyond existing works in three ways: (i) It inherently supports data sets with heterogeneous types of attributes and a large portion of missing values (i.e., high sparsity). (ii) It inherently supports and leverages complex schemata of high dimensionality. (iii) It achieves state-of-the-art results without requiring any labelled data. Our contributions are the following:

- We propose D-HAT, an automated end-to-end, clustering-based framework for deduplicating high-dimensional data sets with heterogeneous attribute types and missing values. Its matching algorithm uses as features a comprehensive set of signals, coupling them with a novel greedy clustering method that defines as matches the records with mutually maximum matching scores.
- We conduct experiments on established real benchmark data sets, showing that: (i) In terms of effectiveness, D-HAT outperforms the state-of-the-art supervised and unsupervised baseline methods. (ii) In terms of time efficiency, D-HAT has an undeniable advantage over the baseline methods.
- We have publicly released all data and code used in our experiments through https://github.com/Loujainl/D-HAT.

## 2   Related Work

The growing research on deduplication reflects its increasing importance, with numerous methods tackling various aspects [4,6,8].

One of deduplication's main challenges is its quadratic complexity: in the worst case, it examines all possible pairs of records. *Blocking* is typically used to alleviate this complexity and to scale deduplication to voluminous data sets [5,19]. Blocking puts together similar records in groups called blocks by applying blocking schemes or functions. A blocking function extracts signatures from every record, dividing the input data set into a set of overlapping blocks – comparisons are reduced to *candidates*, i.e., pairs of records sharing at least one block, reducing the computational cost to a significant extent. Yet, the higher time efficiency comes with the risk of missing potential matches [20].

After blocking, *matching* is performed to determine the degree of similarity between the candidate pairs of records. In essence, it applies similarity functions to the values of selected attributes of the candidate records, obtaining numerical matching scores. Next, it determines whether the resulting degree of similarity is sufficient for designating two records as duplicates. We distinguish the matching algorithms into unsupervised and supervised ones.

The former category includes a collection of methods that are provided by JedAI [18,21] and Stringer [12], with *ZeroER* [24] constituting the state-of-the-art unsupervised approach; it represents every candidate pair as a feature vector. Unlike supervised methods, it does not require a training set. Instead, at its core lies the observation that the distribution of the feature vectors for duplicate records differs from that of the non-matching records. Based on this idea, it learns the parameters of the Gaussian distribution of matching vectors by iteratively applying expectation maximization to compute the posterior probability of a matching label given the feature vector. A posterior probability higher than 0.5 is considered as an indication of duplicate records.

Among the supervised methods, the most popular one is *Magellan* [13], a system that combines a variety of features with the main machine learning classifiers, such as decision trees, logistic regression and support vector machines. After providing an annotated sample of candidate pairs $T$, matching is performed by training a classifier over $T$. Magellan also offers a set of blocking methods.

DeepMatcher [17] is a space of matching solutions based on neural networks with three modules: i) attribute embedding, ii) attribute similarity representation, and iii) a classification module. In most cases, the first module relies on pre-trained fastText embeddings [1] to convert every token to a vector. EMTransformer [2] and DITTO [15] go beyond DeepMatcher by leveraging attention-based transformers like BERT [7], and RoBERTa [16]. These solutions perform well on textual data, outperforming Magellan in terms of accuracy [2,15,17]. We disregard them, as they require large training sets and many hours of training [17] in order to fine-tune hundreds of thousands of parameters [23].

## 3  Preliminaries

A *data set* $T$ is a collection of records. A *record* is an object description denoted by $r_i$, where $i$ is a unique identifier. Records are defined by their *attributes*. The set of attributes in $T$ is denoted by $T.A$, while the value of a specific attribute $a$ in record $r_i$ is symbolized as $r_i.a$; $r_i.a = N/A$ indicates that $r_i$ lacks a value for $a$, i.e., there is a missing or a null value. Two records, $r_i$ and $r_j$, that describe the same real-world object are *matching*, i.e., *duplicates*, a situation denoted by $r_i \equiv r_j$. A data set is called *clean* if it does not contain any duplicates.

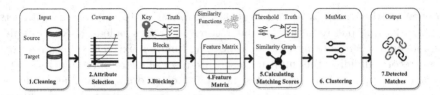

**Fig. 1.** The end-to-end pipeline of D-HAT.

*Deduplication* is the task of identifying and linking duplicate records. A characteristic of this task is that the number of duplicate records scales linearly with the size of the input, unlike its computational cost, which increases quadratically [11]. As a result, Deduplication constitutes a heavily imbalanced task and its effectiveness is measured with respect to the following measures:

1. *Recall*, the portion of existing duplicates that are detected, i.e., $Re = \frac{TP}{TP+FN}$.
2. *Precision*, the portion of record pairs characterized as duplicates that are indeed matching, i.e., $Pr = \frac{TP}{TP+FP}$.
3. *F-Measure*, the harmonic mean of Recall and Precision, $F1 = 2 \times \frac{Pr \times Re}{Pr+Re}$,

where TP stands for the true positive pairs, FP for the false positive ones, and FN for the false negative ones.

In this context, Deduplication can be formally defined as follows:

*Problem 1 (Deduplication).* Given a data set $T$, detect the set of duplicate pairs of records, $D = \{r_i, r_j \in T : i \neq j \wedge r_i \equiv r_j\}$, such that Recall, Precision and F-Measure are maximized.

## 4   Our Approach

We now delve into our framework, whose pipeline is illustrated in Fig. 1.

**Step 1: Data Cleaning.** The first step prepares the input by determining the core characteristics of the attributes describing the given data set(s)[1], i.e., it calculates the number of unique values and the data type per attribute. Attributes that have two unique values are converted to boolean to obtain a more precise degree of similarity. Attributes with very few unique values (<10) are treated as categorical variables. Numerical attributes are identified through regular expressions that detect quantities, possibly accompanied by an optional unit of measurement. E.g., an attribute value width = ''42.8 in'' is transformed into width = 42.8 and is marked as a numeric data type. Min-max normalization is then performed on the values of numeric attributes:

**Step 2: Attribute Selection.** Attributes with a majority of missing values lack valuable information for deduplication and, thus, can be disregarded. The *coverage of an attribute* $a$ expresses the portion of non-empty values in $a$ across

---

[1] In the case of Record Linkage, we assume aligned schemata.

all input records; the fewer missing values there are, the higher is the coverage. We formally define the coverage $c$ of each attribute as: $c(a) = 1 - \frac{|r_i.a=N/A:r_i \in T|}{|T|}$.

This step discards the attributes with a coverage below a specific threshold. Preliminary experiments demonstrated that 0.1 constitutes an effective value.

**Step 3: Blocking.** This step is critical because it determines two things:

1. Time efficiency, because the processing time of the following steps is determined by the number of candidates in the resulting blocks.
2. Effectiveness, because the recall of D-HAT is bounded by the recall of blocking; the false negative pairs of records, which have no block in common, cannot be detected by the subsequent steps, and are excluded from the final output.

Therefore, it is crucial that blocking balances these two competing goals: the reduced search space and the high effectiveness. D-HAT is generic enough to accommodate any blocking method that meets this requirement. Preliminary experiments indicated that Magellan's [13] *overlap blocker* is a robust approach for creating blocks of high performance (see Sect. 5 for more details). It defines as candidate pairs those sharing at least one token in the values of a specific attribute. D-HAT applies the overlap blocker to all textual attributes in the given data sets and opts for the one minimizing the number of candidates, while maximizing coverage – high coverage implicitly signals high recall after blocking.

**Step 4: Feature Matrix.** Similar to supervised approaches, D-HAT represents each pair of records as a feature vector by applying type-specific normalized similarity functions to selected attributes. Unlike supervised approaches, these vectors are unlabelled. In more detail, after detecting the type of every attribute in Step 1, D-HAT creates a feature vector $V_{i,j}$ for each candidate pair of records $(r_i, r_j) \in B$, where $B$ is the set of blocks produced by the previous step and the $k^{th}$ feature/dimension in $V_{i,j}$, $V_{i,j}^k$, stems from a similarity function that is compatible with the type of the $k^{th}$ attribute, $a_k$. If the value of either record for $a_k$ is empty or incorrect (i.e., incompatible with the type of $a_k$), $V_{i,j}^k = $'N/A', which stands for a missing feature. Note that this step does not require any domain knowledge from the user. D-HAT automatically detects the attribute type and applies the appropriate similarity functions in order to create the features.

In particular, the following functions are used by D-HAT:

- For boolean and categorical attributes, the equality operator.
- For numerical attributes, four similarity functions are used:
  1. The equality operator,
  2. The Euclidean similarity, $V_{i,j}^k = 1 - EucDist(r_i.a_k, r_j.a_k)$.
  3. The relative similarity, $V_{i,j}^k = 1 - \frac{|r_i.a_k - r_j.a_k|}{max(r_i.a_k, r_j.a_k)}$.
  4. The normalized Manhattan similarity, $V_{i,j}^k = \frac{|r_i.a_k - r_j.a_k|}{max(r_i.a_k, r_j.a_k)}$.
- For textual attributes, the following functions are used:
  (i) *Syntactic similarity measures.*
      D-HAT distinguishes textual attributes into <u>short strings</u>, if their average value entails less than five words, and <u>long strings</u> otherwise. For both types, it employs the following functions:

1. Jaccard similarity: $V_{i,j}^k = \frac{|token\_set(r_i.a_k) \cap token\_set(r_j.a_k)|}{|token\_set(r_i.a_k) \cup token\_set(r_j.a_k)|}$.

2. Generalized Jaccard, which extends the previous measure to consider the bags of tokens: $V_{i,j}^k = \frac{|bag(r_i.a_k) \cap bag(r_j.a_k)|}{|bag(r_i.a_k) \cup bag(r_j.a_k)|}$.

3. Overlap Coefficient: $V_{i,j}^k = \frac{|token\_set(r_i.a_k) \cap token\_set(r_j.a_k)|}{min(|token\_set(r_i.a_k)|,|token\_set(r_j.a_k)|)}$.

4. Bag: $V_{i,j}^k = 1 - \frac{max(|bag(r_i.a_k)-bag(r_j.a_k)|,|bag(r_j.a_k)-bag(r_i.a_k)|)}{max(|(r_j.a_k)|,|(r_i.a_k)|)|)}$

5. Dice Similarity: $V_{i,j}^k = 2 \times \frac{|token\_set(r_i.a_k) \cap token\_set(r_j.a_k)|}{|token\_set(r_i.a_k)|+|token\_set(r_j.a_k)|}$.

Additionally, D-HAT uses two similarity functions for short strings:

- Levenshtein similarity, the minimum number of edit operations (insert, delete or substitute) required to transform one string to another.
- Hamming, similar to Levenshtein except that it allows only substitution.

(ii) *Semantic similarity measures.* D-HAT exploits pre-trained embedding representations of textual data. Two types of representations are actually used:

  a) *Word-based models* like word2vec and GlobalVectors (GloVe). They substitute each token (word) by a meaningful numeric vector that is learnt from training a shallow feedforward neural network on large, external, un-annotated textual corpora, such as Google News and Wikipedia. In these models, words with contextual similarity have linearly related vector representations. However, they cannot produce vector representations for words that are *out-of-vocabulary.*

  b) To address this limitation, *skipgram models* like fastText [1] represent each word by the sum of the vector representations of its bag of characters. Thus, they are capable of learning a recurrent neural network that yields vector representations for words, independently of their occurrence in the training data.

  To extract numeric features/dimensions from the three pre-trained embeddings (i.e., word2vec, GloVe and fastText), D-HAT applies three similarity functions to the vectors of two records: the cosine, the Euclidean and the word mover's similarity [14]. For the last two functions, the homonymous distance function $d$ is transformed into a similarity value $sim$ as follows: $sim = \frac{1}{1+d}$.

(iii) *Hybrid similarity measures.* This configuration combines the aforementioned syntactic similarity measures with the semantic ones, given that they capture complementary matching evidence.

Overall, D-HAT creates one feature per boolean and categorical attributes, four per numeric ones as well as nine semantic features and up to seven syntactic ones per textual attribute.

**Step 5: Matching Scores.** The goal of this step is to estimate the matching likelihood for each pair of candidates based on the feature matrix of the previous step. This is carried out in two steps:

(i) *Binarizing the feature vectors.* In essence, D-HAT treats each feature as a vote for a "match" (1) or a "non-match" (0) decision. The dimensions of boolean and categorical attributes are already binary. The dimensions of numerical and textual attributes are defined in $[0, 1]$, with higher values indicating a higher matching likelihood. To binarize them, D-HAT employs a similarity threshold $\theta \in [0, 1]$, common to all dimensions, such that all numeric scores above $\theta$ are converted into "match" votes (1), while the rest become "non-match" votes (0). All dimensions with a "N/A" value are ignored.

(ii) *Score estimation.* To calculate the matching score $m_{i,j}$ for two candidate records, $r_i$ and $r_j$, we aggregate the dimensions of their binary feature vector $\hat{V}_{i,j}$ into a single value through their mean, i.e., $m_{i,j} = \sum_{k=1}^{N} \hat{V}_{i,j}^k / (N - n)$, where $N$ is the total number of features, $n$ is the number of missing ones and $\hat{V}_{i,j}^k \in \{0, 1\}$.

At the end of these two steps, the matching scores of all pairs are calculated and stored in a matrix $M$. The records and the matrix define a weighted graph $G(V, M)$, where the set of nodes $V$ represent the input records, and $M$ is the adjacency matrix of weights. $G(V, M)$ is referred to as the *similarity graph.*

**Step 6: MutMax Clustering.** The final step receives as input the similarity graph $G(V, M)$ and partitions it into a set of disjoint clusters, such that every cluster corresponds to a unique entity, containing all duplicate records describing it. The partitioning is performed by **MutMax**, a greedy approach that defines as duplicates the pairs of records with mutually maximum scores. More specifically, MutMax operates as follows: For each record $r_i$, all candidates are sorted in decreasing matching scores and the top one $r^i_{max} = r_j$ is selected as the potential match. If $r_i$ was set as the potential match for $r_j$, the records $r_i$ and $r_j$ are designated as matches. The rest of the candidate pairs are ignored.

**Overall approach.** D-HAT algorithm is outlined in Fig. 2. Step 1 (Data cleaning) is applied first (Line 1). Step 2 (Attribute selection) is performed given threshold $c_{min}$ (Lines 2–7). The overlap blocker is applied to each attribute (Lines 8–10). A performance score is computed per attribute by multiplying the coverage of attribute $a$ with the reduction ratio [5]: $getScore(B_a, a) = c(a) \cdot RR(B_a, T)$, where $|B_a|$ denotes the total number of candidate pairs in the blocks $B_a$. The attribute with the highest score is selected (Lines 11–14), and is applied to retrieve the final set of blocks (Line 16).

The next loop simultaneously applies Steps 4 and 5. It builds a two-dimensional array $M$ with a score for each pair of compared records. In more detail, $F \cap a$ is the set of functions applicable for attribute $a$. For each feature higher than $\theta$, the overall similarity is incremented by one matching vote (Lines 23–25). The average score is finally estimated for the current pair of candidates (Line 29).

Finally, MutMax is applied to $M$ (Lines 31–36). For each record, the most similar candidate is specified and stored in array $O$ (Line 31). Using $O$, D-HAT identifies the record pairs that are mutually most similar (Lines 32–33), adding them to the output (Line 34). Note that $D$ is a set and that each output pair is

---

**Algorithm 1** D-HAT

---

**Input:** A data set with duplicates in itself $T$, a set of features $F$,
a threshold on the minimum coverage per attribute $c_{min}$, and
a threshold on the minimum similarity score for binarization $\theta$

**Output:** The set of duplicate records, $D$

1: Clean $T$ {Step 1}
2: **for all** $a \in T.A$ **do** {Step 2: attribute selection}
3:     calculate coverage $c(a)$
4:     **if** $c(a) < c_{min}$ **then**
5:         $T.A \leftarrow T.A - \{a\}$
6:     **end if**
7: **end for**
8: $best\_attribute = ""$, $best\_score = 0$ {Step 3: blocking}
9: **for all** $a \in T.A$ **do**
10:     $B_a \leftarrow$ overlap_blocker$(T, a)$
11:     **if** $best\_score < getScore(B_a, a)$ **then**
12:         $best\_attribute = a$
13:         $best\_score = getScore(B_a, a)$
14:     **end if**
15: **end for**
16: $B \leftarrow$ overlap_blocker$(T, best\_attribute)$, $M \leftarrow \{\}$ {Steps 4-5}
17: **for all** candidate pair $(r_i, r_j) \in B$ **do**
18:     $sim_{i,j} = 0$, $counter = 0$
19:     **for all** $a \in T.A$ **do**
20:         **if** $r_i.a \neq$ N/A & $r_j.a \neq$ N/A **then**
21:             **for all** $f_{sim} \in F \cap a$ **do**
22:                 $counter + +$
23:                 **if** $\theta \leq f_{sim}(r_i.a, r_j.a)$ **then**
24:                     $sim_{i,j} + +$
25:                 **end if**
26:             **end for**
27:         **end if**
28:     **end for**
29:     $M[i,j] = M[j,i] = sim_{i,j}/counter$
30: **end for**
31: $D = \{\}$, $O = $ **argmax**$(M)$ per $r_i \in T$ {Step 6: clustering}
32: **for all** $r_i \in T$ **do**
33:     **if** $O[i] = j$ & $O[j] = i$ **then**
34:         $D \leftarrow D \cup (i, j)$
35:     **end if**
36: **end for**
37: Return D

---

Fig. 2. The end-to-end algorithm of D-HAT.

formed with the lowest id in the left part (i.e., $i < j$ in $(i, j)$ in Line 34); as a result, no duplicate pairs are returned as output.

In terms of time complexity, the cost of Steps 1, 2 and 3 is linear with the number of attributes in the given data set $T$, i.e., $O(|T.A|)$. For Steps 4 and 5, the cost is $O(|B|)$. For Step 6 no sorting is required. Instead, D-HAT merely iterates once over all cells in the two-dimensional array $M$. A hash table can be used to store the estimated similarities in practice. As a result, both the time and space complexity of Step 6 (and the entire algorithm) are linear with the number of candidate pairs after blocking, i.e., $O(|B|)$.

## 5 Experimental Evaluation

**Setup.** D-HAT is implemented in Python 3.8.5. All experiments were run on an Ubuntu 18.04.5 server with a 12-core Intel Xeon D-2166NT @2GHz, 64 GB of RAM and 300 GB HDD. A single core was employed in all time measurements.

**Benchmark Data Sets.** We employ five established data sets that come from multiple domains: products, bibliography, restaurants, and healthcare.

**Table 1.** Technical characteristics of the benchmark data sets. $|S|$, $|T|$ and $|D|$ stand for the number of source records, target records and duplicate pairs, respectively.

| Data set | $|S|$ | $|T|$ | $|D|$ | #Attributes | #Numerical | #Bool. & Cat. | #Textual | #Selected |
|---|---|---|---|---|---|---|---|---|
| Amazon-Google | 1,363 | 3,226 | 1,298 | 4 | 1 | 0 | 2 | 3 |
| Abt-Buy | 1,081 | 1,092 | 1,095 | 3 | 1 | 0 | 2 | 3 |
| DBLP-ACM | 2,614 | 2,294 | 2,223 | 4 | 1 | 1 | 2 | 3 |
| Fodors-Zagats | 533 | 331 | 112 | 5 | 0 | 0 | 5 | 5 |
| Immucare | 305 | 310 | 305 | 213 | 32 | 6 | 37 | 75 |

Immucare is a healthcare dataset matching two hospital visits of the same patient. The technical details of these data sets [13,24] are summarized in Table 1.

**Baseline Systems.** We compare the performance of D-HAT with Magellan [13] and ZeroER [24]. For the former, we use decision tree as the classification algorithm, while for the latter, no configuration is needed.

**Evaluation Measures.** We use the standard measures of recall, precision, and F1-score, which are defined in Sect. 3. We also report the overall run-time, i.e., the time that intervenes between receiving the data set(s) as input and producing the duplicate pairs as output. We repeat every measurement three times and report the average.

## 5.1 Step 3: Blocking

D-HAT applies Magellan's overlap blocker to all attributes and selects as optimal the one minimizing the number of candidates, while maximizing coverage. The resulting performance appears in Table 2. In all cases, the number of candidate pairs is reduced by whole order of magnitude (i.e., $\gg 90\%$) in comparison to the brute-force approach (i.e., $|S| \times |T|$). The only exception is Abt-Buy, where the candidates drop by 86%, which is a dramatic reduction of the search space, too. Nevertheless, the recall in all cases remains rather high, above 90%. This means that the vast majority of duplicate pairs co-occur in at least one block.

Note that precision after blocking remains very low for most data sets. To raise it to acceptable levels, matching is required. Note also that compared to the

**Table 2.** Blocking performance. Time in Seconds.

| Data set | Key Attribute | #Candidates | Recall | Prec | Time |
|---|---|---|---|---|---|
| Amazon-Google | Name | 131,214 | 0.995 | 0.010 | 7.3 |
| Abt-Buy | Name | 164,072 | 0.994 | 0.007 | 2.6 |
| DBLP-ACM | Authors | 318,404 | 0.993 | 0.007 | 19.4 |
| Fodors-Zagat | Phone | 111 | 0.929 | 0.936 | 0.7 |
| Immucare | Date of Birth | 311 | 1.000 | 0.981 | 26.5 |

**Table 3.** Matching effectiveness of D-HAT, Magellan and ZeroER across all data sets. The best F1 per data set is underlined.

| Data set | D-HAT | | | | | | | | | Magellan | | | ZeroER | | |
|---|---|---|---|---|---|---|---|---|---|---|---|---|---|---|---|
| | Syntactic Features | | | Semantic Features | | | Hybrid Features | | | | | | | | |
| | Pr | Re | F1 | Pr | Re | F1 | Pr | Re | F1 | Pr | Re | F1 | Pr | Re | F1 |
| A-G | 0.904 | 0.479 | 0.626 | 0.828 | 0.349 | 0.534 | 0.925 | 0.532 | <u>0.675</u> | 0.513 | 0.573 | 0.542 | 0.663 | 0.385 | 0.487 |
| A-B | 0.818 | 0.402 | <u>0.539</u> | 0.635 | 0.174 | 0.274 | 0.824 | 0.346 | 0.487 | 0.440 | 0.443 | 0.442 | 0.220 | 0.601 | 0.322 |
| D-A | 0.992 | 0.956 | 0.974 | 0.995 | 0.980 | <u>0.987</u> | 0.997 | 0.974 | 0.985 | 0.980 | 0.983 | 0.981 | 0.936 | 0.945 | 0.940 |
| F-Z | 0.981 | 0.929 | <u>0.954</u> | 0.971 | 0.911 | 0.940 | 0.981 | 0.929 | <u>0.954</u> | 0.939 | 0.969 | <u>0.954</u> | 1.000 | 0.312 | 0.476 |
| CA | 0.993 | 0.987 | <u>0.990</u> | 0.990 | 0.987 | 0.988 | 0.993 | 0.987 | <u>0.990</u> | 0.968 | 1.000 | 0.984 | 1.000 | 0.487 | 0.655 |

overall run-time of D-HAT and the rest of the methods (in Fig. 3), the overhead of blocking is negligible ($< 10\%$ in all cases). The only exception is Immucare, where the overhead of blocking is high, due to the very large number of attributes retained after Step 2 (75).

## 5.2   Steps 4–6: Matching

To ensure fairness, we apply the same blocker to the same key attribute for both baseline systems, (e.g., we use the 'phone' attribute instead of 'name' in Fodors-Zagats). Note that for Amazon-Google, ZeroER could not create its feature matrix within a time limit of 6 h. To complete the assessment, we combined it with the feature vectors created by Magellan instead. As a result, the performance of ZeroER could be slightly different from that reported in [24].

The resulting performance of all algorithms with respect to precision (Pr), recall (Re) and f-measure (F1) appears in Table 3, while the corresponding run-times are reported in Fig. 3. Note that after preliminary experiments, we set $c_{min} = 0.1$ and $\theta = 0.7$ for D-HAT in all cases. Note also that D-HAT is combined with three different groups of features: (i) The syntactic ones, which include only the syntactic similarity functions for textual attributes along with the specialized functions of boolean, categorical and numeric attributes. (ii) The semantic features, which differ from the previous group in that they replace the syntactic similarity functions with the semantic ones. (iii) The hybrid features, which employ all similarity functions for all types of attributes defined in Sect. 4. In this way, we are able to examine the contribution of the two types of textual similarity functions, which account for the majority of features used by D-HAT.

Compared to blocking, precision has actually increased by whole orders of magnitude. This emphasis on precision should be attributed to MutMax clustering, which associates every record only with its most similar candidate.

Comparing the various groups of features between them, we observe that the syntactic ones consistently outperform the semantic ones. The reason is that most data sets contain domain-specific terminology. As a result, especially word2vec and GloVe suffer from a large portion of out-of-vocabulary terms. The only exception is DBLP-ACM, which involves long textual attributes like venue

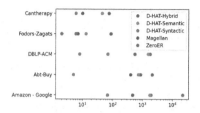

**Fig. 3.** Run-time in seconds.

**Table 4.** The number of features per group.

| Data set | Non-textual | Syntactic | Semantic | Hybrid |
|----------|-------------|-----------|----------|--------|
| A-G | 4 | 14 | 22 | 32 |
| A-B | 4 | 14 | 22 | 32 |
| D-A | 4 | 14 | 22 | 32 |
| F-Z | 0 | 35 | 45 | 80 |
| CA | 78 | 400 | 411 | 733 |

names and publication titles; in these settings, the evidence provided by semantic similarities outperforms the syntactic ones, albeit by just ~2%.

In terms of time-efficiency, the advantage of syntactic similarity functions is clear in all cases, as shown in Fig. 3. The run-time of D-HAT increases by a whole order of magnitude in almost all cases, when replacing the syntactic similarity features with the semantic ones. This is caused by the large number of lookups and computations that are required for converting every attribute value into a high-dimensional embedding vector and a similarity score.

It is interesting to examine whether the combination of syntactic and semantic similarities justifies the lower time efficiency by an increase in effectiveness. This is only true in Amazon-Google, where hybrid features' F1 is higher than the syntactic ones by ~10%. In all other cases, the hybrid features lie between the two other groups of features, usually closer to the top performing one. Hence, *D-HAT should be exclusively combined with the syntactic group of features.*

Compared to ZeroER, Table 3 shows that D-HAT with syntactic features achieves significantly better effectiveness in most cases. Its f-measure is actually higher by 50%, on average, across the five data sets. At the same time, Fig. 3 demonstrates D-HAT is consistently faster than ZeroER by whole orders of magnitude (e.g., 1 min vs 6 hrs over Amazon-Google) – the sole exception is DBLP-ACM, where D-HAT is slower, due to the computation of 10 syntactic similarity functions over textual values. D-HAT takes into account attributes with high level of noises (missing values, heterogeneity of existing values, errors), which inevitably corrupt some matching signals.

Compared to Magellan, in the first two data sets, D-HAT achieves a higher f-measure than Magellan by more than 13%, while in the next three data sets both methods exhibit practically identical performance (i.e., their f-measures differ by less than 1%). The competitive performance of Magellan stems from its supervised functionality: in each dataset, 70% of the candidate pairs are used for training its classification model, leaving only 30% of the pairs as a testing set. In contrast, D-HAT processes all candidate pairs and its performance is bounded by blocking. In terms of time-efficiency, we observe in Fig. 3 that D-HAT takes a clear lead in all cases, as its run-time is lower than Magellan even by a whole order of magnitude (e.g., 35 vs 400 s over Abt-Buy).

Overall, D-HAT typically outperforms the state-of-the-art unsupervised deduplication method to a significant extent in all respects. Compared to the state-of-the-art supervised approach, it exhibits similar effectiveness, if not higher, at a much lower run-time, despite the lack of labelled instances.

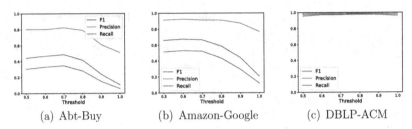

(a) Abt-Buy          (b) Amazon-Google          (c) DBLP-ACM

**Fig. 4.** Performance of D-HAT with syntactic features when varying the threshold $\theta$.

### 5.3 Sensitivity Analysis

The only configuration parameter that is crucial for the performance of D-HAT is the similarity threshold $\theta$, whose value depends on the level of noise and heterogeneity in the data. To assess its impact on the overall performance of D-HAT, we consider all values in the range $[0.5, 1]$ with a step of $0.1$. The results appear in Fig. 4. Due to lack of space, we report three of the five datasets.

We observe that this parameter has no effect on any evaluation measure over DBLP-ACM. The reason is that the pairs identified as matches in these datasets exhibit very high similarity (practically 1.0) for most of the features employed by D-HAT. As a result, the matching decisions of MutMax clustering are not altered by the value of $\theta$. For Abt-Buy and Amazon-Google, we observe that up to 0.7, the performance of D-HAT improves (Abt-Buy) or remains the same (Amazon-Google). For $\theta > 0.7$, *a small increase in the similarity threshold yields slightly lower performance with respect to all measures*. The reason is that both data sets are challenging tasks, because they contain many corner cases, i.e., records that are close to the decision boundary.

Overall, we can conclude that D-HAT is robust with respect to its similarity threshold $\theta$, with $\theta = 0.7$ constituting a reliable default value.

## 6    Conclusions

We presented D-HAT, an efficient, fully automated clustering-based end-to-end deduplication system. D-HAT can process high dimensional data sets with heterogeneous attribute types and missing values without requiring user intervention or any labelled data. The thorough experimental study on benchmark data sets demonstrates that our system achieves high accuracy across different benchmark tasks, and outperforms supervised and unsupervised baselines. The main benefit of D-HAT over unsupervised methods is the high accuracy on all standard tasks, whereas compared to supervised methods, D-HAT eliminates the extra time and effort needed from domain experts to annotate a training set. It also saves the time required to find and train an efficient classification model. In the future, we plan to parallelize D-HAT on top of Apache Spark in order to scale it to huge data sets with millions of records.

# References

1. Bojanowski, P., et al.: Enriching word vectors with subword information. Trans. Assoc. Comput. Linguist. **5**, 135–146 (2017)
2. Brunner, U., Stockinger, K.: Entity matching with transformer architectures - a step forward in data integration. In: EDBT, pp. 463–473 (2020)
3. Chen, M., Mao, S., Liu, Y.: Big data: a survey. Mob. Netw. Appl. **19**(2), 171–209 (2014). https://doi.org/10.1007/s11036-013-0489-0
4. Christen, P.: The data matching process. In: Data Matching. Data-Centric Systems and Applications. Springer, Heidelberg (2012). https://doi.org/10.1007/978-3-642-31164-2_2
5. Christen, P.: A survey of indexing techniques for scalable record linkage and deduplication. IEEE Trans. Knowl. Data Eng. **24**(9), 1537–1555 (2012)
6. Christophides, V., Efthymiou, V., Palpanas, T., Papadakis, G., Stefanidis, K.: An overview of end-to-end entity resolution for big data. ACM Comput. Surv. **53**(6), 1–42 (2021)
7. Devlin, J., et al.: BERT: pre-training of deep bidirectional transformers for language understanding. In: NAACL-HLT, pp. 4171–4186 (2019)
8. Dong, X.L., Srivastava, D.: Big data integration. Synth. Lect. Data Manag. **7**(1), 1–198 (2015)
9. Fan, W., Ma, S., Tang, N., Yu, W.: Interaction between record matching and data repairing. J. Data Inf. Qual. **4**(4), 1–38 (2014)
10. Fellegi, I.P., Sunter, A.B.: A theory for record linkage. J. Am. Stat. Assoc. **64**(328), 1183–1210 (1969)
11. Getoor, L., Machanavajjhala, A.: Entity resolution: theory, practice & open challenges. Proc. VLDB Endow. **5**(12), 2018–2019 (2012)
12. Hassanzadeh, O., et al.: Framework for evaluating clustering algorithms in duplicate detection. Proc. VLDB Endow. **2**(1), 1282–1293 (2009)
13. Konda, P., Das, S., et al.: Magellan: toward building entity matching management systems. Proc. VLDB Endow. **9**(12), 1197–1208 (2016)
14. Kusner, M.J., Sun, Y., Kolkin, N.I., Weinberger, K.Q.: From word embeddings to document distances. In: ICML, vol. 37, pp. 957–966 (2015)
15. Li, Y., Li, J., Suhara, Y., Wang, J., Hirota, W., Tan, W.: Deep entity matching: challenges and opportunities. ACM J. Data Inf. Qual. **13**(1), 1–17 (2021)
16. Liu, Y., et al.: RoBERTa: a robustly optimized BERT pretraining approach. arXiv preprint arXiv:1907.11692 (2019)
17. Mudgal, S., et al.: Deep learning for entity matching: a design space exploration. In: SIGMOD, pp. 19–34 (2018)
18. Papadakis, G., et al.: Three-dimensional entity resolution with JedAI. Inf. Syst. **93**, 101565 (2020)
19. Papadakis, G., et al.: Blocking and filtering techniques for entity resolution: a survey. ACM Comput. Surv. **53**(2), 1–42 (2020)
20. Papadakis, G., et al.: Comparative analysis of approximate blocking techniques for entity resolution. Proc. VLDB Endow. **9**(9), 684–695 (2016)
21. Papadakis, G., et al.: The return of JedAI: end-to-end entity resolution for structured and semi-structured data. Proc. VLDB Endow. **11**(12), 1950–1953 (2018)
22. Wang, J., Kraska, T., Franklin, M.J., Feng, J.: CrowdER: crowdsourcing entity resolution. arXiv preprint arXiv:1208.1927 (2012)
23. Wang, Z., Sisman, B., Wei, H., Dong, X.L., Ji, S.: CorDEL: a contrastive deep learning approach for entity linkage. In: ICDM, pp. 1322–1327 (2020)
24. Wu, R., Chaba, S., Sawlani, S., Chu, X., Thirumuruganathan, S.: ZeroER: entity resolution using zero labeled examples. In: SIGMOD, pp. 1149–1164 (2020)

# Others

# Probing Semantic Grounding in Language Models of Code with Representational Similarity Analysis

Shounak Naik[1,2,3]([⊠]), Rajaswa Patil[1,2,3], Swati Agarwal[1,2][ID],
and Veeky Baths[1,3][ID]

[1] BITS Pilani, K. K. Birla Goa Campus, Goa, India
{f20170835,f20170334,swatia,veeky}@goa.bits-pilani.ac.in
[2] Computational Linguistics and Social Networks Lab, BITS Pilani, Goa, India
[3] Cognitive Neuroscience Lab, BITS Pilani, Goa, India

**Abstract.** Representational Similarity Analysis is a method from cognitive neuroscience, which helps in comparing representations from two different sources of data. In this paper, we propose using Representational Similarity Analysis to probe the semantic grounding in language models of code. We probe representations from the CodeBERT model for semantic grounding by using the data from the IBM CodeNet dataset. Through our experiments, we show that current pre-training methods do not induce semantic grounding in language models of code, and instead focus on optimizing form-based patterns. We also show that even a little amount of fine-tuning on semantically relevant tasks increases the semantic grounding in CodeBERT significantly. Our ablations with the input modality to the CodeBERT model show that using bimodal inputs (code and natural language) over unimodal inputs (only code) gives better semantic grounding and sample efficiency during semantic fine-tuning. Finally, our experiments with semantic perturbations in code reveal that CodeBERT is able to robustly distinguish between semantically correct and incorrect code.

**Keywords:** Language model · Deep learning · Code semantics

## 1 Introduction

Recent development in deep neural network-based (DNN) language modeling has brought in great advancements in various data-driven artificial intelligence (AI) technologies. With the increasing scale of data repositories, the usage of language models in various data-driven AI applications has increased as well. This has brought in a new paradigm of *Language-Model-as-a-Service*. Under this

© The Author(s), under exclusive license to Springer Nature Switzerland AG 2022
W. Chen et al. (Eds.): ADMA 2022, LNAI 13726, pp. 395–406, 2022.
https://doi.org/10.1007/978-3-031-22137-8_29

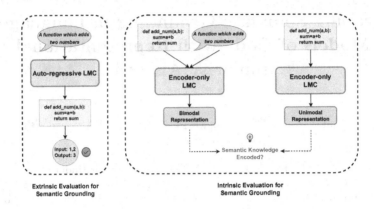

**Fig. 1.** Difference between extrinsic and intrinsic evaluation for semantic grounding in language models of code (LMCs).

paradigm, language models trained on huge web data enable natural language interfaces to various AI applications, as well as provide rich features for various downstream applications such as sentiment analysis, question answering, text completion, etc.

Recent efforts in building AI-based assistive technologies in code and software has seen development of big code datasets and repositories [6,11]. Hence, in parallel to the progress in developing DNN language models for natural language, there has been a recent spurt in developing language models for code (LMC) [3,4]. Representations obtained from LMCs are used as features for a variety of downstream applications. Traditionally, the representations from language models are either evaluated against these downstream tasks (extrinsic evaluation), or are probed for specific knowledge (intrinsic evaluation). A robust language model of code is ideally expected to capture the actual meaning of the code text (semantics) and not just its surface level form-based statistical patterns. As the underlying code repository and databases become large and complex, semantic grounding becomes quite important to avoid spurious outcomes.

Auto-regressive LMCs like Codex (GPT-3) [3] can be probed for such semantic grounding with code-generation tasks as shown in Fig. 1. This can either be achieved by generating code for a given natural language or code prompt and testing it against input-output (IO) specifications [1]. These methods though complete from a semantic evaluation perspective, pose several limitations. Firstly, these are extrinsic evaluation methods which only focus on generating form-level tokens, which does not guarantee true semantic grounding [2]. Secondly, for tasks not involving code-generation and for the tasks that involve scaling to large code databases, the representations obtained from a LMC are more important than the its ability to generate code and text (Fig. 1). Hence, intrinsic evaluation methods become necessary for evaluating the semantic grounding under such settings. Finally, for encoder-only models like Code-

BERT [4], code-generation is not technically feasible, which is a requirement for testing semantic grounding against IO specifications.

In this work we propose using Representational Similarity Analysis (RSA) [7] to probe the semantic grounding in the representations from LMCs. Unlike the other previous intrinsic evaluation methods, RSA is a method from cognitive neuroscience which offers a flexible intrinsic evaluation setting which is agnostic to various properties of the representations like their source, size, structure, modality, etc. Hence, RSA becomes a good choice to evaluate the representations obtained from LMCs against a ground truth semantic representation. In this work we probe CodeBERT with RSA. We use a code's natural language description as its ground truth semantic representation. Through our experiments we show how RSA can be utilized to evaluate various aspects of research and development in using LMCs for various data-driven code intelligence tasks. Specifically, we aim to use RSA-based semantic grounding evaluation to study: **(1)** The localization of semantic knowledge in different layers of CodeBERT; **(2)** The impact of size of the fine-tuning datasets on the semantic grounding in Code-BERT; **(3)** The impact of modality of code repositories (unimodal vs. bimodal) on the semantic grounding in CodeBERT; **(4)** The robustness of CodeBERT model against semantic perturbations in the code.

## 2  Background

LMCs can be broadly classified into two categories: auto-regressive models and encoder-only models. Auto-regressive models like Codex (GPT-3) [3] are pretrained to generate code and text. Whereas, encoder-only models like CodeBERT [4] are pre-trained to encode representations of code, and do not generate any code or text. Depending on their category and their neural architecture, the LMCs are pre-trained with a variety of pre-training objectives by using huge multimodal (natural language and programming language) datasets [6,11].

LMCs have various data-driven applications in programming and software engineering. These include natural language based code search [6], code translation and refinement [9], code repair [9], and bug detection [9]. Representations from pre-trained LMCs are used to enable such applications in practical deployments. LMCs are usually fine-tuned with a relatively smaller dataset for the target downstream application before deployment. This process is usually quite resource-heavy in terms of data annotation costs and fine-tuning computing costs. Hence, getting an estimate of optimal amount of fine-tuning for robust deployments is very important. LMCs are usually evaluated against standard tasks and benchmarks [6,9]. While most such recent LMCs show great performance on such benchmarks, the true code understanding capabilities of these models are still being tested with semantically challenging extrinsic evaluation settings [1,3]. We extend such semantic evaluations into the intrinsic evaluation paradigm by probing the representations from these LMCs for semantic grounding.

**Table 1.** The number of NL-PL sample pairs across the six programming languages used by CodeBERT in the final dataset. *JS: JavaScript

| Data split | Submission type | Go | Java | JS* | PHP | Python | Ruby |
|---|---|---|---|---|---|---|---|
| Test | Incorrect | 10482 | 22213 | 8846 | 6339 | 25322 | 15354 |
| | Correct | 22768 | 25438 | 9729 | 8103 | 25500 | 23658 |
| Training | Correct | 29557 | 62370 | 12377 | 10455 | 67337 | 45433 |

Representational Similarity Analysis (RSA) is a method from cognitive neuroscience, originally invented to compare representations of neural and physiological data and signals from different sources [7]. Recent work in natural language processing research has focused on using RSA for various interpretability studies with language models. Previously, RSA has been used to ground neural representations from language models to that in the human brain [5]. RSA has also been used to probe contextual semantic and syntactic information in language models [8]. RSA has also been proved to be quite useful in studying the effect of fine-tuning on natural language models [10]. Given the versatility of the RSA method, and its successful application in evaluating natural language models, it becomes a great choice to perform intrinsic evaluation of LMCs.

## 3     Experimental Setup

### 3.1     Dataset

We use the IBM CodeNet dataset [11] for all our experiments. The dataset comprises 4000 coding problems with submissions in multiple programming languages. The problem descriptions in the dataset can be used as a natural language (NL) modality. Whereas, the code submissions to these problems can be used as a programming language (PL) modality. Hence each sample in the dataset can be viewed as a NL-PL pair. While the original CodeNet dataset comprises of code samples from over 50 programming languages, we only focus on the six programming languages supported by CodeBERT as shown in Table 1. We sub-sample and clean the CodeNet dataset as per the requirements for our RSA-based semantic grounding probing experiments.

To generate the test data, we filter out 255 problems such that each problem has a correct submission, and a wrong submission in each of the six programming languages under consideration: Go, Java, JavaScript, PHP, Python, and Ruby. In order to generate the training data, we filter out 808 problems (708: Training, 100: Validation), such that each problem has at least one correct submission for each of the six programming languages. Given the noisy metadata files from the CodeNet dataset, we manually extract the problem descriptions from the raw problem description HTML files in the dataset. The extracted set of problem descriptions is multilingual (non-English descriptions) in nature. We translate all the problem descriptions to English using an off the shelf translation tool: DeepL.[1] The translations obtained from DeepL are manually checked for any

---

[1] https://www.deepl.com/translator.

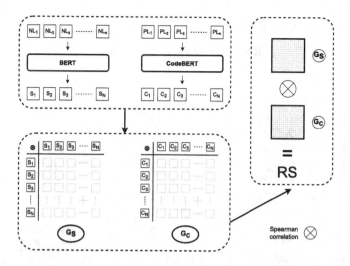

**Fig. 2.** Schematic overview of using the Representational Similarity Analysis method with the code representations from the CodeBERT model.

errors. The final dataset statistics are shown in Table 1. The test data split is used to perform the RSA-based probing for semantic grounding in CodeBERT. On the other hand the training data split is used to fine-tune the CodeBERT model, in order to inspect the role of fine-tuning on its semantic grounding. The original IBM CodeNet dataset can be found here. Our modified version of the IBM CodeNet dataset will be available on request. The code for our experiments is publicly available here.

## 3.2   Representational Similarity Analysis (RSA)

Given $N$ natural language description - code snippet pairs (NL-PL): we first obtain the code representations $\{C_k\}$ from the code snippets with CodeBERT. Similarly, we obtain the semantic representations $\{S_k\}$ from the natural language descriptions with BERT, where $k\epsilon[1, N]$. For ablation purposes, we extract code representations under two different input settings: Unimodal (PL-only) and Bimodal (NL and PL) as supported by the CodeBERT model. Next, we construct the individual representational geometries ($\mathcal{G} \in \mathbb{R}^{N \times N}$) for $\{C_k\}$ and $\{S_k\}$ by computing the pairwise dissimilarities between all the samples in the dataset:

$$\mathcal{G}_C = \{1 - similarity(C_i, C_j)\} \quad (1) \qquad \mathcal{G}_S = \{1 - similarity(S_i, S_j)\} \quad (2)$$

where, $i, j\epsilon[1, N]$

The final representational similarity score ($RS$) between the code and semantic representations can then be obtained by finding the similarity between the two representational geometries:

$$RS_{(C,S)} = similarity(\mathcal{G}_C, \mathcal{G}_S) \tag{3}$$

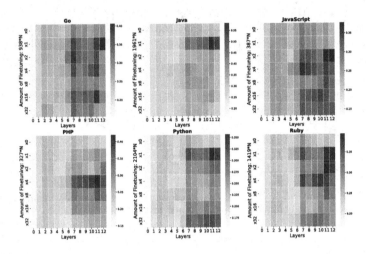

**Fig. 3.** Heatmaps of RSA similarity scores across the six programming languages and 13 layers of the CodeBERT model.

A higher $RS$ value can then be interpreted as higher semantic grounding in the representations of code. We use Spearman correlation coefficient as the *similarity* measure while calculating the pair-wise dissimilarities between the samples, as well as the similarities between the two representational geometries. We obtain statistically significant similarity scores throughout our experiments with correlation p-values $< 0.05$. A schematic overview of using RSA with representations from the CodeBERT model is shown in Fig. 2. While Fig. 2 shows Unimodal setting with CodeBERT (PL-only), we perform RSA with Bimodal setting as well (NL and PL).

## 4   Experiments

Previous work in language model probing has tried to analyze various semantic knowledge localization trends in the layers of a language model [12].[2] This helps understand the model dynamics, as well as in selecting the best representations for downstream applications. Hence, we first begin by evaluating the semantic grounding in the pre-trained CodeBERT model with the default bimodal input setting. We perform RSA with all 13 representations[3] from the pre-trained Code-BERT model. We use the *Correct* code samples from the Test split of the data

---

[2] Localization studies aim to find out which layer and parameters of a neural network capture the maximum amount of specific knowledge.

[3] Embedding (layer 0) + Encoder blocks (layer 1–12).

here. We observe that the semantic grounding is consistently low for the pre-trained model (Fig. 3 x0) across all the layers and programming languages with no particular localization trends. This indicates that the pre-trained CodeBERT model shows low levels of semantic understanding for code data, and directly using the representations from the pre-trained model in downstream practical applications might not yield robust performance. Taking this into consideration, we probe various practically motivated aspects of semantic grounding in LMCs:

## 4.1  Semantic Fine-tuning

We start our experiments by evaluating layer-wise representational similarity scores for CodeBERT model. We visualize the extent of similarity and representative semantic grounding through heatmap plots as shown in Fig. 3. We plot a separate heatmap for each of the six programming languages used by CodeBERT. The X-axis of each heatmap represents the layer of the CodeBERT model. The Y-axis of each heatmap represents the amount of fine-tuning data used to induce semantic knowledge in the CodeBERT model. The heatmap gradient represents the RSA similarity score, which is representative of the amount of semantic grounding as discussed in Sect. 3.2.

Firstly, we observe that the pre-trained CodeBERT model does not show significantly low semantic grounding across all the layers and programming languages (as seen in the first row of each heatmap). Since the original pre-training tasks used by the CodeBERT model do not induce enough semantic knowledge in its representations as seen in Fig. 3, we evaluate how the amount of downstream semantic fine-tuning affects the model's semantic grounding. We divide the training data of each programming language into six splits, where number of samples are increased in the power of 2 at every step: x0, x1, x2, x4, x8, x16, and x32. We use the NL-PL pairs in the training data to fine-tune the model on the semantically relevant task of semantic code search - one of the major downstream applications of code representations. We observe that fine-tuning the model helps with inducing semantic grounding in the representations (Fig. 3). Even fine-tuning on a very small number of samples, significantly increases the semantic grounding. Most of the semantic grounding is localized in the deeper layers of the model (right half of the heatmaps), which is similar to that of previous natural language models [12]. We also observe that the semantic grounding peaks at the pre-final layer. This shows that using representations from the final-layer or from the pre-trained model might not be optimal for data-driven downstream code applications. This also reveals that large multimodal datasets of NL-PL code samples are not required to induce high levels of semantic grounding in the CodeBERT model.

## 4.2  Input Modality

Under practical settings, code repositories can either be unimodal or bimodal in nature. Hence, most LMCs support bimodal NL and PL inputs. Here, we inspect the role of input modality in semantic grounding. Following the fine-tuning done

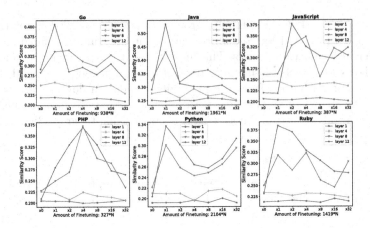

**Fig. 4.** RSA similarity scores after fine-tuning the CodeBERT model with Bimodal (NL and PL) input.

with bimodal data as described in Sect. 4, we repeat the fine-tuning process with unimodal data. We visualize the effect of input modality on the semantic grounding in CodeBERT model with line-chart plots shown in Fig. 4 and Fig. 5. We plot separate sub-plots for each of the six programming languages used by CodeBERT. The X-axis of each sub-plot represents the amount of fine-tuning data used to induce semantic knowledge in the CodeBERT model. The Y-axis of each sub-plot represents the similarity score, which is representative of the amount of semantic grounding as discussed in Sect. 3.2. We report the scores for four layers for each of the six programming languages: {1, 4, 8, 12} - each of which has a separate color-coded trajectory in each of the sub-plots.

We observe that both unimodal and bimodal inputs show increasing semantic grounding in representations with fine-tuning in the deeper layers (8 and 12) as seen in Fig. 5 and Fig. 4 respectively. On the other hand, similar to earlier findings early layers (1 and 4) do not show any significant semantic grounding. Hence, even without natural language descriptions, the deeper layers of the model seem to capture code semantics up to some extent by just looking at the code text. We also observe that representations from bimodal inputs hold significantly more semantic grounding (as high as 500% in languages like Java) than those from unimodal inputs (Fig. 6). Hence, augmenting code repositories with natural language descriptions and comments can help in developing better downstream applications with LMC representations. Bimodal inputs also show better sample efficiency (Fig. 4), where the performance peaks with significantly less amount of fine-tuning as compared to unimodal input which keeps on increasing with an increasing amount of fine-tuning data, while still showing lesser semantic grounding than bimodal input (Fig. 5).

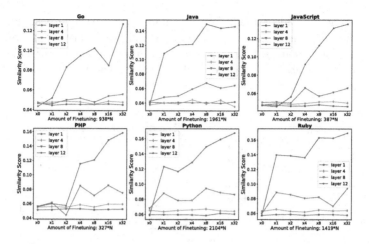

**Fig. 5.** RSA similarity scores after fine-tuning the CodeBERT model with Unimodal (PL-only) input.

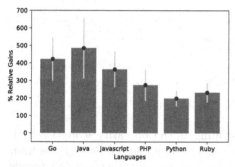

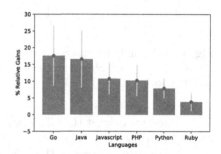

**Fig. 6.** Amount of relative gains in semantic grounding when using bimodal inputs over unimodal inputs.

**Fig. 7.** Amount of relative gains in semantic grounding when using *Correct* code samples over *Incorrect* code samples.

### 4.3 Semantic Perturbations

While all our previous experiments are conducted with the *Correct* code samples from the Test split of data, in this section we focus on using *Incorrect* code samples which are semantically perturbed. Under practical settings, evaluating a code against test specifications provides a complete and strict evaluation of code semantics, where even a small change in code semantics shows error in the outputs. While this is possible with code generation models, intrinsically evaluating code representations is not possible with such a setting. In an attempt to enable such strict evaluation under an intrinsic setting, we compare the representational similarity scores for *Correct* and *Incorrect* submissions in the dataset under a unimodal (PL-only) setting. Using a unimodal setting ensures that cues from the natural language modality do not help the model capture semantics, and its true understanding of code semantics is tested. Overall, we observe that CodeBERT rightly shows significantly higher semantic grounding for *Correct* submissions as compared to *Incorrect* ones across all fine-tuning checkpoints and languages (Fig. 7).

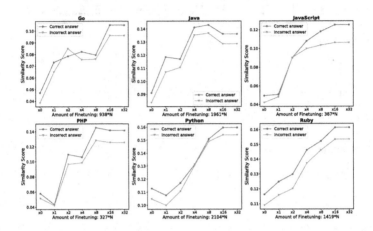

**Fig. 8.** RSA similarity scores for *Correct* and *Incorrect* code samples obtained using Unimodal (PL-only) representations from the best-performing layer (11) of the Code-BERT model.

Further, we inspect the effect of fine-tuning on the similarity scores for both *Correct* and *Incorrect* submissions. Here, we use unimodal representations from the best performing layer 11 (Fig. 3). We visualize the semantic grounding trends in CodeBERT model for *Correct* as well as *Incorrect* submissions with line-chart plots as shown in Fig. 8. We plot separate sub-plots for each of the six programming languages used by CodeBERT. The X-axis of each sub-plot represents the amount of fine-tuning data used to induce semantic knowledge in the CodeBERT model. The Y-axis of each sub-plot represents the similarity score.

We observe that with fine-tuning, the semantic grounding unexpectedly increases equally for both *Correct* and *Incorrect* submissions. This might suggest that similar to other language models, CodeBERT might be optimizing the code search task stochastically on surface-level forms, and not on the code meaning [2]. While active research is still trying to overcome such issues, this can be bypassed by using stricter ground-truth semantic representations derived from code structures like control flow graphs, data flow graphs, etc. and function specifications. Overall, CodeBERT consistently achieves more semantic grounding for *Correct* submissions over the *Incorrect* submissions across all the six programming languages and amounts of fine-tuning data. This suggests that while CodeBERT is unable to penalize *Incorrect* code samples for semantic perturbations, it is able to robustly differentiate between semantically correct and perturbed code samples by assign relatively lower semantic grounding to the *Incorrect* samples. This forms an intrinsic evaluation alternative to the standard practice of evaluating against test specifications in an extrinsic evaluation setting.

## 5    Conclusion and Future Directions

In this work we propose using Representational Similarity Analysis to probe the semantic grounding in language models of code (LMC). Through our experiments with the pre-trained CodeBERT model we show that current pre-training methods do not induce semantic grounding in LMCs. We also show that fine-tuning on semantically relevant tasks helps induce semantic grounding in LMCs, which is localized in the deeper layers of a LMC. Overall, our experiments show that the representations from the pre-final layer of a LMC are most rich in semantic knowledge. Our ablations with the input modalities reveal that different modalities of inputs show different types of semantic grounding and sample efficiency in LMCs, where even a relatively small number of fine-tuning examples is enough to obtain semantically robust performance in downstream applications. Our experiments with semantically perturbed *Incorrect* code samples show that CodeBERT optimizes on form, and not on the meaning. However, it robustly assigns relatively lower semantic grounding to the *Incorrect* code samples as compared to the *Correct* ones across various programming languages and experimental settings.

While in the current work we use natural language descriptions as the ground truth semantic representation, in future more formal semantic representations like program structure and specifications can be used as the ground truth semantic representations. It is also quite imperative to study how insights obtained from RSA-based probing for semantic grounding translates to performance in practical deployments of downstream applications.

## References

1. Austin, J., et al.: Program synthesis with large language models (2021). https://doi.org/10.48550/ARXIV.2108.07732
2. Bender, E.M., Koller, A.: Climbing towards NLU: on meaning, form, and understanding in the age of data. In: Proceedings of the 58th Annual Meeting of the Association for Computational Linguistics, pp. 5185–5198. Association for Computational Linguistics, Online (2020). https://doi.org/10.18653/v1/2020.acl-main.463
3. Chen, M., et al.: Evaluating large language models trained on code. arXiv preprint arXiv:2107.03374 (2021)
4. Feng, Z., et al.: CodeBERT: a pre-trained model for programming and natural languages. In: Findings of the Association for Computational Linguistics: EMNLP 2020, pp. 1536–1547. Association for Computational Linguistics, Online (2020). https://doi.org/10.18653/v1/2020.findings-emnlp.139
5. Gauthier, J., Levy, R.: Linking artificial and human neural representations of language. In: Proceedings of the 2019 Conference on Empirical Methods in Natural Language Processing and the 9th International Joint Conference on Natural Language Processing (EMNLP-IJCNLP), pp. 529–539. Association for Computational Linguistics, Hong Kong, China (2019). https://doi.org/10.18653/v1/D19-1050

6. Husain, H., Wu, H.H., Gazit, T., Allamanis, M., Brockschmidt, M.: CodeSearchNet challenge: evaluating the state of semantic code search (2019). https://doi.org/10.48550/ARXIV.1909.09436

7. Kriegeskorte, N., Mur, M., Bandettini, P.: Representational similarity analysis - connecting the branches of systems neuroscience. Front. Syst. Neurosci. **2** (2008). https://doi.org/10.3389/neuro.06.004.2008

8. Lepori, M., McCoy, R.T.: Picking BERT's brain: probing for linguistic dependencies in contextualized embeddings using representational similarity analysis. In: Proceedings of the 28th International Conference on Computational Linguistics, pp. 3637–3651. International Committee on Computational Linguistics, Barcelona, Spain (Online) (2020). https://doi.org/10.18653/v1/2020.coling-main.325

9. Lu, S., et al.: CodeXGLUE: a machine learning benchmark dataset for code understanding and generation (2021). https://doi.org/10.48550/ARXIV.2102.04664

10. Merchant, A., Rahimtoroghi, E., Pavlick, E., Tenney, I.: What happens to BERT embeddings during fine-tuning? In: Proceedings of the Third BlackboxNLP Workshop on Analyzing and Interpreting Neural Networks for NLP, pp. 33–44. Association for Computational Linguistics, Online (2020). https://doi.org/10.18653/v1/2020.blackboxnlp-1.4

11. Puri, R., et al.: CodeNet: a large-scale AI for code dataset for learning a diversity of coding tasks. In: Thirty-Fifth Conference on Neural Information Processing Systems Datasets and Benchmarks Track (Round 2) (2021). https://openreview.net/forum?id=6vZVBkCDrHT

12. Rogers, A., Kovaleva, O., Rumshisky, A.: A primer in BERTology: what we know about how BERT works. Trans. Assoc. Comput. Linguist. **8**, 842–866 (2020). https://doi.org/10.1162/tacl_a_00349

# Location Data Anonymization Retaining Data Mining Utilization

Naoto Iwata[1]([✉]), Sayaka Kamei[1][iD], Kazi Md. Rokibul Alam[2],
and Yasuhiko Morimoto[1][iD]

[1] Hiroshima University, Hiroshima, Japan
{m223151,s10kamei,morimo}@hiroshima-u.ac.jp
[2] Khulna University of Engineering and Technology, Khulna, Bangladesh
rokib@cse.kuet.ac.bd

**Abstract.** Location information, such as customers' home addresses, is essential for data mining tasks. On the other hand, it is sensitive private information. In most countries, when requesting a third party for data mining, it is legally required to anonymize personal information such as home addresses in the database. However, conventional anonymization methods significantly lose the usefulness that location information has inherently. In this paper, we proposed an anonymization method of location retaining important locational features. In the proposed method, each address is replaced with a ranking value of the distance from each facility that is important in terms of location, such as a station or a supermarket. We examined our method and confirmed that the important rules mined from non-anonymized data could also be mined from our anonymized data.

**Keywords:** Database anonymization · Address · Location · Database utilization · Privacy preserving data mining

## 1 Introduction

The evolution of information and communication technology has made it possible to collect vast amounts of diverse data, which we call "Big Data." Data mining can discover useful regularities and unnoticed insights from big data. Data mining used to be done by a closed organization in which databases were kept inside the organization.

However, recent big data has become too big and diverse to utilize without specialized technologies and skills. Therefore, most organizations must ask a third-party agent for data mining to utilize big data. In addition, current machine learning methods require a large amount of training data whose size is beyond that a single organization can collect. Thus, there is an increasing need to share or outsource databases.

Supported by KAKENHI (20K11830) Japan.

Besides, we must be aware of individuals' privacy when analyzing big data. In Japan and most countries, when requesting a third party for data mining, it is legally required to anonymize personal information in the database [10]. Location information such as customers' home addresses is essential for data mining tasks. However, it has to be anonymized since it is sensitive private information.

## 1.1 Conventional Database Anonymization

Figure 1 is an example of a typical ID-POS database, which contains a POS record table and a customer table. When we ask a third-party agent for data mining, we have to anonymize the database. Figure 2 is the anonymized ID-POS database, in which (1) we remove or replace identifiable values and attributes, (2) we remove or replace foreign key (link) values, and (3) we remove or replace peculiar/specific values.

ID-POS record

| ID | TID (Trans. ID) | Shop ID | Date | Time | Item | Item Class | Price |
|---|---|---|---|---|---|---|---|
| 5579 | 1906150015 | 023 | 20220615 | 1230 | 4902105021590 | 1008 | 138 |
| 5579 | 1906150015 | 023 | 20220615 | 1230 | 4901330523145 | 1012 | 88 |
| | | | | | | | |

Customers

| ID | Name | Gender | D.O.B. | Address | Phone |
|---|---|---|---|---|---|
| 5579 | Aya | F | 19850914 | Kagami 1–7–1, Hiroshima | 0824771111 |
| 5580 | Jiro | M | 19650101 | Shitami 3–4–2, Hiroshima | 0824771111 |
| | | | | | |

**Fig. 1.** An example of ID-POS database

(1) remove/replace identifiers values/attributes
(2) remove/replace foreign key (link) values
(3) remove/replace peculiar values

ID-POS record

| ID (1)(2) | TID (Trans. ID)(2) | Shop ID | Date | Time (3) | Item | Item Class | Price |
|---|---|---|---|---|---|---|---|
| 4918 | 1906155022 | 023 | 20220615 | 123X | 4902105021590 | 1008 | 138 |
| 4918 | 1906155022 | 023 | 20220615 | 123X | 4901330523145 | 1012 | 88 |
| | | | | | | | |

Customers

| ID(1)(2) | Name(1) | Gender | D.O.B.(3) | Address (1)(2) | Phone (1) |
|---|---|---|---|---|---|
| 4918 | | F | 30s | Kagami 1–x–x, Hiroshima | |
| 6466 | | M | 50s | Shitami 3–x–x, Hiroshima | |
| | | | | | |

**Fig. 2.** Anonymized ID-POS database

As for the home address information, the home address string "Kagami 1-7-1, Hiroshima" indicates a specific position on the map as in Fig. 3 (a). To preserve privacy, we generalize, mask, or perturb values, which may lead to the identification of an individual. We usually generalize or mask because perturb may mislead incorrect location insights. The generalization and mask mean an expansion of the area of address information. For example, if the home address "Kagami 1-7-1, Hiroshima" is masked to "Kagami 1-X-X Hiroshima," the masked information in the map becomes somewhere inside the red polygon in Fig. 3 (b). As we can see on the map, conventional anonymization methods significantly lose the usefulness that location information has inherently. As a result, we cannot mine specific location insights from the anonymized database.

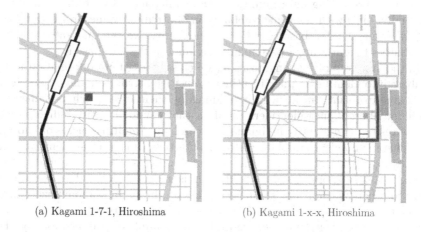

(a) Kagami 1-7-1, Hiroshima          (b) Kagami 1-x-x, Hiroshima

**Fig. 3.** Home address information on the Map

## 1.2  Utility Retained Location Anonymization

In this paper, we proposed a new anonymization method of locations, such as home addresses. Unlike conventional methods, it retains essential locational features for data mining. We replace each address with a distance ranking from each essential facility, such as a station, a bus stop, and a supermarket. For example, an address string, "Kagami 1-7-1, Hiroshima," is replaced with a list of a ranking that implies a place whose location is the $X$-th to the nearest station, the $Y$-th to the nearest bus stop, the $Z$-th to the nearest supermarket, and so forth. Together with the ranking, we also publish statistical information on distance values so that we can approximately infer distance values from ranking information.

Locational features of an address, such as how close to the station, bus stop, and supermarket, are essential for analyzing a database that contains addresses. On the other hand, we cannot disclose home address information to a third-party agent since it is sensitive private information. The proposed anonymization method can preserve privacy while retaining essential locational features.

We examined our method through intensive experiments and confirmed that the crucial rules mined from non-anonymized address data could also be mined from our anonymized data.

The rest of the paper is organized as follows. We discuss related works in Sect. 2. Section 3 describes the proposed anonymization method in detail. Next, we examined the utilities of the proposed anonymization in Sect. 4. Then, we conclude this paper in Sect. 5.

## 2    Related Works

The idea of $k$-anonymity [12] is frequently used for database anonymization and privacy-preserving data mining. $K$-anonymity means that there exist $k$ or more individuals with the same combination of attribute values in the database. By processing the database to satisfy $k$-anonymity, we can guarantee that the probability of identifying a particular individual in the database is $\frac{1}{k}$.

In the first step of $k$-anonymization, we classify attribute information in the database into identifiers, quasi-identifiers, sensitive attributes, and others. Identifiers are data that can identify individuals by themselves, such as names and unique numbers. Quasi-identifiers are data that cannot identify individuals by themselves but can identify individuals by combining multiple pieces of information. There is no clear standard as to which attribute is a quasi-identifier since the nature of the database determines it, but in general, age, gender, and address fall under this category. A sensitive attribute is information one does not want to reveal to others, such as one's annual income or illness. In $k$-anonymization, we assume that the attacker knows the quasi-identifiers and processes them so that their combination is not unique. Figure 4 is an example of $k$-anonymization. For example, suppose that an attacker has the background knowledge that "A" is in the database, and she/he knows "A"'s age and address. In this case, the attacker cannot distinguish whether "A" is line 1 or 2 in the anonymized database.

The idea of $k$-anonymity has been enhanced so that we can reduce the privacy risk [6,8]. The idea and its variants have been used to anonymize location data such as home addresses and GPS records in privacy-aware location-based and sensing services [1,4,5,9]. Though we can reduce the identification risk, conventional location anonymization methods based on grouping significantly lose location information's potential data mining utility.

Database perturbation methods for privacy-preserving data mining have been studied, which add randomized noises so that sensitive values of individuals cannot be identified. Some database perturbation works consider methods to perturb users' location data, which are collected in location-based services [11].

Such privacy-preserving data mining works are kinds of database anonymization methods retaining data mining utility. Though we can retain statistical properties and find insights derived from aggregated values, we cannot retain linkages between individuals' sensitive values. As for location data, we can find insights derived from aggregated values in each pixel grid. For example, we can find a rule: "traffic accidents tend to occur close to crossings." (A grid containing

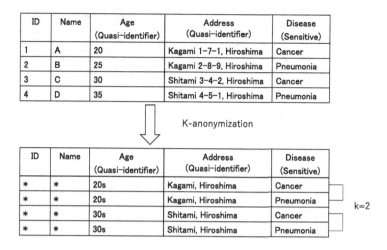

**Fig. 4.** Example of $k$-anonymity

a crossing has more traffic accidents than a grid without a crossing.) However, we cannot find rules related to quantitative location proximity without individual locations and their linkages (distances). We cannot find a rule, for example, "the radius from a crossing that maximizes the ratio of traffic accidents is 0.8 mile" because of the loss of linkages (distances) between individuals' locations. As a result, we cannot construct classification and regression tree models, which are essential prediction models and contain important location proximity insights in area marketing.

## 3   Location Anonymization

In this section, we assume all location information is represented as two-dimensional coordinates such as latitude and longitude. Figure 5 is a tiny example with five sensitive locations (e.g., the customer's home), which are represented as red squares on a map. The map has three kinds of facilities, stations (purple circles), bus stops (blue diamonds), and supermarkets (green triangles), which affect the location.

To anonymize the sensitive locations (red squares), we first calculate distance to the closest station, bus stop, and supermarket. The dot lines in the figure are the corresponding nearest distances. Table 1 is the calculated distance. We, then, replace distance values with ranking for each station, bus stop, and supermarket. The number in parenthesis is the ranking value. For example, the red square 1 is represented as $(3, 4, 1)$, which implies the third, the fourth, and the first in the distance ranking to the nearest station, bus stop, and supermarket, respectively.

Quantitative proximity, i.e., distance values, is one of the most meaningful information in spatial data mining tasks. We also publish statistical information on distance values to infer approximate distance values from the distance

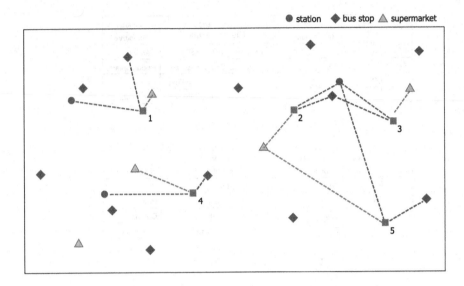

**Fig. 5.** Tiny example with 5 sensitive locations (Color figure online)

ranking. In our proposal, we publish percentile distance values. If the number of location data is not too small like the tiny example, we publish the top 10%, 20%,..., 90% distance values. Even for such percentile representative values, we had better not publish the exact value in a privacy-aware environment. So, if $X$ percentile ranking is $X_r$, we calculate the average distance value of $(X_r - n)$-th, ..., $(X_r - 1)$-th, $(X_r)$-th, $(X_r + 1)$-th, ..., $(X_r + n)$-th and use the average value as $X$ percentile distance value. Table 3 in the next section is an example of the percentile distance value. In the example, we can infer that the top 10% distance to the nearest station is around 157.61.

**Table 1.** Distance and rank for each facility

| ▣ Sensitive Location | ● Dist. & Rank to closest stn | ◆ Dist. & Rank to closest bs | △ Dist. & Rank to closest sup |
|---|---|---|---|
| 1 | 45 (3) | 35 (4) | 10 (1) |
| 2 | 30 (1) | 25 (2) | 25 (3) |
| 3 | 40 (2) | 40 (5) | 20 (2) |
| 4 | 55 (4) | 15 (1) | 35 (4) |
| 5 | 90 (5) | 30 (3) | 85 (5) |

# 4   Experiments

We examined the data mining utility of the proposed anonymization. We perform decision tree analysis using anonymized data and non-anonymized location data. We also evaluate data security from the viewpoint of k-anonymity. Here, we examine how many anonymized locations are in the candidate areas selected from the rankings and percentiles.

## 4.1   Data Set

We used two databases containing real locations. One is a real estate evaluation data set provided by the UCI Machine Learning Repository [3], and the other is the LIFULL HOME'S Monthly Data of Rentals and Sales [7], which is a database of monthly rent for an apartment provided by Japanese rental housing agency.

**UCI Real Estate Evaluation Data Set.** This dataset is the market historical data set of real estate evaluation collected from Sindian Dist., New Taipei City, Taiwan. This data for the regression analysis consists of 414 records. We use three attributes, "house price," "latitude," and "longitude." There are 412 non-duplicate records on the projected three dimensions. We randomly generated the location of six facilities: station, bus stop, supermarket, hospital, market, and school. We decide the number of each facility based on the density of corresponding facilities in the city. The number of each facility is summarized in the second column of Table 2.

**Table 2.** Number of facility

| Facility (UCI * Taipei) | Number | Facility (HOMES:Kyoto) | Number |
|---|---|---|---|
| Station | 13 | Station | 34 |
| Bus stop | 296 | Primary School | 20 |
| Supermarket | 171 | Junior high school | 13 |
| Hospital | 51 | Convenience store | 137 |
| Market | 10 | Supermarket | 28 |
| School | 33 | General hospital | 26 |

Using the location of six kinds of facilities, we replace each location information (the latitude and longitude) with a six-dimensional distance ranking to the nearest station, bus stop, supermarket, hospital, market, and school. Together with the location anonymization, we publish percentile representative distance values for every 10%. Since the size of the location database is not large enough, we divide the database into just two (the minimum number of divisions) and calculate the percentile representative distance values for each. Table 3 is the average percentile distances of the anonymized location database.

Table 3. Average percentile distance table (UCI:Taipei)

|    | to station | To bus stop | To supermarket | To hospital | To market | To school |
|----|-----------|-------------|----------------|-------------|-----------|-----------|
| 10 | 157.61 | 54.83 | 55.36 | 85.48 | 325.04 | 158.75 |
| 20 | 251.60 | 74.22 | 81.03 | 122.60 | 403.86 | 212.62 |
| 30 | 330.04 | 93.97 | 113.98 | 153.52 | 561.75 | 302.16 |
| 40 | 391.71 | 117.94 | 140.28 | 211.36 | 761.79 | 366.35 |
| 50 | 492.23 | 140.30 | 178.46 | 278.10 | 912.62 | 464.26 |
| 60 | 641.72 | 154.89 | 205.18 | 328.71 | 1037.98 | 527.19 |
| 70 | 1164.84 | 174.14 | 242.42 | 398.11 | 1157.61 | 611.83 |
| 80 | 1783.18 | 195.97 | 276.04 | 460.21 | 1303.33 | 730.69 |
| 90 | 2645.22 | 234.24 | 374.69 | 993.36 | 1766.23 | 860.04 |

**LIFULL HOME'S Monthly Data of Rentals and Sales.** This dataset is the monthly rent for approximately 66,360,000 apartments LIFULL HOME'S. In this experiment, we use data from July 2015. We extracted 7,121 apartments (4,424 apartments in Nakagyo Ward and 2,697 apartments in Shimogyo Ward) in Kyoto City, which have no duplicate on projected four dimension ID, monthly rent, latitude, and longitude. We chose the city because it is an average urban city in Japan.

Next, we collected real locations of six facilities: station, primary school, junior high school, convenience store, supermarket, and general hospital from a digital map. The number of each facility is summarized in the fourth column of Table 2. Using the facility locations, we anonymize real apartment locations and then make a similar average percentile distance table, shown in Table 4.

Table 4. Average percentile distance table (HOMES:Kyoto)

|    | to pschool | To jhschool | To conveni | To super | To hospital | To station |
|----|-----------|-------------|------------|----------|-------------|------------|
| 10 | 160.26 | 197.38 | 48.93 | 110.19 | 196.46 | 179.97 |
| 20 | 232.91 | 268.88 | 76.68 | 162.91 | 311.60 | 232.04 |
| 30 | 286.14 | 357.56 | 100.13 | 210.91 | 404.48 | 277.86 |
| 40 | 337.03 | 435.35 | 126.87 | 251.81 | 493.41 | 322.30 |
| 50 | 393.78 | 505.99 | 152.08 | 289.97 | 572.67 | 372.77 |
| 60 | 447.07 | 579.39 | 176.03 | 328.59 | 675.59 | 413.52 |
| 70 | 517.95 | 672.46 | 201.30 | 390.25 | 768.93 | 456.06 |
| 80 | 591.62 | 756.73 | 225.17 | 460.23 | 891.24 | 508.11 |
| 90 | 684.56 | 907.27 | 259.84 | 543.71 | 1013.25 | 589.00 |

## 4.2   Decision Tree and Regression Tree

To examine the data mining utility, we constructed two regression trees of the house price for the UCI data; one is by using the non-anonymized original data, and the other is by using the anonymized data. Figure 6 is the tree from the non-anonymized data and Fig. 7 is the one from the anonymized data. Table 5 compares the top-3 significant branch conditions of the two regression trees. In the table, rank 263.5 among 412 locations in the "to station" attribute means that the rank is between 60% (approx. 247th) and 70% (approx. 288th). Based on the percentile values in Table 3, the estimated value becomes between 641.72 and 1164.84. As for rank 123.5 is *around* 30% (approx. 124th). In this case, we estimate the range between 290.82 (average of 251.60 (20%) and 330.04 (30%)) and 360.88 (average of 330.04 (30%) and 391.71 (40%)). Similarly, for rank 379.5, which is further than 90% (approx. 371th), we estimate the range between 2645.22 (90%) and $\infty$.

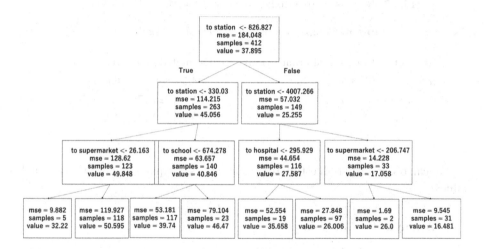

**Fig. 6.** Regression tree from non-anonymized locations (UCI:Taipei)

**Table 5.** Comparison of two models (UCI:Taipei)

| Attribute | Conditional branch (ranking) | Estimated value | Actual value |
|-----------|------------------------------|-----------------|--------------|
| to station | 263.5 | [641.72, 1164.84] | 826.827 |
| to station | 123.5 | [290.82, 360.88] | 330.03 |
| to station | 379.5 | [2645.22, ∞] | 4007.266 |

Similarly, we constructed regression trees of monthly rent for the HOME'S data. Table 6 compares the top-3 significant branch conditions of the two trees.

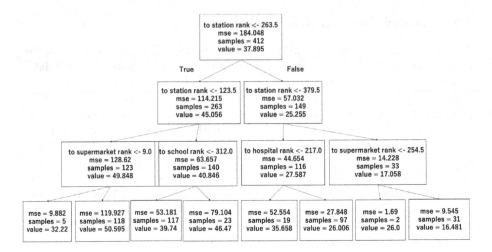

**Fig. 7.** Regression tree from anonymized locations (UCI:Taipei)

**Table 6.** Comparison of two models (HOMES:Kyoto)

| Attribute | Conditional branch (ranking) | Estimated value | Actual value |
|---|---|---|---|
| to pschool | 6513.5 | $[684.56, \infty]$ | 696.444 |
| to conveni | 5127.5 | $[201.30, 225.17]$ | 220.251 |
| to super | 5037.5 | $[390.25, 460.23]$ | 401.356 |

We can observe that all estimated value ranges adequately contain the actual values.

### 4.3 Anonymity Analysis

One can observe six-dimensional distance ranking and percentile statistics in our proposed anonymization in the two location databases. Assume distance ranking $(sta, bus, sup, hos, mkt, sch) = (60, 60, 230, 20, 140, 100)$ in 412 UCI data as an example. According to the percentile statistic in Table 3, the $1st$ column's ($60th$) estimated distance range is between $157.61(41st = 10\%)$ and $251.60(82nd = 20\%)$, the $2nd$ column's ($60th$) estimated distance range is between $54.83(41st = 10\%)$ and $74.22(82nd = 20\%)$, ..., and the 6th column's ($100th$) estimated distance range is between $212.62(82nd = 20\%)$ and $302.16(124th = 30\%)$. If one knows the locations of all 412 sensitive data, she/he can identify the location by the query using the estimated distance conditions. It is an unrealistic situation. However, some kinds of background knowledge may cause identification risks.

We examined how many sensitive locations are in an area that satisfies the estimated distance condition inferred from a six-dimensional distance ranking for each. We found that 185 records are identical by corresponding six-dimensional distance ranking among 412 sensitive data in the UCI data. Similarly, there are 571 identical records among 7, 121 sensitive data in the HOMES data. (Note that facility locations in the UCI data are randomly generated and uniformly distributed in the city. On the other hand, facility locations in the HOMES data are existing facilities in the city. We think it causes the difference in the unique record ratio between the two location databases.)

These identical records are 1-anonymity and are risky in the sense of the $k$-anonymity idea. Suppose Bob has background knowledge that Alice's record exists in the database and knows Alice's home. He can calculate distances from Alice's home to the nearest station, bus stop, ..., and school. Using the calculated distances, he can get a list of estimated distance rankings. If the estimated distance rankings match such an identical record, he may identify Alice's anonymized record in the database.

To prevent such a compromise from critical background knowledge, we should not provide both a distance ranking table and the corresponding percentile table together to a third-party organization. Note that data mining concerning quantitative proximity (distance) can be done without a percentile table. So, a third-party organization that is provided only a distance ranking analyzes and gets data mining results based on rankings. A database owner with the corresponding percentile table can interpret the data mining results analyzed by the third-party organization into those based on distance. So, we can claim that the proposed anonymization preserves location privacy.

On the other hand, small ranking values, such as the nearest, the second nearest, and so on, are themselves risky. For example, assume that we disclose that a location record has a distance rank 1st from a station, which implies that the record's location is the nearest one from a station. If there is one residence/apartment very close to a station in the city, one can easily imagine that the location is that residence/apartment. In such a case, we should replace the $k$ exact ranking, say, "1st," "2nd," ..., "$kth$" with "top-$k$" for small $k$.

## 5 Conclusions

In this paper, we proposed a location anonymization method in which we replace sensitive locations such as the home addresses of users with a list of distance rankings to the nearest facilities. Intensive experimental results show that we can find informative quantitative proximity insights from the distance rankings by using percentile statistics of distance values. For future works, we will try to apply this anonymization to other spatial data mining functions such as clustering, association rule mining and so forth. We should also consider analyzing the proposed anonymization's theoretical foundations of privacy risks.

**Acknowlendgement.** In this paper, we used "LIFULL HOME'S Dataset" provided by LIFULL Co., Ltd. via IDR Dataset Service of National Institute of Informatics.

# References

1. Cornelius, C., Kapadia, A., Kotz, D., Peebles, D., Shin, M., Triandopoulos, N.: AnonySense: privacy-aware people-centric sensing. In: Proceedings of the 6th International Conference on Mobile Systems, Applications, and Services, pp. 211–224. ACM (2008)
2. Du, W., Zhan, J.Z.: Using randomized response techniques for privacy-preserving data mining. In: Proceedings of the ACM SIGKDD International Conference on Knowledge Discovery and Data Mining, pp. 505–510. ACM (2003)
3. Dua, D., Graff, C.: UCI machine learning repository (2019). http://archive.ics.uci.edu/ml. School of Information and Computer Science, University of California, Irvine. Yeh, I.C., Hsu, T.K.: Building real estate valuation models with comparative approach through case-based reasoning. Appl. Soft Comput. **65**, 260–271 (2018)
4. Huang, K.L., Kanhere, S.S., Hu, W.: Towards privacy-sensitive participatory sensing. In: Proceedings of International Conference on Pervasive Computing and Communications, pp. 1–6. IEEE (2009)
5. Kazemi, L., Shahabi, C.: TAPAS: trustworthy privacy-aware participatory sensing. Knowl. Inf. Syst. **37**(1), 105–128 (2013)
6. Li, N., Li, T., Venkatasubramanian, S.: t-closeness: Privacy beyond k-anonymity and l-diversity. In: Proceedings of IEEE International Conference on Data Engineering (ICDE), pp. 106–115 (2007)
7. LIFULL Co., Ltd.: LIFULL HOME'S monthly data of rentals and sales. Informatics Research Data Repository, National Institute of Informatics (dataset) (2015). https://doi.org/10.32130/idr.6.3
8. Machanavajjhala, A., Kifer, D., Gehrke, J., Venkitasubramaniam, M.: l-diversity: privacy beyond k-anonymity. ACM Trans. Knowl. Discov. Data (TKDD) **1**(1), 3-es (2007)
9. Mano, M., Ishikawa, Y.: Anonymizing user location and profile information for privacy-aware mobile services. In: Proceedings of the ACM SIGSPATIAL International Workshop on Location Based Social Networks, pp. 68–75. ACM (2010)
10. The Personal Information Protection Commission JAPAN. https://www.ppc.go.jp/en/index.html
11. Quercia, D., Leontiadis, I., McNamara, L., Mascolo, C., Crowcroft, J.: SpotME if you can: randomized responses for location obfuscation on mobile phones. In: Proceedings of International Conference on Distributed Computing Systems, pp. 363–372 (2011)
12. Sweeney, L.: k-anonymity: A model for protecting privacy. Internat. J. Uncertain. Fuzziness Knowl.-Based Syst. **10**(5), 557–570 (2002)

# A Distributed SAT-Based Framework for Closed Frequent Itemset Mining

Julien Martin-Prin[1] , Imen Ouled Dlala[1] , Nicolas Travers[1(✉)] ,
and Said Jabbour[2]

[1] Pôle Universitaire Léonard de Vinci, Research Center, Paris La Défense, Paris,
France
{julien.martin-prin,imen.dlala,nicolas.travers}@devinci.fr
[2] CRIL CNRS UMR 8188, Université d'Artois, 62307 Lens Cedex, France
jabbour@cril.fr

**Abstract.** Frequent Itemset Mining is an essential part of data mining. SAT-based approaches that extract frequent itemsets in big data encounter significant challenges when computing power and storage capacity are limited. This paper proposes an efficient distributed SAT-based framework for the Closed Frequent Itemset Mining problem (CFIM) which minimizes communications throughout the distributed architecture and reduces bottlenecks due to shared memory. Moreover, it enhances scalability and fault tolerance. This approach makes use of a Computation-Distributed Paradigm to efficiently enumerate the set of all closed itemsets, by reducing the processing time. To the best of our knowledge, this paper presents the first attempt towards a distributed SAT-based approach for CFIM. An extensive empirical evaluation on various real-word datasets shows the efficiency of the approach.

**Keywords:** Data mining · Closed frequent itemset mining ·
Propositional satisfiability · Big data · Distributed computing

## 1 Introduction

Frequent Itemset Mining (FIM, for short) [1,4] is a fundamental problem in Data Mining (DM, for short), knowledge discovery and data analysis. This problem aims at finding all sets of items that frequently occur in a given transaction database. There are plenty of various real-word applications to FIM, including marketing, scientific analytics, e-commerce, etc.

Recently, approaches that extract useful knowledge from data have appeared [24]. This research trend has endowed in a natural and flexible way the combination of complex constraints. Therefore, in contrast to traditional approaches, user preferences can be easily handled without rewriting the algorithm from scratch. In this context, several studies based on symbolic AI, including Constraint Programming (CP), Propositional Satisability (SAT) and Answer Set

© The Author(s), under exclusive license to Springer Nature Switzerland AG 2022
W. Chen et al. (Eds.): ADMA 2022, LNAI 13726, pp. 419–433, 2022.
https://doi.org/10.1007/978-3-031-22137-8_31

Programming (ASP), were conducted to deal with different DM tasks, such as FIM [18], Closed Frequent Itemsets Mining (CFIM, for short) [19,25], frequent sequence mining [14,22], association rules mining [5], etc. Those AI-based techniques work well in practice on typical data sets. However, they are still less efficient than classical approaches and not suitable for larger datasets. Specifically, their efficiency decreases significantly when the dataset increases in size or the support threshold turns to be low (*i.e.*, a large number of itemsets). This issue can be explained by the huge size of constraint networks/propositional formulas encoding the itemset mining task.

Some efforts have been brought to speed up the efficiency of specialized algorithms for FIM as well as SAT solvers by running them in parallel and in distributed environments. We provide a deep overview on it in Sect. 2.

The introduction of the first parallel SAT-based framework for FIM has reduced the performance gap between declarative and specialized approaches [10]. However, even if this framework significantly pushed forward the performance of SAT-based itemset mining frameworks, there are still improvements to be made. In this paper, we propose an extensible distributed framework `distriSATMiner` that aims at distributing the computation to scale up. It extends the `paraSATMiner` framework presented in [10] to distributed systems. The main idea is to use a distributed architecture where the main server is in charge of preparing the dataset as well as the computation instructions to SAT based solvers avoiding communications between solvers. Each distributed SAT solver is responsible for finding and sending models (*i.e.*, frequent itemsets) back to the main server. A fault tolerance work is done during the mining strategy distribution to check if a solver is down or if the main server itself has failed.

## 2   Literature Review

In this section we propose a literature review on distributed FIM solvers using specialized approaches and distributed SAT solvers.

*Distributed Frequent Itemset Mining Approaches.* There have been many challenges regarding the development of distributed FIM (DFIM) [13] using specialized approaches. In this context, distributed algorithms were proposed to mine itemsets and association rules on distributed databases [6,7]. Later, the emergence of Hadoop, a platerform based on the MapReduce framework [9], and Spark [26], a new in-memory data flow platform, led to the introduction of new frameworks for DFIM based on traditional algorithms. As an example, PFP [20] proposed a new parallel form of FP-Growth [15] adapted to MapReduce [9]. PFP [20] works with distributed datasets and uses the performances of MapReduce [9] to perform parallel tasks in distributed environments. More recently, ParallelCharMax [12], a FIM algorithm based on Charm [27] introduced a master-slave architecture based on MapReduce framework [9] and Hadoop to solve the maximal frequent itemsets problem. However, the common feature of those distributed environments is to distribute the computation by using data distribution. However, while all these specialized approaches addressed various challenges, they still have a major drawback in terms of flexibility. Indeed, they are faster than the frameworks based on a declarative approach but faces real challenges when new constraints come into play. A complete redesign is needed to extract a different itemset with specialized solvers, which is not the case for declarative approaches where adding a new constraint is sufficient.

*Distributed SAT Solvers.* Distributed SAT solvers rely on two strategies: divide-and-conquer and portfolio algorithms. In portfolio, the same formula can be assigned to concurrent workers with different strategies to find a solution. Divide-and-conquer solvers operate in partitioning the search space to evaluate among several concurrent workers.

**Portfolio SAT Solvers.** Distributed portfolio SAT based solvers introduce features such as hierarchical parallelism and decentralized design [3,23]. *Horde-Sat* [3] proposed a modular and decentralized design facilitating the implementation. The decentralized design offers the possibility to have all its node equivalent in the parallel system and leaderless. It enables more scalability as well as simplifying the algorithm. However, this paradigm is not suited for enumeration since all slaves will search for every model, introducing a workload redundancy.

**Divide-and-Conquer SAT and #SAT Solvers.** Several distributed divide-and-conquer SAT and #SAT solvers have been proposed [8,16,17]. The master slave architecture is widely used which allows reliability, facility of deployement and extensibility to various problems. In [8], the master is responsible for reading the problem file and generating the final output. Only servants are assigned part of the search space to investigate.

To reduce the computation time, some approaches propose scalable distributed learning and adaptive resource scheduling [8], clause learning [17], looking ahead [16] and partitioning based on scattering [16] (DPLL-based partitioning). In [17], two schemes are used for maintaining the sets of learned clauses: i) a database of partitioned constraints and independent learned clauses, ii) each node supports a limited set of unary learned clauses specific to that node. The VSIDS branching heuristic [21] and unit propagation look ahead are used to partition the formula and to always branch on the most propagating literal.

*Grid-SAT* [8], a solver based on Chaff [21] introduces a smart backtracking. It uses zChaff [2] as a solver core but in a new parallel form and tackling resource sharing issues by implementing a distributed clause database subsystem that can acquire and release memory from a grid-resource pool. However, while this technique might be useful in this case, the bottleneck induced by the access to memory when sharing it makes memory-sharing inefficient on our solver.

**Hybrid SAT Solvers.** As several contributions to distributed SAT frameworks tend to focus on distributing existing parallel approaches, a new kind of hybrid solver has been introduced in [11,20]. Those frameworks combine the divide-and-conquer and the portfolio approaches.

As for the divide-and-conquer solver, the *Dolius* framework introduces in [11] with a master-slave model. The master divides the SAT formula through guiding paths and sends it to workers. The laters process the formula with potentially different types of SAT solvers. To avoid connection bottlenecks, the authors proposed a tree architecture where a slave could be a master and uses its own sub-slaves to solve its assigned problem which relies on a load balancing technique.

*PaInleSS* [20] is a distributed hybrid SAT framework which strategy, as for Dolius, implements a tree architecture. However, the core of the PaInleSS architecture is formed by the three core concepts of a typical parallel SAT solver: a sequential engine, parallelization and a sharing strategy.

# 3   SAT and FIM Problem Formalism

We provide here some preliminaries and notations about propositional satisfiability problems, itemset mining problem and SAT-encoding for itemset mining.

## 3.1   Propositional Logic and SAT Problem

Consider a propositional language $\mathcal{L}$ defined over a finite set of propositional variables $Var = \{p, q, r, \ldots\}$, logical constants $\bot, \top$, and logical connectives ($\neg$, $\wedge, \vee, \rightarrow$, and $\leftrightarrow$). Greek letters $\Phi, \Psi$, etc. denote propositional formulas. $Var(\Sigma)$ refers to the set of propositional variables occurring in the formula $\Sigma$.

A formula is called in *Conjunctive Normal Form* (CNF) if it is a conjunction ($\wedge$) of clauses, where a *clause* is a disjunction ($\vee$) of literals. A *literal* is a propositional variable ($p$) or its negation ($\neg p$). A *Boolean interpretation* (or world) $\mu$ of a formula $\Sigma \in \mathcal{L}$ is a total function from $Var(\Sigma)$ to $\{0, 1\}$.

$\mu$ is a *model* of $\Sigma$ if $\Sigma$ is satisfied under $\mu$ ($\mu(\Sigma) = 1$). Then, $\Sigma$ is said *satisfiable* if there exists a model of $\Sigma$. $models(\Sigma)$ denotes the set of models of a formula $\Sigma$. Finally, the propositional satisfiability problem SAT is NP-complete that consists in deciding whether a given CNF formula is satisfiable or not.

The all models enumerating problem for a CNF formula is a variant of the propositional satisfiability one; a significant issue in many practical SAT application domains such as data mining, unbounded model checking, etc.

## 3.2   Itemset Mining Problem

Let $\Omega = \{a_1, a_2, \ldots, a_m\}$ be a set of $m$ distinct items, an itemset $\mathcal{I}$ is a set of items such that $\mathcal{I} \subseteq \Omega$. We use $2^\Omega$ to denote the set of all possible itemsets over $\Omega$. A transaction database $\mathcal{D}$ is a set of subsets of $\Omega$, denoted by $\mathcal{D} = \{t_1, t_2, \ldots, t_n\}$, where each transaction $t_i \subseteq \Omega (1 \leq i \leq n)$ is defined as a couple $(tid_i; \mathcal{I}_i)$, where $tid_i$ is the transaction identifier and $\mathcal{I}_i \subseteq \Omega$ an itemset. Now, let we give some basic definitions and notations of frequent itemset mining problems.

Given a transaction database $\mathcal{D}$, the cover of an itemset $\mathcal{I} \subseteq \Omega$ in $\mathcal{D}$ is defined as $Cover(\mathcal{I}, \mathcal{D}) = \{tid \mid (tid, \mathcal{J}) \in \mathcal{D} \text{ and } \mathcal{I} \subseteq \mathcal{J}\}$ and its support in $\mathcal{D}$, is defined as $\mathcal{S}(\mathcal{I},\mathcal{D}) = |Cover(\mathcal{I},\mathcal{D})|$. An itemset is said to be closed if all its supersets have strictly inferior support. Formally, an itemset $\mathcal{I}$ is closed if for all $\mathcal{J} \supset \mathcal{I}$, $\mathcal{S}(\mathcal{I},\mathcal{D}) > \mathcal{S}(\mathcal{J},\mathcal{D})$.

The problem of frequent itemsets is that of finding sets of items that appear many times in a given a transaction database $\mathcal{D}$ i.e., $FIM(\mathcal{D},\theta) = \{\mathcal{I} \subseteq \Omega| \mathcal{S}(\mathcal{I},\mathcal{D}) \geq \theta\}$. We denote by $CFIM(\mathcal{D},\theta)$ the set of all closed frequent itemsets. Note that an itemset $\mathcal{I}$ is frequent if and only if all its subsets are frequent. This property is called the anti-monotonicity. More formally, let $\mathcal{I}$ and $\mathcal{J}$ be two itemsets such that $\mathcal{I} \subseteq \mathcal{J}$. If $\mathcal{S}(\mathcal{J},\mathcal{D}) \geq \theta$ then $\mathcal{S}(\mathcal{I},\mathcal{D}) \geq \theta$.

### 3.3   SAT-Encoding for Closed Frequent Itemset Mining

Here, we review the SAT encoding scheme of the FIM problem proposed in [18]. Formally, the frequent itemsets mining problem is encoded into propositional logic as a set of variables and constraints. Those encoding the itemsets named $x_a$, for all $a \in \Omega$ and those representing transactions $y_i$, $1 \leq i \leq m$ i.e., $y_i$ is true if the itemset appears in transaction $T_i$.

The following constraints ensure the required mapping. More precisely, Constraint $(\Phi^{cov})$ captures the cover of an itemset i.e., each item belonging to the itemset exclude the transaction which does not contain it to be part of the solution. The one of $(\Phi^{freq})$, allows enforcing the itemset to be frequent. Finally $(\Phi^{clos})$ ensures the closeness of the itemset.

$$\Phi^{cov} = \bigwedge_{i=1}^{n} (\neg y_i \leftrightarrow \bigvee_{a \in \Omega \setminus \mathcal{I}_i} x_a) \qquad \Phi^{freq} = \sum_{i=1}^{n} y_i \geq \theta$$

$$\Phi^{clos} = \bigwedge_{a \in \Omega} (( \bigvee_{(tid_i, \mathcal{I}_i) \in \mathcal{D},\ a \notin \mathcal{I}_i} y_i) \vee x_a)$$

The set of the models of $\Phi^{cfim} = \Phi^{cov} \wedge \Phi^{freq} \wedge \Phi^{clos}$ corresponds to the closed frequent itemsets of $\mathcal{D}$. From each model $\mu$, the corresponding closed frequent itemset is $X_\mu = \{a \in \Omega,\ x_a \in \mu\}$.

Even though the polynomial encoding space complexity of FIM into propositional logic, i.e., $O(|\Omega| \times |\mathcal{D}|)$, in practice, the solving is intractable especially for large databases. To overcome such issue, a decomposition strategy has been successfully applied in [10]. A common motivation for using decomposition is to avoid dealing with the encoding of the whole database by making its solving equivalent to the one of many but of reasonable size sub-problems.

More formally, assume $\Omega = \{a_1, \ldots, a_m\}$ and let $\Delta(\mathcal{D},\theta)$ denotes the set of closed frequent itemsets of $\mathcal{D}$. This latter can be partitioned as follows:

$$\Delta(\mathcal{D},\theta) = \biguplus_{1 \leq i \leq m} \Delta_i \qquad (1)$$

where $\Delta_i$ is the set of closed frequent itemsets involving $a_i$ but not $a_j$ for $j < i$, i.e., $\Delta_i = \{X \in \Delta(\mathcal{D},\theta),\ a_i \in X \text{ and } \forall j < i, a_j \notin X\}$

From propositional point of view, each subset $\Delta_i$ can be computed by considering the models of $\Phi_i$ defined as a constrained $\Phi^{cfim}$ defined as follows:

$$\Phi_i = \Phi^{cfim} \wedge x_{a_i} \wedge \bigwedge_{1 \leq j < i} \neg x_{a_j} \tag{2}$$

Since each itemset in $\Delta_i$ involves $a_i$, the formula $\Phi^{cfim}$ in $\Phi_i$ can be substituted by the one restricted to transactions containing $a_i$ to avoid generating the whole encoding over $\mathcal{D}$. The sub-formula $\Psi_i = x_{a_i} \wedge \bigwedge_{1 \leq j < i} \neg x_{a_j}$ added to $\Phi^{cfim}$ is then a kind of guiding path [28]. Then the required models are those resulting from each $\Phi_i$. This decomposition method allows to considerably improve the performance of the enumeration. Moreover, as the set of models of $\Phi_i$, $1 \leq i \leq m$ are disjointed, the solving of the sub-problems $\Phi_1, \ldots, \Phi_m$ can be performed in parallel as detailed in [10] where each core solves is assigned with a subset of tasks (sub-problems). Even if balancing workload between cores enhance parallelism, the shared memory is a bottleneck when the core number grows. To overcome this issue, we propose an extensible distributed framework that aims at distributing the model enumeration to scale up.

## 4    Distributed SAT-Based Approach for CFIM

Thanks to the decomposition and the encoding, each task is independent and distributable without further communications between servers. The distriSAT-Miner framework is designed as a scaling up architecture with fault tolerance.

### 4.1    Architecture Presentation

The goal of our approach is to distribute enumeration of models between the main server and solvers through guiding paths and to reduce the synchronization steps. Each solver encodes subsets of models $\Delta_i$ from $\mathcal{D}$ using the threshold $\theta$.

The global architecture is presented in Fig. 1. We identify the main node, the broker service, the distributed database service and solver nodes.

The *Main Node* has three different jobs: 1) storing the transaction database $\mathcal{D}$ with corresponding parameters ($\Omega$), 2) scheduling the tasks among the solvers with a mining strategy based on the guiding path, and 3) launching the fault tolerance job to check if any solver is down or if the main node itself has failed.

The *Data Transfer Layer* relies on a distributed database that contains the transaction database $\mathcal{D}$ accessed by the solvers and the enumerated models of each $\Phi_i \in \Phi^{cfim}$. In fact, this storage procedure allows for further computations which reduces the communication transfers (warm start computation).

The *Instruction Layer* relies on a broker service that roots tasks $\Psi_i$ from the guiding path to the corresponding solvers. It manages a distributed pool of tasks allocated to each solver which enables fault tolerance and load balancing. The fault tolerance module interacts with the broker service to check if a solver node has failed and redistributes corresponding tasks using the mining strategy.

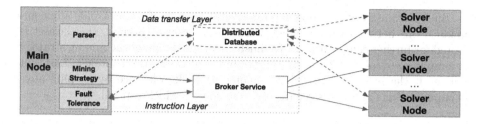

**Fig. 1.** Architecture schema.

To finish with, a *Solver Node* is composed of a solver component of `paraSAT-Miner` and connectors to the *data transfer layer* and the *instruction layer*.

Input: $\mathcal{D}$: transaction database,
    $\theta$: min support threshold,
    $\mathcal{B}$: list of brokers
/* Parser Step                  */
1  @$\mathcal{D} \leftarrow$ db_address($\mathcal{D}$);
2  **if** *empty(@$\mathcal{D}$)* **then**
3     |  $\Omega \leftarrow$ getSortedItems($\mathcal{D}$);
4     |  $\Omega \leftarrow$ filterItems($\Omega, \theta$);
5     |  db_Store(@$\mathcal{D}, \Omega, \mathcal{D}$);
6  **else**
7     |  $\Omega \leftarrow$ db_getItems(@$\mathcal{D}$);
8     |  $\Omega_p \leftarrow$
          db_processedItems(@$\mathcal{D}, \theta$);
9     |  $\Omega_B \leftarrow$ broker_queuedItems($\mathcal{B}$);
10    |  $\Omega \leftarrow \Omega - \Omega_p - \Omega_B$ ;
11 **end**
12 **foreach** $b \in \mathcal{B}$ **do**
13   |  broker_sendConfig(b, @$\mathcal{D}, \theta$);
14 **end**
/* Mining step               */
15 MiningStrategy($\Omega', \mathcal{B}$);
16 FaultTolerance($\mathcal{B}, \theta$);

**Algorithm 1:** Main Node

Input: $\Omega$: List of items,
    $\mathcal{B}$: list of brokers
1 **foreach** $i \in \{1, ..., |\Omega|\}$ **do**
2  |  $b \leftarrow \mathcal{B}[i\%|\mathcal{B}|]$;
3  |  broker_sendItem(b, i);
4 **end**

**Algorithm 2:** MiningStrategy

/* Initialization           */
1  (@$\mathcal{D}, \theta) \leftarrow$ broker_getConfig();
2  $\Omega \leftarrow$ db_getItems(@$\mathcal{D}$);
3  $\mathcal{D} \leftarrow$ db_getDB(@$\mathcal{D}$);
/* Mining guiding paths   */
4  $i \leftarrow$ broker_nextItem();
5  **while** $i \neq null$ **do**
6   |  $\Psi_i \leftarrow$ encodeGuidingPath($\Omega$, i);
7   |  $\Phi_i \leftarrow$ encodeDB($\mathcal{D}, \theta, \Psi_i$);
8   |  $\Delta_i \leftarrow PSATMiner(\Phi_i, \theta)$;
9   |  db_addModels(@$\mathcal{D}, \theta, i, \Delta_i$);
10  |  broker_ackItem(i);
11  |  $i \leftarrow$ broker_nextItem();
12 **end**

**Algorithm 3:** Solver Node

Input: $\theta$: min support threshold,
    $\mathcal{B}$ : list of brokers
1  **while** *!empty(broker_queue($\mathcal{B}$))* **do**
2   |  $s \leftarrow$ getDeadSolver();
3   |  **if** $s \neq null$ **then**
4     |  $b \leftarrow \mathcal{B}[s]$;
5     |  $\Omega_b \leftarrow$ broker_queuedItems(b);
6     |  MiningStrategy($\Omega_b, \mathcal{B} - b$);
7   |  **end**
8   |  wait();
9  **end**
10 **foreach** $b \in \mathcal{B}$ **do**
11  |  broker_sendItem(b, $null$);
12 **end**

**Algorithm 4:** FaultTolerance

Such a framework is interesting for several aspects. First, it avoids any communications between solvers which work autonomously. Second, the dataset is stored for further computations. To finish with, the design of the mining strategy is simplified by only defining the distribution of tasks in the broker. Last but not least, the broker service allows managing fault tolerance by keeping alive communications and pools separately from solvers which can fail.

## 4.2    The distriSATMiner Algorithm

The distriSATMiner algorithm is composed of three main tasks which are 1) the *treatment distribution* with database manipulations, the resources' allocation and the mining strategy, 2) the *production of models* which computes given tasks $\Phi_i$, and 3) the *fault tolerance* of the process. Algorithms 1 and 3 detail the computation led on main and solver nodes while Algorithm 2 defines the resource allocation strategy and Algorithm 4 deals with solvers' fault tolerance.

We must notice that main and solver nodes are designed in an asynchronous architecture to avoid time loss during the process. Since every task is made to be independent, no strong synchronization is required (except waiting for tasks).

**Main Node.** Algorithm 1 details instructions processed in the Main Node whose goal is to distribution tasks among the solvers. As seen in Fig. 1, it is composed of a *parser* and a *mining strategy*.

The *parser step* starts by allocating an identifier to the database @$\mathcal{D}$ and checks if it has already been stored (lines 1–2). It is used to check if an instance is already in progress (fault tolerance at *Main node* level). If not, it sorts increasingly the list of items $\Omega$ according to their frequencies from $\mathcal{D}$ (line 3) in order to ease the guiding path while encoding, and stores the list of items $\Omega$ and the database $\mathcal{D}$ in the distributed database (line 4) for further manipulations. Once the database is stored, the SAT problem configuration is sent to all solvers (lines 11–13) that shares the database address and the threshold to be computed.

The *mining step* treats the whole mining process by taking into account both the distribution of tasks and the fault tolerance (detailed below). First, it filters the sorted list of items by the given threshold $\theta$ (line 14). Then, this new list is then treated in the mining strategy (line 15) with the corresponding items and available brokers. To finish with, the fault tolerance module is launched (line 16).

**Mining Strategy.** This pluggable module in Algorithm 2 aims at allocating resources (items) among the available solvers through corresponding brokers. It distributes all items $i \in \Omega$ to solvers (line 1). For this, it relies on the broker service which associates to each solver a distributed pool (denoted by $b \in \mathcal{B}$). Thus, asynchronously the main node sends an item $i$ to a given solver node $b$ (line 3). Notice that this simple strategy distributes items uniformly to solvers (line 2). Other strategies will be studied in future works.

**Solver Node.** Algorithm 3 details the mining tasks executed in each solver node. First, it waits for the configuration of the mining problem. For this, it asks the broker service (*i.e.*, the associated pool) the database address @$\mathcal{D}$ and the threshold $\theta$ used for mining tasks (line 1). Then, it acquires encoded items $\Omega$ and transactions $\mathcal{D}$ (lines 2 and 3).

The following steps consist in mining frequent itemsets with $\theta$ and $\Omega$ and $i^{th}$ tasks. For each new item $i \in \Omega$ from the broker (line 4), it encodes the guiding path $\Psi_i$ (line 6) by keeping items from $\Omega$ to true if their occurrence is higher or equal to $i$. $\Psi_i$ is combined with the threshold $\theta$ to encode the guiding path $\Phi_i$ simultaneously with both $\phi^{cfim}$ and $\Psi_i$ (line 7) from Definition 2. Then, $\Phi_i$ can be used to mine $\Delta_i$ in the solver (line 8). Finally, it stores the mined models $\Delta_i$ (line 8) with corresponding information (unicity on the triplet: $\mathcal{D}$, $\theta$ and $i$). Once the models are mined and stored the solver can acknowledge the $i^{th}$ task to the broker (line 10) and proceed to the next item (line 11). Thus, the process is repeated until no more items are required to be processed (lines 5 and 11).

**Fault Tolerance.** We need to take into account the fault tolerance to avoid a full restart. Different failures are identified in the process: *Main Node* (parsing (Algorithm 1, lines 1–14), mining strategy (Algorithm 2), and the fault tolerance itself (Algorithm 4)), *Solver Node* (initialization (Algorithm 3, lines 1–4), encoding (lines 6–8), iii) mining (lines 8–9), and communications with the broker (lines 10–11)), the *Broker service*, and the *Distributed database*.

We must notice that *brokers* and the *distributed database* already deal with failures byt replicating both the database $\mathcal{D}$, processed models ($\Delta_i$ linked by @$\mathcal{D}, \theta, i$), and pools of tasks in progress $\mathcal{B}$. The Fault Tolerance is then handled for both the *Main Node* (Algorithm 1) and *Solver Nodes* (Algorithm 4).

If a breakdown occurs at the *Main Node* level (Algorithm 1), when it restarts (or another instance), it requires to check the tasks list $\Omega$. In that case, the database has already been stored (line 4) and we can get $\Omega$ from the database (line 6). Then, it checks if items have been processed $\Omega_p$ (line 7) and also if some are still to be mined $\Omega_B$ in brokers (line 8). Consequently, the list of missing processed items $\Omega$ is the set of items from the database without $\Omega_p$ and $\Omega_B$ (line 9). The following steps of the Algorithm 1 proceed with the mining strategy.

At solver nodes (Algorithm 4), if one fails the *Main Node* will distribute the corresponding tasks to remaining nodes through brokers. For this, while it remains items to be processed (line 1) it checks if any solver is down (lines 2 & 3). If so, it gets none-processed items $\Omega_b$ (lines 4 & 5) and apply the mining strategy on it (line 6). Recall that if the failure occurs while processing a task, items are removed from the pool only once computed models are stored. Thus, if it is not stored, it remains in the pool and will be processed by Algorithm 4.

To finish with, once all items are processed (*i.e.*, all brokers are empty) an "ending" message is sent to solvers (lines 10–12).

## 5    Experimental Results

In this section, we instantiated our approach `distriSATMiner` and evaluate the performances of our solution. Our approach uses a distributed architecture without any direct communication between main and solver nodes (see Fig. 1).

### 5.1    Technical Architecture

All the experiments were done on a kubernetes cluster equipped with an Intel Xeon E5-2698V4 with 20 cores at 2.2 GHz base frequency and 250 GB of RAM. To evaluate our distributed architecture, we scale up the experiments by either deploying the `paraSATMiner` [10] on a single VM with up to 24 cores/threads and a cluster of 24 VMs for solver nodes (`distriSATMiner`).

To instantiate our distributed architecture (Fig. 1), each node corresponds to a distinct virtual machine. The *data transfer layer* and its distributed database are handled by a `MongoDB` cluster which stores and replicates all configuration data and computed models in dedicated collections shared by the main and solver nodes. The *broker service* is provided by a `Kafka` which manages, distributes and replicates pools for each solver node. Notice that both *mongodb* and *kafka* guarantee fault tolerance at the storage level. Thus, our solution only requires to handle computation at fault tolerance level (Algorithm 1 and 4).

The whole architecture setup with the main and solver nodes implementation are available on Github with the deployment scripts[1].

As mentioned the set of closed frequents itemsets corresponds to formulas' models $\Phi$. To enumerate them, we followed the approach of [10] that adapts the MinSAT solver by disabling the restart and conflict analysis components. It showed that using a DPLL-based procedure is more suitable than a CDCL-based one. Since the number of models is usually huge, adding a blocking clause to avoid having the same model each time will slow down the unit propagation. To tackle this issue, each time a model is found, a chronological backtracking is performed while inhibiting the restart component to ensure completeness. Similarly a backtracking to the last decision is performed when a conflict occurs. Let us note that the cardinality constraint ($\Phi^{freq}$) is managed inside the solver.

### 5.2    Performance Analysis

We carried out our experimental evaluation using different real-world data mining datasets. We selected different datasets from the FIMI[2] repository.

We limit our comparison to the parallel SAT-based approach, called ParaSATMiner, proposed in [10]. It showed that the algorithm witnesses better performances than existing declarative methods in literature for mining frequent closed itemsets. Experiments were done using three different support thresholds

---

[1] https://github.com/leonard-de-vinci/Distributed_SAT_Approach_FIM.

[2] http://fimi.ua.ac.be/data/.

$\theta$, respectively representing a light, medium and heavy computation task. Performances are compared by testing the scale up of the number of solver nodes (**D**) *vs.* the number of cores (**P**) in a local ParaSaTMiner from 1 to 24 instances.

We must notice that the framework stores the dataset to be shared with solver nodes. Thus, for any experiment the global computation time can benefit from avoiding further precomputations of the dataset. However, to be fair with the parallel approach we count precomputing time (*i.e.,* "*cold start*").

Figure 2 shows the parallel and distribution effect on the solving time by varying the number of cores (P - dashed lines) and the number of solver nodes (D - plain lines). As shown, on most experiments, the distributed approach obtains better performances. In fact, the bottleneck in a parallel version occurs on the main memory access which makes this solution converging faster than the distributed version, especially when the number of cores exceeds 16. This effect can be seen for Retail (Fig. 2a) with a low threshold of 10 and 20, Pumsb (Fig. 2d) with $2.5 \times 10^4$ Accident (Fig. 2f) with $6 \times 10^4$ and $4 \times 10^4$.

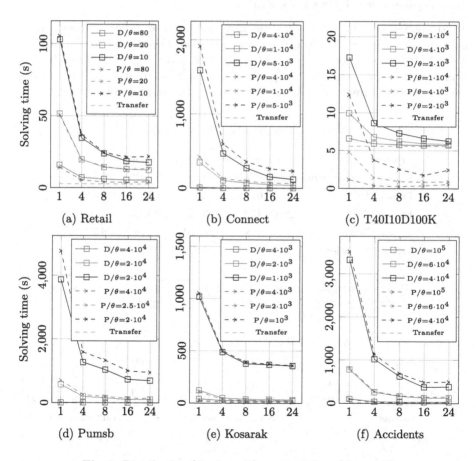

**Fig. 2.** Distribution (#solvers D) vs parallelism (#cores P)

T40I10D100K's dataset witnesses a very different behavior. In fact, the obtained solving time is about 5 times longer in the distributed environment. This effect is explained by the fact that the dataset when stored is pretty small but most of all, the number of enumerated models is very low for this dataset. Thus, the time spent to data transfer is too high compared to the gain on processing time. This transfer time is shown on every experiment by the dashed constant gray line (negligible in Figs. 2b, 2d, 2e and 2f). In Fig. 2c we can remark that the transfer time is higher than the processing time, but for others it's negligible w.r.t. to the processing time. In a "warm start" execution, this transfer time disappears and the distribution outperforms the parallel version.

As expected, the lower the threshold $\theta$ is the higher the running time becomes. In fact, this is due to the large number of models that can be found. The convergence obtained by ParaSaTMiner is all the more noticeable. Thus, the distributed approach can still continue to benefit from a complete parallelism.

## 5.3 Aggregation Effect on CFIM

To study further the effect of scalability, we propose a new dataset based on touristic visits over France from comments on *TripAdvisor* as depicted in Table 1. It provides for each tourist, the set of geolocalized locations he has commented. Such a study will help tourist stakeholders to understand interconnections between frequent destinations and propose adapted tour operators.

**Table 1.** Touristic data characteristics

| Instance | #Transactions | #Items | Density | Size |
|---|---|---|---|---|
| France (Department) | 2,131,781 | 96 | 3.82% | 26.5 M |
| France (District) | 2,529,011 | 350 | 1.22% | 41.4 M |
| France (City) | 2,700,450 | 3,664 | 0.13% | 58 M |
| France (Town) | 2,750,818 | 36,612 | 0.014% | 86.7 M |
| France (Location) | 3,128,134 | 240,911 | 0.003% | 129.3 M |

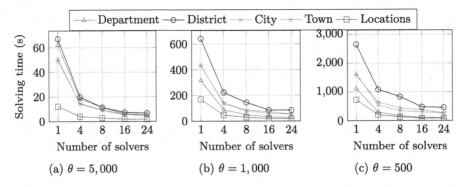

(a) $\theta = 5,000$     (b) $\theta = 1,000$     (c) $\theta = 500$

**Fig. 3.** TripAdvisor France at various aggregation scales and different thresholds

Since the data contains geographical features, we can mine the dataset by extracting frequently visited zones by aggregating those locations on their common area. Thus, by exploiting the linked administrative area (GADM database[3]) we can group locations on towns, cities, districts and departments, in order to analyze correlations between zones in a country. Another interesting aspect of this dataset is to study the effect of aggregation on CFIM with a same dataset; higher density when aggregated since locations belong to the same area[4].

To study the effect of various aggregation, we run the experiments with three different support thresholds and we compared the datasets against each other. Recall that we use the same dataset (at location level) aggregated on shared properties "area" leading to Town, City, District and Department.

Figure 3 gives the solving time of these different scales on the Tripadvisor dataset with three different thresholds and by varying the number of solvers. It is interesting to notice that the increasing order of aggregation is not clearly correlated to the computation time. In fact, the department scale with fewer items (Table 1) takes less time than Districts with more items (3 times) while expecting to have fewer models. This is due to the fact that data are skewed; for instance, lots of visits in Paris lead to more extracted models between Districts in Paris while fewer in other cities. This effect can also be seen with Town and City scales that cost less than district scale for every thresholds. This means that the District scale produces more computation to find models than other levels of aggregation since many backtracking are necessary to find new models.

Moreover, changing the threshold has especially an impact on the computation time of all models except at Department scale. We can see that mining models at a high threshold took more time for the Department aggregation level, while it takes less time for lower thresholds. This means that most itemsets are highly frequent and detected with very frequent items but it does not change as much as other scales when trying lower thresholds.

It is interesting to notice that City and Town scales witness a same-solving time. Thus, trips of users have a same logic while visiting France on destinations (Cities). The low time obtained for Location means that few models are found and it becomes harder to find common trips between users (Table 2).

---

[3] GADM https://gadm.org/download_country.html.

[4] Datasets are anonymized, formatted and available on Github: https://github.com/leonard-de-vinci/Distributed_SAT_Approach_FIM/tree/main/data.

**Table 2.** distriSATMiner *vs.* paraSAT-Miner on FIMI datasets

| Instance (#Transactions, #Items) | $\theta$ | DSAT M-1s | PSAT M-1c | DSAT M-4s | PSAT M-4c | DSAT M-8s | PSAT M-8c | DSAT M-16s | PSAT M-16c | DSAT M-24s | PSAT M-24c | Sending time | #Models |
|---|---|---|---|---|---|---|---|---|---|---|---|---|---|
| Retail (#88 162 #16 470) | 80 | 13.14 | 14.06 | 4.48 | 5.4 | 3.38 | 4.05 | 2.79 | 3.79 | 2.79 | 4.39 | 2.69 | $> 8.10^3$ |
| | 60 | 17.66 | 19.1 | 6.15 | 6.95 | 4.82 | 5.67 | 4.06 | 5.25 | 3.78 | 5.56 | | $> 1.10^4$ |
| | 40 | 26.28 | 27.82 | 9.42 | 10.63 | 6.77 | 8.55 | 5.94 | 7.52 | 5.84 | 7.98 | | $> 2.10^4$ |
| | 20 | 48.72 | 50.54 | 16.89 | 19.35 | 11.64 | 14.79 | 10.29 | 12.57 | 9.76 | 13.58 | | $> 5.10^4$ |
| | 10 | 100.34 | 105.34 | 31.99 | 36.86 | 21.18 | 24.57 | 15.87 | 21.26 | 15.01 | 21.88 | | $> 1.10^5$ |
| | 40000 | 5.19 | 7.13 | 1.51 | 2.09 | 0.83 | 1.56 | 0.51 | 1.28 | 0.56 | 1.57 | 4.09 | $> 7.10^4$ |
| | 20000 | 40.49 | 52.58 | 14.45 | 19.23 | 10.99 | 15.45 | 10.39 | 14.64 | 10.36 | 15.02 | | $> 5.10^5$ |
| Connect (#67 558 #129) | 10000 | 340.13 | 417.96 | 99.72 | 129.68 | 67.98 | 92.12 | 43.22 | 69.76 | 39.66 | 66.97 | | $> 3.10^6$ |
| | 5000 | 1586.39 | 1915.19 | 463.96 | 595.14 | 267.91 | 354.73 | 149.52 | 262.23 | 116.94 | 232.05 | | $> 1.10^7$ |
| | 10000 | 1.02 | 1.16 | 0.35 | 0.37 | 0.17 | 0.27 | 0.11 | 0.38 | 0.13 | 0.56 | 5.58 | $\cong 1.10^2$ |
| T40I10D100K (#100 000 #942) | 8000 | 1.64 | 1.72 | 0.51 | 0.57 | 0.25 | 0.38 | 0.15 | 0.47 | 0.12 | 0.63 | | $> 1.10^2$ |
| | 6000 | 2.62 | 2.98 | 0.84 | 0.84 | 0.43 | 0.57 | 0.26 | 0.64 | 0.19 | 0.9 | | $> 2.10^2$ |
| | 4000 | 4.34 | 4.79 | 1.23 | 1.44 | 0.68 | 0.92 | 0.37 | 0.86 | 0.28 | 0.91 | | $> 4.10^2$ |
| | 2000 | 11.69 | 12.34 | 3.05 | 3.72 | 1.72 | 2.52 | 1.06 | 1.76 | 0.7 | 2.44 | | $> 1.10^3$ |
| Pumsb (#49 046 #2113) | 40000 | 2.84 | 3.64 | 0.92 | 1.26 | 0.57 | 0.84 | 0.43 | 1329.55 | 0.43 | 1.24 | 5.39 | $> 2.10^4$ |
| | 35000 | 17.14 | 22.23 | 5.55 | 7.42 | 4.46 | 6.07 | 4.19 | 6.14 | 4.47 | 6.47 | | $\cong 2.10^5$ |
| | 30000 | 77.68 | 97.64 | 37.05 | 47.08 | 20.37 | 26.86 | 18.74 | 26.15 | 20.23 | 26.63 | | $\cong 9.10^5$ |
| | 25000 | 564.96 | 700.07 | 192.69 | 252.83 | 144.48 | 191.5 | 96.33 | 142.67 | 99.22 | 140.44 | | $\cong 6.10^6$ |
| | 20000 | 3858.91 | 4764.99 | 1259.72 | 1586.33 | 1028.38 | 1329.55 | 726.24 | 999.13 | 688.0 | 948.15 | | $> 3.10^7$ |
| Kosarak (#990 002 #41267) | 4000 | 24.81 | 26.44 | 8.79 | 9.82 | 6.34 | 8.12 | 5.51 | 7.74 | 5.46 | 8.81 | 15.88 | $> 2.10^3$ |
| | 3000 | 38.25 | 39.94 | 12.89 | 15.24 | 10.07 | 12.44 | 8.78 | 11.42 | 8.31 | 12.97 | | $> 4.10^3$ |
| | 2000 | 105.64 | 109.85 | 32.98 | 35.96 | 20.57 | 25.76 | 18.88 | 23.84 | 15.67 | 23.28 | | $> 3.10^4$ |
| | 1000 | 1003.55 | 1049.89 | 472.52 | 499.12 | 361.28 | 389.97 | 352.43 | 373.03 | 340.58 | 359.44 | | $\cong 5.10^5$ |
| Accidents (#340 183 #468) | 100000 | 86.54 | 92.76 | 24.45 | 29.53 | 16.65 | 18.9 | 14.54 | 18.96 | 15.21 | 20.2 | 13.69 | $\cong 1.10^5$ |
| | 80000 | 233.2 | 350.55 | 72.47 | 81.03 | 49.72 | 57.09 | 42.36 | 54.53 | 42.99 | 54.96 | | $\cong 4.10^5$ |
| | 60000 | 768.65 | 813.26 | 243.93 | 274.24 | 151.34 | 177.18 | 113.5 | 147.78 | 111.07 | 149.36 | | $> 1.10^6$ |
| | 40000 | 3307.0 | 3512.26 | 1003.75 | 1132.71 | 607.95 | 690.34 | 361.03 | 484.3 | 366.59 | 492.24 | | $\cong 6.10^6$ |

# 6   Conclusion

We proposed a distributed framework for Closed Frequent Itemset Mining using propositional satisfiability. It relies on a Main Node that defines the mining strategy by sending pieces of guiding paths to Solver Nodes. The distributed architecture is fault tolerant and outperforms traditional parallel mining since it avoids bottlenecks access to main memory. We also provided a new mining dataset with various aggregation scales which brings new questions on mining optimization using correlations between geographic areas.

For future works, we plan to enhance the mining strategy by finding better heuristics allowing to improve the load balancing. In fact, most Solver Nodes work in parallel and witness good performances. However it can occur that one task is longer than others and other nodes wait it to finish.

Another search direction concerns the dynamic decomposition to avoid the idleness of the workers. Moreover, we plan to exploit our distributed architecture for other mining tasks such as maximal frequent itemset and top-k prefered ones.

# References

1. Agrawal, R., Imielinski, T., Swami, A.N.: Mining association rules between sets of items in large databases. In: SIGMOD 1993, Washington, USA, pp. 207–216 (1993)
2. Audemard, G., Hoessen, B., Jabbour, S., Piette, C.: Dolius: a distributed parallel SAT solving framework. In: Workshop POS 2014. @ SAT, vol. 27, pp. 1–11 (2014)
3. Balyo, T., Sanders, P., Sinz, C.: Hordesat: a massively parallel portfolio SAT solver. CoRR abs/1505.03340 (2015)
4. Borgelt, C.: Frequent item set mining. Wiley Interdisc. Rev. Data Min. Knowl. Disc. **2**(6), 437–456 (2012)
5. Boudane, A., Jabbour, S., Sais, L., Salhi, Y.: A SAT-based approach for mining association rules. In: IJCAI 2016, New York, USA, pp. 2472–2478 (2016)

6. Cheung, D.W.L., Han, J., Ng, V.T.Y., Fu, A.W.C., Fu, Y.: A fast distributed algorithm for mining association rules. In: ICPADS 1996, pp. 31–42 (1996)
7. Cheung, D.W., Ng, V.T.Y., Fu, A.W., Fu, Y.: Efficient mining of association rules in distributed databases. IEEE TKDE **8**(6), 911–922 (1996)
8. Chrabakh, W., Wolski, R.: Gridsat: a chaff-based distributed SAT solver for the grid. In: ACM SC 2003, Phoenix, AZ, USA, p. 37. ACM (2003)
9. Dean, J., Ghemawat, S.: Mapreduce: a flexible data processing tool. Commun. ACM **53**(1), 72–77 (2010)
10. Dlala, I.O., Jabbour, S., Raddaoui, B., Sais, L.: A parallel SAT-based framework for closed frequent itemsets mining. In: CP 2018, vol. 11008, pp. 570–587 (2018)
11. Frioux, L.L., Baarir, S., Sopena, J., Kordon, F.: PaInleSS: a framework for parallel SAT solving. In: SAT 2017, Melbourne, Australia, vol. 10491, pp. 233–250 (2017)
12. Gahar, R.M., Arfaoui, O., Hidri, M.S., Hadj-Alouane, N.B.: ParallelCharMax: an effective maximal frequent itemset mining algorithm based on mapreduce framework. In: AICCSA 2017, Hammamet, Tunisia, pp. 571–578 (2017)
13. Gan, W., Lin, J.C., Chao, H., Zhan, J.: Data mining in distributed environment: a survey. Wiley Interdisc. Rev. Data Min. Knowl. Disc. **7**(6), e1216 (2017)
14. Gebser, M., Guyet, T., Quiniou, R., Romero, J., Schaub, T.: Knowledge-based sequence mining with ASP. In: IJCAI 2016, New York, USA, pp. 1497–1504 (2016)
15. Han, J., Pei, J., Yin, Y., Mao, R.: Mining frequent patterns without candidate generation: a frequent-pattern tree approach. DMKD **8**(1), 53–87 (2004)
16. Hyvärinen, A.E.J., Junttila, T.A., Niemelä, I.: Partitioning SAT instances for distributed solving. In: Fermüller, C.G., Voronkov, A. (eds.) LPAR 2017, Yogyakarta, Indonesia, vol. 6397, pp. 372–386 (2010)
17. Hyvärinen, A.E.J., Junttila, T.A., Niemelä, I.: Grid-based SAT solving with iterative partitioning and clause learning. In: CP 2011, vol. 6876, pp. 385–399 (2011)
18. Jabbour, S., Sais, L., Salhi, Y.: The top-k frequent closed itemset mining using top-k SAT problem. In: ECML PKDD 2013, vol. 8190, pp. 403–418 (2013)
19. Lazaar, N., et al.: A global constraint for closed frequent pattern mining. In: CP 2016, Toulouse, France, vol. 9892, pp. 333–349 (2016)
20. Li, H., Wang, Y., Zhang, D., Zhang, M., Chang, E.Y.: PFP: parallel fp-growth for query recommendation. In: RecSys 2008, Lausanne, Switzerland, pp. 107–114 (2008)
21. Moskewicz, M.W., Madigan, C.F., Zhao, Y., Zhang, L., Malik, S.: Chaff: engineering an efficient SAT solver. In: DAC 2001, pp. 530–535 (2001)
22. Négrevergne, B., Guns, T.: Constraint-based sequence mining using constraint programming. CoRR abs/1501.01178 (2015)
23. Ngoko, Y., Trystram, D., Cérin, C.: A distributed cloud service for the resolution of SAT. In: $SC^2$ 2017, Kanazawa, Japan, pp. 1–8 (2017)
24. Raedt, L.D., Guns, T., Nijssen, S.: Constraint programming for itemset mining. In: SIGKDD 2008, Las Vegas, Nevada, USA, pp. 204–212 (2008)
25. Schaus, P., Aoga, J.O.R., Guns, T.: CoverSize: a global constraint for frequency-based itemset mining. In: CP 2017, vol. 10416, pp. 529–546 (2017)
26. Zaharia, M., et al.: Resilient distributed datasets: a fault-tolerant abstraction for in-memory cluster computing. In: NSDI 2012, pp. 15–28 (2012)
27. Zaki, M.J., Hsiao, C.: Efficient algorithms for mining closed itemsets and their lattice structure. IEEE TKDE **17**(4), 462–478 (2005)
28. Zhang, H., Bonacina, M.P., Hsiang, J.: PSATO: a distributed propositional prover and its application to quasigroup problems. J. Symb. Comp. **21**(4), 543–560 (1996)

# Index Advisor via DQN with Invalid Action Mask in Tree-Structured Action Space

Yang Wu[1,2], Yong Zhang[1(✉)], and Ning Li[2]

[1] BNRist, DCST, Institute of Precision Medicine, Institute of Internet Industry,
Tsinghua University, Beijing, China
wu-y22@mails.tsinghua.edu.cn, zhangyong05@tsinghua.edu.cn
[2] School of Computer Science, Northwestern Polytechnical University, Xi'an,
Shaanxi, China
lining@nwpu.edu.cn

**Abstract.** Indexes are essential for increasing query speed. Traditional databases require database administrators to manually tune indexes based on knowledge and their experience. In recent years, AI techniques have been successfully applied to many areas including automatic index recommendation. Reinforcement Learning (RL) methods such as Deep Q-Network (DQN) can find better indexes than traditional methods, but still suffer from the huge action space. Previous RL methods tried to solve it by pre-narrowing action space to several candidate indexes, which may omit some useful indexes. This paper focuses on offline Index Selection Problem (ISP) and tries to solve the problem via invalid action mask in a tree-structured action space. First, we use Double DQN and Dueling DQN to replace traditional DQN to get better estimation of Q-values. Then we propose a novel index recommendation approach DQN-AMTAS that collects all possible indexes in a tree and recommends multi-column indexes from left to right via invalid action mask based on the Leftmost Prefix Rule. We conduct extensive experiments on TPC-H and TPC-DS datasets. The experimental results show the superiority of our proposed DQN-AMTAS compared with state-of-the-art index recommendation algorithms.

**Keywords:** Index Selection Problem · Deep Q-Network · Invalid action mask

## 1 Introduction

Research on automatic index recommendation started as early as 1970 and is often referred to as Index Selection Problem (ISP) in the literature. It is about selecting what indexes to build for given database tables, data and workload queries and has been proven to be NP-hard [13] with a huge solution space. Therefore, it is hard to naively enumerate all possible index combinations. How to search optimal or near-optimal solutions effectively is the major difficulty.

Both non-machine learning methods [5,8] and machine learning methods [10, 15] have been proposed. We mainly try to remedy the potential weakness of heuristic rules in Lan's work  [10] that uses Deep Q-Network (DQN) [11] to search optimal index configuration in offline scenarios where database tables, data and queries are given and fixed. Our contributions can be summarized as follows:

- We replace the traditional DQN with Double DQN and Dueling DQN in the index recommendation framework to get better estimation of action values.
- We propose DQN with Invalid Action Mask in Tree-Structured Action Space (DQN-AMTAS) that collects all combination of indexes in a tree-structured action space and uses invalid action mask during training.
- In the experiments, we first find that Double DQN and Dueling DQN are better than traditional DQN. Then we conduct experiments on TPC-H and TPC-DS with different scale factors and memory budgets demonstrating our DQN-AMTAS can find the best indexes that make query execution time the shortest compared with baseline methods.

This paper is organized as follows. In Sect. 2, the related work is discussed. The architecture design is described in Sect. 3. Then we replace the traditional DQN with Double DQN and Dueling DQN, and propose DQN-AMTAS in Sect. 4. The experiment results are given in Sect. 5. We conclude the paper in Sect. 6.

## 2   Related Work

Algorithms for automatic ISP were firstly published in the early 1970s s and various methods vary in implementation and complexity. Many commercial DBMSs have provided automatic index recommendation tools, such as Anytime [3], DB2advis [16]. These algorithms can be divided into two categories: traditional methods and machine learning-based methods.

Traditional index recommendation algorithms can be categorized from two aspects. From the aspect of searching, some methods start from an empty set and gradually add indexes (DB2advis [16], Anytime [3], Extend [14]), while some other methods start from a large set of indexes (all combinations) and gradually remove indexes (Drop [4], Relaxation [2]). From the aspect of formulating ISP as some well-researched problems, researchers formulate ISP as Linear Programming problems [5,12] and variants of 0–1 knapsack problem [8] because of the similarity between these problems.

Machine Learning methods can be further divided into traditional Machine Learning methods and Deep Learning methods (mainly Deep RL). There are traditional Machine Learning methods such as genetic algorithm [6] and least-square policy iteration [1]. Deep RL methods are mainly proposed after Sharma's first try to apply Deep RL approach [15] to index recommendation in 2018. Since then, many researches on ISP utilize Deep RL, from Sharma's method which can only recommend single-column indexes to others recommending multi-column indexes, from only B+ tree index to other index types.

# 3    RL-Based Index Advisor

In this section, we first give the definition of Index Selection Problem, then explain how to model ISP as a RL problem. After illustrating the general framework, we describe our design of state, action, value function and neural network.

## 3.1    Index Selection Problem

**Definition 1.** *Given a database $D = \{T_1, T_2, ..., T_d\}$, where $T_d$ represents the $d_{th}$ table in the database, with a known workload $W = \{q_1, q_2, ..., q_n\}$, the aim is to find a set of index configurations $I^* = \{i_1, i_2, ..., i_k\}$ that $I^* = argmax_I Cost_I(W)$ under constraints $B = \{b_1, b_2, ..., b_j\}$ such as index number of index space. $Cost_I(W)$ is the execution cost of workload $W$ under index configuration I. It is the sum of execution time of all queries in the workload, calculated as $Cost_I(W) = \sum_{j=1}^{Q}(cost_j)$.*

Suppose we want to build a multi-column index of width c, the table has n columns, there are n choices for the first column in the index, n-1 choices for the second column, and so on. When the database contains hundreds of tables, there are too many possible index configurations to consider. This requires the design of efficient algorithms to automatically find the optimal solution in such a large search space.

## 3.2    General Framework of the Index Advisor

The candidate set of ISP is too large to directly find the optimal configuration at a time. It is often necessary to use an iterative and greedy search algorithm to select and try index one by one. RL is used as a solution to the sequential decision making problem to search for the optimal index configuration. Under given constraints, RL algorithm can automatically learn how to maximize target value in interaction with the environment in episodes. As shown in Fig. 1, at each step of an episode, the agent (index advisor) interacts with the environment (DBMS). The agent selects an action (i.e. recommending an index) based on the current state and policy, and the environment returns a reward value (decreasing in the execution time, e.g. Formula 2 ) as value estimation of the chosen action from the state. When the index size exceeds space constraint, the episode terminates, indexes recommended in this episode are recorded, the environment is reset by deleting all recommended indexes and setting the state to the initial state, and a new episode starts. Among these episodes, we choose indexes with the shortest workload execution time as the final recommendation.

We will explain in detail the design of states, actions, value function and neural network in the following subsections.

## 3.3    States

We arrange all candidate indexes as an array, represented by one-hot encoding schema: "1" represents the index at corresponding position is built and "0"

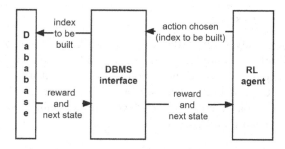

**Fig. 1.** The framework for index recommendation

otherwise. This array indicates current state. Candidates can be single-column or multi-column indexes generated by enumerating all possible combination of columns. We can also pre-extract some candidates by heuristic rules as Lan et al. did.

### 3.4   Actions

An action in ISP is an index chosen by the agent to be built. In DQN, the one-dimensional array representing the current state is input into the neural network, and the array of the same dimension is the output. Each value of the output array represents the value function of selecting the corresponding index of that position, that is, the Q value. Using the greedy strategy, the action corresponding to the maximum Q value is selected:

$$A = argmax_\pi E(\sum_{t=0}^{T} \lambda^t R(s_t, a_t)|s_0 = s, a_0 = a) \tag{1}$$

### 3.5   Value Functions

At every step, the agent selects an index from the candidate index set, adding to the current index configuration to reduce the execution time of workload queries without violating the constraints on the number of indexes or storage space. Let $X_i$ denote the index configuration after the i-th step, and $X_0$ the initial index configuration(empty set). The reward function is defined as the ratio of the cost reduction and the initial cost [10] as Formula 2 shows.

$$R(X_t) = \frac{Cost(W, X_{t-1}) - Cost(W, X_t)}{Cost(W, X_0)} \tag{2}$$

where T represents the maximum number of steps not to exceed the constraints.

## 3.6   Neural Network

We use three-layer fully connected Neural Network(NN) as shown in Table 1. Note that the number of candidate indexes is equal to the dimension of the state space as well as the dimension of the action space. The output array is an array of Q-values. Each Q-value is the value of choosing the corresponding action from the input state. The parameters of each layer are normalized with a Gaussian distribution with mean 0 and variance 1.

**Table 1.** Structure of Neural Network

| Layer | Input dim | Output dim |
|---|---|---|
| 1 | Number of candidates | 1024 |
| 2 | 1024 | 1024 |
| 3 | 1024 | Number of candidates |

# 4   DQN-AMTAS

In this section, we describe the benefit of replacing traditional DQN with Double DQN and Dueling DQN. Then we illustrate the idea and implementation of our proposed DQN-AMTAS, with an example on REGION table .

## 4.1   Improvements on Traditional DQN

**Double DQN.** Traditional DQN [11] tends to overestimate Q-value because every time the maximum Q-value of next state is added as show below:

$$Q(S, A) \leftarrow Q(S, A) + \alpha(R + \gamma max_{a'} Q(S', a') - Q(S, A)) \tag{3}$$

The solution is to use the eval-net (parameters always being updated) to choose action with the highest Q-value, and use the target-net (same structure with eval-net, parameters fixed but periodically synchronized with eval-net) to evaluate Q-value for the chosen action [17]:

$$Q(S, A; \theta) = R + \gamma Q(w', argmax_{a'} Q(S', A'; \theta); \theta') \tag{4}$$

**Dueling DQN.** Dueling DQN divides the Q-value into two parts in Fig. 2: the state value function V(s) and advantage function A(s,a) to get better estimation of state-action value [18]. The Q-value is the sum of V(s) and A(s,a):

$$Q(s, a) = V(s) + A(s, a) \tag{5}$$

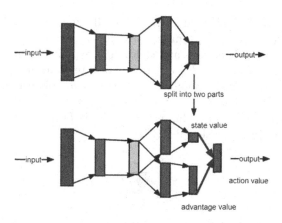

**Fig. 2.** The interaction between the agent and the environment

## 4.2 DQN-AMTAS

In this section, we first introduce the basic idea of our method and then describe the implementation.

**Idea.** Take TPC-H as an example, there is a total of 51 columns in all tables and 59 candidate indexes will be generate by Lan's heuristic rules [10]. If all columns in each table are arranged in permutations, there will be 3551 candidate indexes.

Since recommending single-column indexes is not good enough, we need to consider multi-column indexes. Lan et al. [10] design heuristic rules to pre-narrow candidate set. However, we find that indexes recommended by Extend that make queries faster contain some indexes not in the 59 candidate indexes extracted by the heuristic rules. This means that when the heuristic rules are used, better candidate indexes are excluded early before searching.

We then try all 3551 permutations as candidate set, but unfortunately the agent is bewildered by so many actions and fails to find appropriate indexes, even resulting in longer execution time than single-column indexes.

Now we need to devise a way to both make all combinations possible and make the agent focus on more beneficial actions. Considering the leftmost matching principle of a multi-column index, when building a multi-column index, it is expected that the columns on the left of it are also helpful for speeding up data retrieval. When recommending a multi-column index, we always first recommend single-column indexes, and then choose an existing index, add a column to its right as a new wider index. In DQN, the action space should be fixed, and we could not add or delete indexes. Therefore, we use invalid action mask [7] to make some action valid and other invalid in each step. In this way, valid action space of each step is greatly reduced and all permutation of columns are selectable. In this way the RL agent is able to find appropriate indexes during its trial-and-error interaction with DBMS.

**Implementation.** To collect information of all candidates and enable convenient update of state, action and mask, we design a tree to store all permutations of indexes as well as their parents and children as explained above. Taking TPC-H(a decision support benchmark consisting of a suite of business oriented ad-hoc queries and concurrent data modifications) as an example, the root of the tree has 8 children nodes in its first layer (representing 8 tables in TPC-H). Nodes in the second layer are all single-column indexes in its father table node. The tree structure of table Region is shown in Fig. 3.

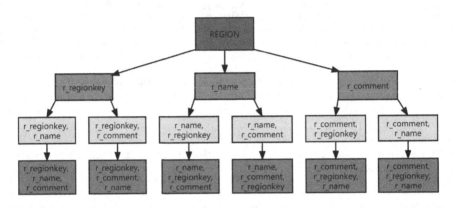

**Fig. 3.** The tree-structured action space (REGION table as an example)

Since the input to the neural network is the current state and needs a one-dimensional representation, we convert this tree into a one-dimensional array by pre-order traversal. Each element in the array has the record of the columns contained in the index and positions of its parent node and child nodes. Position of a node is the node's index in the array, which is convenient for updating valid action information when the agent chooses an action and moves to a new state.

At the beginning of each episode, only single-column indexes are valid, as shown in Fig. 4(a). At this time, a mask tensor is generated. The corresponding position of a single-column index is True, and the others are False. The output of the neural network is masked with the mask tensor and the Q value of the invalid action is set to a negative number with a very large absolute value, so that the invalid action will not be selected.

After the agent selects an action and executes it, the action is set to be invalid and cannot be recommended again. At this time, the parent node of the chosen index is deleted from the current index set because the left part of the newly chosen index can be used for retrieval according to the leftmost matching principle; the actions corresponding to the child nodes of the chosen index are marked as valid, possible to be chosen in later steps, and so on.

Suppose the agent chooses r_regionkey for the first time, then r_regionkey itself becomes invalid, and its children become valid as shown in Fig. 4(b).

Assuming r_name and (r_name, r_regionkey) are selected in the next two steps, the state changes to Fig. 4(c).

| REGION | r_regionkey | r_regionkey, r_name | r_regionkey, r_name, r_comment | r_regionkey, r_comment | r_regionkey, r_comment, r_name | r_name | r_name, r_regionkey |
|---|---|---|---|---|---|---|---|
| NULL | Valid:True Index: No | Valid: False Index: No | Valid: False Index: No | Valid: False Index: No | Valid: False Index: No | Valid:True Index: No | Valid: False Index: No |
| **r_name, r_regionkey, r_comment** | **r_name, r_comment** | **r_name, r_comment, r_regionkey** | **r_comment** | **r_comment, r_regionkey** | **r_comment, r_regionkey, r_name** | **r_comment, r_name** | **r_comment, r_regionkey, r_name** |
| Valid: False Index: No | Valid: False Index: No | Valid: False Index: No | Valid:True Index: No | Valid: False Index: No | Valid: False Index: No | Valid: False Index: No | Valid: False Index: No |

(a) initial state (no index built, only single-column indexes to recommend)

| REGION | r_regionkey | r_regionkey, r_name | r_regionkey, r_name, r_comment | r_regionkey, r_comment | r_regionkey, r_comment, r_name | r_name | r_name, r_regionkey |
|---|---|---|---|---|---|---|---|
| NULL | **Valid:False Index: Yes** | **Valid: True Index: No** | Valid: False Index: No | **Valid: True Index: No** | Valid: False Index: No | Valid:True Index: No | Valid: False Index: No |
| **r_name, r_regionkey, r_comment** | **r_name, r_comment** | **r_name, r_comment, r_regionkey** | **r_comment** | **r_comment, r_regionkey** | **r_comment, r_regionkey, r_name** | **r_comment, r_name** | **r_comment, r_regionkey, r_name** |
| Valid: False Index: No | Valid: False Index: No | Valid: False Index: No | Valid:True Index: No | Valid: False Index: No | Valid: False Index: No | Valid: False Index: No | Valid: False Index: No |

(b) state after choosing r_regionkey

| REGION | r_regionkey | r_regionkey, r_name | r_regionkey, r_name, r_comment | r_regionkey, r_comment | r_regionkey, r_comment, r_name | r_name | r_name, r_regionkey |
|---|---|---|---|---|---|---|---|
| NULL | **Valid:False Index: Yes** | **Valid: True Index: No** | Valid: False Index: No | **Valid: True Index: No** | Valid: False Index: No | Valid:False Index: No | Valid: False Index: Yes |
| **r_name, r_regionkey, r_comment** | **r_name, r_comment** | **r_name, r_comment, r_regionkey** | **r_comment** | **r_comment, r_regionkey** | **r_comment, r_regionkey, r_name** | **r_comment, r_name** | **r_comment, r_regionkey, r_name** |
| Valid: True Index: No | Valid: True Index: No | Valid: False Index: No | Valid:True Index: No | Valid: False Index: No | Valid: False Index: No | Valid: False Index: No | Valid: False Index: No |

(c) state after choosing r_regionkey, r_name and (r_name,r_regionkey) successively

**Fig. 4.** State Illustration of DQN-AMTAS

Mask of next state also needs to be saved to replay buffer in addition to <s,a,r,s'>. The mask is used to calculate $max_{a'}Q(S', a')$ when updating the parameters of the networks. Finally, after the episode ends, the environment needs to be reset, masks set to only single columns recommendable. In addition, because of $\epsilon$ greedy strategy, the algorithm can select a multi-column index whose left part is not recommended but effective on the whole.

We call this algorithm DQN with Invalid Action Mask in Tree-structured Action Space (DQN-AMTAS). The pseudocode is shown in Algorithm 1.

---

**Algorithm 1:** DQN-AMTAS

---

**input**  : Database with several tables $D = \{T_1, T_2, ..., T_d\}$, workload of several queries $W = \{q_1, q_2, ..., q_n\}$
**output:** Selected indexes $I^* = \{i_1, i_2, ..., i_k\}$

1  Randomly initialize eval-net
2  Randomly initialize target-net
3  Generate tree-structured action space
4  **for** *each episode* **do**
5  |    State $s_0 \leftarrow$ initial state
6  |    **for** *each step* **do**
7  |    |    With probability $\epsilon$ select random action, otherwise choose $a_t = argmax Q(s, a)$ from only valid actions
8  |    |    $s_{t+1} \leftarrow$ execute action, update the tree-structured action space and get new state
9  |    |    Compute reward $R_t$
10 |    |    Store transition$< s_t, a_t, R_t, s_{t+1} >$ in replay memory
11 |    |    Sample experience from replay memory
12 |    |    Update eval-net with stochastic gradient descent
13 |    |    Update target-net frequently

---

## 5   Experimental Results

### 5.1   Baselines

We compare DQN-AMTAS with the following baselines on the platform provided by Kossmann et al [9]:

- *Extend*: Extend [14] greedily chooses an action to maximize the ratio of cost decrease and storage increase, either selects one column as a new index or adds it to the right of an existing index.
- *DB2advis*: DB2advis [16] gradually adds indexes to index configuration according to the ratio of cost decrease and storage increase from large to small.
- *Relaxation*: Relaxation [2] starts from full optimal index set, gradually reduces space until the storage constraint is satisfied.
- *Anytime*: Anytime [3] first recommends indexes for each query and then combines them for the total workload.
- *Lan's DQN*: Lan's DQN [10] uses heuristic rules to pre-collect candidate indexes before searching. We re-implement it in the platform [9] and name it 'DQN with heuristics'.

## 5.2    Experimental Results

Our experiments are conducted on de-facto industry standard benchmarks TPC-H and TPC-DS datasets. Scale factor determines how large data are generated, e.g. 0.1 GB when scale factor=0.1. Overall cost is measured by the execution time of all queries in the workload. We use the total cost returned by the 'explain' command.

**Improvements of Double DQN and Dueling DQN.** Table 2 shows the results of different algorithms. From the table, we can see that the performance of index recommendation using Double DQN and Dueling DQN algorithms is better than that of traditional DQN.

**Table 2.** Improvements of Double DQN and Dueling DQN on TPC-H (scale factor = 0.1 memory budget = 50 MB)

| Algorithm | Overall cost | Index space (MB) | Index number |
|---|---|---|---|
| Lan's DQN | 289261.76 | 45703168 | 11 |
| Lan's DQN + Double DQN | 288203.86 | 48005120 | 12 |
| Lan's DQN + Double DQN + Dueling DQN | 288018.61 | 46358528 | 10 |

**Table 3.** Comparison results on TPC-H (scale factor = 0.1, memory budget = 50 MB, only single-column index)

| Algorithm | Overall cost | Index space (MB) | Index number |
|---|---|---|---|
| Extend | 288701.58 | 45596672 | 7 |
| DB2advis | 300405.44 | 49930240 | 18 |
| DQN with heuristics | 303822.04 | 48775168 | 14 |
| DQN-AMTAS | 287227.59 | 49889280 | 19 |

**Comparison of DQN-AMTAS with Baselines** Table 3 shows the results of different algorithms recommending only single-column indexes in 0.1 GB TPC-H dataset under 50 MB constraint. Our DQN-AMTAS's overall cost is 287227.59 that is smaller than Extend. In the same experimental setting but recommending multi-column indexes, the leftmost of Fig. 5 shows indexes recommended by DQN-AMTAS further reduce query execution time to 280258, shorter than Extend's 283254.

As shown in Fig. 5 and Table 4, DQN-AMTAS has lower cost than other algorithms. Only on 1GB TPC-H dataset under 100MB storage constraint, the result of DQN with heuristics is similar to the result of DQN-AMTAS.

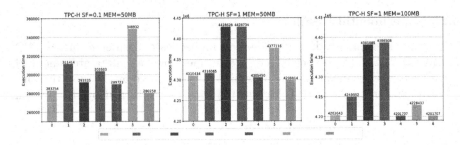

**Fig. 5.** Comparison of DQN-AMTAS with baselines

**Table 4.** Comparison results on TPC-DS (scale factor $= 1$, memory budget $= 50$ MB)

| Algorithm | Overall cost | Index space (MB) | Index number |
|-----------|-------------|------------------|--------------|
| Extend | 12382736.36 | 43515904 | 10 |
| Anytime | 12467281.36 | 49979392 | 20 |
| DB2advis | 12608707.42 | 49905664 | 57 |
| DQN-AMTAS | 12347613.18 | 49053696 | 12 |

## 6    Conclusions

In this paper, we apply improved DQN (Double DQN and Dueling DQN) with invalid action mask in tree-structured action space to Index Selection Problem. Including all useful candidates, our DQN-AMTAS can find better indexes than other baseline algorithms.

In the future work, we will turn to online case of index recommendation and adapt to constantly changing data and workload. After fully pre-training the model on enough tables, data and queries, the model is expected to recommend indexes in a single run and be faster than any search-based algorithms.

**Acknowledgements.** This study was supported by the Natural Science Foundation of Shaanxi Province of China (Grant No.2021JM068).

## References

1. Basu, D., et al.: Regularized cost-model oblivious database tuning with reinforcement learning. In: Hameurlain, A., Küng, J., Wagner, R., Chen, Q. (eds.) Transactions on Large-Scale Data- and Knowledge-Centered Systems XXVIII. LNCS, vol. 9940, pp. 96–132. Springer, Heidelberg (2016). https://doi.org/10.1007/978-3-662-53455-7_5
2. Bruno, N., Chaudhuri, S.: Automatic physical database tuning: a relaxation-based approach. In: Proceedings of the 2005 ACM SIGMOD International Conference on Management of Data, pp. 227–238 (2005). https://doi.org/10.1145/1066157.1066184

3. Chaudhuri, S., Narasayya, V.: Anytime algorithm of database tuning advisor for microsoft sql server (2020)
4. Choenni, S., Blanken, H., Chang, T.: Index selection in relational databases. In: Proceedings of ICCI 1993: 5th International Conference on Computing and Information, pp. 491–496. IEEE (1993). https://doi.org/10.1109/ICCI.1993.315323
5. Dash, D., Polyzotis, N., Ailamaki, A.: Cophy: a scalable, portable, and interactive index advisor for large workloads. arXiv preprint arXiv:1104.3214 (2011). https://doi.org/10.48550/arXiv.1104.3214
6. Fotouhi, F., Galarce, C.E.: Genetic algorithms and the search for optimal database index selection. In: Sherwani, N.A., de Doncker, E., Kapenga, J.A. (eds.) Great Lakes CS 1989. LNCS, vol. 507, pp. 249–255. Springer, New York (1991). https://doi.org/10.1007/BFb0038500
7. Huang, S., Ontañón, S.: A closer look at invalid action masking in policy gradient algorithms. arXiv preprint arXiv:2006.14171 (2020). https://doi.org/10.48550/arXiv.2006.14171
8. Ip, M.Y.L., Saxton, L.V., Raghavan, V.V.: On the selection of an optimal set of indexes. IEEE Trans. Softw. Eng. 2, 135–143 (1983). https://doi.org/10.1109/TSE.1983.236458
9. Kossmann, J., Halfpap, S., Jankrift, M., Schlosser, R.: Magic mirror in my hand, which is the best in the land? an experimental evaluation of index selection algorithms. In: Proceedings of the VLDB Endowment, vol. 13, no. 12, pp. 2382–2395 (2020). https://doi.org/10.14778/3407790.3407832
10. Lan, H., Bao, Z., Peng, Y.: An index advisor using deep reinforcement learning. In: Proceedings of the 29th ACM International Conference on Information & Knowledge Management, pp. 2105–2108 (2020). https://doi.org/10.1145/3340531.3412106
11. Mnih, V., et al.: Playing atari with deep reinforcement learning. arXiv preprint arXiv:1312.5602 (2013). https://doi.org/10.48550/arXiv.1312.5602
12. Papadomanolakis, S., Ailamaki, A.: An integer linear programming approach to database design. In: 2007 IEEE 23rd International Conference on Data Engineering Workshop, pp. 442–449. IEEE (2007). https://doi.org/10.1109/ICDEW.2007.4401027
13. Piatetsky-Shapiro, G.: The optimal selection of secondary indices is np-complete. ACM SIGMOD Rec. 13(2), 72–75 (1983). https://doi.org/10.1145/984523.984530
14. Schlosser, R., Kossmann, J., Boissier, M.: Efficient scalable multi-attribute index selection using recursive strategies. In: 2019 IEEE 35th International Conference on Data Engineering (ICDE), pp. 1238–1249. IEEE (2019). https://doi.org/10.1109/ICDE.2019.00113
15. Sharma, A., Schuhknecht, F.M., Dittrich, J.: The case for automatic database administration using deep reinforcement learning. arXiv preprint arXiv:1801.05643 (2018). https://doi.org/10.48550/arXiv.1801.05643
16. Valentin, G., Zuliani, M., Zilio, D.C., Lohman, G., Skelley, A.: Db2 advisor: an optimizer smart enough to recommend its own indexes. In: Proceedings of 16th International Conference on Data Engineering (Cat. No. 00CB37073), pp. 101–110. IEEE (2000). https://doi.org/10.1109/ICDE.2000.839397
17. Van Hasselt, H., Guez, A., Silver, D.: Deep reinforcement learning with double q-learning. In: Proceedings of the AAAI Conference on Artificial Intelligence, vol. 30 (2016). https://doi.org/10.1609/aaai.v30i1.10295
18. Wang, Z., Schaul, T., Hessel, M., Hasselt, H., Lanctot, M., Freitas, N.: Dueling network architectures for deep reinforcement learning. In: International Conference on Machine Learning, pp. 1995–2003. PMLR (2016)

# A Hybrid Model for Demand Forecasting Based on the Combination of Statistical and Machine Learning Methods

Fadoua Ouamani[1(✉)], Asma Ben Fredj[1,2], Mohamed Rayen Fekih[2], Anwar Msahli[2], and Narjès Bellamine Ben Saoud[1]

[1] RIADI Research Laboratory, National School of Computer Science, University of Manouba, Manouba, Tunisia
fadoua.ouamani@ensi-uma.tn
[2] COGNIRA, Imm. PREMIUM, Les Berges du Lac, Tunis, Tunisia
https://cognira.com/

**Abstract.** Demand Forecasting (DF) is nowadays a key component of successful businesses in retailing field. In fact, accurate customer's demand forecasts and insights into the reasons driving the forecasts may increase confidence, assist decision-making and therefore boost's the retailer's profit. It is then crucial for an accurate DF model to not only understand the retail time-series repeated patterns but also the impacts of factors such as the promotions on the data behavior. The literature review of existing research works has shown that statistical models gave good results in accurately detecting time-series components such as seasonality or trend but they fail when it comes to detecting external factors or causal effects compared to machine learning models. Moreover, the combination of both models either focused only on the trend component and neglected the seasonality or considered both of them but used sophisticated neural networks, which are computationally expensive. To this end, in this paper, we propose an approach that combines statistical and machine learning models to take advantages of their aforementioned properties. We used first Multiple Linear Regression (served as the baseline model as well) and linear interpolation to remove the promotions effect from the data and compute promotional multipliers. Then, each resulting data was fed to two statistical models (Prophet and Exponential Triple Smoothing). Finally, the combination step consisted in reintegrating the promotions effect into the forecasting results of each statistical model. Quantitative and qualitative evaluations of the hybrid models' performance showed that the hybrid models outperformed the baseline model.

**Keywords:** Retailing · Data mining · Demand forecasting · Time-series · Statistical models · Machine learning · Hybrid model · Prophet

## 1 Introduction

In today's fast-paced world, data collection, understanding and mining has become the key to staying ahead of the competition, achieving success and boosting profits in many

W. Chen et al. (Eds.): ADMA 2022, LNAI 13726, pp. 446–458, 2022.
https://doi.org/10.1007/978-3-031-22137-8_33

business fields. In retailing for example, it helps retailers gathering the most useful information about their products and customers, forecasting the customer's demand, understanding the reasons behind sales actions and behaviors and making of the right business decision. In retailing, the data is daily recorded and is organized into three-dimensional hierarchies: time hierarchy (year, half of year, month, week, and day), product hierarchy (: group, department, class, subclass and Stock Keeping Unit-SKU) and location hierarchy (chain, area, region, district and store). Moreover, Retail data is usually time-series data which is a collection of data sorted chronologically by a time index. It has three components: 1) trend component ($T(t)$) which shows the overall tendency of the data to increase, decrease or remain stable through long periods, 2) seasonal component ($S(t)$) which represents regularly spaced fluctuations having almost the same pattern and magnitude during the same period every year and 3) irregular component ($I(t)$) which refers to the unpredictable sudden changes in data. Time-series ($Y(t)$) may be additive ($Y(t) = T(t) + S(t) + I(t)$) or multiplicative ($Y(t) = T(t)*S(t)*I(t)$). When the magnitude of the seasonal pattern evolves with the trend then the time series is multiplicative.

Demand forecasting make then use of this data history to forecast the amount of goods or services that will be sold in the future or in response to particular causal-effects such as the promotions. In real world scenarios, the demand forecasting is based on the Eq. (1) where the $Y$ is the demand to forecast, Level is the average selling unit in the absence of seasonality and Promotional_mutliplier is the ratio representing the increase in sales within the presence of a promotion.

$$Y = \text{Level} \times \text{Seasonality} \times \text{Promotional\_multiplier} \qquad (1)$$

Due to the particularity of retail data, it is crucial for a demand forecasting model to not only understand the time series repeated patterns (trend and seasonality) but also the impact of factors such as the promotions on the behavior of the data. However, the existing research works on retail time-series forecasting propose statistical models, machine learning models or a combination of both models that do not satisfy the two aforementioned requirements simultaneously. In fact, statistical models focus on detecting the time series components such as seasonality and trend, whereas machine learning models have proven high effectiveness in handling causal effects. To this end, this paper proposes a new approach to retail demand forecasting that combines both models to handle causal effects and detect time series components with low complexity and easily interpretable models.

The remainder of the paper is then organized as follows: Sect. 2 will be dedicated to the review of existing research works on retail time-series forecasting and discussion of the findings. This will lead us to the paper's contributions described in Sect. 3 in terms of a revisited KDD (Knowledge Discovery from Databases) that goes through different steps: from data collection and preprocessing to model evaluation. Section 4 will be then devoted to the presentation of the efforts made to carry out the first step. Once data is ready, it's fed to two machine learning techniques (Multiple Linear regression and Linear Interpolation) to remove the promotions effect and compute promotional multipliers; which will be described in Sect. 5. Section 6 will present the statistical models used, namely Prophet and Exponential Triple Smoothing (ETS), their configurations and results of their trainings. Section 7 will finally present the model combination process and comparison results of the performance of the different hybrid models based on both quantitative and qualitative evaluations.

## 2  Related Work

Several retail time-series forecasting models have been proposed in the literature by following two different approaches: single model-based approach and model combination based approach. In this section, we will briefly present the research works using each approach, which help us to draw important conclusions and come up with our contribution to the field.

### 2.1  Statistical Models

Different types of statistical models were utilized for retail time-series forecasting. MA(Moving Average), ARIMA (Auto Regression Moving Average), SARIMA(Seasonal ARIMA), ETS(Exponential Triple Smoothing), Prophet, to cite but a few. In [1], the authors have used ARIMA which is an optimized version of MA, as MA cannot handle all time-series components and thus the high dimensionality variable's space and business related information about promotion and price fluctuation. ARIMA was deployed for demand forecasting as an important component of supply chain management process but it was incapable of capturing the seasonality pattern within the time-series data. In an attempt to solve this problem, researchers in [2] have used SARIMA for red lentils market price forecasting. Reference [3] proposes a ETS model to forecast the monthly retail sales of consumer goods. The model has shown good results with seasonal data. In [4], the researchers focused on forecasting monthly sales of different items using Prophet and creating a product portfolio based on the forecast reliability. Reference [5] has compared the performance of Prophet to ARIMA based on the MAPE (Mean Absolute Percentage Error) evaluation metric calculation. The performance of Prophet was slightly better than ARIMA in forecasting stock prices of a unique individual. In [6], the authors compared classical Prophet model to its optimized version by modifying the default Fourier order used to detect the seasonality to meet the particularity of the training data. Both models were used for daily sales forecasting.

### 2.2  Machine Learning Models

An alternative to statistical methods for retail time-series forecasting, are machine learning method, including deep learning methods. In [7], the authors compared the forecasting results of prophet and SARIMA to machine learning model LightGBM (Light Gradient Boosting Machine) on daily sales of Walmart stores. It has been shown that the latter model outperformed the former models based on the calculation of RMSE (Root Mean Squared Error) evaluation metric. The authors relate this difference to the fact that the data used is highly affected by external factors, which statistical models failed to capture. Whereas in [8], SARIMA outperformed ARIMA, Holt-Winters and ANN when forecasting Amazon's daily sales. The authors brought these results to the fact that the date is characterized by a strong seasonality pattern and was less affected by external factors. More sophisticated deep learning techniques were also proposed in [9], in which the authors compared LSTM (Long-Short Term Memory), RNN (Recurrent Neural Networks) and CNN (Convolutional Neural Network) models' performance on stock price prediction. As CNN use the information of the particular instant when forecasting unlike the others models and since stock markets are subjects to sudden changes, it captured the effect of external factors and outperformed the other models.

### 2.3 Combined Models

Reference [10] provides a literature review on retail sales forecasting, in which single models and combined models were compared to show better results among combined models. In [11], the authors propose an ensemble model for demand forecasting that combines Regression, Exponential smoothing, Holt-Winters and ARIMA models. The method examines the performance of each model over time and combines the weighted forecasts of the best performing models. In [12], the researchers used both MA and ANN models. MA was used to remove the trend effect from the sale time-series data to show only the differences in values from the trend. This has enhanced the performance of ANN model as it will be able to understand and capture accurately causal effects. Moreover, in [13], the authors have also proposed an hybrid model that uses LSTM and RF (Random Forest) for demand forecasting in multi-channel retail. The authors started by applying LSTM but they found that a residual caused by non-temporal explanatory causes remained. Thus, they applied RF on these unexplained components which allow them to outperform the single models applied on the same data.

To sum up, the research works studied have shown that statistical models gave good results in accurately detecting time-series data components, namely trend and seasonality but they fail when it comes to detecting external factors or causal effects compared to machine learning models. The reason why some research works ([12] and [13]) have combined them. However in [12], the authors focused only on the trend component and neglected the seasonality component. Whereas in [13], the authors have considered all the time-series data components but they have used sophisticated Neural Networks, which are computationally expensive.

In fact, for optimization and interpretability sake, and as stated by the authors of [14, 15], and due to the huge scale of item-level sales in the retail business, it is crucial to apply simple forecast methods and time-applicable approaches based on a decomposable models that can be updated gradually with only a limited amount of new sales data. Based on these findings, we propose a new approach for retail demand forecasting that will ensure a trade-off between the following three objectives: Handling causal effects, detecting all time-series data components and proposing a decomposable approach with low complexity and explainable results. To the best of our knowledge, this is the first investigation in such an approach that satisfies simultaneously these three criteria.

## 3   The Proposed Approach

The proposed approach (see Fig. 1) is based on a hybrid model that will allow us to capture both time-series data components and causal effects, and mainly the promotion effects on demand forecasting. The hybrid model will combine machine learning models and statistical models. We will then use Multiple Linear Regression (MLR) Model, not only for its capability to detect causal effects, but also for its easily interpretable outputs. This model will serve also as a baseline model. Linear interpolation will be the second model that we will use to eliminate the promotion's effect. The idea is to remove the actual value of the data in presence of the promotions; then linearly link the rest of the data points in order to have an approximation of the data with absence of promotion. Statistical models like ETS, Prophet and SARIMA will be then used on the resulting data (without promotion's effect). The choices were based on the findings from the literature review and motivated by the fact that these models were commonly used and have given accurate results when handling seasonality and trend components of time-series data where external factors are removed.

After preparing the data, we will first apply MLR to get the coefficients that will be used to compute the promotions multipliers. The multipliers will be then used to remove the promotion effect from the data. We will also apply Interpolation to get another version of the data without promotional effect. The resulting data will be fed to statistical models after being preprocessed accordingly to meet the requirements of each model. The combination step will reintegrate the promotional effect into the time-series data by multiplying it with the promotion multipliers whenever there is a promotion to reconstitute the predicted time-series with the correct magnitude. Finally, the results of each combination will be compared to the baseline model.

## 4   Data Generation, Preprocessing and Exploration

The real data was inaccessible due to confidentiality issues. That being said, the data on which the process was applied is a generated data that represents a simulation of a grocer.

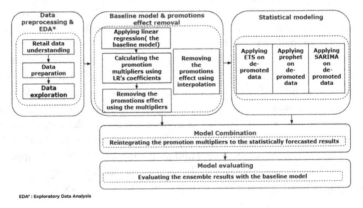

**Fig. 1.** The proposed approach for demand forecasting

## 4.1 The Dataset

The generated data is described by 5 parquet files as follows:

1. The product file: data about 100 products, 12 classes, 24 subclasses and 5 different departments;
2. The stores file: data about 30 stores, 4 regions in 1 country;
3. The customers file: data about 15000 customers and their purchasing frequency;
4. The calendar file: data about the time hierarchy of 1093 days (order of the week, the month in a year, the name of the year, the days of weeks and the weeks of years (WOY))
5. The Tlog file: data about the transactions (4702056 transactions) containing the action of buying a product by a customer from a store in a certain week within a promotion category: $'promo\_cat'$ and its discount $'promo\_discount'$.

## 4.2 Data Preprocessing

A data pipeline of three steps (joining, aggregation and cleaning) was applied on the raw data to transform it into exploitable one. Specific additional techniques were used depending on the model to meet its requirements (see Subsects. 6.1 and 6.2).

In the first step, we joined the transactional data in the Tlog file with their corresponding time information from the calendar file, the region and country information, and the products data. Then, we aggregated the transactions having the same combination of SKU, store, and week to the same row and filled it with the corresponding features in addition to a sales feature containing the sum of these transactions.

The common resulting input for the next steps is composed of categorical features and numerical features. We have as categorical features, the promotion category, at the product level, we have subclasses and 100 SKU, at the location level, we have 4 Regions an 30 stores, at the time level, we have the year (2018 to 2020), the WOY (52) and the Week (from 01-01-2018 to 12-28-2020). The numerical features are the target variable which is the Units and the feature variable which is the promotion discount.

## 4.3 Exploratory Data Analysis (EDA)

In an attempt to gain insight into time-series data components and their behaviors a plot is proposed (see Fig. 2.). The seasonality peaks in the plot occurred in the same period, a pattern that was observed in most SKUs. Same plots for the other subclasses have shown almost the same results. The seasonality phenomenon was observable at the subclass level and it is additive as its magnitude doesn't change significantly over time. Whereas, the trend component was absent.

The promotion effect (compared to a baseline plot that represents the median of non-promoted sales; green line in Fig. 2) was observable through the variation of behavior of the time-series data, with a peak in the fourth week of February. In fact, promotion effect varies according to the promotion categories (BXGY, BXGX, 50%_off, 25%_off, 10%_off). The average of estimated promotion multipliers has shown that 50%_off had the biggest effect on sales. This allows us to conclude that there is a positive correlation between the estimated promotion multiplier and the promotion category.

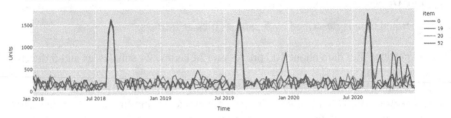

**Fig. 2.** Plot showing the sum of sales for the products in subclass 211

# 5  Removing the Promotions Effect Using MLR and Linear Interpolation

Since statistical models are univariate and unable to model causal effects, the promotions effect was first removed from data using two different techniques: MLR and linear interpolation. Customized data preparation was done before feeding the data to the MLR model. First, data was sorted by time to assure a non-erroneous forecast. Then, the discount was transformed into a discount elasticity using the Eq. (2)

$$discount\_elasticity = 1 - discount \tag{2}$$

After that, log based data scaling was applied to reduce the vast range of the data and to fit it to the additive model using the Eq. (3). Finally, one-hot encoding was applied to the categorical features to make them understandable by the model.

$$log((\widehat{y_t})) = \beta_1 log(x_{1t}) + \beta_2 log(x_{2t}) + \beta_3 log(x_{3t}) + \cdots + \beta_n log(x_{nt}) + log(e_t) \tag{3}$$

$$(\widehat{y_t}) = e^{(\beta_1 log(x_{1t}) + \beta_2 log x_{2t} + \beta_3 log(x_{3t}) + \cdots + \beta_n log(x_{nt}) + log(e_t))}$$

## 5.1  Multiple Linear Regression Coefficients Generation and Promotions Effect Removal

MLR has more than one predictor variable as shown by the Eq. (4) unlike simple Linear regression.

$$y_t = \beta_0 + \beta_1 x_{1,t} + \beta_2 x_{2,t} + \cdots + \beta_n x_{n,t} + \varepsilon_t \tag{4}$$

$\beta_1, \beta_2, \beta_3,..., \beta_n$ are the coefficients for the predictor variables $x_{1,t}, x_{2,t}, x_{3,t},..., x_{n,t}$. For $\varepsilon_t$, it represents the unobserved random variable

Before training the model, we selected the following three parameters: 1) fit_intercept: True ($\beta_0$ must be calculated for each execution of the model); 2) normalize: False (data is already scaled using the log); 3) positive: False (the coefficients can be negative or positive). The model was applied with subclass pooling. Prediction results (see Fig. 3) using MLR showed infinite values prediction (after july 2020). This

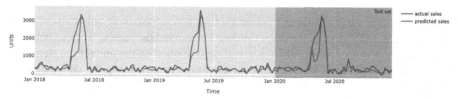

**Fig. 3.** Excerpt from the actual and predicted aggregated sum of sales within the subclass 200 using MLR with subclass pooling

is due to the correlation between input features as depicted by EDA. The problem was solved by using only three values for promo_cat: percentage, BXGX and BXGY.

The MLR coefficients obtained were then used to compute the promotion multipliers. The promotion multipliers will be used in turn to remove the promotions effect from data by dividing the sales data by their corresponding ones. Figure 4. Shows that the higher the discount value is, the stronger the promotional effect gets as well as its coefficient. The coefficients are capturing accurately the effects and are leading to accurate promotion multipliers; thus an effective promotions effect removal.

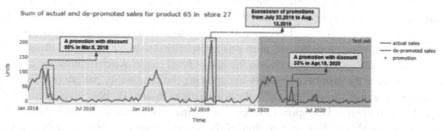

**Fig. 4.** Excerpt from the actual time-series and time-series without promotions effect using MLR coefficients for the product 65 in store 27

## 5.2   Interpolation Based Promotion Effect Removal

The promotion effect removal was also applied using the linear interpolation. The results of the two techniques look similar. However, on the one hand, using only linear interpolation is risky since it is basically linking linearly the data without understanding the category, discount, or any other information about the promotion effect. On the other hand, this linear linking can be useful as it outperforms the MLR based approach in some cases where the coefficients are relatively not accurate. Moreover, there are no specific metrics to evaluate the two approaches. Thus, we decided to use both of them in the next steps and integrate them into the final benchmark.

# 6 Statistical Modelling

Statistical models require as input a univariate time series containing only two columns, one for the ordered dates and the other for their corresponding values. To do so, the data is first sorted by week feature. Then, log scaling is applied on the units if the statistical model is additive. Finally, time-series data version is created using the data and units columns.

## 6.1 Prophet Model

Prophet model is capable of handling trend, seasonality, and holiday effects, which occur on an irregular basis over a day or a period of days. To do so, it decomposes the time-series y(t) into trend (g(t)), seasonality (s(t)), holiday (h(t)) and error using the Eq. (5)

$$y(t) = g(t) + s(t) + h(t) + e_t \tag{5}$$

The model was then configured as follows: Seasonality_mode was set to additive (default value) as we previously detected additive seasonality within the data; Yearly_seasonality was set to True as the presence of repeated pattern of seasonality will be detected in each year; growth was set to its default value (Linear) as the data do not follow a logistic growth; changepoint_range was set also to its default minimum value (0.6) as our data doesn't contain the trend component, so the model don't need to be flexible to it; prior_scale_changepoints was set to 0.05 as the model will focus less on trend component; changepoint_num was not specified to make the model investigate by itself the number of changepoint in the above fixed range. After training the model, the results show the prophet capability in capturing the seasonality pattern (see Fig. 5). However, they also show a case of under-forecasting. In fact, Prophet gives significant importance to the trend component when forecasting. When in the first period of sales, the seasonal peak was higher than that of the second period; the model assumed the presence of a decreasing trend. The altitude of this trend was high enough to cause an under-forecasting in the test set.

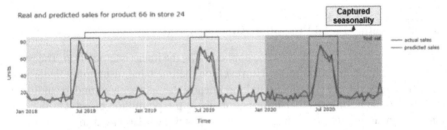

**Fig. 5.** Excerpt from the Prophet forecasting results on the resulting data after promotion effect removal using MLR coefficient for the product 66 in store 24

### 6.2  Triple Exponential Model (ETS)

Known as Holt-Winters Exponential Smoothing, ETS presents an improved version of classical Smoothing models (single and double exponential smoothing) since it explicitly accounts for seasonality in the time series. It allows the configuration of the same parameters: $\alpha$, $\beta$, Trend Type, Dampen Type and Damping coefficient $\phi$; and adds a new parameter $\gamma$ to control the seasonal component's influence.

The ETS model requires specific parameters for both the model definition and fitting steps. For model fitting, the parameter optimized will be set to True so that the model will estimate automatically the best fitting parameters ($\alpha$, $\beta$, $\phi$, $\gamma$, Trend Type, Dampen Type). The parameters chosen for model definition were based on the EDA findings and are summarized as follows: seasonal was set to additive; trend was set to none as there is no trend in our data, freq was set to W as data is weekly, seasonal_period was set to 52 (52 weeks each year). After training the model, the results have shown that ETS was capable of capturing the seasonal pattern accurately (see Fig. 6). However in another excerpts, an over-forecasting case was spotted out as by default the model gives more importance to the recent values when predicting.

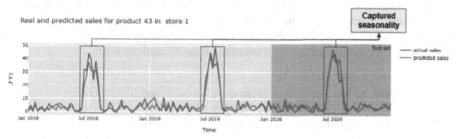

**Fig. 6.** Excerpt from the results of ETS's forecasting on data without promotions effect using MLR coefficients for the product 43 in store 1.

## 7  Model Combination and Results

### 7.1  Model Combination Process

The model combination process will combine the promotion coefficients and the predictions. Based on the Eq. (1) (see Sect. 1), the process will encompass four combination approaches as depicted by Fig. 7.

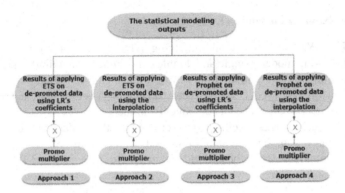

**Fig. 7.** The model combination process

## 7.2 Benchmarking Results

The comparison of the four approaches was based on quantitative and qualitative evaluation. A first evaluation effort was based on the use of SMAPE, MAE and RMSE statistical metrics as summarized in Table 1. The four experiments had similar results and all of them outperformed the baseline model. However for approach 3, according to the SMAPE metric, the ensemble model is relatively underperforming the baseline model. Thus, we cannot make a clear assumption based on these results.

**Table 1.** Results of the quantitative evaluation of models performance

| Models | SMAPE | MAE | RMSE |
| --- | --- | --- | --- |
| Baseline model | 64.394 | 9.570 | 5.411 |
| Approach 1 | 55.616 | 7.190 | 4.775 |
| Approach 2 | 61.450 | 7.262 | 4.781 |
| Approach 3 | 64.480 | 8.314 | 4.849 |
| Approach 4 | 57.200 | 7.175 | 4.730 |

Qualitative evaluation was then carried out since it shows an overview of the model performance on all the SKUs. For the improvement cases, the SKUs and their interpretation are the same for the four approaches as they were capable of detecting both the seasonal component and the promotion effect with almost the same performance. However, for the deterioration cases, they were similar per statistical model. In layman's words, approaches 1 and 2 that used ETS have their common deteriorated SKUs and interpretations. Whereas approaches 3 and 4 which used Prophet have their particular deterioration causes. But, for the majority of deteriorated cases, the main reason is the non-seasonality of the data. This result is understandable since our hybrid models were not designed to accurately forecast this type of data.

# 8 Conclusion

Through this paper, we have introduced a hybrid modeling approach that combines in different ways MLR and Linear interpolation with Prophet and ETS. The evaluation results of the approach showed that our solution was able to accurately forecast the demand based on both time-series data components and the promotion effect while using a decomposable low complexity models. Even though, the generated data was a good asset to conduct this work, a future work will consider the application of the approach on real world data to obtain more generalized performance evaluation.

# References

1. Fattah, J., Ezzine, L., Aman, Z., El Moussami, H., Lachhab, A.: Forecasting of demand using ARIMA model. Int. J. Eng. Bus. Manag. **10**, 1847979018808673 (2018). https://doi.org/10.1177/1847979018808673
2. Divisekara, R.W., Jayasinghe, G.J.M.S.R., Kumari, K.W.S.N.: Forecasting the red lentils commodity market price using SARIMA models. SN Bus. Econ. **1**(1), 1–13 (2020). https://doi.org/10.1007/s43546-020-00020-x
3. Aohan, L.: An empirical analysis of total retail sales of consumer goods based on holt-winters. In: Proceedings of the 2020 2nd International Conference on Big Data and Artificial Intelligence, pp. 54–57. Association for Computing Machinery, New York (2020). https://doi.org/10.1145/3436286.3436298
4. Žunić, E., Korjenić, K., Hodžić, K., Đonko, D.: Application of facebook's prophet algorithm for successful sales forecasting based on real-world data. Int. J. Comput. Sci. Inf. Technol. **12**, 23–36 (2020). https://doi.org/10.5121/ijcsit.2020.12203
5. Garlapati, A., Krishna, D.R., Garlapati, K., Yaswanth, N.M.S., Rahul, U., Narayanan, G.: Stock price prediction using facebook prophet and arima models. In: 2021 6th International Conference for Convergence in Technology (I2CT), pp. 1–7 (2021). https://doi.org/10.1109/I2CT51068.2021.9418057
6. Liço, L., Enesi, I., Jaiswal, H.: Predicting customer behavior using prophet algorithm in a real time series dataset. Eur. Sci. J. ESJ. 17, 10 (2021). https://doi.org/10.19044/esj.2021.v17n25p10
7. Jiang, H., Ruan, J., Sun, J.: Application of machine learning model and hybrid model in retail sales forecast. In: 2021 IEEE 6th International Conference on Big Data Analytics (ICBDA), pp. 69–75 (2021). https://doi.org/10.1109/ICBDA51983.2021.9403224
8. Singh, B., Kumar, P., Sharma, N., Sharma, K.P.: Sales forecast for amazon sales with time series modeling. In: 2020 First International Conference on Power, Control and Computing Technologies (ICPC2T), pp. 38–43 (2020). https://doi.org/10.1109/ICPC2T48082.2020.9071463
9. Selvin, S., Vinayakumar, R., Gopalakrishnan, E.A., Menon, V.K., Soman, K.P.: Stock price prediction using LSTM, RNN and CNN-sliding window model. In: 2017 International Conference on Advances in Computing, Communications and Informatics (ICACCI), pp. 1643–1647 (2017). https://doi.org/10.1109/ICACCI.2017.8126078
10. Aras, S., Kocakoç, İD., Polat, C.: Comparative study on retail sales forecasting between single and combination methods. J. Bus. Econ. Manag. **18**, 803–832 (2017). https://doi.org/10.3846/16111699.2017.1367324
11. Akyuz, A.O., Uysal, M., Bulbul, B.A., Uysal, M.O.: Ensemble approach for time series analysis in demand forecasting: ensemble learning. In: 2017 IEEE International Conference on INnovations in Intelligent SysTems and Applications (INISTA), pp. 7–12 (2017). https://doi.org/10.1109/INISTA.2017.8001123

12. Nunnari, G., Nunnari, V.: Forecasting monthly sales retail time series: a case study. In: 2017 IEEE 19th Conference on Business Informatics (CBI), pp. 1–6 (2017). https://doi.org/10.1109/CBI.2017.57

13. Punia, S., Nikolopoulos, K., Singh, S.P., Madaan, J.K., Litsiou, K.: Deep learning with long short-term memory networks and random forests for demand forecasting in multi-channel retail. Int. J. Prod. Res. **58**, 4964–4979 (2020). https://doi.org/10.1080/00207543.2020.1735666

14. Beheshti-Kashi, S., Karimi, H.R., Thoben, K.-D., Lütjen, M., Teucke, M.: A survey on retail sales forecasting and prediction in fashion markets. Syst. Sci. Control Eng. **3**, 154–161 (2015). https://doi.org/10.1080/21642583.2014.999389

15. Seaman, B.: Considerations of a retail forecasting practitioner. Int. J. Forecast. **34**, 822–829 (2018). https://doi.org/10.1016/j.ijforecast.2018.03.001

# A Boosting Algorithm for Training from Only Unlabeled Data

Yawen Zhao[1], Lin Yue[2], and Miao Xu[1(✉)]

[1] The University of Queensland, Brisbane, QLD 4072, Australia
{yawen.zhao,miao.xu}@uq.edu.au
[2] The University of Newcastle, Callaghan, NSW 2308, Australia
Lin.Yue@newcastle.edu.au

**Abstract.** Unlabeled-unlabeled (UU) learning was proposed to cope with the high cost of data annotation and some realistic cases, in which we cannot get labeled data. It allows us to train a classifier with only unlabeled data. State-of-the-art (SOTA) UU methods with good performance based on neural networks (NN) have been proposed; however, there is a lack of studies on boosting algorithms for learning from only unlabeled data, even though boosting algorithms sometimes perform very well with simple base classifiers. We propose a novel boosting algorithm for UU learning: Ada-UU, which compares against neural networks. The proposed method follows the general procedure of AdaBoost while the classification error is estimated with two sets of unlabeled (U) data. We empirically demonstrate that Ada-UU outperforms neural networks on several large-scale benchmark UU datasets and has comparable performance on a small-scale benchmark dataset.

**Keywords:** Boosting · Weakly supervised learning · UU learning

## 1 Introduction

In the past few decades, mining insights from big data has become a sought-after research objective owing to the rapid development of machine learning techniques. Supervised learning is one of the most popular genres among various machine learning categories, where an intelligent system learns predictions from labeled input-output pairs. However, researchers and stakeholders have progressively realized that acquiring a vast amount of labeled data is exceedingly difficult and expensive. This exists in numerous fields such as medicine, biometrics and social media analysis. There are two typical difficulties in labeling the data. The first one is the expense, while another is the label's uncertainty. For example, in social media sentiment analysis, the amount of data produced by Twitter's users could easily hit terabytes per day. It is impossible to label every record as positive, negative or neutral. For the label's uncertainty, taking disease diagnosis as an example, we can only label the confirmed patients provided by the hospital with the corresponding disease. A patient with mild or asymptomatic symptoms cannot be labeled since they have not been diagnosed yet.

W. Chen et al. (Eds.): ADMA 2022, LNAI 13726, pp. 459–473, 2022.
https://doi.org/10.1007/978-3-031-22137-8_34

To satisfy these real-world demands, positive-unlabeled (PU) learning [2–4] has been widely studied, where a binary classifier can be trained with a few labeled positive (P) data and sufficient unlabeled (U) data without knowing negative (N) data. However, PU learning still requires a few labeled data. There still exists a cost to label data, and there is no guarantee that we can get accurate labels for a dataset [22]. Therefore, UU learning [12] was proposed with a stricter setting, where it only learns from unlabeled (U) data, which is closer to a real-world scenario.

Unlike clustering, UU learning does not rely on geometric assumptions of the classes, such as that one cluster corresponds to one class. It does not need to introduce additional assumptions upon which the learning objectives are built [10,20] either. So far, most SOTA UU methods are based on neural networks (NN) [11–14]. However, NN may not always be the most suitable model for classification tasks. As mentioned in [12,14], NN-based UU methods require heavy hyper-parameter tuning, sophisticated correction functions and designed network structures. Besides, large computational resources are required for training NN-based models.

A boosting method for PU learning has been proposed and proved to be effective [23], which generates a sequence of base classifiers and combines them with corresponding weights. The same as traditional boosting methods for supervised learning, [23] iteratively adjusts data weights during the boosting procedure. The experimental results show that NN-based methods might not always be the most suitable for a classification task. So far, no boosting algorithm has been proposed to learn directly from only unlabeled data yet, even though the boosting algorithms are the most effective in some of the practical machine learning tasks on tabular data [1]. Inspired by [23], we propose a boosting algorithm Ada-UU for UU learning, which does not require heavy parameter tuning or large GPU resources. There remain two critical challenges in designing a boosting method for training from only unlabeled data.

- How do we get the ideal base classifiers and their weights?
- How should we update the data weights in each iteration?

To address these challenges, four different data weights of two sets of U data are maintained and updated in each iteration in Ada-UU. The classification error is estimated with the two sets of U data and further contributes to the weight of each base classifier. The experimental results show that the proposed algorithm, Ada-UU, achieves higher performance than most baseline methods. The main contributions of this work include:

- We innovatively propose Ada-UU, which is the first boosting method for UU learning.
- We solve the two challenges of obtaining the optimal base classifiers with corresponding weights and updating the data weights with only U data during the boosting process.
- Compared with NN-based UU learning methods, our proposed algorithm does not require extensive tuning of hyperparameters or a large number of expen-

sive GPU devices. It can be counted as an efficient and green approach to solving the UU problem.

## 2  Formulation and Background

### 2.1  UU Learning

Let $X \in \mathbb{R}^d$ and $Y \in \{+1, -1\}(d \in \mathbb{N}^+)$ be the input and output random variables, respectively. $\pi_p = p(Y = +1)$ is the class-prior probability and $\pi_n = p(Y = -1) = 1 - \pi_p$. We use $\pi$ to represent $\pi_p$ in the following for convenience. In UU learning, the training dataset $X$ is composed of two sets of U data, $U_1$ with size $n_1$ and $U_2$ with size $n_2$. $U_1$ and $U_2$ are both sampled from $p(x)$. Therefore, we have $\{x_1, x_2, \cdots, x_{n_1}\} \sim p_{U_1}(x)$ and $\{x_1, x_2, \cdots, x_{n_2}\} \sim p_{U_2}(x)$.

$$p_{U_1}(x) = \theta_1 p_p(x) + (1 - \theta_1)p_n(x), \ \ p_{U_2}(x) = \theta_2 p_p(x) + (1 - \theta_2)p_n(x), \quad (1)$$

where $\theta_1$ and $\theta_2$ are the class priors of $U_1$ and $U_2$ respectively.

Following the convention [12,14], we assume $\pi$, $\theta_1$ and $\theta_2$ as known and $\theta_1 > \theta_2$ throughout the paper. Based on the given data, our objective is to learn a binary classifier $g : \mathbb{R}^d \to \{+1, -1\}$. Let $\ell : \mathbb{R} \times \{+1, -1\} \to \mathbb{R}$ be a loss function that $\ell(g(x), y)$ means the loss occurred when the classifier $g$ outputs the prediction when the label of $x$ is $y$. Then the risk of the classifier $g$ is

$$R(g) = \pi \mathbb{E}_p[\ell(g(x), +1)] + (1 - \pi)\mathbb{E}_n[\ell(g(x), -1)]. \quad (2)$$

Here $\mathbb{E}_p[\cdot]$ means $\mathbb{E}_{X \sim p_p}[\cdot]$ and $\mathbb{E}_n[\cdot]$ means $\mathbb{E}_{X \sim p_n}[\cdot]$, where $p_p(x) = p(x|Y = +1)$ and $p_n(x) = p(x|Y = -1)$. In a PN learning scenario, we can approximate $R(g)$ by:

$$\widehat{R}_{pn}(g) = \frac{\pi}{n_p} \sum_{i=1}^{n_p} \ell(g(x_i), +1) + \frac{(1 - \pi)}{n_n} \sum_{j=1}^{n_n} \ell(g(x_j), -1), \quad (3)$$

where $n_p$ is the size of P data and $n_n$ is the size of the N data.

However, we cannot directly estimate $\mathbb{E}_p[\ell(g(x), +1)]$ or $\mathbb{E}_n[\ell(g(x), -1)]$ in UU learning, since we do not have P data or N data. More specifically, given only $U_1$ and $U_2$, we cannot approximate Eq. (2) by Eq. (3). Inspired by [15–17], we can rewrite $R(g)$ to make it possible to be approximated based on the given $U_1$ and $U_2$. Recall that $\theta_1$ and $\theta_2$ are the class priors of $U_1$ and $U_2$, and we have Eq. (1). Thus we can rewrite the risk of classifier $g$ as follows.

$$
\begin{aligned}
R_{uu}(g) =& \mathbb{E}_{p_{U_1}}[a\ell(g(x), +1) + b\ell(g(x), -1)] + \mathbb{E}_{p_{U_2}}[c\ell(g(x), +1) + d\ell(g(x), -1)] \\
=& \theta_1 \mathbb{E}_p[a\ell(g(x), +1) + b\ell(g(x), -1)] \\
&+ (1 - \theta_1)\mathbb{E}_n[a\ell(g(x), +1) + b\ell(g(x), -1)] \\
&+ \theta_2 \mathbb{E}_p[c\ell(g(x), +1) + d\ell(g(x), -1)] \\
&+ (1 - \theta_2)\mathbb{E}_n[c\ell(g(x), +1) + d\ell(g(x), -1)].
\end{aligned}
$$

$$(4)$$

By setting Eq. (2) and Eq. (4) to be equal, we can get

$$a = \frac{(1 - \theta_2)\pi}{\theta_1 - \theta_2}, \quad b = -\frac{\theta_2(1 - \pi)}{\theta_1 - \theta_2}, \quad c = -\frac{(1 - \theta_1)\pi}{\theta_1 - \theta_2}, \quad d = \frac{\theta_1(1 - \pi)}{\theta_1 - \theta_2}, \quad (5)$$

Therefore, similar to Eq. (3), the risk in Eq. (4) can be approximated by:

$$\widehat{R}_{\mathrm{uu}}(g) = \frac{a}{n_1} \sum_{i=1}^{n_1} \ell(g(x_i), +1) + \frac{b}{n_1} \sum_{i=1}^{n_1} \ell(g(x_i), -1)$$
$$+ \frac{c}{n_2} \sum_{j=1}^{n_2} \ell(g(x_j), +1) + \frac{d}{n_2} \sum_{j=1}^{n_2} \ell(g(x_j), -1), \quad (6)$$

Equation (6) has been proved to be unbiased [12]. Therefore, the proposed UU method, which uses the unbiased risk estimator, is called UU-Unbiased [12]. However, UU-Unbiased suffers from severe overfitting and the overfitting usually happens when the empirical risk goes negative, which is not legitimate. In order to solve the overfitting problems, a family of consistently corrected risk estimators

$$\widehat{R}_{\mathrm{cc}}(g) = f_1(a\widehat{R}^+_{\mathrm{U}_1}(g) + c\widehat{R}^+_{\mathrm{U}_2}(g)) + f_2(b\widehat{R}^-_{\mathrm{U}_1}(g) + d\widehat{R}^-_{\mathrm{U}_2}(g)) \quad (7)$$

have been proposed [14], where $f_1$ and $f_2$ can be any consistent correction function. [14] introduced three typical types of correction functions and named the methods with those functions as: UU-ABS, UU-ReLU and UU-LReLU. These methods have been studied in [14] and proved to be effective in solving overfitting problems and improving model performance. Denote $a\widehat{R}^+_{\mathrm{U}_1}(g) + c\widehat{R}^+_{\mathrm{U}_2}(g)$ as $\widehat{R}_{\mathrm{uu}_p}$ and $b\widehat{R}^-_{\mathrm{U}_1}(g) + d\widehat{R}^-_{\mathrm{U}_2}(g)$ as $\widehat{R}_{\mathrm{uu}_n}$. UU-ReLU uses the ReLU function that

$$\widehat{R}_{\mathrm{uu-ReLU}}(g) = \max\{0, R_{\mathrm{uu}_p}\} + \max\{0, R_{\mathrm{uu}_n}\}$$
$$= \max\{0, a\widehat{R}^+_{\mathrm{U}_1}(g) + c\widehat{R}^+_{\mathrm{U}_2}(g)\} + \max\{0, b\widehat{R}^-_{\mathrm{U}_1}(g) + d\widehat{R}^-_{\mathrm{U}_2}(g)\}. \quad (8)$$

Using ReLU function as correction function to prevent overfitting problems has also been discussed in [9]. UU-ReLU uses the absolute function that

$$\widehat{R}_{\mathrm{uu-abs}}(g) = |R_{\mathrm{uu}_p}| + |R_{\mathrm{uu}_n}|$$
$$= |a\widehat{R}^+_{\mathrm{U}_1}(g) + c\widehat{R}^+_{\mathrm{U}_2}(g)| + |b\widehat{R}^-_{\mathrm{U}_1}(g) + d\widehat{R}^-_{\mathrm{U}_2}(g)|. \quad (9)$$

UU-LReLU uses the generalized leaky ReLU function, i.e. $f_1(x) = f_2(x) = \mathbb{I}_{\{x \geq 0\}}x + \mathbb{I}_{\{x \leq 0\}}\lambda x$, where $\lambda \leq 0$. The parameter $\lambda$ controls the weights of the negative risks. In this way, the model can learn from all the training data, since it does not completely ignore the negative part as UU-ReLU and UU-ABS do. Note that the absolute function and the ReLU function are special cases of the generalized leaky ReLU function [14].

## 2.2 AdaBoost

Boosting is an ensemble method, that generates a sequence of base classifiers and combines them into a strong model. AdaBoost (adaptive boosting) is one of the most classic boosting algorithms [5,6]. It has been widely studied and applied to different areas, such as facial recognition [19] and financial fraud detection [18]. It has also been combined with PU learning together [23] and achieved good performance, which shows the potential of the algorithm in weakly supervised learning. However, there is still a lack of study on boosting algorithms for UU learning as far as we know.

In this paper, we review AdaBoost from the view of [7], following [24]. Let us set the number of base classifiers to $T$. Then the algorithm will generate $T$ hypotheses $\{h_t(x) : t = 1, \ldots, T\}$ sequentially and combine them with corresponding weights $\alpha_t$. The final output of the algorithm can be written as $H(x) = \text{sign}(\sum_{t=1}^{T} \alpha_t h_t(x))$. At each iteration, the data distribution $\mathcal{D}_t$ of training samples is updated based on the output of the algorithm that the weights of the misclassified data will be increased while the weights of the correctly classified data will be decreased. In other words, the later classifiers focus more on the instances misclassified by the earlier classifiers. In this way, AdaBoost can achieve better performance than a single base classifier by combining the base classifiers with proper weights.

Now we review the details of the algorithm. The algorithm tries to minimize an exponential loss

$$R_{\text{exp}}(h) = \mathbb{E}_{(x,y)}[e^{-yh(x)}] \tag{10}$$

in each round when a new hypothesis $h$ is generated. Assume we are in iteration $t$ of the algorithm, where we already have a set of hypotheses and their weights. Denote the weighted combination of the generated hypotheses in the previous $t-1$ iterations as $H$ and the hypothesis we generate in the current round as $h$. The current exponential loss can be written as

$$R_{\text{exp}}(H + \alpha h) = \mathbb{E}_{(x,y)}[e^{(-y(H(x)+\alpha h(x)))}]. \tag{11}$$

Denote $\epsilon = \mathbb{E}_{x \sim \mathcal{D}}[y \neq h(x)]$, we can get

$$\alpha = \frac{1}{2} \ln \frac{1 - p(y \neq h(x))}{p(y \neq h(x))} = \frac{1}{2} \ln \frac{1 - \epsilon}{\epsilon},$$

where $\alpha$ is the weight of $h$ that greedily minimizes $R_{\text{exp}}(H + \alpha h)$. This is how we decide the weight of each generated hypothesis in AdaBoost.

Now we consider how to generate $h$ and update the data distribution. Since it will not lose generalization, we expand Eq. (11) to second order with respect to $h(x) = 0$ and fix $\alpha = 1$,

$$R_{\text{exp}}(H + h|x) \approx \mathbb{E}_y[e^{-yH(x)}(1 - yh(x) + y^2 h(x)^2/2)|x]$$
$$= \mathbb{E}_y[e^{-yH(x)}(1 - yh(x) + 1/2)|x],$$

which is further minimized by

$$h^*(x) = \arg\max_h \mathbb{E}_y[e^{-yH(x)}yh(x)|x].$$

Therefore, to get the optimal classification performance, the distribution is updated in each iteration as

$$\mathcal{D}_{(t+1)}(x) = e^{-y(H(x)+\alpha h(x))}p(y|x) = \mathcal{D}_t(x) \cdot e^{-\alpha yh(x)}.$$

## 3    Proposed Method: Ada-UU

As mentioned in Sect. 1, we solve two challenges in this paper: i) How do we obtain the ideal base classifiers with corresponding weights? ii) How do we update the data weights as the training process goes on? We explain the details of our proposal in the following two subsections.

### 3.1    Estimation of the Combination Weight

In this part, we are going to explain how we decide the weights of the base classifiers. Note that with only U data available, our goal is still to minimize the exponential risk in Eq. (10). Assume we are at round $t$ and we already have the previous generated hypotheses $h_i$ and their weights $\alpha_i$, $i \in \{1, 2, \cdots, t-1\}$. A new hypothesis $h$ has just been generated. The same as Sect. 2.2, we denote the weighted combination of the generated hypotheses as $H$, then we can represent the risk as

$$R_{\exp}(H + \alpha h) = e^{-yH(x)}(e^{-\alpha} \cdot p(y = h(x)) + e^{\alpha} \cdot p(y \neq h(x))) \quad (12)$$

when using exponential loss as the surrogate loss. Let the derivation of Eq. (12) equal to zero, then we can get the weight of the hypothesis

$$\alpha = \frac{1}{2}\ln\frac{p(y = h(x))}{p(y \neq h(x))} = \frac{1}{2}\ln\frac{1 - p(y \neq h(x))}{p(y \neq h(x))}, \quad (13)$$

which minimizes the risk. Note that we only have U data in UU learning, so we need to estimate $P(y \neq h(x))$ based on the U data. Because we have

$$p(y \neq h(x)) = \pi\mathbb{E}_p[\ell_{01}(h(x), +1)] + (1 - \pi)\mathbb{E}_n[\ell_{01}(h(x), -1)], \quad (14)$$

inspired by the procedure of rewriting the classification risk in UU learning introduced in Sect. 2.1, given $U_1$ with class prior $\theta_1$ and $U_2$ with class prior $\theta_2$, we can rewrite Eq. (14) as

$$\begin{aligned} p(y \neq h(x)) = &\theta_1\mathbb{E}_p[a\ell_{01}(g(x), +1) + b\ell_{01}(g(x), -1)] \\ &+ (1 - \theta_1)\mathbb{E}_n[a\ell_{01}(g(x), +1) + b\ell_{01}(g(x), -1)] \\ &+ \theta_2\mathbb{E}_p[c\ell_{01}(g(x), +1) + d\ell_{01}(g(x), -1)] \\ &+ (1 - \theta_2)\mathbb{E}_n[c\ell_{01}(g(x), +1) + d\ell_{01}(g(x), -1)], \end{aligned} \quad (15)$$

whose empirical estimation is

$$\widehat{p}_{UU}(y \neq h(x)) = \frac{a}{n_1} \sum_{i=1}^{n_1} \ell_{01}(g(x_i), +1) + \frac{b}{n_1} \sum_{i=1}^{n_1} \ell_{01}(g(x_i), -1)$$
$$+ \frac{c}{n_2} \sum_{j=1}^{n_2} \ell_{01}(g(x_j), +1) + \frac{d}{n_2} \sum_{j=1}^{n_2} \ell_{01}(g(x_j), -1), \tag{16}$$

where

$$a = \frac{(1-\theta_2)\pi}{\theta_1 - \theta_2}, \quad b = -\frac{\theta_2(1-\pi)}{\theta_1 - \theta_2}, \quad c = -\frac{(1-\theta_1)\pi}{\theta_1 - \theta_2}, \quad d = \frac{\theta_1(1-\pi)}{\theta_1 - \theta_2}. \tag{17}$$

Note that in AdaBoost, the weights of the training examples will be updated in each iteration. In the following paper, we use $w$ to represent the data weights. In UU learning, we maintain four different kinds of U data as shown in Eq. (16). Following [9,23], we denote them as $U_1^+$, $U_1^-$, $U_2^+$ and $U_2^-$. Denote the data weights in iteration $t$ as $w_t$, we define

$$w_1(x) = \begin{cases} \frac{a}{n_1}, & \text{for } x \in U_1^+ \\ \frac{b}{n_1}, & \text{for } x \in U_1^- \\ \frac{c}{n_2}, & \text{for } x \in U_2^+ \\ \frac{d}{n_2}, & \text{for } x \in U_2^- \end{cases}. \tag{18}$$

In iteration $t$, after we get the current hypothesis $h$ and the previous combined classifiers $H_{t-1}$, the weighted error estimated in the $t$-th iteration can be written as

$$\widehat{\epsilon}_t(U_1, U_2) = \sum_{x \in U_1^+} [w_1(x) \exp(-H_{t-1}(x)) \ell_{01}(h(x), +1)]/Z_1 + \tag{19}$$

$$\sum_{x \in U_1^-} [w_1(x) \exp(H_{t-1}(x)) \ell_{01}(h(x), -1)]/Z_2 +$$

$$\sum_{x \in U_2^+} [w_1(x) \exp(-H_{t-1}(x)) \ell_{01}(h(x), +1)]/Z_3 +$$

$$\sum_{x \in U_2^-} [w_1(x) \exp(H_{t-1}(x)) \ell_{01}(h(x), -1)]/Z_4,$$

where $Z_1$, $Z_2$, $Z_3$ and $Z_4$ are normalization constants that
$Z_1 = \sum_{x \in U_1^+} w_1(x) \exp(-H_{t-1}(x))/a$, $Z_2 = \sum_{x \in U_1^-} w_1(x) \exp(-H_{t-1}(x))/b$,
$Z_3 = \sum_{x \in U_2^+} w_1(x) \exp(-H_{t-1}(x))/c$, $Z_4 = \sum_{x \in U_2^-} w_1(x) \exp(-H_{t-1}(x))/d$,
and

$$\alpha_{UU}^t = \frac{1}{2} \ln \frac{1 - \widehat{\epsilon}_t(U_1, U_2)}{\widehat{\epsilon}_t(U_1, U_2)}. \tag{20}$$

Similar to Eq. (8) and Eq. (9), we denote

$$\sum_{x \in U_1^+} [w_1(x) \exp(-H_{t-1}(x)) \ell_{01}(h(x), +1)]/Z_1 +$$

$$\sum_{x \in U_2^+} [w_1(x) \exp(-H_{t-1}(x)) \ell_{01}(h(x), +1)]/Z_2$$

as $\widehat{\epsilon}_p$, and

$$\sum_{x \in U_1^-} [w_1(x) \exp(H_{t-1}(x)) \ell_{01}(h(x), -1)]/Z_3 +$$

$$\sum_{x \in U_2^-} [w_1(x) \exp(H_{t-1}(x)) \ell_{01}(h(x), -1)]/Z_4$$

as $\widehat{\epsilon}_n$ for convenience.

## 3.2   Updating the Data Weight

Since we do not know the labels in UU learning, we cannot update the data weights in the same way as AdaBoost in PN learning, where only a set of data weights needs to be updated for all the data. In our proposal, four different sets of data weights are updated simultaneously in each round. In this part, we are going to explain the details of how we update the data weights. The same as [23], we omit the conditional dependence on $x$ in the original derivation as it does not impact our learning objective.

We start from our original goal, which is minimizing the exponential loss in Eq. (10). Given generated hypotheses at iteration $t$, we already know their weighted combination $H_{t-1}$ and the corresponding weight of each one $\alpha_t$. We fix $\alpha_t = 1$ without losing generalization. Then we have

$$R_{\exp}(H_{t-1} + h) = \mathbb{E}_{(x,y)}[e^{-y(H_{t-1}(x)+h(x))}]$$

$$\approx \mathbb{E}_{(x,y)}[e^{-yH_{t-1}(x)}(1 - yh(x) + \frac{y^2 h(x)^2}{2})]$$

$$= \mathbb{E}_{(x,y)}[e^{-yH_{t-1}(x)}(1 - yh(x) + \frac{1}{2})]$$

$$= \mathbb{E}_{(x,y)}[\frac{3}{2}e^{-yH_{t-1}(x)} - yh(x)e^{-yH_{t-1}(x)}]. \qquad (21)$$

We can find that minimizing Eq. (21) is equal to maximizing

$$\mathbb{E}_{(x,y)}[e^{-yH_{t-1}(x)}yh(x)]. \qquad (22)$$

Denote Eq. (22) as $E_t$, then the empirical estimation of Eq. (22) is

$$\widehat{E}_t = \frac{a}{n_1} \sum_{i=1}^{n_1} e^{-H_{t-1}(x_i)} h(x_i) + \frac{b}{n_1} \sum_{i=1}^{n_1} -e^{H_{t-1}(x_i)} h(x_i)$$

$$+ \frac{c}{n_2} \sum_{j=1}^{n_2} e^{-H_{t-1}(x_j)} h(x_j) + \frac{d}{n_2} \sum_{j=1}^{n_2} -e^{H_{t-1}(x_j)} h(x_j). \qquad (23)$$

In practice, we maximize $\widehat{E}$ to achieve our objective of maximizing Eq. (22). Since $H_t(x) = H_{t-1}(x) + \alpha_{UU}^t h_t(x)$, we have $e^{H_t(x)} = e^{H_{t-1}(x)} e^{\alpha_{UU}^t h_t(x)}$ for $t \geq 1$. With the $w_1(x)$ defined in Eq. (18), if we further define

$$
w_{t+1}(x) = \begin{cases}
w_t(x)e^{-\alpha_{UU}^t h_t(x)}, & \text{for } x \in U_1^+ \\
w_t(x)e^{\alpha_{UU}^t h_t(x)}, & \text{for } x \in U_1^- \\
w_t(x)e^{-\alpha_{UU}^t h_t(x)}, & \text{for } x \in U_2^+ \\
w_t(x)e^{\alpha_{UU}^t h_t(x)}, & \text{for } x \in U_2^-
\end{cases}
\tag{24}
$$

then we can rewrite Eq. (23) as

$$
\begin{aligned}
\widehat{E}_t &= \sum_{x \in U_1^+} w_1(x)e^{-\sum_{j=1}^{t-1}\alpha_{UU}^j h_j(x)} h_t(x) - \sum_{x \in U_1^-} w_1(x)e^{\sum_{j=1}^{t-1}\alpha_{UU}^j h_j(x)} h_t(x) \\
&+ \sum_{x \in U_2^+} w_1(x)e^{-\sum_{j=1}^{t-1}\alpha_{UU}^j h_j(x)} h_t(x) - \sum_{x \in U_2^-} w_1(x)e^{-\sum_{j=1}^{t-1}\alpha_{UU}^j h_j(x)} h_t(x) \\
&= \sum_{x \in U_1^+} w_t(x)h_t(x) - \sum_{x \in U_1^-} w_t(x)h_t(x) + \sum_{x \in U_2^+} w_t(x)h_t(x) - \sum_{x \in U_2^-} w_t(x)h_t(x) \\
&= \sum_{x \in U_1^+ \cup U_2^+} w_t(x)h_t(x) - \sum_{x \in U_1^- \cup U_2^-} w_t(x)h_t(x).
\end{aligned}
\tag{25}
$$

From Eq. (25), we can see that four sets of data weights needs to be updated as defined in Eq. (24) in each iteration. Re-weighting the training examples based on Eq. (24) and maximizing our objective $\widehat{E}_t$ could lead to a desired base classifier $h_t^*$ in iteration $t$. We can rewrite $\widehat{\epsilon}_t(U_1, U_2)$ in Eq. (19) with a simpler form as

$$
\begin{aligned}
\widehat{\epsilon}_t(U_1, U_2) &= \sum_{x \in U_1^+} [w_t(x)\ell_{01}(h(x), +1)]/Z_1 + \sum_{x \in U_1^-} [w_t(x)\ell_{01}(h(x), -1)]Z_2 \\
&+ \sum_{x \in U_2^+} [w_t(x)\ell_{01}(h(x), +1)]/Z_3 + \sum_{x \in U_2^-} [w_t(x)\ell_{01}(h(x), -1)]/Z_4.
\end{aligned}
$$

### 3.3   Algorithm

In the previous two subsections, we introduced how we obtain the base classifiers with corresponding weights and how we update the data weights in the training process in our proposal. In this section, we summarize the procedure of the algorithm Ada-UU and introduce the implementation details.

Ada-UU has a similar procedure as AdaBoost as shown in Algorithm 1 and Algorithm 2. We first initialize the data weight as Eq. (18). Then in each iteration, we generate the desired classifier with the largest $\widehat{E}$ calculated by Eq. (23) and update the data weight as Eq. (24). Below are some implementation details.

- We skip the stump whose classification error on the training examples is larger than 0.5, since a weak learner should be slightly better than random guessing.
- We skip the stump with non-positive $\widehat{\epsilon}_p$ and non-positive $\widehat{\epsilon}_n$ to ensure they are legitimate according to [14].
- We set the parameter $\beta$ additionally based on the phenomenon we observed in experiments that a proper $\beta$ improves the performance of Ada-UU. Details are discussed in Sect. 4.4.

---

**Algorithm 1** Ada-UU StumpGenerator

---

**Input:** Class prior $\pi$; Data weight $w_t(\cdot)$;
       Training data $U_1$ and $U_2$; Number of steps $S$.
**Output:** Generated stump $h_t^*$, weight of the generated stump $\alpha_{UU}^t$.

**procedure:** StumpGenerator
    Initialize $\widehat{E}_{max} = -\infty$, $\epsilon_{min} = \infty$
    **for** feature $f$ in all features **do**
        **for** $s = 1, \cdots, S$, **do**
            Randomly select a value $v$ in the range of $f$
            Use $v$ as the split point of the stump and get $h$
            Calculate $\widehat{\epsilon}_t$, $\widehat{\epsilon}_p$ and $\widehat{\epsilon}_n$ based on Eq. (19)
            **if** $\widehat{\epsilon}_t \geq 0.5$ or $\widehat{\epsilon}_p < 0$ or $\widehat{\epsilon}_n < 0$ **then**
                Skip the stump and continue
            **else**
                Calculate $\widehat{E}_t$ by Eq. (23)
                **if** $\widehat{E}_t > \widehat{E}_{max}$ **then**
                    $\widehat{E}_{max} = \widehat{E}_t; h_t^* = h; \epsilon_{min} = \widehat{\epsilon}_t$
                **end if**
            **end if**
        **end for**
    **end for**
    $\alpha_{UU}^t = \frac{1}{2} \ln \frac{1 - \epsilon_{min}}{\epsilon_{min}}$
**end procedure**

---

## 4  Experiments and Results

In this section, we study the empirical performance of our proposed method Ada-UU from the following two aspects.

- Comparing the performance of Ada-UU with SOTA NN-based UU methods under different class prior settings.
- Studying the impact of different $\beta$ on the performance of Ada-UU under different class prior settings.

---

**Algorithm 2** Ada-UU

---

**Input:** Class prior $\pi$; Training data $U_1$ and $U_2$;
      Number of training rounds $T$; $S$; $\beta$.
**Output:** $F(x)$.

**procedure:** Ada-UU
    Initialize $H_0 = 0$; Calculate the initial data weight $w_1(\cdot)$ based on Eq. (18),
    **for** $t = 1, \cdots, T$ **do**
        $h_t^*, \alpha_{UU}^t = $ StumpGenerator $(\pi, w_t, U_1, U_2, S)$
        Calculate $w_{t+1}$ based on Eq. (24)
        $H_t = \beta H_{t-1} + \alpha_{UU}^t h_t^*$
    **end for**
    $F(x) = \text{sign}\left(\sum_{t=1}^{T} \alpha_{UU}^t h_t^*(x)\right)$
**end procedure**

---

We first introduce the datasets we use and the methods we compare with. Then we give the experimental results and discussion.

### 4.1  Dataset

We study the performance of our proposed method on three public datasets. The details of the datasets are listed below.

- Epsilon[1] is a binary text classification dataset with $+1$ and $-1$ labels for Parscal large scale learning challenge in 2008 [21]. It has $400,000$ training data points and $100,000$ test data points, with $2,000$ features for every data point. We randomly sample $40,000$ data points as training data and $10,000$ data points as test data. The class prior $\pi$ of this dataset is 0.50.
- Breast Cancer dataset[2] is a binary classification data set with label 0 and $+1$. The label 0 represents malignant breastcancer and the label $+1$ represents the benign cancer. We preprocessed label 0 into label $-1$. It has 455 training data points and 113 test data points. Each data point has 30 features. The class prior $\pi$ of this dataset is 0.59.
- UNSW-NB15 Dataset[3] is a large-scale dataset in cyber security with label 0 and label $+1$. The label 0 represents normal logs and $+1$ represents abnormal logs. We preprocessed the label 0 into label -1. This datset has $175,340$ training points and $82,331$ test data points. Each data point has 39 features. The class prior $\pi$ of this dataset is 0.68.

Following [12,14], we preprocess the training data into two unlabeled datasets $U_1$ and $U_2$ with the same size. The class priors of the two datasets $\theta_1$ and $\theta_2$ are set into $(0.8, 0.2)$, $(0.7, 0.3)$ and $(0.6, 0.4)$ respectively for the experiments.

---

[1] https://www.csie.ntu.edu.tw/~cjlin/libsvmtools/datasets/binary.html.
[2] https://goo.gl/U2Uwz2.
[3] https://research.unsw.edu.au/projects/unsw-nb15-dataset.

## 4.2    Methods

We compare our proposed Ada-UU with the NN-based SOTA UU methods introduced in Sect. 2.1. The details of each method we compare with are summarized as follows.

- UU-unbiased, the UU learning method with unbiased risk estimator described in Eq. (4) without correction function. The overfitting issue occurs when the empirical risk goes negative.
- UU-ABS, the unbiased UU learning method using absolute function described in Eq. (9) as correction function.
- UU-ReLU, the unbiased UU learning method using ReLU function described in Eq. (8) as correction function.
- UU-LReLU, the unbiased UU learning method using generalized leaky ReLU function described in Sect. 2.1 as correction function.

The model structure and the optimizer are the same as [14], where they showed the efficiency of the SOTA UU methods with experimental results. More specifically, the model structure is 5-layer multilayer perceptron (MLP) with ReLU as the activation function. The batch size and learning rate were set to 128 and $5e-5$ respectively. The resulting objectives were minimized by Adam [8] with the default momentum parameters $\beta_1 = 0.9$ and $\beta_2 = 0.999$ as in [14].

## 4.3    Performance of Ada-UU

The empirical performance of SOTA UU methods and our proposed Ada-UU are shown in Table.1. We can see that when $\theta_1$ and $\theta_2$ are getting closer, in most cases, the performance of both NN-based methods and Ada-UU decrease. This finding is consistent with [12,14]. Intuitively, as $\theta_1$ and $\theta_2$ are getting closer, $U_1$ and $U_2$ are becoming more similar, which provides less information to the learning process.

The results in Table.1 show that Ada-UU outperforms the other methods on the large-scale dataset (i.e. Epsilon and UNSW-NB15), especially when the difference of $\theta_1$ and $\theta_2$ gets larger. Ada-UU does not outperform NN-based methods on the small-scale dataset (i.e. Breastcancer), but still has comparable performance. Note that Ada-UU performs significantly better than NN-based methods on the large-scale cyber security dataset UNSW under each different class prior setting.

## 4.4    Impact of $\beta$

Apart from comparing Ada-UU with other methods, we also discovered the impact of $\beta$ on the prediction performance. The $\beta$ in Ada-UU controls the contribution of historical stumps to the final prediction in the training phase. In other words, adjusting $\beta$ results in the change in how much the decisions made by existing stumps affect the final prediction while there is a new stump. The experiments were conducted with the dataset introduced in Sect. 4.1 under the

**Table 1.** Classification comparison on Epsilon, UNSW-NB15 and BreastCancer. With means and standard deviations from 5 runs.

| Dataset | Model | (0.8, 0.2) | (0.7, 0.3) | (0.6, 0.4) |
|---|---|---|---|---|
| Epsilon | UU-unbiased | 0.7957 (0.0016) | 0.7483 (0.0035) | 0.6063 (0.0038) |
| | UU-ABS | 0.8028 (0.0015) | 0.7446 (0.0035) | 0.6024 (0.0038) |
| | UU-ReLU | 0.8032 (0.0012) | 0.7472 (0.0033) | 0.6336 (0.0026) |
| | UU-LReLU | 0.8036 (0.0012) | 0.7471 (0.0033) | 0.6331 (0.0027) |
| | **Ada-UU** | **0.8069 (0.0020)** | **0.7849 (0.0022)** | **0.7439 (0.0045)** |
| UNSW | UU-unbiased | 0.7467 (0.0003) | 0.7469 (0.0005) | 0.7472 (0.0008) |
| | UU-ABS | 0.7466 (0.0003) | 0.7469 (0.0004) | 0.7477 (0.0008) |
| | UU-ReLU | 0.7466 (0.0004) | 0.7469 (0.0005) | 0.7475 (0.0009) |
| | UU-LReLU | 0.7466 (0.0004) | 0.7469 (0.0004) | 0.7477 (0.0008) |
| | **Ada-UU** | **0.8309 (0.0122)** | **0.8127 (0.0164)** | **0.8058 (0.0106)** |
| Breastcancer | UU-unbiased | 0.9439 (0.0043) | 0.9421 (0.0181) | 0.9140 (0.0428) |
| | UU-ABS | **0.9456 (0.0035)** | **0.9456 (0.0195)** | 0.9070 (0.0456) |
| | UU-ReLU | **0.9456 (0.0035)** | 0.9439 (0.0181) | 0.9175 (0.0391) |
| | UU-LReLU | **0.9456 (0.0035)** | 0.9439 (0.0181) | **0.9175 (0.0349)** |
| | Ada-UU | 0.9368 (0.0187) | 0.9193 (0.0151) | 0.9018 (0.0321) |

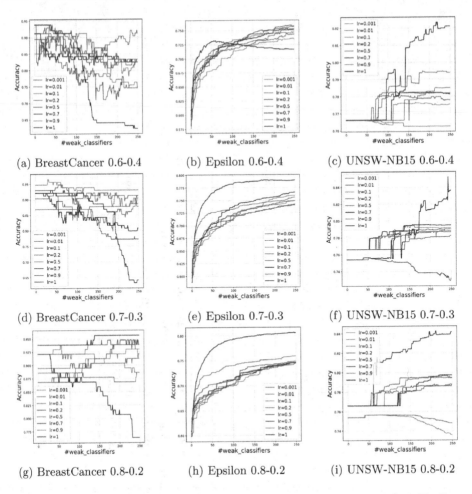

**Fig. 1.** Experimental results given $\beta \in \{0.001, 0.01, 0.1, 0.2, 0.5, 0.7, 0.9, 1\}$ on Epsilon, UNSW-NB15 and BreastCancer. $\beta$ is denoted by lr in the figures.

three different class prior settings the same as the experiments in Sect. 4.3. Moreover, we experimented $\beta \in \{0.001, 0.01, 0.1, 0.2, 0.5, 0.7, 0.9, 1\}$. The experimental results are presented in Fig 1, where we can clearly see that the variation in $\beta$ leads to remarkable differences in the model's performance. Note that we use lr to represent $\beta$ in the figures.

## 5    Conclusion

In this paper, we propose a novel boosting method for UU learning Ada-UU. We solve the problem of obtaining the base classifiers with corresponding weights and updating the data weight with only U data in the boosting procedure. We show the efficiency of our proposed method empirically. In the future, we will prove that the theoretical properties of AdaBoost can be guaranteed in UU learning.

# References

1. Chen, T., Guestrin, C.: XGBoost: a scalable tree boosting system. In: KDD (2016)
2. De Comité, F., Denis, F., Gilleron, R., Letouzey, F.: Positive and unlabeled examples help learning. In: Watanabe, O., Yokomori, T. (eds.) ALT 1999. LNCS (LNAI), vol. 1720, pp. 219–230. Springer, Heidelberg (1999). https://doi.org/10.1007/3-540-46769-6_18
3. Denis, F.Ç.: PAC learning from positive statistical queries. In: Richter, M.M., Smith, C.H., Wiehagen, R., Zeugmann, T. (eds.) ALT 1998. LNCS (LNAI), vol. 1501, pp. 112–126. Springer, Heidelberg (1998). https://doi.org/10.1007/3-540-49730-7_9
4. Denis, F., Gilleron, R., Letouzey, F.: Learning from positive and unlabeled examples. Theoret. Comput. Sci. **348**(1), 70–83 (2005)
5. Freund, Y., Schapire, R.: A decision-theoretic generalization of on-line learning and an application to boosting. J. Comput. Syst. Sci. **55**(1), 119–139 (1997)
6. Freund, Y., Schapire, R., Abe, N.: A short introduction to boosting. J. Japan. Soc. Artif. Intell. **14**(771–780), 1612 (1999)
7. Friedman, J., Hastie, T., Tibshirani, R.: Additive logistic regression: a statistical view of boosting. Ann. Stat. **28**(2), 337–407 (2000)
8. Kingma, D.P., Ba, J.: Adam: a method for stochastic optimization. In: ICLR (2015)
9. Kiryo, R., Niu, G., du Plessis, M.C., Sugiyama, M.: Positive-unlabeled learning with non-negative risk estimator. In: NeurIPS (2017)
10. Krause, A., Perona, P., Gomes, R.: Discriminative clustering by regularized information maximization. In: NeurIPS (2010)
11. Lu, N., Lei, S., Niu, G., Sato, I., Sugiyama, M.: Binary classification from multiple unlabeled datasets via surrogate set classification. In: ICML (2021)
12. Lu, N., Niu, G., Menon, A.K., Sugiyama, M.: On the minimal supervision for training any binary classifier from only unlabeled data. In: ICLR (2019)
13. Lu, N., Wang, Z., Li, X., Niu, G., Dou, Q., Sugiyama, M.: Federated learning from only unlabeled data with class-conditional-sharing clients. In: ICLR (2021)
14. Lu, N., Zhang, T., Niu, G., Sugiyama, M.: Mitigating overfitting in supervised classification from two unlabeled datasets: a consistent risk correction approach. In: AISTATS (2020)
15. Natarajan, N., Dhillon, I.S., Ravikumar, P.K., Tewari, A.: Learning with noisy labels. In: NeurIPS (2013)
16. Patrini, G., Rozza, A., Menon, A.K., Nock, R., Qu, L.: Making deep neural networks robust to label noise: a loss correction approach. In: CVPR (2017)
17. van Rooyen, B., Williamson, R.C.: A theory of learning with corrupted labels. J. Mach. Learn. Res. **18**(1), 8501–8550 (2017)
18. Su, G., Chen, W., Xu, M.: Positive-unlabeled learning from imbalanced data. In: IJCAI (2021)
19. Viola, P., Jones, M.: Robust real-time face detection. In: IJCV (2004)
20. Xu, L., Neufeld, J., Larson, B., Schuurmans, D.: Maximum margin clustering. In: NeurIPS (2004)
21. Yuan, G.X., Ho, C.H., Lin, C.J.: An improved GLMNET for L1-regularized logistic regression. J. Mach. Learn. Res. **13**(64), 1999–2030 (2012)
22. Zampieri, M., Malmasi, S., Paetzold, G., Specia, L.: Complex word identification: challenges in data annotation and system performance. In: NLPTEA (2017)
23. Zhao, Y., Zhang, M., Zhang, C., Chen, T., Ye, N., Xu, M.: A boosting algorithm for positive-unlabeled learning. arXiv (2022)
24. Zhou, Z.-H.: Machine Learning. Springer, Singapore (2021). https://doi.org/10.1007/978-981-15-1967-3

# A Study of the Effectiveness
# of Correction Factors for Log Transforms
# in Ensemble Models

Ray Lindsay[1]([✉])[iD] and Yanchang Zhao[2][iD]

[1] Smarter Data, Australian Taxation Office, Canberra, Australia
ray.lindsay@ato.gov.au
[2] Data61, CSIRO, Canberra, Australia
yanchang.zhao@csiro.au

**Abstract.** We consider the problem of making predictions about long-tailed interval variables. Such variables are commonplace in revenue prediction, where a small part of the population has very large positive or negative values. Log transforms are often used in such problems, and when modeled in a log space, a correction factor is required when converting back to the original space. In this work, we study the effectiveness of two different approaches of applying correction factors at the individual model level and the whole model level. Particularly, we consider ensembles of simpler models (decision trees) with individual correction factors, compared with XGBoost, an averaging model using an overall correction factor. We show that the ensembles of simple models outperform XGBoost, when the correction factors are applied separately to each component model.

**Keywords:** Machine learning · Ensemble models · Log transformation · Long-tailed variables

## 1 Introduction

In the world of finance and revenue agencies, it is often the case that there are a small number of very large values and many smaller ones. An example is a charity donations dataset used in the KDD98 Cup [13], which has such a pattern of many small donations and a few much larger ones. If linear regression methods are used ignoring the long tail, the residuals in the model tend to be proportional to the size (i.e., heteroscedastic) and so predictions are least reliable at the larger values – where the predictions are most important. Log transformation [15] is often used and then regression is more reliable. The KDD98 data and many other finance/revenue datasets have missing values in some of their explanatory variables, and these are seldom Missing Completely At Random (MCAR) [6] which makes approaches like complete case analysis or single imputation of questionable use or validity. Many of the decision tree (DT) and variant algorithms [7]

---

Work undertaken when RL was a Data Fellow at CSIRO's Data61.

used in machine learning are robust to missing values, so lend themselves to these problems. Decision trees also will only ever predict within the range of the target variable since each prediction is an average of values in the data. Decision trees are known to have high variance in that a small change in an input can cause a large change in output. One way round that is to use an ensemble approach, averaging over many decision trees with slightly different sets of rows and columns. More recent algorithms (which internally are averaging over many trees) such as XGBoost [1,14] and LightGBM [4,5] have been shown to be more accurate than older methods. However, in the case of a log-transformed target this is not always so, as these algorithms do not 'know' the target has been transformed. In this study we show that an ensemble of simpler DT methods can outperform either a single or ensemble of more sophisticated methods – because the ensembles of DT make explicit allowance for the different correction factors required for each component model when a target has been log-transformed.

## 2    Problem Statement

A typical example of a long-tailed interval variable in financial and revenue data is shown in Figs. 1 and 2. Figure 1 is the cumulative distribution function (CDF) on the original scale and is essentially very close to a step function at zero and Fig. 2 shows it on a scale closer to zero. From Fig. 2 a proportion of the values are zero and there is a suggestion of a log-normal distribution on both sides. We modelled the (absolute value of) negative and positive parts separately, excluding zeros in both cases. Note that in some populations there is no negative part. In this work, we address the challenging problem of predicting such long-tailed variables using appropriate correction factors for log transforms.

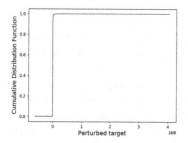

**Fig. 1.** Cumulative Distribution Function showing full range: note step function at 0

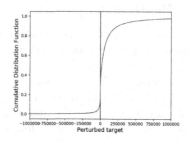

**Fig. 2.** Cumulative Distribution Function showing range closer to zero: note distribution near zero

## 3    Methodology

When using predictive models, such as decision trees and random forest, to predict a long-tailed variable, the log of the target variable has been modelled, and therefore, a correction factor needs to be applied when back transforming – this is derived from the expected value of the mean of the log-normal distribution. A seminal reference is Miller [8]. The correction factor is based on the variance, so is always greater than 0 in log space and 1 in original space respectively. The correction factor (in log space) is defined as

$$\alpha = \exp(\sigma^2/2), \tag{1}$$

where $\sigma$ is the standard error of the regression and is derived from the fit to the training data.

Miller's work [8] shows that this transformation has been used in wide variety of fields, including allometry [12], river loads [2], flood frequency [10], extreme value time series [9], and psychological data [3]. Smith notes that not using the correction factor results in too low predictions of fossil body mass [12]. Pandey and Nguyen note that linear models had higher bias and Root Mean Squared Error and under predicted floods [10].

As the transformation is affine and the rank-order is important to our results, the transformation would not matter if only one model was considered. However as we are explicitly looking at ensembles of models each of which will have a different correction factor, we do need to use the individual correction factors for each model. To study the effectiveness of two different ways of applying correction factors, we conducted an experimental study of ensembles of simpler models (decision trees) with individual correction factors, compared with XGBoost, an averaging model using an overall correction factor, and report our experimental results in next section.

## 4    Experiments

### 4.1    The Dataset

The data used in our experiments are sourced from a government revenue office for predicting revenue or income of two types of entities. As each type of entity has both positive and negative parts, there are 4 populations to model. For each population the data has about 100,000 rows, randomly sampled from a much larger population. The explanatory variables number about 90 interval and 50 nominal (character and numeric) – these being from a larger pool of about 1,000 explanatory variables. Many of the interval variables are themselves log transformed. The positive and negative parts of each interval variable are treated separately, yielding two new variables.

The nominal variables are transformed using one-hot encoding to derive a design matrix. The numbers of levels per nominal variable vary from 2 to about 20.

The data has been sanitised so that the transformed data had nearly the same mathematical properties but was no longer identifiable or sensitive. These transformations were:

– Identifiers were replaced by random integers;

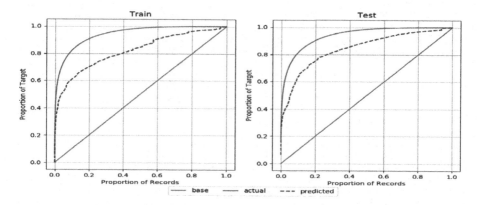

**Fig. 3.** C-statistic plots for Train and Test data

- The target variable was perturbed by an amount proportional to its size;
- All interval variables were converted to ranks, retaining missing values and given uninformative names; and
- Nominal variables (character and numeric) were given uninformative levels and names and missing values retained.

The interval variables contain missing values which XGBoost can deal with though the Python *sklearn* package does not. Hence the missing values were imputed for the Decision Tree modelling. Missing nominal variables define a new level.

## 4.2   Evaluation Measure

For an interval variable with a small number of very large values (i.e., long tailed), a widely-accepted measure for model evaluation is the *c-statistic*, which measures how well the rank-order is predicted. To calculate this, the data are sorted in descending order of the actual or predicted value, and the area under the curve calculated as proportions of cases ordered from zero to one. The c-statistic is quite sensitive to the most extreme values in a population, so the ratio of predicted to actual is reported—an example plot is shown in Fig. 3.

Note that for targets which are always strictly greater or less than zero the c-statistic is always less than 1. For populations with both positive and negative parts, it can exceed 1. The ratio of prediction to actual is nearly always less than 1.

## 4.3   Experimental Settings

At the start, the data were randomly split into two subsets, Train (70%) or Test (30%) and kept that role throughout. The numbers of ensembles tested were 1, 5, 10, 25 and 50.

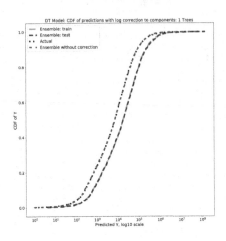

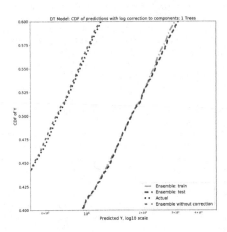

**Fig. 4.** CDF for actual and predicted – single Decision Tree (Color figure online)

**Fig. 5.** Close up of CDF – single Decision Tree (Color figure online)

The specific learners used were the BaggingRegressor from Python *sklearn* [11]. The default base estimator (Decision Tree) was used for the DecisionTree modelling and XGBRegressor was used for the XGBoost modelling.

### 4.4   Decision Trees Vs XGBoost

Considering first a single decision tree, Fig. 4 shows the actual CDF (see black line) of the target variable, the Ensemble without correction (green line) and the predictions corrected (yellow and blue lines). Because the corrections are greater than 1, the CDF of the predictions are shifted to the right. However the ensemble predictions for Train and Test (yellow and blue lines, respectively) are much closer to the actual data. Figure 5 zooms in on part of the plot and here the 'lumpy' nature of DT predictions is apparent.

For 50 ensembles the equivalent plots are Figs. 6 and 7. In both, there is a band of predictions (grey lines) for each model, which follow essentially the same curve and lie to the right of the actual (black line). However averaging 50 predictions results in Train and Test CDFs (yellow and blue lines, respectively) follow the actual much more closely.

The predictions from a XGBoost model with 1 and 50 ensembles are shown in Figs. 8, 9, 10 and 11. As expected, the individual prediction curves are much smoother for these models. The averaged predictions and individual ones overlay each other.

Figures 12 and 13 show histograms of the correction factors for 50 models, respectively with DT and XGBoost. The DT ones (see Fig. 12) are larger than XGBoost (see Fig. 13) because the errors on individual models are larger. Looking at the variability of the factors for the XGBoost models suggests all factors lie within [2.02, 2.055] whereas those for DT lie within a larger range [1.3, 2.6]

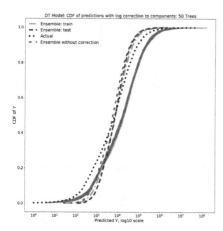

**Fig. 6.** Ensembles – Decision Trees (Color figure online)

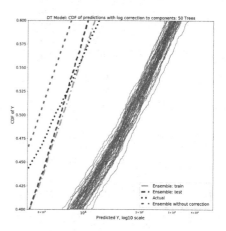

**Fig. 7.** Close up of ensembles – Decision Trees (Color figure online)

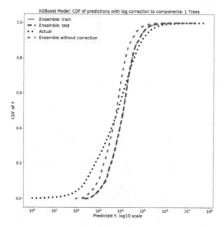

**Fig. 8.** CDF of actual and predicted for single XGBoost model

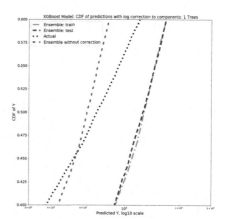

**Fig. 9.** Close up of single XGBoost model

In Table 1, we compare the model performances for the positive part for entities of type 1. For a single model the c-statistic ratio for decision tree (DT) models is slightly worse than XGBoost models for the Train data (0.8537 vs 0.8576) and noticeably worse for the Test data (0.8065 vs 0.853). As the number of ensembles increases, the DT model's performance increases more than that of XGBoost, so that at 50 ensembles the DT (0.8899) is better than XGBoost (0.8616). For the Test data, the XGBoost results are still better (0.850 for DT vs 0.855 for XGBoost).

Tables 2, 3 and 4 repeat this pattern for other populations: for 1 ensemble the XGBoost outperforms DT, but as the number in the ensemble increases, the XGBoost improves only slightly while the DT improves markedly, being better

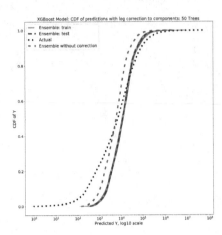

**Fig. 10.** CDF of actual and predicted for Ensembles – XGBoost

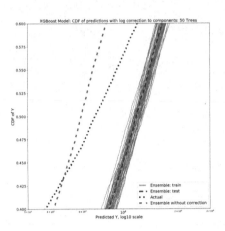

**Fig. 11.** Close up of Ensembles – XGBoost

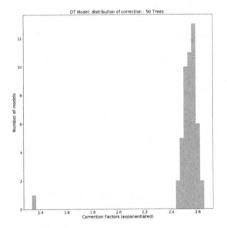

**Fig. 12.** Histogram of correction factors for Decision Tree models

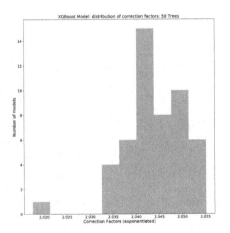

**Fig. 13.** Histogram of correction factors for XGBoost models

than XGBoost usually by 5 ensembles for the Train data. In all scenarios a single DT performs worse that a single GB, but as the number of ensembles increases, the GB performance does not improve much and the DT becomes in some cases superior to the GB for the Train data though not for the Test data. Although GB remains best, its advantage is relatively small in most tables.

This model performance needs to sit alongside the computational performance: the DT models are of the order of 7–8 times faster to run than XGBoost, so an ensemble of 5–10 will be of a similar order of computation effort as a single XGBoost model. The XGBoost models sometimes failed for problems with more than 300 columns, likely due to the limit of computer memory.

**Table 1.** Positive part of type 1 ratios of C-statistics

| Ensemble | DT | DT | GB | GB |
|---|---|---|---|---|
| Size | Train | Test | Train | Test |
| 1 | 0.8537 | 0.8064 | 0.8576 | 0.8536 |
| 5 | 0.8797 | 0.8389 | 0.8609 | 0.8552 |
| 10 | 0.8855 | 0.8473 | 0.8615 | 0.8555 |
| 25 | 0.8891 | 0.8497 | 0.8615 | 0.8552 |
| 50 | 0.8899 | 0.8500 | 0.8616 | 0.8551 |

**Table 2.** Negative part of type 1 ratios of C-statistics

| Ensemble | DT | DT | GB | GB |
|---|---|---|---|---|
| Size | Train | Test | Train | Test |
| 1 | 0.7391 | 0.7780 | 0.8010 | 0.8775 |
| 5 | 0.8475 | 0.8566 | 0.8039 | 0.8797 |
| 10 | 0.8627 | 0.8676 | 0.8037 | 0.8782 |
| 25 | 0.8721 | 0.8688 | 0.8049 | 0.8797 |
| 50 | 0.8762 | 0.8768 | 0.8067 | 0.8793 |

**Table 3.** Positive part of type 2 ratios of C-statistics

| Ensemble | DT | DT | GB | GB |
|---|---|---|---|---|
| Size | Train | Test | Train | Test |
| 1 | 0.8711 | 0.8429 | 0.8852 | 0.8894 |
| 5 | 0.9080 | 0.8753 | 0.8874 | 0.8908 |
| 10 | 0.9140 | 0.8861 | 0.8885 | 0.8919 |
| 25 | 0.9183 | 0.8918 | 0.8885 | 0.8918 |
| 50 | 0.9192 | 0.8875 | 0.8877 | 0.8909 |

**Table 4.** Negative part of type 2 ratios of C-statistics

| Ensemble | DT | DT | GB | GB |
|---|---|---|---|---|
| Size | Train | Test | Train | Test |
| 1 | 0.8297 | 0.7719 | 0.8731 | 0.8369 |
| 5 | 0.8796 | 0.7944 | 0.8694 | 0.8356 |
| 10 | 0.9093 | 0.8236 | 0.8728 | 0.8406 |
| 25 | 0.9220 | 0.8351 | 0.8734 | 0.8398 |
| 50 | 0.9216 | 0.8326 | 0.8711 | 0.8396 |

## 4.5   Effect of Not Log Transforming

Figures 14 and 15 show boxplots of residuals from models with respectively no transformations and log transformed target variables. Such plots are useful in determining whether or not heteroscedasticity is present in a regression model. There are 20 boxes grouped by rank of the predicted value. Without transformation (Fig. 14) the largest ranked residuals are roughly 107 larger than the other groups. With transformation (Fig. 15), the residuals are all of a similar scale. As the predictions at the highest values are those of most interest, not transforming has resulted in predictions that are least accurate where it matters most.

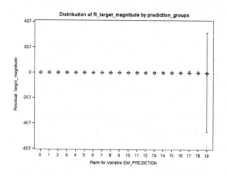

**Fig. 14.** Residuals of models without log transformations

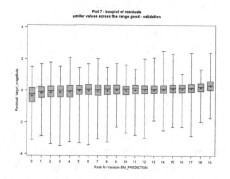

**Fig. 15.** Residuals of models with log transformations

## 5  Conclusions

We have studied the problem of predictive modeling for long-tailed interval variables and have experimentally investigated two different methods of applying correction factors for log transforms in ensemble models. Our experimental results on real-world data show that using an ensemble of simpler models can outperform when building machine learning models to predict long-tailed interval target variables. Alternatively, at least give a very similar performance to a sophisticated method such as XGBoost. This is because many sub-models that make up an XGBoost model do not have individual correction factors—they are only available at the whole model level. Future work will further evaluate the individual correction factors approach with more machine learning algorithms and more datasets from other domains.

**Acknowledgements.** Author RL thanks the ATO and CSIRO for the opportunity to pursue this topic while as a Data Fellow at CSIRO's Data61.

## References

1. Chen, T., Guestrin, C. : XGBoost: a scalable tree boosting system. In: Proceedings of the 22nd ACM SIGKDD International Conference on Knowledge Discovery and Data Mining, New York, NY, USA, pp. 785–794. ACM. (2016) https://doi.org/10.1145/2939672.2939785
2. Ferguson, R.I.: River loads underestimated by rating curves. Water Resour. Res. **22**(1), 74–76 (1986)
3. Judd, C.M., McClelland, G.H., Culhane, S.E.: Data analysis: continuing issues in the everyday analysis of psychological data. Annu. Rev. Psychol. **46**, 433–465 (1995)
4. Ke, G., et al.: LightGBM: a highly efficient gradient boosting decision tree. In: Advances in Neural Information Processing Systems (NIPS 2017), vol. 30, pp. 3149–3157 (2017)
5. LightGBM. https://lightgbm.readthedocs.io/en/latest/
6. Little, R.J.A., Rubin, D.: Statistical Analysis with Missing Data, 2nd edn. Wiley, Hoboken (2002)
7. Loh, W.-Y.: Classification and regression tree methods. In: Ruggeri, Kenett and Faltin (eds.) Encyclopedia of Statistics in Quality and Reliability, pp. 315–323. Wiley (2008)
8. Miller, D.M.: Reducing transformation bias in curve fitting. Am. Stat. **38**(2), 124–126 (1984). http://www.jstor.com/stable/2683247
9. Mudelsee, M.: Extreme value time series. In: Climate Time Series Analysis. AOSL, vol. 51, pp. 217–267. Springer, Cham (2014). https://doi.org/10.1007/978-3-319-04450-7_6
10. Pandey, G.R., Nguyen, V.T.V.: A comparative study of regression based mods in regional flood frequency analysis. J. Hydrol. **225**(1–2), 92–101 (1999)
11. Sklearn   Documentation.   https://scikit-learn.org/stable/modules/generated/sklearn.ensemble.BaggingRegressor.html
12. Smith, R.J.: Logarithmic transformation bias in allometry. Am. J. Biol. Anthropol. **90**(2), 215–228 (1993)

13. The KDD Cup 1998 Data. https://kdd.ics.uci.edu/databases/kddcup98/kddcup98.html
14. XGBoost Documentation. https://xgboost.readthedocs.io/en/latest/
15. Xiao, X., White, E.P., Hooten, M.B., Durham, S.L.: On the use of log-transformation vs. nonlinear regression for analyzing biological power laws. Ecology **92**, 1887–1894 (2011). https://doi.org/10.1890/11-0538.1

# Author Index

Printed in the United States
by Baker & Taylor Publisher Services